D1043138

THE CHEMICAL FORMULARY

The
Chemical Formulary

*A Collection of Valuable, Timely, Practical
Commercial Formulae and Recipes for
Making Thousands of Products in
Many Fields of Industry*

VOLUME VII

Editor-in-Chief

H. BENNETT

1945

CHEMICAL PUBLISHING CO., INC.

BROOKLYN, N. Y., U. S. A.

PREFACE

Chemistry as taught in our schools and colleges is confined to synthesis, analysis and engineering—and rightly so. It is part of the proper foundation for the education of the chemist.

Many a chemist on entering an industry soon finds that the bulk of the products manufactured by his concern are not synthetic or definite chemical compounds but are mixtures, blends or highly complex compounds of which he knows little or nothing. The literature in this field, if any, may be meager, scattered or antiquated.

Even chemists, with years of experience in one or more industries, spend considerable time and effort in acquainting themselves on entering a new field. Consulting chemists, similarly, have problems brought to them from industries foreign to them. A definite need has existed for an up-to-date compilation of formulae for chemical compounding and treatment. Since the fields to be covered are many and varied, an editorial board was formed, composed of chemists and engineers in many industries.

Many publications, laboratories, manufacturing companies and individuals have been drawn upon to obtain the latest and best information. It is felt that the formulae given in this volume will save chemists and allied workers much time and effort.

Manufacturers and sellers of chemicals will find in these formulae new uses for their products. Non-chemical executives, professional men and others, who may be interested, will gain from this volume a "speaking acquaintance" with products which they may be using, trying, or with which they are in contact.

It often happens that two individuals using the same ingredients in the same formula get different results. This may be the result of slight deviations or unfamiliarity with the intricacies of a new technique. Accordingly, repeated experiments may be necessary to get the best results. Although many of the formulae given are being used commercially, many have been taken from patent specifications and the literature. Since these sources are often subject to various errors and omissions, due regard must be given to this factor. Wherever possible it is advisable to consult with other chemists or technical workers regarding commercial production. This will save time and money and avoid "headaches."

It is seldom that any formula will give exactly the results which one requires. Formulae are useful as starting points from which to work out one's own ideas. Formulae very often give us ideas which may help us in our specific problems. In a compilation of this kind, errors of omission, commission and printing may occur. We shall be glad to receive any constructive criticism.

To the layman, it is suggested that he arranges for the services of a chemist or technical worker familiar with the specific field in which he is interested. Although this involves an expense, it will insure quicker and better formulation without waste of time and materials.

H. BENNETT

PREFACE TO VOLUME VII

Additional new formulae have been gathered to compile a seventh volume of the *Chemical Formulary*—an addition which will broaden and bring up-to-date the contents of volumes I, II, III, IV, V and VI.

Schools and colleges in increasing numbers seem to find it advisable to use the *Chemical Formulary* as an aid in promoting a practical interest in chemistry. By its use, students learn to make cosmetics, inks, polishes, insecticides, paints and countless other products. The result is that chemistry becomes an extremely interesting practical and useful subject. This interest often continues even when the students reach the theoretical or more difficult phases of chemistry.

Since some mature users of this book have not had the good fortune to have had previous training or experience in the art of chemical compounding, the simple introductory chapter of directions and advice has been repeated. This chapter should be studied carefully by all beginners (and some more experienced workers) and some of the preparations given in it should be made before attempting to duplicate the more complex formulae in the succeeding chapters.

An enlarged directory of sources of chemicals and supplies is included. This should prove useful in locating new as well as old materials and products.

It is a sincere pleasure to acknowledge the valuable assistance of the members of the board of editors and others who have given of their time and knowledge in contributing the special formulae which have made this volume possible.

H. BENNETT

NOTE

All the formulae in volumes I, II, III, IV, V, VI and VII (except in the introduction) are different. Thus, if you do not find what you are looking for in this volume, you may find it in one of the others.

A *cumulative index* for the first six volumes is now available. Many will find this a useful, time-saving adjunct.

TABLE OF CONTENTS

                                                                    PAGE

PREFACE   vii

PREFACE TO VOLUME VII   ix

ABBREVIATIONS xxxi

I. INTRODUCTION   1

II. ADHESIVES   17
Water-Soluble or Dispersible
Labeling; Non-Warping; Billposter's; Cold Water;
Envelope; Wall Paper; Water-Resistant; "Iceproof";
Sizes; Non-Warping Paper; Binder for Cork; Tin
Labeling; Starch; Veneer; Potato Starch; Sweet Po-
tato; Dextrin; Gummed Paper; Paper to Metal;
Powdered; Casein; Laminating; Dried Blood; Soy-
bean; Sodium Silicate; Ceramics; High Temperature;
Marble and Alabaster; Bonding for Abrasive Wheel;
Gum Arabic; Waxed Paper; Mucilage; Magazine and
Catalog; Bookbinders'; Moistureproof Cellophane;
Wood; Plywood; Paper Board; Calcium Sucrate;
Synthetic Resin; "Protectoid"; Decalcomania; Tire
Tube Air Seal; Latex; Paper.

Water-Insoluble Adhesives
Synthetic Rubber; Neoprene; Buna N; Buna S;
Thiokol; Photo Mounting; Mixed; Rubber to Metal;
Leather Driving Belts; Insole; Metal to Metal; Cel-
luloid to Hard Rubber; Rosin; Sticky Rubber-Rosin;
Permanently Tacky-Pressure; Non-Webbing Rubber;
Crockery; Rosin Linoleum; Electrical Sealing; Rub-
ber-Resin; Bakelite Coated Cloth; Polyvinyl Chloride
Plastics; Coating; Ethyl Cellulose; Cellulose Acetate;
Cellulose Acetate Foil; Wood Ply Lamination; Lam-
inated Wood; Laminating; Plywood; Veneer; Ther-
moplastic; Flexible Thermoplastic; Thermoplastic
High-Melting; Transparent Resin; Stainless Steel
Pipe; Pipe Joint; Plastic Heatproof; Pipe Joint
Lute; Pipe Joint Calking; Metal Joint Seal; Plastic
Sealing; Can Seam Seal; Container Joint Thermo-
setting; Aluminum Thread Joint; Thermosetting;

Crack Sealer; Plate Glass Setting; Lantern Slide;
Lens; Glass Sealing; Glass to Metal; Glass; High-
Vacuum; Universal; Quick-Setting Waterproof; Pett-
man; Lacquer; Asphalt Emulsion; Sound-Deadening
Pads; Pressure-Sensitive; Tape; Masking; Removing
Surgical Tape; Strapping; Surgical Waterproof;
Grafting Wax; Wood Filler; Furniture Filler; Metal
to Glass; Metal to Wood; Metal to China; Metal to
Horn; Rubber to Glass; Rubber to Stone; Rubber
to Metal; Metal to Linoleum; Metal to Stone; Brass
to Marble; Metal to China; Glass to Glass; Glass to
Wood; Marine; Metal to Celluloid; Stone to Stone;
Stone to Horn; Stone to Hard Rubber; China to
China; White Lead; Alberene Stone Filler; Stone-
ware; Refractory; Metal Filler and Calking; Calking; Boat Calking; Heat-Sealable; Rectifier; Denture; Preparation of Surfaces for Cementing Leather
with Gutta-Percha; Leak-Sealing; Refrigerator; Oil-
Well Water; Bonding Aluminum to Steel.

III.   FLAVORS AND BEVERAGES   51
Pure Fruit Concentrates: Strawberry; Raspberry;
Grape; Cherry. *Nectars:* Orange; Lime. *Imitation
Nectars:* Grape; Cherry; Strawberry; Raspberry;
Walnut Extract; Pineapple Extract; Almond Ex-
tract; Spearmint Extract; Lemon Extract. *Imita-
tion Spices:* Allspice; Nutmeg; Sage Oil; Root
Beer Extract; Quick Dissolving Syrup; Sugar-
less Syrup Substitutes; Hop Extract Emulsion;
Emulsifier and Stabilizer; Marmalade from Orange
Peels; Rye Bread; Chocolate Syrup; Lemon Oil;
Orange Oil; Orangeade Syrup; Orange Juice
Syrup; Ice Cream or Soda Fountain Crushed Fruits;
Prepared Lemon Beverage.

IV.   COSMETIC AND DRUG PRODUCTS   59
Creams: Cleansing; Hydrogenated Vegetable Oil;
Liquid; Emulsifying Base; Toilet; Ointment Base;
Washable Ointment Base; Emulsifier; Sulfathiazole
Urea; Hormone; Camouflage; Glycerin; Acid; Four-
Purpose; Concentrated; Cosmetic Base; Tissue; Mod-
ern Tissue; Cold; Suntan Oil; Suntan Lotion; Sun-
burn Ointment; Suntan Cream; Anti-Sunburn Oint-
ment; Milky Suntan Oil; Milk of Almond; Face
Lotion; Hand Lotion; Vanishing, Hand, Founda-

tion; Alginate Hand Lotion Base; Invisible Glove; Lithographers' Protective Hand Wash; Hand Protective; Oil-Resistant; Nail; Non-fatty; Honey-Gel; Vitamin F.

Nail Polish; Solvents; Oily Nail Polish Remover; Antiseptic Dusting Powder; Astringent Dusting Powder; Douche Powder; Face Powder; Solid Perfume; Cream Cologne; Eau de Cologne; Spanish "Paste"; Mouth Wash; Glycerin Thymol; Germicidal; Gargle Powder; Antiseptic Dentifrice; Pumice-Bentonite Dental Paste; Salt-Lime Dentifrice; Sodium Oleate Dentifrice; Epithelial Solvent; Trench Mouth Treatment; Toothache Drops; Dental Desensitizer; Topical Anesthetics; Cavity Cement Lining; Denture Adhesive; Styptic Dental Cotton; Dental Instruments Sterilizing Fluid; Dental Pulp Devitalizer; Dental Antiseptic; Denture-Cleaner; Hair Tonic; Hair Lotion; Permanent-Waving Compositions; Chemical-Heating Powder; Hair-Waving Self-Heating Composition; Cold Hair-Waving Solution; Wave Set (Hair-Waving) Fluid; Bandoline; Wavy Hair Pomade; Hair Creams; Washing and Bathing Composition; Soapless Shampoo; "Crude" Oil for Shampoo; Head Lice Lotion; Bluing for White Hair; Electric Shaving Aid; Brushless Shaving Cream; After-Shaving Lotion; Throat and Nasal Spray; Nasal Inhalant; Vapor Inhalant; Nasal Constrictor; Ephedrine Nose Drops; Isotonic Nose Drops; Asthma Spray; Hay-Fever Treatment; Asthma Nose Drops; Asthma Smoke Powder; Hay-Fever Asthma Inhalants; Non-Irritating Eye Drops; Hay-Fever Eye Wash; Cake Mascara; Kohl (Eye-Brow Black) Neutral Eye Wash; Treating Hives on Eyelids; Eye-Burn Treatment; Ointment Bases; Absorption Base; Healing Salve; Stainless Ointment Base; Analgesic Ointment; Cod-Liver Oil Ointments; Lanolin Substitute; Disinfectant; Germicidal Cream; Germicidal Bombs; (Aerosols) Germicide; Antiseptic Solution; Disinfecting Shoes; Silver Fluoride Soap; Antiseptic Hand Soap; Aluminum Chloride Soap; Anti-Perspirant Liquid; Under-Arm Cream; Anti-Perspirant and Body Deodorant; Control of Foot Perspiration; Deodorant Cream; Deodorant-Fungicide; Liniment; Bay Rum; Menthol Liniment; Athlete's Rub Liniment; Isopropyl Rubbing Alcohol; Foot Bath; Antiseptic

Military Foot Powder; Athlete's Foot Powder; Soaking Bath for Athlete's Foot; Athlete's Food Remedy; Callous Softener and Remover; Corn Remover; Corn Collodion; Callous Skin Remover; Antacid Mixture; Colloidal Aluminum Hydroxide; Anti-Acid Powder; "Acetic" Acid Tablets; Medicinal Barium Sulfate Suspension; Acne Lotions; Chloracne Treatment; Waterproofing Skin Cream; Skin Treatment Cream; Stable Zinc Peroxide Ointment; Coal-Tar Skin Ointment; Ulcer Ointment; Dermatitis Lotions; Cable Rash Ointment; Mucilage of Quince Seed; Sulfathiazole Suspension; Medicine and Salves for Chromic Acid Skin Injuries; Dermatitis Salve for Metal Poisoning; Metal Poisoning Salve; Poison Ivy and Oak Remedies; Poison Ivy Treatment; Scabies Preparations; Antipruritic Lotion; Itch Remedy; Chlorophyll Impetigo Ointment; Impetigo and Wound Powder; Sulfa-Zinc Peroxide Ointments; Sulfa-Drug Surgical Film; Wound and Burn Film Treatment; Burn Ointment; Mustard-Gas Ointment; Anaerobic Wound-Infection Germicide; Aseptic Wax; Tannic Acid Burn Emulsion; Burn Cream; Insect-Repellant Cream; Treatment of Chigger Bites; Cedar Chest Compound; Hemorrhoidal Suppository; Vaginitis Suppository; "Sulfa" Vaginal Suppository; Rectal Suppository; Cocoa-Butter Suppository; Suppository Mold Lubricant. *Sulfa Drug Compositions:* Foot Fungus Ointment; Vaginal Ointment; Washable Jelly; Ointment Emulsion. Medicinal-Vanishing Creams; Ophthalmic-Ointment Base; Sugarless-Pharmaceutical Syrups; Drug-Penetrating Aid; Anthelmintic Tablet; Sedative; Non-Toxic Bitter-Tasting Product; Elixir of Raspberry; Vitamin B Elixir; Pharmaceutical-Flavoring Vehicles; Surgical-Lubricating Jelly; Flexible-Adhesive Surgical Dressing; Surgical-Cast Plaster; Capsule-Filling Hints; Pharmaceutical-Tablet Coating; Enteric-Pill Coating; Pill Excipient; Liquid-Petrolatum Emulsion; Methyl Cellulose; Vegetable Physic; Cod-Liver Oil, Orange Juice, Malt Extract, Emulsion. Substitutes for Glycerin; Depilatory; Cosmetic Stocking; Feminine-Hygiene Preparations; Embalming Fluid; Chapped Lip Stick; Mustache Wax; Bubble Bath; Air Odor-Neutralizer; Air Deodorant; Air Sterilization; Telephone-Mouthpiece Antiseptic.

PAGE

V. EMULSIONS AND COLLOIDS 111

Paraffin Wax; Ceresin Wax; Ouricury Wax; Cetyl
Alcohol; Dichloroethyl Ether Spray; Benzyl Benzoate; Neat's Foot Oil; Castor Oil; Cod-Liver Oil; Peppermint Oil; Polyvinyl Chloride Polymerization;
Cumar; Polyisobutylene Polymer; Pentaerythritol
Abietate; "Staybelite" Ester; Colloidal Sulphur;
Oleoresin Capsicum; Sulfur; Cottonseed Pitch; Asphalt; Aqueous Colloidal Graphite; Breaking Emulsions; Fluorspar Flotation Frother.

VI. FARM AND GARDEN SPECIALTIES 115

Soilless-Growth Plant Foods; Tomato-Plant Nutrient;
Plant-Growth Regulator; Root-Growth Stimulant;
Indolebutyric Acid; Granulated Fertilizer; Conserv-
ing Agent for Plants; Cut-Flower Preservative; Pre-
venting Dropping of Apples; Protecting Potatoes
against Decay; Grafting Wax; Tree Gall Treatment.
General Insecticides: Mosquito Larvicide; Insecti-
cidal Aerosol; D.D.T. Fly Spray; D.D.T. Insect
Powder; Vermin Exterminator; Silver-Fish Poison;
Roach Powder; Powdered Pyrethrum; Sowbug Poi-
son Bait; Dry Insecticide; Rotenone Derris Substi-
tutes; Disinfectant Spray. *Agricultural Insecticides:*
Cutworm; Lawn Grub and Japanese Beetle Larvae;
Subterranean Grass Caterpillar; Slug; Tomato Moth;
Tomato Pinworm; Boxwood Leaf Miner; Potato
Leafhopper; Cherry Fruit Fly; Parasiticide Spray;
Control of Pea Aphids; Codling Moth Spray; Codling
Moth Ovicide and Larvicide; Water Hyacinth Spray;
Horticultural Spray Base; Thrip, Aphid, Mite and
Mealy Bug Control; Gladiolus Thrip Spray; Paris
Green Sprays; Ornamental Plant Insecticide Spray;
Shade Tree Insecticides; General Agricultural Spray;
Kerosene Spray Emulsion; Agricultural Spray
Spreader; Dry Agricultural Spray; Bordeaux Mix-
ture; Nicotine Bentonite Insecticide; Nicotine De-
coctions; Nicotine Dust; Tobacco Blue Mold Control;
Rust-Fungus Disinfectant Dust; Disinfecting Seeds;
Control of Potato Ring Rot and Scab; Poultry Louse
Powder; Disinfectant and Insecticide for Poultry
Houses; Tetrachlorethylene Anthelmintic Emulsion;
Phenothiazine Suspension; Sheep Anthelmintic;
Sheep Tick Dip; Cattle Louse Dust; Cattle Grub Con-
trol; Control of Hide Beetles; Cattle Horn Fly Spray;

Treating Fungous Diseases in Pet Fish; Weed Killers; Honeysuckle-Weed Killer; Poison-Ivy Killer; Mole-Cricket Poison Bait; Solution for Cleaning Eggs; Feather and Animal Hair Depilatory; Dehusking Carob Beans; Artificial Honeycomb. *Poultry Mash:* Broiler Feed; Chick Starting and Growing; Chick Starters. Poultry Feed; Grain and Mash; Broiler All-Mash; Breeding Mash; Turkey Starters; Turkey-Growing Mashes; Calf Meal Feed; Dry Calf Feed; Reinforced Calf-Starter Mixture; Preserving Green Feed (Fodder); Cattle Salt Iodine Blocks.

VII. FOOD PRODUCTS 139
Canned Fruit Salad; Canned Tomato Aspic; New Fruit Desserts; Syrup for Pickling Fruits; Preventing Discoloration of Cut Fruit; Grape Shipment Protection; Fruit Protective Coatings; Non-Discoloring Peeled Potatoes; Vegetable Preservation; Pea-Soup Cubes; Bleaching Nut Shells; Removing Nut Skins; Orange Marmalade; Strawberry Jam or Preserves; Grape Jelly; Wine Jelly; Cocoa Jelly; Pectin Solution; Pectin Candy Formula. *Jelly Candy:* Tart Fruit Flavors; Mildly Tart Pieces; Pieces with Chocolate Flavor; High-Grade Licorice Flavored Piece; Pieces with Orange Pulp and Juice; Pieces with Grapefruit Pulp and Juice; Pieces with Fig Pulp; Apple Confection. Apple Leather Candy; Turkish Paste; Fondant Icing Sugar; Butterscotch Sauce; Starch-Albumen Marshmallow; Reducing Viscosity of Chocolate; Glycerin-Gelatin Bases. Heat-Resistant Candy and Cake Cocoa Coating. *Bakers' Icings, Glazes and Washes:* Pectin Syrup; Coffee Cakes, Danish Pastry, Sweet Rolls, etc.; Icing; Glazed Topping for Nut Butter Coffee Cake; Fruit and Berry Glazes; Strawberry and Other Short Cakes; Pecan Bun Glaze; Piping Jelly. *Icings:* Cream; Chocolate; Boiled Chocolate; Pineapple Fluff or Fruit; Fondant Type; White; Coffee Cake; Marshmallow; Peach Fluff; Peach Base; Malted Chocolate; Egg Whip. Whipped Cream; Buttercream Filling; Cookie Flat; Buttercream Icing; "Buttercream"; Doughnut Icing and Glaze; Boiled Icing; Chocolate Fudge. *Biscuit Coatings:* Lemon; Malt Milk; Orange; Dark Chocolate; White; Custard Icing; Boiled Icing or Base; Orange Crumb Topping; Spe-

cial Icing; French Custard Cream Filling; Lemon Cream Pie. Bakers' Flavored Pectin Jelly Powder; Powdered Pie Fillings; Pectin Solutions; Lemon Cheese; Meringue; Custard-Pie Base; Chocolate-Pie Filler; Fruit-Flavor Pie Filler. Spreading Jelly for Bakers; Lemon-Pie Filler; Fig Slab Jellies; Firm Jelly for Bakers; Orange-Jelly Paste; Yeast-Raised Doughnuts with Potatoes; Non-Staling Baked Products; Sugar Wafers; Filler for Sugar Wafers; Sugar-Wafer Shell; Frankfurter Roll; Prepared Pancake Flour; Prepared Biscuit Flours; Prepared Waffle Flours; Bleaching and Maturing Flour; Baking Powder; Dry Lecithin Shortening; Malto-Dextrin; Liquid Red Food Color; Liquid Green Food Color; Liquid Yellow Food Color; Substitute for Cocoanut Oil in Chocolate Icing. Soft Curd Milk; Artificial Cream; Substitute for Coffee Cream; Cream Whipping Aids; Vanilla Ice Cream; Chocolate Ice Cream; Water Ice; Sherbet; Tangerine Water Ice; Stabilizer for Sherbets and Water Ices; Pineapple Sherbet; Ice-Cream Bar Coating; Reducing Churning Time in Butter; Non-Melting Butter; Fat-Free Butter Substitute; Non-Weeping Oleomargarine; Stabilized Homogenized Butter Spread; Cultured Buttermilk; Reducing the Tendency of Blowing in Cheese; Non-Molding Cheese Coating; Cheese Coating Wax; French Dressing; Composition of Dry Mixed Flavoring Material for French Dressing; Mayonnaise; Curry Powder; Celery Salt; Red Pepper Substitute; Capsicine-Free Red Pepper. New-England-Style Pressed Ham; Pork Luncheon Loaf; Blood and Tongue Loaf; Baked Hamburger Loaf; Pro-Lac Sandwich Special; Hamburger Style Patties; Roast-Beef Loaf; Cooked-Ham Loaf; Corned-Beef Loaf; Cooked Corned-Beef and Spaghetti Loaf; Liver and Bacon Loaf; Baked-Veal and Pork Loaf; Luncheon Meat Loaf; Utility Meat Loaf; Defense Meat Loaf; Luxury Meat Loaf; Southern Meat Loaf; Corsica-Style Meat Loaf; Cumberland-Style Meat Loaf; Meat, Pickle and Pimento Loaf; Celery Meat Loaf; Dutch Meat Loaf; Southern Peanut Meat Loaf; Delectable Meat Loaf; Mosaic Liver Loaf; Barbecue-Style Pork Loaf; Liver Loaf; Translucent Liver Loaf; Minced Luncheon Loaf; Skim-Milk Meat Loaf; Tongue and Cheese Loaf; Liver and Cheese Loaf; Meat, Macaroni

PAGE

and Cheese Loaf; Hamburger and Cheese Loaf;
Glaze for Meat Loaves; Barbecue Sauce for Hams;
Scrapple; Breakfast Pork and Apple Patties; Chili
Con Carne; Smoked Roast Chicken; Nutrition Sau-
sage; Cervelate Sausage; Marbled Sausage; Morta-
della-Style Sausage, etc.; Salami Cotto; Skinless
Frankfurters; High-Grade Frankfurter; Vienna-
Style Sausage; Mayence-Style Sausage; Liver Sau-
sage; Braunschweiger-Style Liver Sausage; Marble
Liverwurst; Smoked Metwurst; Bockwurst; Blood
Pudding; Liver Pudding; Liverwurst; Ring Bologna;
Bologna with Soya Flour; Cure for Sausage Meat;
Meat Curing Salt; Meat Preservative; Pork-Sausage
Seasoning; Preserving Lard; Dried Salted Meat; Pro-
tein Food; Oxtail-Type Soup Cubes; Preservation of
Crab Meat; Fish-Preserving Ice; Fish Deodorizer.

VIII. INKS AND MARKING SUBSTANCES 215

Fluorescent; Sympathetic; Detecting Invisible-Ink
Messages; Black Writing; Semi-Gallate Writing;
Hectograph; Duplicating; Quick-Drying Printing;
Black Stencil; Black Marking; Printing; Printing
Offset Composition; Clock Numeral; Glass Mark-
ing; China and Glass; Glass Etch; Marking Porce-
lain; Ceramic Stenciling; Laboratory; India-Ink
Thinner; Erasing Fluid for Tracing Cloth; Reju-
venating Typewriter Ribbons; Transfer Carbon
Paper; Red Marking Crayon; Luminous Crayon;
Hot-Metal Marking Crayon.

IX. SKINS—LEATHER AND FUR 218

Egg-Yolk Substitute for Use in Tanning; Tanning
Extract; Softening Dry Hides; Control of Hide and
Skin Beetles; Dehairing Hides; Waterproofing
Leather Emulsion; Sole Leather Waterproofing;
Preventing Mold Growth on Leather; Oil for Soften-
ing Leather Goods; Leather Belt Stuffing; Box Calf
Oil; Fine-Leather Dressing; Shoe-Bottom Filler;
Wax for Leather Strips; Shoe Box Toe Stiffening;
Shoe Stiffener; Coloring Used Shoes; Black Leather
Dye Solution; Shoe-Sole Stain; Bactericidal Bristles;
Washing Sheepskins; Curing Hairy Sheepskins; Fur
Carroting Solution; After-Chrome for Fur Felt Hats;
Dyeing Fur without Dyeing Skin; Wax Finish for
Furs.

PAGE

X. LUBRICANTS AND OILS 229
Wire Drawing; Metal Can; Powder-Metallurgy Die;
Gasoline Line and Airplane Parts; Gun; Corrosionless
Bearing; Non-Rusting Turbine; Air-Pump; Anti-
Seize Thread; Anti-Seize Paste; Solid; Lubricating-
Oil Corrosion Inhibitor; Anti-Corrosive Spindle;
Clock; Non-Flowing; Oil-Base Well-Drilling Fluid;
Plastic-Molding; Rubber-Mold; Ethyl Cellulose
Molding; Soluble; Metal-Cutting; Forging Tool and
Die; Emulsive; Metal-Quenching; Leather-Packing
and Gasket; Belt-Dressing; Leather-Belt; GR-S Rub-
ber-Belt Dressing; Penetrating Oil; Stopcock Lubri-
cant; High-Temperature Stopcock Grease; Hydro-
carbon-Resistant Stopcock Lubricant; Extracting
Fish Liver Oil; Bleaching Oils and Fats; Prevent-
ing Discoloration of Higher Fatty Acids.

XI. MATERIALS OF CONSTRUCTION 234
Plasticizing Concrete; Concrete Made with Sea Water;
Concrete Road Protective; Hydraulic Cement; Plastic
Building Cement; Building Cement; Oxychloride
Cement; Increasing Fluidity of Cement Mix;
Masons' Mortar; Artificial Building Stone; Preparing
Cement Floors for Painting; Waterproofing Cement
Flooring; Cement Floor Hardener; White "Black"-
Board; Gasoline- and Kerosene-Resistant Plaster;
Waterproofing Cast Gypsum; Catalyst for Quick-
Setting Anhydride Plaster; Retarding the Harden-
ing of Plaster of Paris; Separating Fluid for Dental
Plaster of Paris; High-Strength Brick Tile; Fire-
proof Thermal Insulation; Sound Insulation; Re-
fractory; Clay Refractory; Zircon Refractory; Fusion
Cast Refractory; Refractory Crucible; Hard Refrac-
tory Non-Acid Brick; Refractory; Dental Invest-
ment; Refractory Coating for Metal or Ceramics;
Electric Furnace Luting Lining; Firebrick; Steatite
Body; Ceramic Switch Composition; Electrical Cera-
mic Insulator; High-Frequency Ceramic Insulation;
Ceramic Binder Lubricants; Multiple Ceramic Flux;
Form for Slip Casting Ceramics; Ceramic Pigment;
High-Temperature-Resistant Ceramic Glaze; Pink
Ceramic Underglaze Stain; Clear Vitreous Enamel
for Metal; Vitreous Undercoat for Enamelled Iron;
Permanent Marking of Mica; Ceramic Denture Base;
Clear, Colorless Glass Batch; Low-Expansion Glass;

Optical Glass Batch; Ruby Glass; High-Silica Glass
Batch; Boro-Silicate Enamel; Cork Substitute; Asbestos Roofing Paper; Light-Weight Asphalt Aggregate; Improving Adhesion of Asphalt to Wet Surfaces; Salt-Water Pit Lining; Wood Preservative;
Fireproofing Wood.

XII. METALS AND ALLOYS 242

Iron and Steel Pickling; Stainless Steel Pickling
Solution; Pickling Solution for Chrome; Pickling
Agent for Copper-Beryllium; Passivation of Stainless Steel Wire; Pickling Corrosion Inhibitor; After-
Treatment of Pickled Metals; Rust Inhibitor or Reinhibitor for Internal Combustion Engines; Rendering Stainless Steel Resistant to Sea Water; Prevention of Corrosive Effect of Brine; Cleaning and Protecting Metal; Rustproofing; Phosphatizing Bath;
Purification of Phosphating Baths; Prevention of
Corrosion of Metal in Contact with Chloroform; Pre-
venting Moisture Absorption and Corrosion; Weight
of Silica Gel Required to Prevent Corrosion;
Corrosionproofing Magnesium Alloys; Bright Mag-
nesium Finish; Gray Finish for · Magnesium;
Pre-Painting Treatment for Magnesium; Non-Aque-
ous Metal Cleaning Bath; Stripping Oxide Film from
Aluminum; Pre-Galvanizing Coating; Aluminum
Pre-Painting Treatment; Silver Tarnish Inhibitor;
Repairing Damaged Galvanized Coatings; Tinplate
Protective Coating; Coloring Cadmium Brown;
Coloring Stainless Steel; Black Coating of Stain-
less Steel; Bluing Steel; Cold Tinning Compounds;
Hot Tinning; Tinning Copper; Hot Lead Coating
Alloy; Lead Alloy for Hot Coating Steel; Hot Metal
Coating on One Side Only; Metallic Spray Coating;
Immersion Tin-Plating of Cast Iron; Chromium-
Plating Cast Iron; Chromium-Plating Bath; Pal-
ladium-Plating Bath; Aluminum-Plating; Coating
Aluminum; Nickel-Plating Bath; Nickel-Plating on
Magnesium; Rhenium-Plating; Bright Zinc-Plating;
Brush Electroplating; Manganese-Plating; Simple
Copper-Plating; Silver Mirrors on Glass; Stainless
Steel Anodic Polishing; Electrolytic Metal Polishing;
Electropolishing Steel; Metal Electrocleaning Bath;
Copper-Deplating of Steel; Stereotype Surfacing;
Recovering Silver from Stripping Solutions; Silver-

Plating Magnesium; Silver-Plating without Electricity; Macroscopic Brass Etching; Metallographic Etchant for Silver Solders; Etching Solution for Beryllium; Non-Ferrous Metallographic Etch; Etching Magnesium Metallographic Specimens; Acid Etching Resist; Testing for Traces of Grease or Finger Prints on Metals; Identifying Cadmium, Tin and Zinc Coatings; Powdered Lead Coated Iron; Powdered Iron; Powdered Metals; Sponge Iron Reducing Agent; Metal Carbides; Pulverizing Cemented Tungsten Carbide; Diamond Embedding Composition; Calorizing Metal; Case-Hardening Iron or Steel; Case-Hardening Compound; High-Speed Steel-Hardening Bath; Metal-Tempering Compound; Case-Hardening Small Articles; Resurfacing Cold-Rolled Low-Carbon Steel; Hardened Aluminum Alloy Casting; Permanent Magnet Alloy; Non-Corrosive Bearing; Hardened Lead Bearing Alloy; Cadmium Bearing Alloy; Iron Bearing Alloy; Lead Alloy; Non-Corrosive Lead Alloy; Non-Discoloring Tin Foil; Casting and Rolling Magnesium Alloy; Electric Thermostat; Electric Fuse Alloy; Electrical Contact Alloy; Spark Plug Electrode Alloy; Resistor Alloy; Electric Resistance Alloy; Electric Wire Resistance Alloy; Heat-Hardenable Copper Alloy; Hardening Copper; Corrosion-Resisting Coating for Copper; Cleaning and Covering for Molten Copper Alloys; Tin-Free Gear Bronze; Refining White Metal Alloys; Hard Platinum Alloy; Hard Non-Corrosive Gold Jewelry Alloy; Dental Copper Amalgam; Dental Filling Alloy; Dental Casting Alloy; Dental Model Alloy; Dental Inlay Alloy; Corrosion-Resistant Denture Alloy; Dental Clasp Alloy; Solder for Zinc-Aluminum Alloys; Solder for Aluminum Bronze; Solder for Aluminum Foil; Powdered Solder for Aluminum and Stainless Steels; Aluminum Solders; Hard Solder; Brazing Solder; Gas Meter Solder; Soft Solder for Tinplate; Tinless Soft Solder; Low-Tin Solder; Soft Solder; Soldering Flux; Hard Soldering Flux; Soldering Flux for Galvanized Iron; Foundry Mold Composition; Foundry Mold and Core; Magnesium Alloy Foundry Sand; Mold-Facing Sand; Casting Mold Coating; Ingot Mold Coating; Insulating Compounds for Molten Cast Steel; Dental Investment; Foundry Mold Wash; Mold

Composition for Making Complicated Precision Castings of Aluminum or Brass; Parting Compound to Be
Sprayed on Patterns before Pouring Plaster; Non-
Slip Stair Tread Coating; Welding Rod; Brazing Alloy; Brazing Alloy for Brass, Copper and Steel; Braz-
ing Alloy for Welding to Iron; Copper Brazing
Wire; Welding Rod for Red Brass; White Metal Weld-
ing Composition; Welding Electrode Coating; Alumi-
num Welding Rod Coating; Welding Rod Flux Coat-
ing; Brazing Flux; Nickel-Chrome Alloy Welding
Flux; Gold Melting Flux; Copper Welding Flux;
Flux for No. 3 Welding Copper-Base Materials; Alu-
minum or Magnesium Welding Flux; Flux for
Welding Steel Contaminated with Sulfur; Flux for
Welding Iron-Base Alloys; Non-Adhering Weld
Spatter Coating; Flux for Melting Non-Ferrous
Metals; Non-Corrosive Magnesium Flux; Magnesium-
Refining Flux; Aluminum Scrap Melting Flux.

XIII. PAINT, VARNISH, LACQUER AND OTHER COATINGS . .   271
Exterior House Paints: Outside White House; White
Primer-Undercoat; House Base for Tints; Brick and
Stucco; Brick and Stucco Primer; White Brick and
Stucco Paint; Brick and Stucco Paint Base for Tints;
Exterior Trim and Trellis Paints; Black Trim and
Trellis; Wagon and Tractor Enamels; Traffic Paints;
Congo Resin Traffic Paints; Exterior Metal Paints;
Metal Protection; Pickling and Rustproofing for
Iron; Red Lead Primer; Galvanized Iron Primer;
Zinc Dust-Zinc Oxide Primer; Zinc Chromate
Primer; Zinc Tetroxy Chromate-Iron Oxide Primer;
Zinc Chromate Primer; Alkyd Resin Primer; Trans-
parent Flexible Metal Primer; Red Lead Paint for
Structural Steel; Carbon Black Paint for Steel; Heat-
Resistant Black Enamel; Black Bridge Paint; Metal-
Protective Synthetic Coating; Anti-Corrosive Paint
for Structural Steel; Black Coating Steel; Black Bak-
ing Enamel; Enamel Liquid. *Marine Paints:* Anti-
Corrosive Shipbottom; Anti-Fouling Shipbottom;
Light-Gray Deck; Red Deck; Hull Black; Hull
Light-Gray; Marine Outside White; Navy Chromate
Primer; Fire-Resisting Canvas Preservative; Marine
Interior Flat; Marine Interior Enamel. *Camouflage
Coatings:* Quick-Drying Enamel; Lusterless Sand-
ing Enamel; Lusterless Olive Drab Enamel; Congo

Resin Ammunition Paint; Congo Resin Vehicle for
Gasoline-Soluble and Gasoline-Removable Paint.
Waterproof Coatings: Thermosetting for Raincoats,
Hospital Sheeting and Tents; Vinylite Cement; Flex-
ible Coating for Cloth; Awning; Moistureproof Coat-
ing; Flame-Resisting Paints; Water Emulsion Paints
for Exterior Use; Emulsion Stone Paint; Waterproof-
ing for Cement and Stucco; Sealing and Waterproof-
ing Porous Masonry; Water Paint; Cement-Water
Paint; Congo Resin Camouflage Paints; Oleo-
Resinous Emulsifiable Paint; Congo Resin Fire-
Retardant Paint. *Exterior Varnishes:* Spar; Chemi-
cal-Resistant Spar; Clear Automobile; Traffic
Paint; Synthetic Resin. *Aluminum Vehicles:* Ready-
Mixed Aluminum Alkyd for Brush, Spray or Dip;
Clear Varnish for Heat-Resistant Aluminum Paint;
Wall Coating Vehicle; Congo Resin Treatment and
Congo Vehicles; "Slack Melt" Congo Resin; "Fine
Melt" Congo Resin; Congo Resin Varnishes and Ve-
hicles for Paints and Enamels; Congo Linseed Oil;
Congo Resin Sealer; Interior Flat Wall Finishes;
Sealer; Undercoater; Flat Wall Paint; Interior Semi-
Gloss Finish; Semi-Gloss White Enamel and Base
for Tints. *Interior Gloss Paints:* Varnish KV-91;
Interior Quick-Dry Enamels; Non-Yellowing White
Quick-Dry Enamel; Quick-Dry Enamel Base for
Tints; Quick-Dry Red Enamel; Quick-Dry Green
Enamel; Black Enamel; Blue Quick-Dry Enamel.
Floor Enamels: Liquid Wood Sealer; Green; Brown;
Gray; Interior Emulsion-Type Finishes; Water-Re-
ducible Oil-Type Paste Paint; Congo Resin Emulsion
Paint. *Interior Varnishes:* High-Grade General-Pur-
pose; Floor and Deck; Low-Cost Floor and Trim;
Penetrating Sealer for Floors; Pale-Gloss for Art
Work; Furniture-Rubbing; Synthetic Rubbing;
Wall-Sizing; Insulating Impregnating; White Arch-
itectural Enamel Liquid; Alkyd Blending Varnish;
General-Purpose Grinding Oil, Mixing Varnish and
Paint Oil. *Special Paints and Coatings:* Cold Wall
Glaze; Foundry-Mold; Asphalt Drums; for Iron
Pipes; Protective for Aluminum; Acoustic Sound-
Deadening; Protective for Lithographic Plates; Cor-
rosion and Abrasionproof for Petroleum Refinery
Equipment; Non-Leafing Aluminum Paste Paint;
for Ice Trays; Hot-Melt; Moistureproof; Paint or

PAGE

Lacquer "Stop-Off"; Oilproof Barrel; Oilproof for
Paper Containers; Gilder's Size; Semi-Gloss Polish-
ing Paint; Non-Stick for Asphalt Containers; As-
phalt Drum Lining; Fusible Coating Composition;
Wine Tank Coating; Protective Coatings for Stone
Top Tables; Casein Powder Paints; Protective Fruit
Coatings; Cleaning Paint Brushes; Filling Scratches
on Varnished Furniture; Preventing Skinning of
Paint in Cans; Fire-Resistant Coating; Wood Filler;
Wax Wood Filler; Wax Furniture Filler; Paint-
ing Glazed Tile; Olive Drab Enamel; Black
Enamel; Phosphorescent Paints; Clear Protec-
tive Coat for Phosphorescent Paint; Wood Stains.
Lacquers: White Automobile or Refrigerator; Clear
Furniture; Flat; for Metal or Glass; Transparent
Silver; Clear Metal; Artificial Leather Coating;
Book; Brushing Linoleum; Wood; Cellulose Acetate;
Cheese-Wrapper Foil; Avoiding Stickiness of Lac-
quer-Coated Paper. For Moistureproofing Cello-
phane; Gold; Flat; Lacquer Thinner; Crystallizing;
Frosting Glass; Filling and Binding; Flame-Resist-
ant Coating; Perspirationproof Lacquer Base; Protec-
tive Lacquer for Lenses; Dipping Lacquers for Glass
Identification; Clear, Transparent Metal Lacquer for
Exterior Exposure; Stable Zein-Rosin Solution. Pro-
lamine Plastic Lacquers; Imitation Shellac Lacquer;
Adhesive Scratch-Resisting Primer; Paint Remover;
Finish for Laboratory Table Tops.

XIV. PAPER 325

Tub Sizing for Paper; Rosin Size; Alumina Sol;
Lime in Making Paper Size; Defoamer for Pulp and
Paper Stock; Pest-Resistant Paper; Anti-Mist Paper
for Glass; Paper Transparentizing Solution; Non-
Tarnish Paper; Paper Softening; Paper Glaze Coat-
ing; Non-Slip, Non-Hygroscopic Adhesive-Coated
Paper; Coating or Laminating Paper; Paper Pulp
Oilproofing; Grease-Resistant Paper; Grease-Resistant
Glue Coatings; Oilproofing Paper Containers; Mois-
tureproofing for Cellulose Foil; Paraffin Wax Sizing
and Waterproofing; Wetproof Cigarette Paper; Heat-
Sealable Waxed Paper; Waxed Paper; Improving
Transparency of Waxed Paper; Milk Bottle Cap
Coating; De-inking Newspapers.

PAGE

XV. PHOTOGRAPHY. 329

Sensitized Photographic Paper; Gelatin Composition
for True-to-Scale Prints; Universal Developer; Low-
Contrast Developer; Stable Developer; Economical
Photographic Developing; Non-Dichroic Fog Developer; Fine-Grain Developer; Copper Sulfide Intensification of Film or Prints; Intensifying Baths
for Negatives; Intensifiers; Gelatin-Hardening Solu-
tions; Photographic Hardener; Photographic Fixing
and Hardening Bath; Acid Hardening Fixer; Re-
moval of Yellow Spots on Prints; Restoring Yellow
Prints; Photographic Bleaching Solution; Normaliz-
ing Over-Developed Negatives; Reduction of Over-
Developed Films; Photographic Resist; Eliminating
Abrasion Fog on Photographs; Rapid Drying of
Negatives; Non-Curling Photographic Negatives;
Preventing Drying Spots on Negatives; Preventing
Curl in Photographic Prints; Restoring Shrunken
Photographic Film; Anti-Foam in Photo Solutions;
Fog Prevention; Fog Removal; Improving X-Ray
Photographs; Photographic Etching Bath; Photo-
graphic Print-Out Emulsion; Kallitype Silver Print-
ing; Toning Agents; Reddish-Tone Developer; Blue-
Tone and Blue-Black-Tone Developer; Lithographic
Plate Desensitizing Etch; Lithographic Image Coat-
ing; Photoengraver's Cold Enamel; Albumin Sub-
stitute; Lithographic Paper Coating; Lithographic
Printing Plate Treaments; Photographic Prints on
Oxidized Aluminum; Photographic Substitutes; Re-
storing Faded Documents or Detecting Falsified Docu-
ments; Preventing Silver Nitrate Stains; Blueprint
Paper; Lacquer for Blueprints, Direct Prints, Maps,
etc.; Correction Fluid for Blueprints; Red Writing
Fluid for Blueprints; Correction Fluid for Vandyke
Prints; Developer for B. W. Prints; Recovery of
Silver from Fixing Baths; Photographic Flashlight
Powder.

XVI. POLISHES 341

Paste Wax; Liquid Wax; Auto; Windshield Glass
and Cleaner; Glass; Oven; Silver; Gold and Soft
Metal; Metal Polishing Cloth; Chromium; Leather;
Liquid Leather; Paste Leather; Shoe Paste (for
Tubing); Wax Shoe; Colorless Shoe; Shoe Creams;
Furniture; "Two-Tone" Furniture; Liquid Floor

and Furniture; Floor Wax Remover; Bright Drying
Floor Emulsion; Floor Wax; Diamond Abrasive;
Synthetic Abrasive; Auto-Rubbing Compound;
Buffing Compounds; Cream Buffing Wax.

XVII. PYROTECHNICS AND EXPLOSIVES 349

Ammunition Primer; Non-Combustible Blasting
Primer; Stable Explosive Powder; Blasting Powder;
Oil-Well Explosive Charge; Demolition Explosive;
Torpedo Explosive; Pyrotechnic Flare; Green Pyro-
technic Torch; Colored Rocket Smoke Flares; Colored
Smokes.

XVIII. PLASTICS, RESINS, RUBBER, WAX 351

Waxes: Cartridge; Cheese Coating; Ski; Batik;
Coopers'; Modelling; Ironing (Laundry); Gilders';
Copper Engraving; Cable; Grafting; Gravure; Shoe-
maker's Thread; Saddler's; Inhibiting Discoloration
of Paraffin; Carnauba Substitute; Lanette SX Substitute; Modifiers for Paraffin; Penetration of Modified Paraffin; Dental Basic Plate; Dental Inlay; Dental Adhesive; Dental Carving; Dental Investment;
Dental Duplicating and Impression Compound; Electrolytic Wax Plating. *Plastics:* Sulfur; Self-Hardening; Molded Cork and Cane Stalk; Acrylic Molding
Composition; Polymerization Inhibitor for Methacrylates; Melamine Resin; Ethyl Cellulose; Injection
Molding Powder; Polystyrene Sponge; Stabilizing
Polystyrene; Inhibiting Polymerization of Styrene;
Single Stage Phenol-Formaldehyde Resin; Cast
Phenolic Dye; Molding Powder; Emulsion Polymerization of Vinyl Acetate; Moistureproof Coating
for Acetate Sheet; Tough Bituminous Composition;
Improving Stretch of Cellulose Acetate; Polymeriz-
ing Butadiene; Injection Molding Shellac Com-
pound; Shellac Substitute for Records; Flexible Plas-
tic Piping; Thermoplastic, Tailor's Dummy; Mois-
tureproof Coating for "Cellophane"; Milk Bottle Cap
Coating; Electrical Insulating Filling Compound;
Electrical Insulation; Chemically-Resistant Packing
Gasket; Plastic Stuffing Box Packing; Resin-Im-
pregnated Wood; Bow Rosins; Flesh-Colored Latex
Compound; Preserving Latex; Heat-Sterilizing Fresh
Latex; Solvent for Chlorinated Rubber; Surgeon's
Gloves Dusting Powder; Tire-Puncture Sealing Com-

position; Gasolineproof Synthetic Rubber Composition; Vulcanization of Butadiene Polymers; Emulsion Polymerization; Gas-Expanded Rubber; Reviving "Dead" Rubber Balls; Reclaiming Scrap Rubber; Non-Sticking Rubber; Preventing Adhesion of Rubber; Reducing Viscosity of Rubber Solutions; Gelling Neoprene Latex; GR"S"—Olive Drab Raincoat Material; Self-Curing Lean Cement for Cured GR"S" Coating; Thermosetting Polyvinyl Butyral Coatings for Raincoats; 100% Rubber Reclaim Compounds; Code Wire Insulation; Non-Swelling Thiokol Compound; General-Utility Neoprene Compound; Calendered Raincoat Material; GR"I" Compound for Laminating Fabrics; GR"S" Heel Compound; Highly-Loaded GR"S" for Mechanical Goods; Self-Curing 100% Reclaim Compound; Neoprene Olive Drab Coating; Neoprene Latex Coating; Buna N Oil-Resistant Coating; Men's Rubber Belts; Artificial Sponge; Heat Paper Transfer; Nitrocellulose Coatings; Oil-Resistant Sponge; Ceramic Made without Heat; Molding Clay; Glueless Printers' Roller; Coloring Gelatin Masses.

XIX. SOAP AND CLEANERS   368

Soaps: Home-Made; Rosin; Rosin-Tallow; Non-Efflorescing Bar; Substitute; Lemon; Transparent Glycerin; Fungicidal Borax Soap Washing Mixture; Powdered Cleaner; Detergent Cake; Cleaning Paste; Industrial Hand Soaps and Cleaners; Abrasive Hand Cleaning Powder; Hand Cleaner Requiring No Water; Scouring Powder; Bleaching Soap; Mercury Fulminate Detector; Antiseptic Liquid; Disinfecting Detergent; Detergent Paper Towel; Laundry Blue; Laundry Sour and Bluing; Stabilized Laundry Sour; Spectacle Lens Cleaner and Mist Preventer; Cleaning Solution for Lenses; Fish and Cannery Factory Antiseptic Wash; Cleaning Bottles; Non-Corrosive Laboratory Glass Cleaning Solution; Window and Glassware Cleaners; Saddle Soap; Substitute for Household Ammonia; Leather Cleaners; Antiseptic Dry Cleaning Fluid; Dry Cleaners' Soap; Rug Dry Cleaner; Pile Fabric Cleanser; Rug and Textile Cleaning Powder; Jewelry Cleaning Fluid; Drawing Pen Cleaner; Non-Corrosive Tin Plate Cleaner; Aluminum Cleaner; Pre-Anodizing Metal Cleaner; Uni-

versal Cleaner; Metal-Degreasing Compound; Metal-
Machinery Cleaner; Cylinder Cleaner; Type Cleaner;
Printing-Plate Cleaner; Linotype Matrix Cleaner;
Rust Remover; Brass Cleaner; Brass and Bronze
Cleaner; Removing Dye Stains from the Hands.
Stain Removers: Coffee or Beer; Burnt Sugar;
Blood; Cadmium, Cobalt, Mercury or Nickel Stains;
Chromium; Copper; Manganese; Milk; Mildew;
Picric Acid; Urine; Scorch; Egg Yolk; Grass Stains;
Iodine; Iron; Lead. Floor Sweeping Compound;
Coloring Sweeping Compound.

XX. TEXTILES AND FIBERS 379

Chlorinating Wool; Lusterizing Wool; Textile Dull-
ers for Hosiery; Delustering Shiny Fabric; Deluster-
ing Acetate Silk; Boil-Off of Nylon Fabrics; Thin
Boiling Starch; Textile Finish and Size; Sizing Vis-
cose Ribbon; Sizing Glazed Cotton Yarn; Rayon Size;
Rayon Warp Sizing; Setting Twist in Rayon Knit-
ting Yarns; Creping Fabrics; Scrooping Mercerized
Cotton; Silky-Scroop on Bleached Cotton Yarn;
Scroop on Cotton Yarn; Scroop Effect on Wool;
Wool Finish; Emulsion for Finishing Dyed Rayon;
Thickening Satin Finish for Calicos; Mercerizing
Bath; Unkinking Natural Wool; Softener and Lubri-
cant for Artificial Fibres; Softening and Lubricating
Wool; Stearic Acid Textile Softener; Textile
Weighter; Wool Oiling Emulsions; Textile-Soften-
ing Cream; Textile Scouring Agents; Prefabricating
Silk Treatment; Improving the Wearing Qualities of
Silk Stockings; Bleaching Jute; White Bleach on
Union Goods; Bleaching Textiles; Heat-Stabilization
of Yarn; Crease-Preserving Composition; Jute Twine
Polishing Size; Jute Yarn Dressing; Wool-Like Jute;
Dyeing Sisal and Manila Hemp; Dyeing Straw Hat
Plaits; Dyeing Brush Bristles; Dyeing Grass; Mor-
danting Rayon; Aged Black on Loose Cotton; Dye-
ing Wool Bunting Cloth for Flags; Dyeing Aralac;
Basic Chrome Nitrate-Acetate; Printing Calico with
Catechu; Textile-Printing Paste; Silk-Printing Dis-
charge; Textile-Printing Resist; Improved Textile
Dyeing; Brightening Black Dyeings; Acetate Rayon
"Burnt-Out" Fabrics; Zinc-Dust Resist for Wool;
Reserve for Wool Yarn; Stripping Dye from Wool;
Stripping Dyed Cotton; Flameproof, Waterproof and

Mildewproof Textile Coating; Textile Flameproofing;
Fireproof Cloth; Fireproofing Cotton Goods; Nitrated
Non-Inflammable Lace Fabrics; Wetting and Con-
ditioning Agent for Yarns; Mothproofing Solution;
Mothproofing Emulsion; Rotproofing Fibres; Rot-
proofing Fabrics; Improving Absorbency of Towels;
Porous Waterproofing for Textiles; Waterproofing
Cotton Duck; Textile Waterproofing; Water-Repel-
lent Finish for Ribbons; Making Casein Fibre Resist-
ant to Hot Water; Waterproofing Surgical Dressings;
Increasing Wet Strength of Yarn; Manufacture of
Organdie Cloth; Metallized Yarn; Drum Head and
Banjo Cloth; Asbestos Cloth Substitute.

XXI. MISCELLANEOUS 406

Separation of Butadiene; Non-Caking Cracking Cata-
lyst; Refinery Gas Polymerization Catalyst; Deodoriz-
ing Sulfurous Petroleum; Aviation Fuel; Anti-
Knock Motor Fuel; Stabilizing Tetra Ethyl Lead
Gasoline; Colloidal Fuel; Improving Fuel Oil; Easily
Inflammable Coke; Engine Carbon Solvent; Motor
Gum Solvent; Plugging of Brine Bearing Earth;
Petroleum Desalting and Demulsifying; Removing
Metal Obstructions from Oil Wells; Brine Cooling
Fluid; Preserving Gelatin Solutions; Deodorizing
Isopropyl Alcohol; Filter Aids; Heat-Generating
Composition; Solid Dielectric; Electrical Capacitator;
Temperature Compensating Resistor; Fluorescent
Lamp Coating; Vacuum Tube Getter; Hydraulic
Fluid; Door Check Fluid; Magnetizable Fluid; Man-
ometer Fluid; Anti-Incendiary Material; Magnesium
and Aluminum Powder; Dry Fire Extinguisher; Fire
Foam Extinguisher; Fire Extinguisher Foam Stabi-
lizer; Bonded Fibrous Insulation; Heat-Exchange
Medium; Diamond-Dust Brooch; Clutch Plate Studs;
Brake and Clutch Frictional Material; Sewage Treat-
ment; Water Disinfectant; Carbonaceous Zeolite;
Cation-Exchange Water Softener; Water-Impurity
Coagulant; Coagulant for Hard Water; Water Puri-
fying Compound; Water Softener Briquette; Water
Corrective; Boiler Scale Compound; Processing Tan-
nin for Boiler Water Treatment; Locating Leaks in
Pipe Lines Under Water; Radiator Rusting Inhibi-
tor; Radiator and Water Boiler "Stop Leak"; Auto
Anti-Freeze; Non-Foaming Anti-Freeze; Airplane

De-Icing Coating; Windshield Defroster; Cathode or
X-Ray Tube Screen Coating; X-Ray Protective Com-
position; Animal Tissue Fixative; Regenerating
Cuprous Chloride Solution; Solvent-Resistant Pump
Packings; Settling Fumes; Activated Carbon; Alpha
Cellulose; Lycopodium Substitute; Extracting Sa-
ponin from Soap Bark; Cuprous Oxide; Hydrogen
Peroxide; Potassium Chlorate; Glass Syringe Aids;
Restoring Fraudulent or Faded Documents; Electro-
chemical Recording Solution; Dyestuffs Recom-
mended for the Coloring of Trichlor-ethylene.

TABLES 420

REFERENCES AND ACKNOWLEDGMENTS 427

TRADE NAME CHEMICALS 429

CHEMICALS AND SUPPLIES: WHERE TO BUY THEM 430

SELLERS OF CHEMICALS AND SUPPLIES 444

WHERE TO BUY CHEMICALS OUTSIDE OF THE UNITED STATES . . 454

INDEX 455

ABBREVIATIONS

amp.	ampere
amp./dm²	amperes per square decimeter
amp./sq. ft.	amperes per square foot
anhydr.	anhydrous
avoir.	avoirdupois
Bé.	Baumé
b.p.	boiling point
C.	Centigrade
°C.	degrees Centigrade
cc.	cubic centimeter
c.d.	current density
cm.	centimeter
cm³	cubic centimeter
conc.	concentrated
c.p.	chemically pure
cps.	centipoises
cu. ft.	cubic foot
cu. in.	cubic inch
cwt.	hundredweight
d.	density
dil.	dilute
dm.	decimeter
dm²	square decimeter
dr.	dram
E.	Engler
F.	Fahrenheit
°F.	degrees Fahrenheit
f.f.c.	free from chlorine
f.f.p.a.	free from prussic acid
fl. dr.	fluid dram
fl. oz.	fluid ounce
f.p.	freezing point
ft.	foot
ft.²	square foot
g.	gram
gal.	gallon
gr.	grain
hl.	hectoliter
hr.	hour
in.	inch
kg.	kilogram
l.	liter
lb.	pound
liq.	liquid
m.	meter
min.	minim, minute
ml.	milliliter—cubic centimeter
mm.	millimeter
m.p.	melting point
N.	normal
N.F.	National Formulary
oz.	ounce
pH	hydrogen-ion concentration
p.p.m.	parts per million

pt.	pint
pwt.	pennyweight
q.s.	a quantity sufficient to make
qt.	quart
r.p.m.	revolutions per minute
S.A.E.	Society of Automotive Engineers
sec.	second
sp.	spirits
sp. gr.	specific gravity
sq. dm.	square decimeter
tech.	technical
tinc.	tincture
tr.	tincture
Tw.	Twaddell
U.S.P.	United States Pharmacopeia
v.	volt
visc.	viscosity
vol.	volume
wt.	weight
x.	extra

Chapter I

INTRODUCTION

At the suggestion of a number of teachers of chemistry and home economics the following introductory matter has been included.

The contents of this section are written in a simple way so that anyone, regardless of technical education or experience, can start making simple products without any complicated or expensive machinery. For commercial productions, however, suitable equipment is necessary.

Chemical specialties en masse are composed of pigments, gums, resins, solvents, oils, greases, fats, waxes, emulsifying agents, water, chemicals of great diversity, dyestuffs, and perfumes. To compound certain of these with some of the others requires certain definite and well-studied procedure, any departure from which will inevitably result in failure. The successful steps are given with the formulas. Follow them explicitly. If the directions require that A should be added to B, carry this out literally, and not in reverse fashion. In making an emulsion, the job is often quite as tricky as the making of mayonnaise. In making mayonnaise, you add the oil to the egg, *slowly*, with constant and even and regular stirring. If you do it correctly, you get mayonnaise. If you depart from any of these details: if you add the egg to the oil, or pour the oil in too quickly, or fail to stir regularly, the result is a complete disappointment. The same disappointment may be expected if the prescribed procedure of any other formula is violated.

The next point in importance is the scrupulous use of the proper ingredients. Substitutions are sure to result in inferior quality, if not in complete failure. Use what the formula calls for. If a cheaper product is desired, do not obtain it by substituting a cheaper material for the one prescribed: resort to a different formula. Not infrequently a formula will call for some ingredient which is difficult to obtain: in such cases, either reject the formula or substitute a similar material only after preliminary experiment demonstrates its usability. There is a limit to which this rule may reasonably be extended. In some instances the substitution of an equivalent ingredient may legitimately be made. For example: when the formula calls for *white wax* (beeswax), yellow wax can be used, if the color of the finished product is a matter of secondary importance. Yellow beeswax can often replace white beeswax, making due allowance for color: but paraffin will *not* replace beeswax, even though its light color recommends it above yellow beeswax.

And this leads to the third point: the use of good quality ingredients, and ingredients of the correct quality. Ordinary lanolin is not the same thing as *anhydrous* lanolin: the replacement of one for the other, weight for weight, will give discouragingly different results. Use exactly what the formula calls for: if you are unacquainted with the material and a doubt arises as to just what is meant, discard the formula and use one that you understand. Buy your materials from reliable sources. Many ingredients are obtainable in a number of different grades: if the formula does not designate the grade, it is understood that the best grade is to be used. Remember that a formula and the directions can tell you only a part of the story. Some skill is often required to attain success. Practice with a small batch in such cases until you are sure of your technique. Many instances can be cited. If the formula calls for steeping quince seed for 30 minutes in cold water, your duplication of this procedure may produce a mucilage of too thin a consistency. The originator of the formula may have used a fresher grade of seed, or his conception of what "cold" water means may be different from yours. You should have a feeling for the right degree of mucilaginousness, and if steeping the seed for 30 minutes fails to produce it, steep them longer until you get the right kind of mucilage. If you do not know what the right kind is, you will have to experiment until

1

you find out. Hence the recommendation to make small experimental batches until successful results are arrived at. Another case is the use of dyestuffs for coloring lotions, and the like. Dyes vary in strength: they are all very powerful in tinting value: it is not always easy to state in quantitative terms how much to use. You must establish the quantity by carefully adding minute quantities until you have the desired tint. Gum tragacanth is one of those products which can give much trouble. It varies widely in solubility and bodying power: the quantity prescribed in the formula may be entirely unsuitable for *your* grade of tragacanth. Hence a correction is necessary, which can only be made after experiments to determine *how much* to correct.

In short, if you are completely inexperienced, you can profit greatly by gaining some experience through recourse to experiment. Such products as mouth washes, hair tonics, astringent lotions, need little or no experience, because they are as a rule merely mixtures of simple liquid and solid ingredients, the latter dissolving without difficulty and the whole being a clear solution that is ready for use when mixed. On the other hand, face creams, tooth pastes, lubricating greases, wax polishes, etc., which require relatively elaborate procedure and which depend for their usability on a definite final viscosity, must be made with the exercise of some skill, and not infrequently some experience.

Figuring

Some prefer proportions expressed by weight, volume or in terms of percentages. In different industries and foreign countries various systems of weights and measures are used. For this reason no one set of units could be satisfactory for everyone. Thus diverse formulae appear with different units in accordance with their sources. In some cases, parts instead of percentages or weight or volume is designated. On the pages preceding the index, tables of weights and measures are given. These are of use in changing from one system to another. The following examples illustrate typical units:

Ink for Marking Glass

Glycerin	40	Ammonium Sulfate	10
Barium Sulfate	15	Oxalic Acid	8
Ammonium Bifluoride	15	Water	12

Here no units are mentioned. When such is the case it is standard practice to use parts by weight, using the same system throughout. Thus here we may use ounces or grams as desired. But if ounces are used for one item then ounces must be the unit for all the other items in the particular formula.

Flexible Glue

Glue, Powdered	30.9 %	Glycerin	5.15%
Sorbitol (85%)	15.45%	Water	48.5 %

Where no units of weight or volume but percentages are given then forget the percentages and use the same instructions as given under Example No. 1. Example No. 3

Antiseptic Ointment

Petrolatum	16 parts	Benzoic Acid	1 part
Coconut Oil	12 parts	Chlorthymol	1 part
Salicylic Acid	1 part		

The same instructions as given under Example No. 1 apply to Example No. 3. It is not wise in many cases to make up too large a quantity of material until one has first made a number of small batches to first master the necessary technique and also to see whether it is suitable for the particular outlet for which it is intended. Since, in many cases, a formula may be given in proportions as made up on a commercial factory scale, it is advisable to reduce the proportions accordingly. Thus, taking the following formula: Example No. 4

Neutral Cleansing Cream

Mineral Oil	80 lb.	Water	90 lb.
Spermaceti	30 lb.	Glycerin	10 lb.
Glyceryl Monostearate	24 lb.	Perfume	to suit

Here, instead of pounds, grams may be used. Thus this formula would then read:

Mineral Oil	80 g.	Water	90 g.
Spermaceti	30 g.	Glycerin	10 g.
Glyceryl Monosterate	24 g.	Perfume	to suit

Reduction in bulk may also be obtained by taking the same fractional part or portion of each ingredient in a formula. Thus in the following formula:

Example No. 5

Vinegar Face Lotion

Acetic Acid (80%)	20	Alcohol	440
Glycerin	20	Water	500
Perfume	20		

We can divide each amount by ten and the finished bulk is only 1/10th of the original formula. Thus it becomes:

Acetic Acid (80%)	2	Alcohol	44
Glycerin	2	Water	50
Perfume	2		

Apparatus

For most preparations, pots, pans, china and glassware, such as are used in every household, will be satisfactory. For making fine mixtures and emulsions a "malted-milk" mixer or egg-beater is necessary. For weighing, a small, low priced scale should be purchased from a laboratory supply house. For measuring of fluids, glass graduates or measuring glasses may be purchased from your local druggist. Where a thermometer is necessary a chemical thermometer should be obtained from a druggist or chemical supply house.

Methods

To better understand the products which you intend making, it is advisable that you read the complete section covering such products. Very often an important idea is thus obtained. You may learn different methods that may be used and also avoid errors which many beginners are prone to make.

Containers for Compounding

Where discoloration or contamination is to be avoided (as in light-colored, or food and drug products) it is best to use enameled or earthenware vessels. Aluminum is also highly desirable in such cases, but it should not be used with alkalies which dissolve and corrode this metal.

Heating

To avoid overheating, it is advisable to use a double boiler when temperatures below 212° F. (temperature of boiling water) will suffice. If a double boiler is not at hand, any pot may be filled with water and the vessel containing the ingredients to be heated is placed in it. The pot may then be heated by any flame without fear of overheating. The water in the pot, however, should be replenished from time to time as necessary—it must not be allowed to "go dry." To get uniform higher temperatures, oil, grease or wax is used in the outer container in place of water. Here of course care must be taken to stop heating when thick fumes are given off as these are inflammable. When higher uniform temperatures are necessary, molten lead may be used as a heating medium. Of course, where materials melt uniformly and stirring is possible, direct heating over an open flame is permissible.

Where instructions indicate working at a certain temperature, it is important that the proper temperature be attained—not by guesswork, but by the use of a thermometer. Deviations from indicated temperatures will usually result in spoiled preparations.

Temperature Measurements

In Great Britain and the United States, the Fahrenheit scale of temperature measurement is used. The temperature of boiling water is 212° Fahrenheit (212° F.) ; the temperature of melting ice is 32° Fahrenheit (32° F.).

In scientific work and in most foreign countries the Centigrade scale is used. On this scale of temperature measurement, the temperature of boiling water is 100 degrees Centigrade (100° C.) and the temperature of melting ice is 0 degrees Centigrade (0° C.).

The temperature of liquids is measured by a glass thermometer. The latter is inserted as deeply as possible in the liquid and is moved about until the temperature remains steady. It takes a little time for the glass of the thermometer to come to the temperatures of the liquid. The thermometer should not be placed against the bottom or side of the container, but near the center of the liquid in the vessel. Since the glass of the bulb of the thermometer is very thin, it can be broken easily by striking it against any hard surface. A cold thermometer should be warmed gradually (by holding over the surface of a hot liquid) before immersion. Similarly the hot thermometer when taken out should not be put into cold water suddenly. A sharp change in temperature will often crack the glass.

Mixing and Dissolving

Ordinary solution (e.g. sugar in water) is hastened by stirring and warming. Where the ingredients are not corrosive, a clean stick, bone or composition fork or spoon is used as a mixing device. These may also be used for mixing thick creams or pastes. In cases where most efficient stirring is necessary (as in making mayonnaise, milky polishes, etc.) an egg-beater or a malted-milk mixer is necessary.

Filtering and Clarification

When dirt or undissolved particles are present in a liquid, they are removed by settling or filtering. In the former the solution is allowed to stand and if the particles are heavier than the liquid they will gradually sink to the bottom. The upper liquid may be poured or siphoned off carefully and in some cases is then of sufficient clarity to be used. If, however, the particles do not settle out, then they must be filtered off. If the particles are coarse they may be filtered or strained through muslin or other cloth. If they are very small, then filter paper is used. Filter papers may be obtained in various degrees of fineness. Coarse filter paper filters rapidly but will not, of course, take out extremely fine particles. For the latter, it is necessary to use a very fine grade of filter paper. In extreme cases even this paper may not be fine enough. Here it will be necessary to add to the liquid 1-3% of infusorial earth or magnesium carbonate. The latter clog up the pores of the filter paper and thus reduce their size and hold back undissolved material of extreme fineness. In all such filtering, it is advisable to take the first portions of the filtered liquid and pour them through the filter again as they may develop cloudiness in standing.

Decolorizing

The most commonly used decolorizer is decolorizing carbon. The latter is added to the liquid to the extent of 1-5% and heated with stirring for ½ hour to as high a temperature as is feasible. It is then allowed to stand for a while and filtered. In some cases bleaching must be resorted to. Examples of this are given in this book.

Pulverizing and Grinding

Large masses or lumps are first broken up by wrapping in a clean cloth and placing between two boards and pounding with a hammer. The smaller pieces are then pounded again to reduce their size. Finer grinding is done in a mortar with a pestle.

Spoilage and Loss

All containers should be closed when not in use to prevent evaporation or contamination by dust; also because, in some cases, air affects the material adversely. Many materials attack or corrode the metal containers in which they are received. This is particularly true of liquids. The latter, therefore, should be transferred to glass bottles which should be as full as possible. Corks should be covered with aluminum foil (or dipped in melted paraffin wax when alkalies are present).

Materials such as glue, gums, olive oil or other vegetable or animal products may ferment or become rancid. This produces discoloration or unpleasant odors. To avoid this, suitable antiseptics or preservatives must be used. Too great stress cannot be placed on cleanliness. All containers must be cleaned thoroughly before use to avoid various complications.

Weighing and Measuring

Since, in most cases, small quantities are to be weighed, it is necessary to get a light scale. Heavy scales should not be used for weighing small amounts as they are not accurate for this type of weighing.

For measuring volume (liquids) measuring glasses or cylinders (graduates) should be used. Since this glassware cracks when heated or cooled suddenly it should not be subjected to sudden changes of temperature.

Caution

Some chemicals are corrosive and poisonous. In many cases they are labeled as such. As a precautionary measure, it is advised not to smell bottles directly, but only to sniff a few inches from the cork or stopper. Always work in a well ventilated room when handling poisonous or unknown chemicals. If anything is spilled, it should be wiped off and washed away at once.

Where to Buy Chemicals and Apparatus

Many chemicals and most glassware can be purchased from your druggist. A list of suppliers of all products will be found at the end of this book.

Advice

This book is the result of cooperation of many chemists and engineers who have given freely of their time and knowledge. It is their business to act as consultants and, for a fee, to give advice on technical matters. As publishers, we do not maintain a laboratory or consulting service to compete with them.

Please, therefore, do not ask us for advice or opinions, but confer with a chemist in your vicinity.

Extra Reading

Keep up with new developments of new materials and methods by reading technical magazines. Many technical publications are listed under references in the back section of this book.

Calculating Costs

Purchases of raw materials, in small quantities, are naturally higher in price than when bought in large quantities. Commercial prices, as given in the trade papers and catalogs of manufacturers, are for quantities such as barrels, drums or sacks. For example, a pound of epsom salts, bought at retail, may cost 10 or 15 cents. In barrel lots its price today is about 2 to 3 cents per pound.

Typical Cost Calculation
Formula for Beer or Milk Pipe Cleaner

Soda Ash	25 lb. @	.02½ per lb.	= $0.63
Sodium Perborate	75 lb. @	.16 per lb.	= 12.00

Total 100 lb.	Total $12.63

If 100 lb. cost $12.63, 1 lb. will cost $12.63 divided by 100 or about $0.126 per lb. for raw materials, assuming no loss.

Always weigh the amount of finished product and use *this* weight in calculating costs. Most compounding results in some loss of material because of spillage, sticking to apparatus, evaporation, etc. Costs of making experimental lots are always high and should not be used for figuring costs. To meet competition, it is necessary to buy in larger units and costs should be based on the latter.

ELEMENTARY PREPARATIONS

The recipes that follow have been formulated in a very simple way. Only one of each type is given to avoid confusion. These have been selected because of their importance and because they can be made readily.

The succeeding chapters go into greater detail and give many different types and modifications of these and other recipes for home and commercial use.

Cleansing Creams

Cleansing creams as the name implies serve as skin cleaners. Their basic ingredients are oils and waxes which are rubbed into the skin. When wiped off they carry off dirt and dead skin. The liquefying type of cleansing cream contains no water and melts or liquefies when rubbed on the skin. To suit different climates and likes and dislikes harder or softer products can be made.

Cleansing Cream (Liquefying)

Liquid Petrolatum (White Mineral Oil)	5½ oz.
Paraffin Wax	2½ oz.
Petrolatum	2 oz.

Melt together with stirring in an aluminum or enamelled dish and allow to cool. Then stir in a dash of perfume oil. Allow to stand until a haziness appears and then pour into jars, which should be allowed to stand *undisturbed* overnight.

Cold Creams

The most important facial cream is cold cream. This type of cream consists of a mineral oil and wax which are emulsified in water with a little borax or glycosterin. The function of a cold cream is to furnish a greasy film which takes up dirt and waste tissue which are removed when the skin is wiped thoroughly. Many modifications of this basic cream are encountered in stores. They vary in color, odor, **and** in claims but, essentially, they are not more useful than this simple cream. The latest type of cold cream is the non-greasy cold cream which is of particular interest because it is non-alkaline and, therefore, non-irritating to sensitive skins.

Cold Cream

Liquid Petrolatum (White Mineral Oil)	52 g.
White Beeswax	14 g.

Heat the above in an aluminum or enamelled double boiler (the water in the outer pot should be brought to a boil). In a separate aluminum or enamelled pot dissolve:

Borax	1 g.
Water	33 cc.

and bring this to a boil. Add it in a thin stream, to the melted wax, while stirring vigorously in one direction only. Use a fork for stirring. When the mixture turns to a smooth thin cream, immerse the bottom of the thermometer in it from time to time, stirring continuously. When the temperature drops to 140° F. add ½ cc. of perfume oil and continue stirring until the temperature drops to 120° F. At this point pour into jars where the cream will "set" after a while. If a harder cream is desired, reduce the amount of liquid petrolatum. If a softer cream is wanted increase it.

Cold Cream (Non-Greasy)

White Paraffin Wax	1¼
Petrolatum	1½
Glycosterin or Glyceryl Monostearate	2¼
Liquid Petrolatum (White Mineral Oil)	3

Heat the above in an aluminum or enamelled double boiler (the water in the outer pot should be boiling). Stir until clear. To this slowly add, while stirring vigorously with a fork,

Water (boiling)	10

Continue stirring until smooth and then add, with stirring, a little perfume oil. Pour into jars at 110–130° F. and cover the jars as soon as possible.

Vanishing Creams

Vanishing creams are non-greasy creams, soapy in nature. Some are white and others have a very beautiful pearly appearance. This type of cream depends on the soapiness for its cleansing character and is useful as a powder base.

Vanishing Cream

Stearic Acid	18 oz.

Melt the above in an aluminum or enamelled double boiler (the water in the outer pot must be boiling). To the above add, in a thin stream, while stirring vigorously with a fork, the following boiling solution made in an aluminum or enamelled pot:

Potassium Carbonate	¼ oz.
Glycerin	6½ oz.
Water	5 lb.

Continue stirring until the temperature falls to 135° F., then stir in a little perfume oil and stir from time to time until cold. Allow to stand overnight and stir again the next day. Pack into jars which should be closed tightly.

Hand Lotions

Hand lotions are usually clear or milky liquids or salves which are useful in protecting the skin from roughness and redness because of exposure to cold, hot water, soap and other materials. "Chapped" hands are a common occurrence. The use of a good hand lotion keeps the skin smooth, soft, and in a normally healthy condition. The lotion is best applied at night, rather freely, and cotton gloves may be worn to prevent soiling. During the day it should be put on sparingly and the excess wiped off.

Hand Lotion (Salve)

Boric Acid	1
Glycerin	6

Warm the above in an aluminum or enamelled dish and stir with a clean wooden stick until dissolved (clear). Then allow to cool and work into the following mixture with a potato masher, or rounded stick, adding only a little of the above liquid at a time to the mixture below and not adding a further portion until it is fully absorbed.

Lanolin	6
Petrolatum	8

If it is desired to impart a pleasant odor to this lotion a little perfume may be added and worked in.

Hand Lotion (Milky Liquid)

Lanolin	¼ teaspoonful
Glycosterin or Glyceryl Monostearate	1 oz.
Tincture of Benzoin	2 oz.
Witch Hazel	25 oz.

Melt the first two items together in an aluminum or enamelled double boiler. If no double boiler is at hand improvise one by standing the dish in a small pot containing boiling water. When the mixture becomes clear, remove from the double boiler and add slowly, while stirring vigorously with a fork or stick, the tincture of benzoin and then the witch hazel. Continue stirring until cool and then put into one or two large bottles and shake vigorously. The finished lotion is a beautiful milky liquid comparable to the best hand lotions on the market sold at high prices.

Brushless Shaving Creams

Brushless or latherless shaving creams are soapy in nature and do not require lathering or water. The formula given below is of the latest type being free from alkali and non-irritating. It should be borne in mind, however, that certain beards are not softened by this type of cream and require the old-fashioned lathering shaving cream.

Brushless Shaving Cream
White Mineral Oil	10
Glycosterin or Glyceryl Monostearate	10
Water	50

Heat the first two ingredients together in a Pyrex or enamelled dish to 150° F. and into this run slowly, while stirring with a fork, the water which has been heated to boiling. Allow to cool to 105° F. and while stirring add a few drops of perfume oil. Continue stirring until cold.

Mouth Washes

Mouth washes and oral antiseptics are of practically negligible value. Many, however, insist on their use because of their refreshing taste and deodorizing value.

Mouth Wash
Benzoic Acid	⅝
Tincture of Rhatany	3
Alcohol	20
Peppermint Oil	⅛

Just shake together in a dry bottle until it is dissolved and it is ready. A teaspoonful is used to a small wineglassful of water.

Tooth Powders

Tooth powders depend for their cleansing action on soap and mild abrasives such as precipitated chalk and magnesium carbonate. The antiseptic present is practically of no value. The flavoring ingredients mask the taste of the soap and give the user's mouth a pleasant after-taste.

Tooth Powder
Magnesium Carbonate	420 g.
Precipitated Chalk	565 g.
Sodium Perborate	55 g.
Sodium Bicarbonate	45 g.
Soap, Powdered White	50 g.
Sugar, Powdered	90 g.
Wintergreen Oil	8 cc.
Cinnamon Oil	2 cc.
Menthol	1 g.

Dissolve the last three ingredients together and then rub well into the sugar. Add the soap and perborate mixing in well. Add the chalk with good mixing and then the sodium bicarbonate and magnesium carbonate. Mix thoroughly and sift through a fine wire screen. Keep dry.

Foot Powders

Foot powders consist of a filler such as talc or starch with or without an antiseptic or deodorizer. In the following formula the perborates liberate oxygen when in contact with perspiration which tends to destroy unpleasant odors. The talc acts as a lubricant and prevents friction and chafing.

Foot Powder
Sodium Perborate	3
Zinc Peroxide	2
Talc	15

Shake together thoroughly in a dry container until uniformly mixed. This powder must be kept dry or it will spoil.

Liniments

Liniments usually consist of an oil and an irritant such as methyl salicylate or turpentine. The oil acts as a solvent and tempering agent for the irritant. The irritant produces a rush of blood and warmth which is often slightly helpful.

Sore Muscle Liniment
Olive Oil	6
Methyl Salicylate	3

Shake together and keep in a well stoppered bottle. Apply externally but do not apply to chafed or cut skin.

Chest-Rubs

In spite of the fact that chest-rubs are practically useless countless sufferers use them. Their action is similar to that of liniments and they differ only in that they are in the form of a salve.

"Chest-Rub" Salve

Yellow Petrolatum	1	lb.
Paraffin Wax	1	oz.
Eucalyptus Oil	2	fl. oz.
Menthol	½	oz.
Cassia Oil	⅛	fl. oz.
Turpentine	½	fl. oz.

Melt the petrolatum and paraffin wax together in a double boiler and then add the menthol. Remove from the heat, stir, and cool a little; then stir in the oils, turpentine, and acid. When it begins to thicken pour into tins and cover.

Insect Repellents

Preparations of this type may irritate sensitive skins. Moreover, they will not always work. Psychologically they often are helpful, even though they may not keep insects away, because they give one confidence of protection.

Mosquito Repelling Oil

Cedar Oil	2
Citronella Oil	4
Spirits of Camphor	8

Just shake together in a dry bottle and it is ready for use. This preparation may be smeared on the skin as often as is necessary to repel mosquitoes and other insects.

Fly Sprays

Fly sprays usually consist of deodorized kerosene, perfuming material, and an active insecticide. In some cases they merely stun the flies who may later recover and begin buzzing again.

Fly Spray

Deodorized Kerosene	89 fl. oz.
Methyl Salicylate	1 fl. oz.
Pyrethrum Powder	10 oz.

Mix thoroughly by stirring from time to time; allow to stand covered overnight and then filter through muslin.

Caution: This spray is inflammable and should not be used near open flames.

Deodorant Spray

(For public buildings, sick-rooms, lavatories, etc.)

Pine Needle Oil	2
Formaldehyde	2
*Acetone	6
*Isopropyl Alcohol	20

One ounce of the above is mixed with a pint of water for spraying.

Cresol Disinfectant

†Caustic Soda	25½ g.
Water	140 cc.

Dissolve the above in a Pyrex or enamelled dish and warm it. To this add slowly the following warmed mixture:

†Cresylic Acid	500 cc.
Rosin	170 g.

Stir until dissolved and add water to make 1000 cc.

Ant Poison

Sugar	1 lb.
Water	1 qt.
‡Arsenate of Soda	125 g.

Boil and stir until uniform; strain through muslin; add a spoonful of honey.

Bedbug Exterminator

*Kerosene	90
Clove Oil	5
§Cresol	1
Pine Oil	4

Simply shake and bottle.

Mothproofing Fluid (Non-Staining)

Sodium Aluminum Silico-fluoride	½
Water	98
Glycerin	½
Sulfatate (Wetting Agent)	¼

Stir until dissolved.

Fly Paper

Rosin	32
Rosin Oil	20
Castor Oil	8

Heat the above in an aluminum or enamelled pot on a gas stove with stirring until all the rosin has melted and dissolved. While hot pour on firm paper sheets of suitable size which have been brushed with soap water just before coating. Smooth out the coating with a long knife or piece of thin flat wood and allow to cool. If a heavier coating is desired increase the amount of rosin used. Similarly a thinner coating is obtained by reducing the amount of rosin. The finished paper should be laid flat and not exposed to undue heat.

Household Products

Household Baking Powder

Bicarbonate of Soda	28
Mono Calcium Phosphate	35
Corn Starch	27

* Inflammable.
† Should not touch skin as it is corrosive.
‡ Poison.
§ Corrosive to skin.

Mix the above powders thoroughly in a dry can by shaking and rolling for a half hour. Pack into dry air-tight tins as moisture will cause lumping.

Malted Milk Powder
Malt Extract, Powdered	5
Skim Milk, Powdered	2
Sugar, Powdered	3

Mix thoroughly by shaking and rolling in a dry can. Pack in an air-tight container.

Cocoa Malt Powder
Corn Sugar	55
Malt, Powdered, Mild	19
Skim Milk, Powdered	12½
Cocoa	13
Vanillin	⅛
Salt, Powdered	⅜

Mix thoroughly and then run through a fine wire sieve.

Sweet Cocoa Powder
Cocoa	17½ oz.
Sugar, Powdered	32½ oz.
Vanillin	¾ g.

Mix thoroughly and sift.

Pure Lemon Extract
Lemon Oil U.S.P.	6½ fl. oz.
Alcohol	121½ fl. oz.

Shake together in a gallon jug till dissolved.

Artificial Vanilla Flavor
Vanillin	¾ oz.
Coumarin	¼ oz.
Alcohol	2 pt.

Stir the above in a glass or china pitcher until dissolved. Then stir in the following solution which has been made by stirring in another pitcher.
Sugar	12 oz.
Water	5¼ pt.
Glycerin	1 pt.

Color brown by adding sufficient "burnt" sugar coloring.

Canary Bird Food
Yolk of Eggs, Dried and Chopped	2
Poppy Heads (Coarse Powder)	1
Cuttlefish Bone (Coarse Powder)	1
Granulated Sugar	2
Soda Crackers, Powdered	8

Mix well together.

Writing Ink (Blue-Black)
Naphthol Blue Black	1 oz.
Gum Arabic, Powdered	½ oz.
Carbolic Acid	¼ oz.
Water	1 gal.

Stir together in a glass or enamelled vessel until dissolved.

Laundry Marking Ink (Indelible)
A. Soda Ash	1 oz.
Gum Arabic, Powdered	1 oz.
Water	10 fl. oz.

Stir the above until dissolved.
B. Silver Nitrate	4 oz.
Gum Arabic, Powdered	4 oz.
Lampblack	2 oz.
Water	40 fl. oz.

Stir this in a glass or porcelain dish until dissolved. Do not expose it to strong light or it will spoil. Finally pour into a brown glass bottle. In using these solutions wet the cloth with solution A and allow to dry. Then write on it with solution B using a quill pen.

Marking Crayon (Green)
Ceresin	8
Carnauba Wax	7
Paraffin Wax	4
Beeswax	1
Talc	10
Chrome Green	3

Melt the first four ingredients in any container and then add the last two slowly while stirring. Remove from the heat and continue stirring until thickening begins. Then pour into molds. If other color crayons are desired, other pigments may be used. For example for black, use carbon or bone-back; for blue, Prussian blue; for red, orange chrome yellow.

Antique Coloring for Copper
Copper Nitrate	4
Acetic Acid	1
Water	2

Dissolve by stirring together in a glass or porcelain vessel. Pack in glass bottles.

To use: Wet the copper to be colored and apply the above solution hot.

Blue-Black Finish on Steel
a. Place object in molten sodium nitrate (700–800° F.) for 2–3 minutes. Remove and allow to cool somewhat; wash in hot water; dry and oil with mineral or linseed oil.

b. Place in following solution for 15 minutes:
Copper Sulfate	½ oz.
Iron Chloride	1 lb.
Hydrochloric Acid	4 oz.

Nitric Acid ½ oz.
Water 1 gal.
Then allow'to dry for several hours; place in above solution again for 15 min.; remove and dry for 10 hours. Place in boiling water for ½ hour; dry and scratch brush very lightly. Oil with mineral or linseed oil and wipe dry.

Rust Prevention Compound
Lanolin 1
*Naphtha 2
Mix until dissolved.
The metal to be protected is cleaned with a dry cloth and then coated with the above composition.

Metal Polish
Naphtha 62 oz.
Oleic Acid ⅓ oz.
Abrasive 7 oz.
Triethanolamine Oleate ⅓ oz.
Ammonia (26°) 1 oz.
Water 1 gal.
In one container mix together the naphtha and oleic acid to a clear solution. Dissolve the triethanolamine oleate in water separately, stir in the abrasive, if it is of a clay type, and then add the naphtha solution. Stir the resulting mixture at a high speed until a uniform creamy emulsion results. Then add the ammonia and mix well, but do not agitate as vigorously as before.

Glass Etching Fluid
Hot Water 12 fl. oz.
†Ammonium Bifluoride 15 oz.
Oxalic Acid 8 oz.
Ammonium Sulfate 10 oz.
Glycerin 40 oz.
Barium Sulfate 15 oz.
Warm the washed glass slightly before writing on it with this fluid. Allow the fluid to act on the glass for about two minutes.

Leather Preservative
Neatsfoot Oil (Cold Pressed) 10
Castor Oil 10
Just shake together.
This is an excellent preservative for leather book bindings, luggage and other leather goods.

White Shoe Dressing
Lithopone 19 oz.
Titanium Dioxide 1 oz.
Shellac (Bleached) 3 oz.

* Inflammable—keep away from flames.
† Corrosive.

Ammonium Hydroxide ¼ fl. oz.
Water 25 fl. oz.
Alcohol 25 fl. oz.
Glycerin 1 oz.
Dissolve the last four ingredients by mixing in a porcelain vessel. When dissolved, stir in the first two pigments. Keep in stoppered bottles and shake before using.

Waterproofing for Shoes
Wool Grease 8
Dark Petrolatum 4
Paraffin Wax 4
Melt together in any container. Apply this grease warm but never hotter than the hand can bear.

Polishes
Polishes are usually used to restore the original luster and finish of a smooth surface. As a secondary purpose they are expected to clean the surface and also to prevent corrosion or deterioration. There is no one polish which will give good results on all surfaces.

Most polishes depend on oil or wax for their lustering or polishing properties. Oil polishes are applied easily but the surfaces on which they are used attract dust and show finger marks. Wax polishes are more difficult to apply but are more lasting.

Oil or wax polishes are of two types: waterless and with water. The former are clear or translucent and the latter are milky in appearance.

For use on metals abrasives of various kinds such as tripoli, silica dust or infusorial earth are incorporated to grind away oxide films or corrosion products present.

Shoe Polish (Black)
Carnauba Wax 5½ oz.
Crude Montan Wax 5½ oz.
Melt together in a double boiler (the water in outer container should be at a boil), then stir in the following melted and dissolved mixture:
Stearic Acid 2 oz.
Nigrosine Base 1 oz.
Then stir in
Ceresin 15 oz.
Remove all flames and run in slowly, while stirring
Turpentine 90 fl. oz.
Allow mixture to cool to 105° F. and pour into air-tight tins which should be allowed to stand undisturbed overnight.

Auto Polish (Clear Oil Type)

Paraffin (Mineral) Oil	5	pt.
Raw Linseed Oil	2	pt.
China Wood Oil	½	pt.
*Benzol	¼	pt.
Kerosene	¼	pt.
Amyl Acetate	1	tbsp.

Shake together in a glass jug and keep stoppered.

Auto and Floor Wax (Paste Type)

Yellow Beeswax	1	oz.
Ceresin	2½	oz.
Carnauba Wax	4½	oz.
Montan Wax	1¼	oz.
*Naphtha or Mineral Spirits	1	pt.
*Turpentine	2	oz.
Pine Oil	½	oz.

Melt the waxes together in a double boiler. Turn off the heat and run in the last three ingredients in a thin stream and stir with a fork. Pour into cans; cover and allow to stand undisturbed overnight.

Furniture Polish (Oil and Wax Type)

Thin Paraffin (Mineral Oil)	1	pt.
Carnauba Wax, Powdered	¼	oz.
Ceresin Wax	⅛	oz.

Heat together until all of the wax is melted. Allow to cool and pour into bottles before mixture turns cloudy.

Polishing Wax (Liquid)

Beeswax, Yellow	1	oz.
Ceresin Wax	⅛	oz.

Melt together and then cool to 130° F.; turn off all flames and stir in slowly

*Turpentine	17	fl. oz.
Pine Oil	½	fl. oz.

Pour into cans or bottles which are closed tightly to prevent evaporation.

Floor Oil

Mineral Oil	46	fl. oz.
Beeswax	½	oz.
Carnauba Wax	1	oz.

Heat together in double boiler until dissolved (clear). Turn off flame and stir in

*Turpentine	3	fl. oz.

Lubricants

Lubricants in the form of oils or greases are used to prevent friction and wearing of parts which rub together. Lubricants must be chosen to fit specific

*Inflammable—Keep away from flames.

uses. They consist of oils and fats often compounded with soaps and other unctuous materials. For heavy duty heavy oils or greases are used and light oils for light duty.

Gun Lubricant

White Petrolatum	15
Bone Oil (Acid Free)	5

Warm gently and mix together.

Graphite Grease

Ceresin	7
Tallow	7

Warm together and gradually work in, with a stick

Graphite	3

Stir until uniform and pack in tins when thickening begins.

Penetrating Oil
(For freeing rusted bolts, screws, etc.)

Kerosene	2
Thin Mineral Oil	7
Secondary Butyl Alcohol	1

Shake together and keep in a stoppered bottle.

Molding Material

White Glue	13	lb.
Rosin	13	lb.
Raw Linseed Oil	⅓	qt.
Glycerin	1	qt.
Whiting	19	lb.

This mixture is prepared by cooking the white glue until it is dissolved. Then cook separately the rosin and raw linseed oil until they are dissolved. Add the rosin, oil, and glycerin to the cooked glue, stirring in the whiting until the mass makes up to the consistency of putty. Keep the mixture hot.

Place this putty mass in the die, pressing it firmly into the same and allowing it to cool slightly before removing. The finished product is ready to use within a few hours after removal. Suitable colors can be added to secure brown, red, black or other color.

In applying ornaments made of this composition to a wood surface, they are first steamed to make them flexible; in this condition they can be glued to the wood surface easily and securely. They can be bent to any shape, and no nails are required for applying them.

Grafting Wax

Wool Grease	11
Rosin	22
Paraffin Wax	6

Beeswax	4
Japan Wax	1
Rosin Oil	9
Pine Oil	1

Melt together until clear and pour into tins. This composition can be made thinner by increasing the amount of rosin oil and thicker by decreasing it.

Candles

Paraffin Wax	30
Stearic Acid	17½
Beeswax	2½

Melt together and stir until clear. If colored candles are desired a pinch of any oil-soluble dye is dissolved at this stage. Pour into vertical molds in which wicks are hung.

Adhesives

Adhesives are sticky substances used to unite two surfaces. Adhesives are specifically called glues, pastes, cements, mucilages, lutes, etc. For different uses different types are required.

Wall Patching Plaster

Plaster of Paris	32
Dextrin	4
Pumice Powder	4

Mix thoroughly by shaking and rolling in a dry container. Keep away from moisture.

Cement Floor Hardener

Magnesium Fluosilicate	1 lb.
Water	15 pt.

Mix until dissolved.
In using this, the cement should first be washed with clean water and then drenched with the above solution.

Paperhanger's Paste

Use a cheap grade of rye or wheat flour, mix thoroughly with cold water to about the consistency of dough or a little thinner, being careful to remove all lumps. Stir in a tablespoonful of powdered alum to a quart of flour, then pour in boiling water, stirring rapidly until the flour is thoroughly cooked. Let this cool before using and thin with cold water.

a. White or Fish Glue	4 oz.
Cold Water	8 oz.
b. Venice Turpentine	2 fl. oz.
c. Rye Flour	1 lb.
Cold Water	16 fl. oz.
d. Boiling Water	64 fl. oz.

Soak the 4 oz. of glue in the cold water for 4 hours. Dissolve on a water bath (glue-pot) and while hot stir in the Venice turpentine. Make up c into a batter free from lumps and pour into d. Stir briskly, and finally add the glue solution. This makes a very strong paste, and it will adhere to a painted surface, owing to the Venice turpentine in its composition.

Aquarium Cement

Litharge	10
Plaster of Paris	10
Powdered Rosin	1
Dry White Sand	10
Boiled Linseed Oil	Sufficient

Mix all together in the dry state, and make into a stiff putty with the oil when wanted for use.
Do not fill the aquarium for three days after cementing. This cement hardens under water, and will stick to wood, stone, metal, or glass, and, as it resists the action of sea-water, it is useful for marine aquaria. The linseed oil may have an addition of drier to the putty made up four or five hours before use, but after standing fifteen hours, it loses its strength when in the mass.

Wood Dough Plastic

*Collodion	86
Ester Gum, Powdered	9
Wood Flour	30

Allow first two ingredients to stand until dissolved, stirring from time to time. Then while stirring add the wood flour a little at a time until uniform. This product can be made softer by adding more collodion.

Putty

Whiting	80
Raw Linseed Oil	16

Rub together until smooth. Keep in closed container.

Wood Floor Bleach

Sodium Metasilicate	90
Sodium Perborate	10

Mix thoroughly and keep dry in a closed can. Use 1 pound to a gallon of boiling water. Mop or brush on the floor, allow to stand ½ hour, then rub off and rinse well with water.

* Paint Remover

Benzol	5	pt.
Ethyl Acetate	3	pt.
Butyl Acetate	2	pt.
Paraffin Wax	½	lb.

Stir together until dissolved.

* Inflammable.

Soaps and Cleaners

Soaps are made from a fat or fatty acid and an alkali. They lather and produce a foam which entraps dirt and grease which is washed away with water. There are numerous kinds of soaps depending on the uses to which they are to be put.

Cleaners consist of solvent such as naphtha with or without a soap. Abrasive cleaners are soap pastes containing powdered pumice, stone, silica, etc.

Liquid Soap (Concentrated)

Water	11
†Caustic Potash (Solid)	1
Glycerin	4
Red Oil (Oleic Acid)	4

Dissolve the caustic in water, add the glycerin and bring to a boil in an enamelled pot. Remove from heat, add the red oil slowly while stirring. If a more neutral soap is wanted, use a little more red oil.

Saddle Soap

Beeswax	5
†Caustic Potash	0.8
Water	8

Boil for 5 minutes while stirring. In another vessel heat

Castile Soap	1.6
Water	8

Mix the two with good stirring; remove from heat and add

Turpentine	12

while stirring.

Mechanic's Hand Soap Paste

Water	1.8	qt.
White Soap Chips	1.5	lb.
Glycerin	2.4	oz.
Borax	6	oz.
Dry Sodium Carbonate	3	oz.
Coarse Pumice Powder	2.2	lb.
Safrol	enough to scent	

Dissolve the soap in ⅔ of the water by heat. Dissolve the last three in the rest of the water. Pour the two solutions together and stir well. When it begins to thicken, sift in the pumice, stirring constantly till thick, then pour into cans. Vary amount of water, for heavier or softer paste (water cannot be added to the finished soap).

Dry Cleaning Fluid

Glycol Oleate	2
Carbon Tetrachloride	60
Varnoline (Naphtha)	20
Benzine	18

† Should not touch skin as it is corrosive.

An excellent cleaner that will not injure the finest fabrics.

Wall Paper Cleaner

Whiting	10 lb.
Magnesia, Calcined	2 lb.
Fuller's Earth	2 lb.
Pumice Powder	12 oz.
Lemenone or Citronella Oil	4 oz.

Mix well together.

Household Cleaner

Soap Powder	2
Soda Ash	3
Trisodium Phosphate	40
Finely Ground Silica	55

Mix well and put up in the usual containers.

Window Cleanser

Castile Soap	2
Water	5
Chalk	4
French Chalk	3
Tripoli Powder	2
Petroleum Spirits	5

Mix well and pack in tight containers.

Straw Hat Cleaner

Sponge the hat with a solution of

Sodium Hyposulfite	10
Glycerin	5
Alcohol	10
Water	75

Lay aside in a damp place for 24 hours and then apply

Citric Acid	2
Alcohol	10
Water	90

Press with a moderately hot iron after stiffening with gum water if necessary.

Grease, Oil, Paint & Lacquer Spot Remover

Alcohol	1
Ethyl Acetate	2
Butyl Acetate	2
Toluol	2
Carbon Tetrachloride	3

Place garment with spot over a piece of clean paper or cloth and wet with the above fluid; rub with clean cloth toward center of spot. Use a clean section of cloth for rubbing and clean paper or cloth for each application of the fluid. The above product is inflammable and should be kept away from flames. Use of cleaners of this type should be out-of-doors or in well-ventilated rooms as the fumes are toxic.

Paint Brush Cleaner

Mix (1)

Kerosene	2
Oleic Acid	1

Mix (2)

Strong Liquid Ammonia, 28%	¼
Denatured Alcohol	¼

Slowly stir 2 into 1 until a smooth mixture results. To clean brushes, pour into a can and stand the brushes in it overnight. In the morning, wash out with warm water.

Rust & Ink Remover

Immerse portion of fabric with rust or ink spot alternately in Solution A and B, rinsing with water after each immersion.

Solution A

Ammonium Sulfide Solution	1 oz.
Water	19 oz.

Solution B

*Oxalic Acid	1 oz.
Water	19 oz.

Javelle Water (Laundry Bleach)

Bleaching Powder	2 oz.
Soda Ash	2 oz.
Water	5 gal.

Mix well until reaction is completed. Allow to settle overnight and siphon off the clear liquid.

Laundry Blue (Liquid)

Prussian Blue	1
Distilled Water	32
*Oxalic Acid	¼

Dissolve by mixing in a crock or wooden tub.

"Glassine" Paper

Paper is coated with or dipped in the following solution and then hung up to dry.

Gum Copal	10 oz.
Alcohol	30 fl. oz.
Castor Oil	1 fl. oz.

Dissolve by letting stand overnight in a covered jar and stirring the next day.

Waterproofing Paper and Fiberboard

The following composition and method of application will render uncalendered paper, fiberboard, and similar porous material waterproof and proof against the passage or penetration of water.

Paraffin Melting Point about 130° F.)	22.5
Trihydroxyethylamine Stearate	3.0
Water	74.5

* Poisonous.

The paraffin wax is melted and the stearate added to it. The water is then heated to nearly boiling and then vigorously agitated with a suitable mechanical stirring device while the above mixture of melted wax and emulsifier is slowly added. This mixture is cooled while it is stirred.

The paper or fiberboard is coated on the side which is to be in contact with water. This is then quickly heated to the melting point of the wax, which then coalesces into a continuous film that does not soak into the paper which is preferentially wetted by the water. This method works most effectively on paper pulp molded containers and possesses the advantages of being much cheaper than dipping in melted paraffin as only about a tenth as much paraffin is needed. In addition, the outside of the container is not greasy, and can be printed upon after treatment which is not the case when treated with melted wax.

Waterproofing Liquid

Paraffin Wax	⅖ oz.
Gum Dammar	1⅕ oz.
Pure Rubber	⅛ oz.
Benzol	13 oz.
Carbon Tetrachloride to make	1 gal.

Dissolve rubber in benzol; add other ingredients and allow to dissolve. (Inflammable.)

The above is suitable for wearing apparel and wood. It is applied by brushing on two or more coats, allowing each to dry before applying another coating. Apply outdoors as vapors are inflammable and toxic.

Waterproofing Heavy Canvas

Raw Linseed Oil	1 gal.
Beeswax, Crude	13 oz.
White Lead	1 lb.
Rosin	12 oz.

Heat the above, while stirring, until all lumps are gone and apply warm to upper side of canvas, wetting the canvas with a sponge on the underside before applying.

Cement Waterproofing

Chinawood Oil Fatty Acids	10 oz.
Paraffin Wax	10 oz.
Kerosene	2½ gal.

Stir until dissolved. This is painted or sprayed on cement walls, which must be dry.

Oil and Greaseproofing Paper and Fiberboard

This solution applied by brush, spray, or dipping will leave a thin film which is impervious to oils and grease. Applied to paper or fiber containers, it will enable them to retain oils and greases. All the following ingredients are by weight:

Starch	6.6
Caustic Soda	0.1
Glycerin	2.0
Sugar	0.6
Water	90.5
Sodium Salicylate	0.2

The caustic soda is dissolved in the water and then the starch is made into a thick paste by adding a portion of this solution. This paste is then added to the water. This mixture is placed in a water jacket and heated to about 85° C. until all the starch granules have broken and the temperature maintained for about half an hour longer. The other substances are then added and thoroughly mixed and the composition is completed and ready for application. A smaller water content may be used if applied hot and a thicker coating will result. Two coats will result in a very considerable resistance to oil penetration.

Fireproof Paper

Ammonium Sulfate	8	oz.
Boric Acid	3	oz.
Borax	1¾	oz.
Water	100	fl. oz.

Mix together in a gallon jug, by shaking, until dissolved.

The paper to be treated is dipped into this solution in a pan, until uniformly saturated. It is then taken out and hung up to dry. Wrinkles can be prevented by drying between cloths in a press.

Fireproofing Canvas

Ammonium Phosphate	1	lb.
Ammonium Chloride	2	lb.
Water	½	gal.

Impregnate with above; squeeze out excess and dry. Washing or exposure to rain will remove fireproofing salts.

Fireproofing Light Fabrics

Borax	10	oz.
Boric Acid	8	oz.
Water	1	gal.

Impregnate; squeeze and dry. Fabrics so impregnated must be treated again after washing or exposure to rain as the fireproofing salts wash out easily.

Dry Fire Extinguisher

Ammonium Sulfate	15
Sodium Bicarbonate	9
Ammonium Phosphate	1
Red Ochre	2
Silex	23

Use powdered materials only; mix well and pass through a fine sieve. Pack in tight containers to prevent "lumping."

Fire Extinguishing Liquid

Carbon Tetrachloride	95
Solvent Naphtha	5

The inclusion of the naphtha minimizes production of toxic fumes when extinguishing fires.

Fire Kindler

Rosin or Pitch	10
Sawdust	10 or more

Melt, mix, and cast in forms.

Solidified Gasoline

*Gasoline	½	gal.
White Soap (Fine Shaved)	12	oz.
Water	1	pt.
Household Ammonia	5	oz.

Heat the water, add soap, mix and when cool add the ammonia. Then slowly work in the gasoline to form semi-solid mass.

Boiler Compound

Soda Ash	87
Trisodium Phosphate	10
Starch	1
Tannic Acid	2

Use powdered materials, mixing well and then pass through a fine sieve.

Anti-Freezes

The materials listed below are the basic ingredients used in all good antifreeze liquids. Of these, alcohol is the only one that evaporates. Radiators containing alcohol should be tested from time to time to be sure of protection. A hydrometer for testing alcohol solution strength can be bought from sellers of denatured alcohol.

* Inflammable.

Anti-Freeze Liquids
Pints of anti-freeze per gallon of water for protection at:

	+10° F.	0° F.	—10° F.	—20° F.
Denatured Alcohol 180° proof	3.4	4.9	6.5	8.3
Denatured Alcohol 188° proof	3.3	4.7	6.0	7.7
Glycerin 95%	3.3	5.3	7.1	9.0
Radiator Glycerin 60%	10.0	18.7	39.0	106.5
Ethylene Glycol 95%	2.7	4.0	5.1	6.5

Specific gravity for protection at:

	+10° F.	0° F.	—10° F.	—20° F.	—30° F.
Denatured Alcohol	0.968	0.959	0.950	0.942	0.921
Glycerin	1.090	1.112	1.131	1.147	1.158
Ethylene Glycol	1.038	1.048	1.056	1.064	1.069

Soldering Flux (Non-corrosive)

Rosin, Powdered	1 oz.
Denatured Alcohol	4 oz.

Soak overnight and mix well.

Photographic Solutions
Developing Solution
Stock Solution A
Dissolve the following, separately, in glass or enamel dishes.

Pyro	4 oz.
Sodium Bisulfite, Pure	280 gr.
Potassium Bromide	32 gr.
Distilled Water	64 oz.

Stock Solution B

Sodium Sulfite, Pure	7 oz.
Sodium Carbonate, Pure	5 oz.
Distilled Water	64 oz.

To use take the following proportions:

Stock Solution A	2
Stock Solution B	2
Distilled Water	16

At a temperature of 65° F. this developer requires about 8 minutes.

Acid Hardening Fixing Bath

A. Sodium Hyposulfite	32 oz.
Distilled Water	8 oz.

Stir until dissolved and then add the following chemicals in the order given below, stirring each until dissolved:

B. Distilled Water (Warm)	2½ oz.
Sodium Sulfite, Pure	½ oz.
Acetic Acid (28%), Pure	1½ oz.
Potassium Alum Powder	½ oz.

Add Solution B to A and store in dark bottles away from light.

CHAPTER II

ADHESIVES

Water-Soluble or Dispersible

Labeling Paste
Formula No. 1

Rye Flour	4	oz.
Alum	½	oz.
Water	8	oz.
Glycerin	1	oz.
Clove Oil	2	drops

Dissolve the alum in the water and rub in the rye flour until a smooth paste is obtained. Pour paste into one pint boiling water and heat until thick. Add 1 oz. of glycerin and 2 drops of oil of cloves and stir until homogeneous.

No. 2

This paste is equally suitable for sticking labels on glass, metals, and wood. It has unusually good adhesion to tin plate. The paste has a good odor and will not show mold growth or other decomposition over long periods of time.

Bone Glue	8	oz.
Corn Starch	60	oz.
Water	190	fl. oz.
Moldex		
(Preservative)	⅜	oz.
Clovel		
(Odorant)	½	fl. oz.

Yield 208 fl. oz.

Soak glue overnight in double boiler in 12 fl. oz. water. Melt slowly and add 114 fl. oz. water heated to boiling. Mix starch and 64 fl. oz. water and heat very slowly over slow fire, stirring constantly. When it begins to thicken, add a little of the glue solution and continue heating and stirring until it thickens again. Add more glue solution, thicken, and continue in this manner until all glue is in. Finally, add Moldex and Clovel, stir thoroughly, and fill into cans at once. It should be a heavy paste free of lumps.

Non-Warping Paste

Water (added at the start)	35.0
Sorbitol (85% Syrup)	10.0
Moldex	0.1
Ammonia Alum	0.3
Glucose (43° Bé)	20.0
Flour	
(Soft Winter Wheat)	19.6
Water	15.0

The equipment used in making this paste should be a regular paste making machine fitted with a mechanical agitator and with steam connections to allow cooking by the direct entrance of steam into the mixture.

The first five ingredients mentioned in the formula are placed in the machine and the steam turned on till the glucose dissolves. The agitator is then started and the flour gradually added. Cooking is continued until the paste becomes quite thick. The paste is drawn off into containers and allowed to stand until used. It may be thinned with water as required.

Billposter's Paste

Alum	1
Flour	666

The alum is dissolved in cold water and the flour is gradually added, the mixture is stirred until the paste is creamy and free of lumps. Use enough water to get a thick paste that will brush properly. This paste will not hold in damp or wet weather. After sticking the poster, to improve its moisture resistance, it is washed with soap water or a dilute lead acetate solution.

Cold Water Paste
Australian Patent 8259

Wheat Flour	8
Alum	1
Water	8

Mix until smooth, evaporate to dryness and grind.

Envelope Adhesive
U. S. Patent 2,159,613

Water	10–18
Dextrinized Starch	50–62
Acetic Acid	10–20
Sorbitol Syrup	3–10

Wall Paper Adhesive
U. S. Patent 2,284,800

Corn Gluten Meal	3–8
Rosin	2–7
Ammonia	1–8
Water	30–70

Mix at 80–100°C.
Add:

Hydrogen Peroxide (30%)	0.1–1%

Water-Resistant Glue
Swiss Patent 192,582

Starch	20
Sodium Naphthalene Sulfonate	10
Glue	80
Water	100
Formaldehyde	4
Sodium Bisulfite	2

The glue is soaked in water overnight. The next day, the glue is warmed on a water bath and stirred until dissolved. The starch is then mixed in and the mass is gently heated until completely gelatinized. The sodium naphthalene sulfonate is added next. When cold, add the formaldehyde and sodium bisulfite with rapid stirring.

"Iceproof" Glue
(For labels on containers kept in ice water)
U. S. Patent 2,308,185
Formula No. 1

Animal Glue	120
Ammonium Thiocyanate	50
Tapioca Starch	120
Water	259
Phenol	1

No. 2

Animal Glue	105
Sodium Thiocyanate	52
Sago Starch	105
Water	267
Lanolin	20
Phenol	1

The animal glue is mixed with the water and allowed to stand for one hour without heating. The mixture is then raised to a temperature of about 55 to 60°C. to complete the dispersion of the glue, and the liquefying agent, the thiocyanate, is added to the mixture at about this temperature. The starch which may have been previously moistened with a portion of the water, is then added to the mixture at the elevated temperature. The mixture is then agitated and heated to a temperature of 65 to 70°C. and the remaining ingredients of the composition are added. The resultant adhesive is then strained, if necessary, and cooled. In an alternate procedure, the starch is added in a dry state providing the liquefied glue mixture is cooled from its 55 to 60°C. temperature to about 45°C. to inhibit "lumping."

Glue Sizes

Flexible glue is used for sizing or finishing specialty papers, various textiles, hat materials, airplane cloths, carpet and auto upholstery and the like to make the material less porous, or to give it "glaze," or to add a certain amount of strength as in rayon warp sizing. Flexible glues of this type are generally cast in cake or block form. When the cake or block is to be used, a definite quantity is added to a fixed amount of water and remelted. Thus the actual sizing as used has a high water content, and the film of solid size laid down on the paper or cloth is quite thin. Formulas Nos. 1, 2 and 3 are examples of the final diluted compositions.

Formula No. 1

Glue	12
Arlex	13.2
Anti-Foam (Foamex)	0.8
Water	74.0

Formula No. 1 is intended for jobs like paper sizing or as a coating in heat-sealing paper sacks and in applications where an appreciable film of flexible glue must be laid down on the material. A high grade 450–465 gram glue is recommended.

Arlex raises the melting point more than an equal quantity of glycerin or diethylene glycol, so that the glue grabs quicker and resists blocking at higher temperatures and humidities. The film remains water-soluble and can be heat-sealed at any time.

No. 2

Gelatin	25	lb.
Glycerin	10	lb.
Penetrant (Sulfatate)	0.5	lb.
Water	50	gal.

No. 3

Gelatin	20	lb.
Arlex	7.5	lb.
Penetrant (Sulfatate)	0.5	lb.
Water	50	gal.

Formula No. 3 was developed in the laboratory and has proved in the mill to be an economical substitute for No. 2 in rayon warp sizing. The stronger bond between Arlex and the gelatin used in these formulas, and the greater uniformity imparted to the warp, together with the lower gelatin and softener contents, give a more economical and efficient operation.

Non-Warping Paper Glue

A grade of Arlex containing 70% sorbitol solids, 10% glucose and 20% water, is used as a softener in certain of the above applications where some tanning (insolubilizing) action is required. This grade is known as "Sorbitol B." Its glucose content provides the tanning action:

Formula No. 1

Glue	9.1
"Sorbitol B"	15.6
Anti-Foam (Foamex)	0.8
Water	74.5

No. 2

Glue	20
"Sorbitol B"	20
Water	60

Glue Binder for Cork

Flexible glue is used as an adhesive binder for cork granules in the manufacture of products such as blocks, cylinders, sheets, bottle caps, gaskets, etc.

Formula No. 1

Glue	22.2
Arlex	44.4
Water	33.4
Formaldehyde	To suit

No. 2

Glue	26.7
Glycerin	33.3
Water	40.0
Formaldehyde	To suit

These formulas have about the same viscosity range at 150–160°F., the temperature at which the ground cork is poured into the composition. Of course, the moisture content eventually reaches an equilibrium value so that the water contents shown are only those of the compositions as originally prepared. Generally, a high grade glue is used. With proper operating technique it has been found that No. 1 gives a binder possessing greater tensile strength, greater and more permanent flexibility, greater resistance to organic solvents such as hydrocarbons, and greater stability to changes in atmospheric humidity.

Tin Labeling Paste

Tin plate has a thin coating of oil which makes it very difficult to attach labels with paste. This paste has been developed to overcome this difficulty. It has been found to be extremely satisfactory. No cleaning of the metal is necessary for the label to adhere perfectly.

Corn Starch	40	oz.
76% Flake Caustic Soda	9	oz.
Water	4¾	gal.
Carbolic Acid (88%)	1¼	fl. oz.

Yield 5 gal.

Dissolve starch by boiling over slow fire (water or steam bath preferred) in 4 gal. water. It will become thick and translucent. Pour into earthenware crock warmed in water. Dissolve caustic soda in 3 quarts cold water in enameled pan and add to the crock. Stir, add acid, and stand overnight. Fill into cans. Finished product is a thick, translucent liquid.

Starch Adhesives

Formula No. 1

Tapioca Starch	320
Water	500
Sodium Bicarbonate	0.65
Sodium Hydroxide (36° Bé)	80.0
Formaldehyde (30%)	6.0
Turkey Red Oil (Ammonium Salt)	0.65

The sodium bicarbonate is dissolved in the water and the tapioca is added slowly with constant stirring. The sodium hydroxide is diluted with an equal weight of water and run in slowly. The temperature is maintained at 15–20°C.

throughout the reaction. The stirring is continued for 12 hours. At the end of that time, the formaldehyde and Turkey red oil are added. The stirring is continued for another twenty minutes after which the batch is run off.

No. 2
U. S. Patent 1,020,655

Cassava Starch	100
Water	100
Sulfuric Acid (66° Bé)	2–3

The acid is added to the water and the starch is stirred in gradually. The suspension is then heated to 55°C. and maintained at that temperature from four to six hours. After cooling, the starch is neutralized with caustic soda solution and dried. The starch may be sold in this form and prepared for use as an adhesive in the manner described below.

Acid Treated Cassava

Starch	200
Water	225 or less

Sodium Hydroxide	20
Water	30

Suspend the acid treated cassava starch in water and stir thoroughly. Add the sodium hydroxide dissolved in water very slowly, with constant stirring over a period of 15–20 minutes.

No. 3

Starch	96
Ammonium Persulfate	4

Mix and heat 2–3 hours at 45–50°C.

Cheap Starch Adhesive

Water	150
Starch	100
Caustic Soda (36° Bé)	25
Water	25

Water	550
Borax	0.14
Hydrochloric Acid (22° Bé)	5.0
Water	50.0

Neutral Starch Adhesive

Water	240
Starch	160
Sodium Hydroxide (30° Bé)	40
Water	320
Hydrochloric Acid (22° Bé)	31
Water	200
Formaldehyde (30%)	5
Water	5

The temperature is maintained at 15–20°C. and the stirring with the alkali is continued for about 90 minutes, until the mass is homogeneous. After diluting the paste with water, the acid is run in very carefully.

Veneer Adhesives
Formula No. 1

Tapioca Starch	84.4
Barium Peroxide	0.5
Soda Ash	0.1
Whiting	5.0
Urea	10.0
Water	120.0
Sodium Hydroxide	2.5

No. 2
U. S. Patent 1,020,656

Cassava Starch	100
Water	100
Sodium Peroxide	0.25– 1.0
Water	5.0–20.0
Sodium Hydroxide	0.25– 1.0
Water	5.0–20.0

Suspend the starch in water and stir thoroughly.

Add the sodium peroxide very cautiously, while stirring, to the water in a stainless steel pail. Sodium peroxide is very corrosive and also tends to spatter. Add the solution of sodium peroxide immediately to the starch suspension. The sodium hydroxide dissolved in water may be added immediately before or after the sodium peroxide.

Do not heat the mass but stir for twelve hours. Filter the starch and dry it at a low temperature.

No. 3

U. S. Patent 1,200,488

Raw Cassava Starch	
(Medium Quality)	250
Water	800
Sodium Hydroxide	25
Water	75

Suspend the starch in water using a strong, efficient agitator. Stir ½ hour. Add the sodium hydroxide dissolved in water gradually over a period of about ½ hour.

This gives a colloidal solution of starch in caustic soda, but it is too viscous. By stirring for several hours, the viscosity of the solution is gradually reduced. After five or six hours, the paste has the desired fluidity.

No. 4

Water	850
Sodium Hydroxide	
(36° Bé)	0.45
Hydrogen Peroxide	
(12 Vol.)	0.35
White Soap	0.50
Cassava Starch	140.0
Sodium Fluoride	2.0
Formaldehyde (30%)	5.0
Water	3.0

The cassava starch is suspended in the water with an efficient agitator. The white soap, sodium hydroxide, and hydrogen peroxide are then added. As in the preceding example, the temperature is raised to 85°C. and maintained until the product has the desired viscosity. Finally, the solution of sodium fluoride and formaldehyde is added and the mass is mixed fifteen minutes longer.

No. 5

Water	700
Sodium Hydroxide	
(36° Bé)	0.45
Hydrogen Peroxide	
(12 Vol.)	3.0
Cassava Starch	280.0
Hydrogen Peroxide	
(12 Vol.)	1.0
Water	3.0
Sodium Fluoride	2.0
Formaldehyde (30%)	5.0
Water	3.0

The cassava starch is suspended in water with an efficient agitator. After adding the hydrogen peroxide and sodium hydroxide, the mass is heated to 85°C. When the reaction has subsided, the second addition of hydrogen peroxide is made and the heating continued until the glue has the desired viscosity. Thereupon the sodium fluoride and formaldehyde solution is added and the mass is stirred fifteen minutes longer.

No. 6

Tapioca Starch	130	lb.
Water	20	gal.

Sodium Hydroxide
(36° Bé) 33 lb.

Water 3.3 gal.

Sodium Silicate
(35° Bé) 43.0 lb.

Borax 0.1 lb.

Water 15.0 gal.

Hydrochloric Acid
(22° Bé) 35.0 lb.

Water 20.0 gal.

Suspend the tapioca starch in water and run in the sodium hydroxide solution, at room temperature. When the mass is clear, add the sodium silicate and follow it with the borax dissolved in water. Add the dilute acid, when the mass is homogeneous. Test the alkalinity of the paste with phenolphthalein solution during the addition of the acid and leave the paste slightly alkaline.

Potato Starch Adhesives
Formula No. 1
German Patent 392,660

Potato Starch 100

Calcium Nitrate 3

Sodium Chloride 1.5

Magnesium Sulfate 1.5

Dissolve the above chemicals in a minimum of water, dry, and grind the dried powder.

No. 2
U. S. Patent 1,677,348

Potato Flour 100 kg.

Water 100 l.

Sodium Phosphate 3 kg.

Mix into a uniform paste. Dry on a hot surface and grind.

No. 3
U. S. Patent 2,124,934

Sweet potatoes are sliced, dried, ground fine and sifted to remove peel and fibers.

Sweet Potato Flour 10

Water 20

Sodium Hydroxide 1

Water 7

Stir the sweet potato flour in the water and heat to 60°C. Add the sodium hydroxide solution with vigorous stirring. This converts the flour into a homogeneous brown gel of excellent adhesive properties.

Salt Glue from Sweet Potatoes

Sweet potatoes are treated with sulfur dioxide, pressed, air dried, ground and sifted.

Sweet Potato Flour 8

Water 8

Calcium Chloride 3

Water 10

The sweet potato flour is mixed with water, and the calcium chloride solution is added. The mass is stirred and heated at 60°C. for one hour. The product is a yellow, homogeneous mass with excellent adhesive properties.

Dextrin Adhesives
Formula No. 1

Water 500

Ordinary Soluble
Dextrin 390

Borax 46

Sodium Bisulfite 4
(30° Bé) 6

Sodium Bisulfite 4

Make a thick paste of the dextrin and half the water and rub until all the lumps are gone. Mix in the rest of the water and heat to 85°C. Add the borax and stir until dissolved. Then add the sodium hydroxide and finally, with the temperature still at 85°C., add the sodium bisulfite.

No. 2

Water	50.0
Yellow Dextrin	
(Very Soluble)	40.0
Borax	40.0
Sodium Hydroxide	2.0
Phenol	0.1
Turkey Red Oil	0.5

Dissolve the dextrin in 45 parts water, using heat, if necessary. Add the borax and stir until dissolved. Dissolve the sodium hydroxide in 5 parts water and add it to the dextrin with rapid stirring. Then mix in the phenol and the Turkey red oil.

No. 3

Water	600
Turkey Red Oil	
(Ammonium Salt)	2
White Dextrin	
(Slightly Soluble)	180
Light Yellow Dextrin	
(Soluble)	150
Cassava Starch	60
Borax	46
Sodium Hydroxide	6
Sodium Bisulfite	4

Make a thick paste of the dextrins and the starch with 300 parts water. Mix in the Turkey red oil and then 280 parts water. Heat to 85°C. with stirring. When the mass is gelatinized, add the borax. When the latter is completely dissolved, add the sodium hydroxide dissolved in 20 parts water. The temperature is maintained at 85°C., the sodium bisulfite is added last and the mass is stirred until it is completely dissolved.

No. 4

Water	370.0

Turkey Red Oil (90%)	
(Ammonium Salt)	1.3
Light Dextrin	
(Water-Soluble)	630.0
Sodium Hydroxide	
(36° Bé)	3.0
Formaldehyde (30%)	6.3

The mixture of water, dextrin and Turkey red oil is heated to 80°C. Add just enough sodium hydroxide (36° Bé) to neutralize the acid in the dextrin. This is determined by testing with litmus paper. About 3 parts are required. The formaldehyde is added last.

The formaldehyde does not darken the solution nor does it destroy any of its desirable characteristics.

No. 5

Water	600.0
Dextrin (Extra	
Soluble)	400.0
Turkey Red Oil (90%)	
(Ammonium Salt)	0.4

No. 6

White Dextrin	10	oz.
Potato Starch	10	oz.
Water	7	oz.
Glycerin	3	oz.
Phenol	2	gr.
Formaldehyde (30%)	⅛	oz.
Sassafras Oil	1	gr.

Form a smooth paste with the dextrin, starch, and half of the water required. Add the rest of the water and heat on a steam bath until smooth. Stir in the glycerin, phenol, formaldehyde, and the sassafras oil.

No. 7

Potato Dextrin	100	g.
Water	150	cc.
Calcium Nitrate	10	g.
Phenol	2	g.

Dissolve the calcium nitrate in the water heated to 75°C. Add the dextrin a little at a time while stirring. When everything is in solution, add the phenol.

No. 8
(Gummed Paper Adhesive)

Water	600.0
Dextrin (Extra Soluble)	400.0
Turkey Red Oil(90%) (Ammonium Salt)	0.4
Formaldehyde (30%)	120.0
Gelatin	20.0
Glycerin	40.0

Soak the gelatin in 100 parts of water overnight. Next day heat it to 80°C. with stirring. Add the glycerin and stir until uniform.

Add the dextrin and Turkey red oil to 500 parts of water and heat to 80°C. with stirring. Add the gelatin, glycerin solution which has been previously prepared and stir 15 minutes longer.

No. 9
(Paper to Metal Foil Paste)

(a)	Dextrin	40
	Glucose	1
	Water	40
(b)	Aluminum Sulfate	1
	Water	30
(c)	Glycerin	3

Dissolve (a) and (b), and mix them. Heat until clear. Add (c) and mix.

No. 10
(Powdered Adhesive)

Dextrin	100.0
Borax	120.0
Sodium Carbonate	1.2

The powders are mixed thoroughly. The resulting powder is mixed with one and a half times its weight of water on the evening before the day it is to be used.

Casein Adhesives
Formula No. 1

Casein	100
Water	280
Sodium Hydroxide	8
Water	20

Stir the casein in the first quantity of water for five to ten minutes. Add the sodium hydroxide, dissolved in the second quantity of water, and stir from thirty minutes to an hour.

No. 2

Casein	100
Water	580
Ammonium Hydroxide (28–29% Ammonia)	13
Water	20

As in the first example, the casein is stirred in the first quantity of water and then the diluted ammonium hydroxide is added. Stirring is continued until a clear solution is obtained.

No. 3

Casein	100
Calcium Hydroxide	7
Sodium Carbonate	To suit

No. 4

Casein	7.5
Sodium Hydroxide	1.0
Water	30.0
Calcium Hydroxide	1.5

The casein is thoroughly wetted with 25 parts of water. After ten minutes, the sodium hydroxide dissolved in 5 parts water is added and the solution is stirred until clear. Finally, the calcium hydroxide is mixed in.

No. 5

Casein	10
Calcium Hydroxide	2
Water	40

The casein and calcium hydroxide are stirred into cold water and the adhesive is ready for use. The last formula gives a glue which is very waterproof.

No. 6

Casein	100.0
Borax	14.7

Grind and mix thoroughly.

No. 7

Casein	100.0
Trisodium Phosphate	12.3

Grind and mix thoroughly.

No. 8

Casein	100.0
Soda Ash	16.0

Grind and mix thoroughly.

In each of the above cases, the powder is mixed with from four to six parts of cold water for use.

No. 9

Casein	100
Water	250
Sodium Hydroxide	11
Calcium Chloride	20

The casein is soaked in 180 parts water with stirring. The sodium hydroxide is dissolved in 50 parts water and is added to the casein. When the casein is completely dissolved, add the calcium chloride dissolved in 20 parts water. Stir for a few minutes. The glue is then ready for use. It remains fluid for seven hours.

No. 10

Casein	100
Calcium Hydroxide	30
Sodium Fluoride	12
Water	250

The first three ingredients are stirred into the cold water and the glue is ready for use in a few minutes.

No. 11

Casein	70
Trisodium Phosphate	10
Calcium Hydroxide	20

Sodium Fluoride	3
Water	200
Pine Oil	2

The casein is soaked in the water for ten minutes and then the other ingredients are added in order, with stirring.

No. 12

Casein	50.0
Calcium Hydroxide	5.0
Calcium Carbonate	10.0
Disodium Acid Phosphate	5.0
Sodium Fluoride	5.0
Mineral Oil	2.5

The mineral oil is added to keep the powder from dusting.

No. 13

The following is a formula for a strong veneer glue that brushes well and does not set too quickly. It is useful for work with large areas:

Casein	69
Calcium Hydroxide	20
Sodium Fluoride	5
Sodium Sulfite	3
Mineral Oil	3

No. 14

Casein	100
Water	350
Calcium Hydroxide	32
Sodium Silicate	70

The casein is soaked and stirred in 250 parts water. The calcium hydroxide is stirred up in 100 parts water and is added to the swollen casein. After stirring until the solution is uniform, add the sodium silicate and stir ten minutes longer. The working life of this glue is 11 hours.

The life of this glue can be prolonged, by the addition of copper chloride as given in the following formula:

No. 15

Casein	100
Water	350
Calcium Hydroxide	22
Sodium Silicate	70
Copper Chloride	2–3

The procedure is the same as in the preceding example except that the copper chloride is dissolved in 30 parts water which is deducted from the water used for soaking the casein. The specifications for the sodium silicate are the same as previously given. The sodium silicate must be added after the calcium hydroxide has reacted with the casein. The copper chloride solution is added last. The glue, at first, turns green when the copper salt is added. After a few minutes, the color turns violet and the glue is ready for use. This product is very resistant to water.

No. 16

Casein	100
Water	350
Calcium Hydroxide	10
Sodium Silicate	15

No. 17

Casein	100.00
Sodium Fluoride	18.75
Zinc Oxide	1.25
China Clay	30.62
Bentonite	30.62
Calcium Hydroxide	43.75
Water	450.00

The dry materials are thoroughly mixed and are added to the water in an efficient glue mixer.

No. 18

Casein	50.0
Calcium Hydroxide	10.0
Trisodium Phosphate	7.0
Sodium Fluoride	3.3
Barytes	10.0
Petroleum	2.0

This glue is prepared for use by adding the mixed powder to about five parts of water in an efficient glue mixer.

No. 19

Casein	100
Calcium Hydroxide	18
Soda Ash	5
Sodium Fluoride	7
Sodium Silicate	43

The dry materials are mixed thoroughly and are stirred into the sodium silicate in an efficient mixer.

No. 20
U. S. Patent 1,886,750

Casein	100
Urea	100
Water	75

Do not heat, as casein dissolves more slowly in a urea solution at elevated temperatures. However, it may be heated cautiously to 60–70°C. for a short time (10–15 min.) to hasten the solution of the casein.

No. 21
(Laminating Adhesive)
U. S. Patent 2,300,907

An adhesive which remains pliable and flexible indefinitely, regardless of the presence of moisture, is formed of casein about 100, triethanolamine or urea about 20–30, to serve as a mildly basic material, and glycerol or additional triethanolamine about 50–70 parts as a plasticizer, and sufficient water to give the desired working character.

No. 22
(Non-setting for 8 hours)

Casein	100
Sodium Fluoride	10

Soda Ash	4
Lime	24
Urea	5–5½
Water	230–240

Dried Blood Glue
U. S. Patent 2,292,624

Water	320
Dried Blood	53
Blood Albumin	30
Casein	10
Sodium Fluoride	2
Disodium Phosphate	3
Pine Oil	2

Mix the above at 20°C. until dissolved; then add:

Slaked Lime	12
Caustic Soda	6
Sodium Silicate (N Grade)	25

Soybean Glue
Formula No. 1
Italian Patent 352,378

(a)	Soybean Flour	18.0
	Soda Ash	0.8
	Water	81.2
(b)	Calcium Oxide	10.0
	Water	50.0
(c)	Soda Ash	4.0
	Water	10.0
(d)	Sodium Silicate	30.0
(e)	Carbon Bisulfide	1.8
(f)	Carbon Tetrachloride	1.2

The solutions are made up individually and mixed in the order given above.

No. 2

Soy Protein	100
Calcium Oxide	15
Water	500
Sodium Silicate	7
Cement	2

The lime (calcium oxide) is slaked with a little of the water. The rest of the water and the soy protein are stirred in. This is followed by the silicate and cement.

No. 3
U. S. Patent 1,994,050

Soybean Flour (Oil-Free)	100.0
Disodium Phosphate	10.0
Sodium Fluoride	5.0
Calcium Hydroxide	7–10.0
Calcium Carbonate	50.0
Copper Sulfate	0.5
Sodium Chloride	2.0
Water	530.0

All of the dry ingredients are mixed in powder form and stirred into the water before use. To insolubilize the glue add 5–10 parts of the following reaction product:

Formaldehyde	35
Ammonium Hydroxide (18%)	105

The ammonium hydroxide is added slowly to the formaldehyde so as to maintain the temperature between 15–20°C. during the addition. This glue remains fluid 6–10 hours and is completely set in twenty-four hours.

Sodium Silicate Adhesives
Formula No. 1

Sodium Silicate (36° Bé)	100
Rosin	5

The rosin must be finely powdered and is added carefully to the sodium silicate, in a mixing pan fitted with a powerful agitator. The resulting product is described as being a pale, transparent solution which is more viscous, more adhesive, quicker drying, and less alkaline than the original silicate.

No. 2
U. S. Patent 2,005,900

Sodium Silicate	50

Water   44
Copper Sulfate
    (12.5% Solution)   6

The copper sulfate solution is run carefully into the diluted sodium silicate solution.

No. 3
U. S. Patent 1,949,914

Zinc Sulfate
    (Heptahydrate)   10
Water   40
Ammonium Hydroxide
    (sp. gr. 0.90)   15
Sodium Silicate
    (Na_2O 3.25 SiO_2)
    (42.5° Bé)   200

The zinc sulfate is dissolved in 20 parts water and the ammonium hydroxide is added to it. The clear solution is added to the sodium silicate diluted with 10 parts water with rapid stirring and, finally, 10 parts more water is added to reduce the viscosity to that of the original sodium silicate.

No. 4
U. S. Patent 1,949,914

Copper Sulfate   14.4
Water   16.5
Ammonium Hydroxide
    (sp. gr. 0.90)   14.9
Sodium Silicate
    (Na_2O 3.2 SiO_2)
    (42.5° Bé)   200.0
Water   14.6

The copper sulfate is dissolved in 16.5 parts water and the concentrated ammonia is added. The deep blue solution is run into the sodium silicate slowly with constant stirring. The 14.6 parts water are then added to reduce the viscosity to that of the original silicate solution.

No. 5
U. S. Patent 1,949,914

(a) Soy Bean Meal   12.70
    Water   22.30
    Sodium Silicate
        (Na_2O 3.2 SiO_2)
        (38.5%)   65.00
(b) Sodium Silicate
        (Same Grade
        as Above)   19.40
    Wood Flour   5.10
    Water   7.60
(c) Copper Sulfate   14.30
    Ammonium Hy-
        droxide
        (sp.gr. 0.90)   15.00
    Water   41.30
(d) China Wood Oil
        (Blown)   13.10
    Petroleum Sulfonic
        Acid, Sodium
        Salt   0.13

The ingredients in (a) are ground in a mill until thoroughly dispersed. (b) is added next with rapid stirring and is followed by (c). Finally (d) is added with thorough, rapid agitation.

No. 6

Starch   50
Water   100
Sodium Silicate
    (Na_2O 3.34 SiO_2)
    (sp.gr. 1.38)   50

The starch and water are stirred together and added to the silicate, and the mixture is stirred and heated on a steam bath until it is practically clear.

Dextrin is sometimes added to improve the tack of the silicate when wet, although the tensile strength suffers as a result.

No. 7
U. S. Patent 2,932,142

Flint (Powdered)   62

Sodium Silicate 14
Water 14
Aluminum Fluoride 10
Mix well and dry at 125°C.
after application.

No. 8
Blood Albumin 45
Water 55
Sodium Silicate
(Na$_2$O 2.9 SiO$_2$) 4
(sp.gr. 1.48)
The ingredients are stirred until
a homogeneous mass is obtained.

No. 9
U. S. Patent 2,175,767
Sodium Silicate 60
Emulsified Asphalt 12
Clay 28

Cement for Ceramics
Lithopone 10
Powdered Quartz mixed
with Sodium Silicate
to a thick dough 10

Iron Cement for High
Temperatures
Borax powdered is mixed
and stirred 1
Zinc Oxide and 5
Pyrolusite are mixed with
Sodium Silicate to a
stiff dough 10
Cement for Marble and
Alabaster
Whiting are mixed and
stirred with 100
Zinc Carbonate and 100
Sodium Silicate 50

Bonding Adhesive for Abrasive
Wheels
U. S. Patent 2,311,271
Sodium Silicate
(<2SiO$_2$:1 Na$_2$O) 350
Sulfonated Castor Oil 1¾

Aluminum Silicate 100
Zinc Oxide 7
Umber 25

Gum Arabic Glues
Formula No. 1
Water 250.0
Calcium Hydroxide 0.2
Glycerin 8.0
Gum Arabic 100.0
The ingredients are added in
the order mentioned to the cold
water, with stirring. When everything is dissolved, the solution is
allowed to stand in order to settle
and clear. Supernatant liquid is
filtered.

No. 2
The following formula illustrates the addition of aluminum
sulfate to increase its adhesive
strength:
Gum Arabic 100.0
Calcium Hydroxide 0.2
Water 300.0
Aluminum Sulfate 10.0
Dissolve the gum arabic in 200
parts of cold water. Let the solution settle and filter the clear top
liquid. Add the aluminum sulfate dissolved in 100 parts of
water. Aluminum sulfate also decreases blotting when gum arabic
glue is used on paper which has
not enough size.

No. 3
(Waxed Paper Adhesive)
U. S. Patent 1,983,650)
Gum Arabic 40
Potassium Hydroxide 34
Water 75
The potassium hydroxide is dis-
solved in the water. The solution,
spontaneously, becomes very hot
and should be allowed to cool to
room temperature before the gum

arabic is added. This glue is very caustic.

Mucilage

This mucilage is suitable for use on paper. It is a clear, thick liquid with a good odor.

Powdered Gum Acacia	24 oz.
Moldex (Preservative)	$\frac{1}{16}$ oz.
Water	40 fl. oz.
Sassafras Oil	5 min.

Yield 56 fl. oz.

Stir acacia to a paste with half the water. Heat the remainder of the water to boiling and stir Moldex into it. Add gum dispersion and heat slowly until clear. When cold add oil sassafras and pour into bottles.

Magazine and Catalog Glue

Formulas used in the binding of magazines and catalogs differ from bookbindery formulas mainly in their being designed for seasonal variations in atmospheric humidity rather than for permanent all-weather binding.

The viscosity of the melted flexible glue mixture at time of application on the machine is very important, and formulas are built around this factor. Generally there is more glue than softener as strength is more important than flexibility. A "middle register" glue is often used for economic reasons. The following group of formulas has proved good under conditions of actual usage.

Formula No. 1
Season—Summer

Glue	35.7
*Arlex	12.8
Glycerin	3.6
Water	47.9

No. 2
Season—Spring and Fall

Glue	33.3
*Arlex	14.9
Glycerin	4.2
Water	47.6

No. 3
Season—Winter

Glue	30.9
*Arlex	18.3
Glycerin	5.1
Water	45.7

No. 4
Season—Winter

Glue	32.1
Glycerin	21.4
Water	46.5

Nos. 3 and 4 have the same viscosity range for machine spreading. The glue employed here is an inexpensive 180 g. glue, since long life is not required.

Arlex gives greater tensile strength than glycerin to flexible glue mixes, even when somewhat less glue is used to keep the same viscosity range.

When thicker periodicals, such as large catalogs or telephone books are to be bound, the flexible glue formulas more nearly approach those used in bookbinding. More flexibility and greater strength is required than for the average monthly periodical. The best way to accomplish this is to use Arlex, a stronger glue, and a somewhat higher softener ratio.

No. 5

Glue	42.0
*Arlex	28.4
Water	29.6

* Sorbitol Syrup.

No. 6

Glue	38.7
*Arlex	36.8
Water	24.5

No. 7

Glue	36.5
*Arlex	43.5
Water	20.0

The composition is cast in greased pans, the cakes are stored at 50–55°F. and strips are cut off as needed. These strips are then melted down with a measured additional amount of water (e.g., 33–50%) so as to give the viscosity required for application.

Bookbinders' Flexible Glue

Gelatin	36
Turkey Red Oil	15
Water	49

The gelatin, or any high grade glue, is soaked overnight in water, and then warmed to not over 130°F., until fully dissolved. The rest of the mixture is added, well stirred, and kept warm until complete dissolution is reached. Depending upon the use to which it is put, some dilution is possible to obtain a good flow and spread. A large batch can be cast in trays and remelted for use as required.

Bookbinders' Size
Formula No. 1
U. S. Patent 2,089,063

Egg Albumin	4–15
Water	35–70
Amyl Acetate	4–14

No. 2

Egg Albumin	4–13
Glucose	3–10
Starch	1– 5
Water	35–85
Acetic Acid (28%)	1– 8

* Sorbitol Syrup

Ammonium Hydroxide (26%)	1– 8

The starch is stirred into enough water to make a thin paste and is heated on a steam bath until a clear solution is obtained. The egg albumin is dissolved in the rest of the water and ammonium hydroxide. When the starch is cold, the glucose and egg albumin are added, with stirring, and finally the acetic acid is run in.

Can Seal
Formula No. 1
U. S. Patent 2,114,308

Latex (40%)	28.0
Barytes	30.0
Casein	1.5
Sulfur	0.5
Diphenylguanidine	0.3
Zinc Oxide	4.0
Water	35.7

The casein is dissolved in a little ammonia before it is added to the other components. This mixture is particularly good because it expands on drying.

No. 2
British Patent 441,877

Colloidal Graphite (20% in Water)	75 lb.
Gum Tragacanth (½–1% Solution)	10 lb.
Latex (40% Solids)	22 gal.

Moistureproof Cellophane Adhesive
U. S. Patent 1,953,104

Latex (45% Rubber)	5
Water-Soluble Agglutinant	15
Glycerin	3
Ethyl Lactate	5

The water-soluble agglutinant is prepared by heating the following until a clear solution is obtained.

Water-Soluble Agglutinant

Corn Starch	10
Dextrin	2
Water	63

Wood Glue Adhesive

Water	500.00
Sodium Bicarbonate	0.65
Cassava Starch	320.00
Hydrogen Peroxide (12 Vol.)	10.00
Sodium Hydroxide (36° Bé)	80.00
Water	80.00
Turkey Red Oil (Ammonium Salt, 90%)	0.65

The cassava starch, water, hydrogen peroxide and sodium bicarbonate are mixed at room temperature. The diluted sodium hydroxide is then added and the mass is stirred for twelve hours. The paste is stirred twenty minutes longer after adding the Turkey red oil.

Paper Board Adhesive
U. S. Patent 2,282,364

Dextrin, Converted	70
Water	90

To above add:

Water	50–100

Heat to 70–80°C.

Sodium Acid Phosphate	0.5–1
Dicyanodiamide	0.03–0.04

Plywood Adhesive
U. S. Patent 2,291,586

Soybean Flour	94
Sodium Hydrogen Phosphate	2½
Soda Ash	2
Sodium Fluoride	½
Pine Oil	1
Water	370
Calcium Hydroxide	10

Sodium Silicate ("N" Grade)	25
Carbon Disulphide Carbon Tetrachloride (1:1 by vol.)	2

Apply to veneers and use pressure of 125–225 lb/in² (hot pressing) or 110–140 lb/10000 ft.² (cold pressing).

Calcium Sucrate-Glue Adhesive

Water	200.0
Calcium Hydroxide	30.0
Sugar	35.0
Glue	100.0
Phenol	0.1
Acetic Acid	0.7

The sugar, calcium hydroxide and 100 cc. water are heated to 80° for eight hours. The glue is soaked in 100 cc. water overnight and heated the next day to dissolve. The calcium sucrate and glue solutions are mixed while warm. Finally the phenol and acetic acid are stirred in.

Synthetic Resin Adhesive

Formaldehyde (40%)	500
Cresylic Acid (Dark)	1000
Bleaching Powder	20
Hide Glue (Ground)	1200
Sodium Bicarbonate	24
Water	2000

Run the formaldehyde into the cresylic acid. Add the bleaching powder gradually and maintain the temperature at 90°C. for 30 minutes. Let the resin stand one hour.

Dissolve the hide glue and sodium bicarbonate in the water and heat the solution in an autoclave at 130°C. for 2½–3 hours.

Adhesive for "Protectoid"

Animal Glue	100
Aquaresin GB	25
Invert Sugar	25
Water	100

Decalcomania Adhesive
U. S. Patent 2,143,868

An adhesive which retains its elasticity and color and is non-bleeding even on oil-soluble bleeding colors, contains:

Hide Glue	14.8
Butyl Cellosolve	10.0
Glycerol	1.2
Water	74.0

Tire Tube Air Seal
U. S. Patent 2,347,925

Water	5
Salt	¼
Alcohol	1
Linseed Meal	4

Latex Adhesives

Can Joint Seal
Formula No. 1
U. S. Patent 2,013,651

| Latex | 60 |
| Alginic Acid | 40 |

The alginic acid is dissolved in ammonium hydroxide before it is added to the latex.

No. 2
Canadian Patent 367,342

The following formula illustrates the use of latex with casein:

(a) Bentonite	1.70
Water	10.30
(b) Ammonium Alginate	0.04
Water	0.96
(c) Casein	1.65
Ammonium Hydroxide (28%)	3.70
Zinc Oxide	0.05
Water	0.40
Ammonium Hydroxide (28%)	0.20
(d) Latex (38% Solids)	64.00
Accelerator	6.50

In (a) the bentonite and water are mixed together and added to the clear solution (b). The ammonium alginate acts as a suspending agent. In (c) the casein is dissolved in the mixture of ammonium hydroxide and water. The zinc oxide and excess ammonia are then added and the mixture combined with (a) and (b). The resulting mixture is stirred into (d).

No. 3
(Paper Adhesive)
U. S. Patent 2,093,105

| Latex (40–60%) | 50 |
| Mica (Powdered) | 50 |

The mica serves to anchor the latex to the paper.

Water-Insoluble Adhesives

Synthetic Rubber Cements

Cement is prepared from synthetic rubber by the procedure used for natural rubber. Crude synthetic rubber receives a preliminary mastication on a rubber mill or in a Banbury mixer. The degree of "breaking down" determines the fluidity of the cement. Dispersion is effected by a paddle mixer or rotating churn. Each synthetic has specific solvent requirements. Thiokol dissolves in ethylene dichloride. Buna N (Ameripol, Chemigum, Hycar O.R.) is soluble in ethylene dichloride, methyl ethyl ketone, or a blend of either with an equal volume of aromatic gasoline (Solvesso, etc.). Neoprene requires toluene or aromatic gasoline. Buna S (GR-S), Butyl, and Vistanex make good cements in gasoline.

Neoprene Cement
Formula No. 1

| Neoprene GN | 454 |
| Toluene | 2970 |

No. 2
U. S. Patent 2,313,039

Neoprene	1.5	lb.
Toluene	1	gal.
Morpholine	3	oz.

Buna N Cement

Ameripol	454
Methyl Ethyl Ketone	1334
Solvesso	1334

Buna S, Butyl, or Vistanex Cement

| GR-S, Butyl, or Vistanex | 454 |
| Gasoline | 2377 |

Thiokol Cement

| Thiokol | 12 |
| Ethylene Dichloride | 88 |

Photo Mounting Cement

Buna S or Neoprene cement containing 1 to 2 pounds of synthetic per gallon of cement makes a suitable photo mounting cement. An additive such as rosin may have to be dissolved in the cement to improve tackiness.

Pure Gum Hycar O.R.–15 Cement
—Low-Temperature-Curing
Formula No. 1
Part A

Hycar O.R.–15	100.0
Zinc Oxide	2.5
Sulfur	3.0

Part B

Hycar O.R.–15	100.0
Zinc Oxide	2.5
Butyl Eight	6.0

Dissolve each part separately in ethylene dichloride—one pound per gallon of cement. Mix equal volumes just before using.

No. 2
(High-Tensile Channel Black Cement with Volatile Softener—Low-Temperature-Curing).

Hycar O.R.–15	100.0
Zinc Oxide	5.0
Benzoic Acid	2.0
AgeRite Resin D	2.0
Easy-Processing Channel Black	50.0
Triacetin	30.0
Sulfur	2.5

Dissolve in chlorobenzene—1½ pounds per gallon of cement. Just before using, stir into each gallon eight grams of diethylamine followed by fifty grams of carbon disulfide.

No. 3
(Tacky Channel Black Cement —Heat Curing).

Hycar O.R.–15	100.0
Zinc Oxide	5.0
Benzoic Acid	2.0
Channel Black	50.0
Nypene Resin	20.0
Dibutyl Phthalate	20.0
"Dibutyl Meta-Cresol"	35.0
Sulfur	2.5
Mercaptobenzothiazole	2.0

Dissolve two pounds in a half gallon of chlorobenzene and make up to one gallon with methyl ethyl ketone (preferably anhydrous).

No. 4
(Semi-Active Black Cement — Low-Temperature-Curing).

| Hycar O.R.–15 | 100.0 |

Zinc Oxide	5.0
Semi-Active Black	75.0
BRT #7	35.0
Dibenzyl Sebacate	15.0
Sulfur	2.0
2-Mercapto 4, 5-dimethyl-thiazole 85%	
2-Mercapto 2-ethyl thiazole 15%	2.0

Dissolve two pounds in a half gallon of nitroethane and make up to one gallon with "Solvesso No. 1." Add four grams of butyraldehyde aniline to each gallon before using.

No. 5
(Smooth Soft-Black Exceptionally Stable Cement—Low-Temperature-Curing).

Hycar O.R.–15	100.0
Zinc Oxide	5.0
Lauric Acid	0.5
P-33 Black	100.0
Tricresyl Phosphate	50.0
Sulfur	2.0
2-Mercaptobenzothiazole	2.0

Dissolve in Sharples dichloropentanes No. 14, 1½ to 2 pounds per gallon of cement. Add four grams of butyraldehyde aniline to each gallon before using.

No. 6
(White Cement for Adhesions to Fabric—Fast Heat-Curing).

Hycar O.R.–15	100.0
Zinc Oxide	5.0
"Silene"	35.0
"Titanox"	35.0
Bakelite Resinoid BR 4036	35.0
Dibutyl Phthalate	15.0
Sulfur	2.0
Heptaldehyde Aniline	0.75

Dissolve in methyl ethyl ketone, 1½ pounds per gallon of cement. If slower evaporation is desired,

diisopropyl ketone could be used. Other useful cements based on the above recipe may be obtained by replacing the "Silene" with fine white clay. The above type of cement can be made in a multitude of colors by adding colored mineral pigments or dyes to the batch stock during milling.

No. 7
(Red Ebonite Cement—A Protective Coating Preparation for Baking on to Metallic Surfaces, etc.).

Hycar O.R.–15	100.0
Red Iron Oxide	90.0
Cadmium Selenide	10.0
Tetramethyl Thiuram Disulfide	2.0
Sulfur	35.0
P-25 Cumar	15.0

Dissolve 1½ to 2 pounds in one quart of nitroethane plus one quart of ethylene dichloride; then dilute to one gallon with "Solvesso No. 1." Apply to a clean surface by brushing; dry thoroughly at room temperature or at slightly elevated temperatures; then bake for three hours at 300°F.

Mixed Cements
Cements from Hycar O.R. lend themselves to blending with other cements, particularly with chlorinated rubber cements. For example, a Hycar O.R.–15 cement of recommended concentration may be mixed with an equal volume of 10% chlorinated rubber cement made from 125 centipoise chlorinated rubber or with half the volume of a 20% chlorinated rubber solution to produce a very useful blended cement. Such

blends can serve four important purposes.

1. Chlorinated rubber has a pronounced stabilizing effect.

2. Chlorinated rubber seems to retard the cure of fast-curing cements during storage, but does not noticeably affect the cure after the cement is spread out in a thin film exposed to air.

3. Chlorinated rubber enhances the adhesion strength of many Hycar O.R. cements.

4. Chlorinated rubber increases the tackiness of many Hycar O.R. cements.

Cements made by blending Hycar O.R. cements with neoprene cements are smooth, tacky mixtures especially useful for bonding cured or uncured Hycar O.R. to neoprene or *vice versa*.

Cements from Hycar O.R. may also be blended with Rezyl, Glyptal, and Bakelite resin cements for special uses.

Hycar O.R. cements do not form stable mixtures when blended with natural rubber cements or cements made from the butadiene-styrene synthetic rubber such as Buna S or GR-S.

Useful cements for special purposes can be prepared from blends of Hycar O.R. with vinyl chloride resins such as Koroseal. This type of cement can be manufactured by either of two methods. The first method involves mixing a Hycar O.R. cement with a vinyl chloride resin cement. The second method involves blending Hycar O.R. and a plasticized vinyl chloride resin on a mill, then, with this blend as the starting material, compounding to produce a cement batch to meet the needs at hand. Solvent generalizations made for Hycar O.R. hold closely for mixtures of Hycar O.R. with vinyl chloride resins.

Cement for Rubber to Metal

Rubber Cement (12% Rubber in Benzol)	100.00
Latex	5.00
Zinc Oxide	0.12
Sulfur	2.40
Accelerator	0.12

Strong, flexible adhesives are required for the manufacture of leather belts for power drives. Such products are frequently based on rubber, rosin and linseed oil.

Cement for Leather Driving Belts
Formula No. 1

Rosin (Light)	30
Rubber (Dry, Waste)	20
Linseed Oil Varnish	20
Benzine (High Boiling)	30

Heat the rosin, rubber and linseed oil together until completely dissolved. Add the benzene, taking precautions against fire.

No. 2

Rubber (Raw)	10
Rosin (Powdered)	20
Linseed Oil	20

Heat together and stir until homogeneous.

Insole Cement

Latex (40% Solids)	100 lb.
Casein Solution (10%)	10–15 lb.
Resin Emulsion (50%)	10 lb.

The casein solution is prepared as follows:

Casein	5 lb.

Water   44 lb.
Ammonium Hydroxide   8 oz.
Phenol   1 oz.

The following directions are used in the preparation of the resin emulsion:

Resin (Ester Gum,
  or Cumar)   3 lb.
Oleic Acid   3 oz.
Ammonium Hydroxide   1 oz.
Water   3 lb.

The oleic acid and ammonium hydroxide are dissolved in the water and the resin is emulsified by heating to the melting point and stirring vigorously.

Metal to Metal Cement
British Patent 439,657

Rubber   2
Ester Gum   2
Gasoline   2
Acetone   10

Dissolve in the cold.

Celluloid to Hard Rubber Adhesive
German Patent 707,659

Celluloid   10
Acetone   70
Ether   5
Amyl Acetate   5
Tar   10

Rosin Adhesive Solution

Castor Oil   17
Rosin   100
Alcohol (Denatured)   50

The rosin and castor oil are heated together until a homogeneous liquid is obtained. The denatured alcohol is then added and the mixture is heated until the solution is clear.

Sticky Rubber-Rosin Adhesive

Benzene   60
Rosin   100
Castor Oil   3
Crepe Rubber   5

Dissolve the rosin in the benzene with heat, and add the rubber. When it is all dissolved, add the castor oil. If the adhesive is wanted in paste form, whiting or precipitated chalk may be stirred in.

Permanently-Tacky Pressure Adhesive
British Patent 556,147

Rubber   80
Rosin   16
Zinc Oxide   4

Non-Webbing Rubber Cement
U. S. Patent 2,270,731

"Cobwebbing" during spraying with rubber cements is prevented by adding 3–10% (calculated on weight of rubber) of bentonite, previously swollen in water.

Rubber Cement for Glass
Formula No. 1

Rubber   1
Mastic Gum   12
Dammar Gum   4
Chloroform   50
Benzene   10

Dissolve in the cold.

No. 2

Rubber   2
Mastic Gum   6
Chloroform   100

Dissolve in the cold.

Brown Cement

Rubber (Gum)   10
Carbon Bisulfide   100

Shellac (Powdered) 2
Alcohol 8

Dissolve the rubber in the carbon bisulfide. Add the alcohol, slowly with stirring, avoiding clots. Add the powdered shellac and heat on a boiling water bath until all the shellac is dissolved and the carbon bisulfide is driven off.

Hard Cement for Crockery

Rubber 1
Shellac 1

Melt together on a boiling water bath.

Rosin Linoleum Cement

(a) Castor Oil 5.0
    Rosin 100.0
    Bentonite
    (Refined) 10.0
(b) Bentonite
    (Refined) 2.5
    Alcohol (Denatured) 75.0
    Water 25.0

The castor oil and rosin are melted together and the bentonite is stirred in. The mass is then permitted to cool to 75–80°C.

The bentonite and alcohol are mixed and then the water is stirred in. The mass is heated to 75–80°C. and added to the rosin solution previously prepared. This paste is very viscous and requires a strong agitator. It should be stirred until cold.

Electrical Sealing Compound
U. S. Patent 2,075,885

Rosin 200
Calcium Hydroxide 10
Methyl Abietate 12

Melt the rosin by heating it to above its melting point. Stir in

the calcium hydroxide and maintain the heat for 20 minutes. Stir in the methyl abietate at 150°C.

Rubber-Resin Adhesives

The following formulations, compounded by milling, are dissolved in carbon tetrachloride.

Formula No. 1

Milled Rubber 160
Staybelite Ester No. 2 160
Zinc Oxide 125
Adhesive strength
  lb./sq. in. 12.3

No. 2

Milled Rubber 240
Staybelite Ester No. 3 80
Zinc Oxide 125
Adhesive strength
  lb./sq. in. 15.3

No. 3

Milled Rubber 160
Staybelite Ester No. 10 160
Zinc Oxide 125
Lanolin 55
Adhesive strength
  lb./sq. in. 24.0

Rubber-Resin Cement
U. S. Patent 2,175,797

Cumarone Resin 20–25
Asphalt 20
Rubber 6–9
Soap 0.8–1.2
Clay 1.3
Water To make 100

Cement for Uncured Vinylite-Bakelite Coated Cloth

Vinylite Resin XYHL
  (Dry Basis) 9
Bakelite Resin XJ-16530
  (Wet Basis) 50
Dibutyl Phthalate 18

"Staybelite" (Hydro-
genated Rosin)   18
Acetone   5

This formulation yields a solution of a pasty consistency suitable
for "finger" spreading. Thinning
with suitable solvents such as alcohols or esters is required where
brushing is to be used. Often,
better cementing is obtained if the
coated cloth is brushed with butanol or ethanol before applying
the adhesive. Butanol-ethanol
mixtures may also be used where
solvent cementing of the uncured
coating is preferred. The cemented seams can be cured along
with the cloth coating during the
final baking operations.

Light Polyvinyl Adhesive
Vinyl Acetate Resin   40
Cellulose Acetate or
Nitrate   4–12
Volatile Solvent   60

Adhesive for Polyvinyl Chloride
Plastics
British Patent 551,412
Polyvinyl Acetate   5–10
Mesityl Oxide or
Methylcyclohexanone   95–90
Dilute with either solvent as de-
sired.

Adhesive and Coating Material
U. S. Patent 2,348,447
Polymeric Vinyl
Isobutyl Ether   20
Benzine   40
Acetone   40

Ethyl Cellulose Adhesives
One of the most useful strong
adhesive combinations results
when Staybelite ester No. 3 is
blended with approximately equal
parts of ethyl cellulose. Such a
blend may be applied as a hot
melt, from solvent solution, as an
oil-in-water emulsion, or as a reversed emulsion in the lamination
of industrial textiles. A suggested
formulation for a reversed emulsion follows:

25% solids high-viscosity
Ethyl Cellulose dis-
solved in an 80:20 ratio
of Toluene and Ethyl
Alcohol   100
Staybelite Ester No. 3   20
Water containing 1 part
28% Ammonia   30

The Staybelite ester No. 3 is
dissolved in the ethyl cellulose
solution. To this resin-modified
solution the water phase is added,
with slow agitation, to produce a
homogeneous stable water-in-oil
reversed emulsion.

Adhesive for Cellulose Acetate
Formula No. 1
Aquaresin GM   25 cc.
Cellulose Nitrate   50 g.
Acetone   45 cc.
Alcohol   180 cc.
Methyl Cellosolve
Acetate   45 cc.
Butyl Acetate   30 cc.
No. 2
Vinylite Resin   50
Aquaresin GM   15

Cellulose Acetate Foil Adhesive
U. S. Patent 2,296,891
Nitrocellulose (20–25
visc.)   100
Methyl Phthalylethyl-
glycolate   80
Acetone   335
Methyl "Cellosolve"   540

Wood Ply Lamination
U. S. Patent 2,290,833

Microcrystalline Paraffin Wax	5–25
Rosin	10–35
Oxidized Petroleum Asphalt	To make 100

Laminated Wood
(Plywood Adhesive)
U. S. Patent 2,290,833

Microcrystalline Petroleum Wax	5–25
Rosin or Cumarone Resin	10–35
Oxidized Petroleum Asphalt (m.p. 107°)	To make 100

Laminating Adhesive

For regenerated cellulose, metal foil, fiberboard, leather, paper, wood, veneer, glass and cellulose ester fibers.

U. S. Patent 2,333,676

Cellulose Nitrate	8.2
Toluol	29.0
Dibutyl Phthalate	6.4
Ethyl Acetate	33.0
*Thermoplastic Resin	13.4

Plywood Glue

Casein	500
Dimethylolurea	500

Veneer Glue
U. S. Patent 2,150,175

A vegetable protein material (e.g., soy-bean flour) is mixed with calcium oxide, strontium oxide, barium oxide, or magnesium oxide, 3–17% of sodium hydroxide, and 1–6% of carbon disulfide to give a waterproof glue. The

* Modified alky, vinyl, or sulfonamide resin.

consistency is controlled by varying the sodium hydroxide: carbon disulfide ratio; casein may be added if desired.

Thermoplastic Adhesive
Canadian Patent 417,155

Ethyl Cellulose	15–27
Castor Oil	7–27
Rosin	To make 100

Flexible Thermoplastic
Adhesive
British Patent 551,398

Gutta Percha	6
Rosin	2
Swedish Pitch	2

Melt and mix slowly until uniform. Then add:

Zinc Oleate	1

Mix until uniform.

High-Melting Thermoplastic
Adhesive
U. S. Patent 2,281,483

Vinyl Acetate	70
Pentaerythritol Abietate	30

Use at 350–450°F.

Transparent Resin Adhesive

Urea	200
Thiourea	52
Trioxymethylene	240
Sodium Acetate	6
Formamide	20

Heat the intimate mixture to 140°C. for 20 minutes. The melt becomes viscous and on cooking solidifies into a hard, white mass which may be powdered and used as a thermoplastic adhesive.

Stainless Steel Pipe Joint
Cement
U. S. Patent 2,324,729

Barium Sulfate	50

Castor Oil   50
Aluminum Powder   ½–10

Pipe Joint Cement
U. S. Patent 2,329,014

Asbestos Fiber   10
Cement, Powdered   70
Soda Ash   8–12
Hydrated Lime   8–12

Add water to form a paste before use.

Plastic Heatproof Jointing
Compound
U. S. Patent 2,162,387

Asbestos Fiber   10
Sodium Silicate   8
Sand   27
Fireclay   50
Mineral Wool   5

Pipe Joint Lute

Chalk   65.0
Kaolin   33.5
Linseed Oil, Boiled   35.0
Litharge   1.5

Ingredients are first dried at 105°C., ground and screened, mixed together, then mixed with the boiled linseed oil and passed through a 3-roller color mill.

Pipe Joint Calking
U. S. Patent 2,329,014

Asbestos Fiber   10
Cement, Powdered   70
Soda Ash   8–12
Lime, Hydrated   8–12

Add water before use, to make a paste.

Metal Joint Seal
British Patent 558,135
Blown Castor Oil   4.5–6
Zinc Stearate   1 –2
Butyl Ricinoleate   0.3–1.5

Expansion Joint Plastic Sealing
U. S. Patent 2,286,018
Petrolatum   40–50
Liquid Asphalt   5–15
Red Lead, Powdered   25–35
Asbestos Fiber, Long   10–20

Crack Sealer
U. S. Patent 2,353,723
Vulcanized Rubber,
  Powdered   6–8
Bentonite   2–5
Asbestos Fiber   6–8
Bitumen Emulsion
                   To make 100

Can Seam Seal
U. S. Patent 2,326,966
Alkyd Resin   100
Cellulose Acetate   10–15
Magnesium Silicate   50–150
Acetone   200–300

Container Joint Thermosetting
Adhesive
U. S. Patent 2,333,676
Vinyl Acetate   70
Pentaerythritol Abietate
  Resin   30
Apply at 350–450°F.
This seal will hold at 190°F.

Aluminum Thread Joint Seal
British Patent 556,834
Lead Stearate, Ground
  Fused   20–35
Mineral Oil   120
Heat the oil to 130–160°F. and add the stearate slowly, while mixing.

Thermosetting Adhesive
Starch   25–30
Polyvinyl Alcohol
  (RH-349) (3%–4%
  solution)   100
The starch is merely stirred into
the polyvinyl alcohol solution un-
til a smooth suspension is obtained.
Apply heat to set.

Plate Glass Setting Composition
  U. S. Patent 2,340,840
Calcined Gypsum   96.9–97.3
Portland Cement   2.5
Potassium Sulfate   0.1–0.5
Calcium Chloride   0.1

Lantern Slide Adhesive
  U. S. Patent 2,324,680
Amyl Acetate   150
Linseed Oil   100
Collodion   3600

Lens Cement
Staybelite   75
n-Butyl Methacrylate
  Polymer   25
Melt the ingredients together
and stir until homogeneous. Strain
through pure silk of finest texture.

Watch Crystal Cement
Paraplex RG2 Resin   17
*Gum Solution   14
Pyroxylin Solution   49
Amyl Acetate   35
Butyl Alcohol   13

Glass Sealing Compound
Durite Resin   261 g.
Rosin   392 g.
Shellac   504 g.
Alcohol   367 cc.

---

* Gum solution 20% Rezyl 12 in
toluol. Cotton Solution 50/50 by weight
acetone and ½ second pyroxylin.

Marble Flour   3178 g.
Malachite Green   1 g.
Mix marble flour and Durite
resin and add the rosin, shellac
and alcohol and work the mixture
until dissolved. Then add the
malachite green and mix thoroughly. Mix entire batch to consistency desired, adding more alcohol if necessary.
The cement is baked to the parts
to be sealed until the green color
begins to fade. The color should
not disappear entirely.

Glass to Metal Adhesive
  Formula No. 1
Beeswax (Bleached)   2
Rosin   4
Venetian Turpentine   1
English Red   4
Melt the wax and rosin and stir
in the English red.
  No. 2
Rosin   46¼
Shellac   46¼
Dibutyl Phthalate   7½
  No. 3
Water-Insoluble Glass Adhesive
Glass Powder   10
Calcium Fluoride   20
Water-Glass   70

High-Vacuum Cement
A. Dissolve 10 parts of shellac
in 20 parts of 95% ethyl alcohol
and 1 part of n-butyl phthalate.
Slight warming will increase the
rate of dissolving.
B. Place in vacuum oven and
evacuate for 4 hours at 1 mm. of
mercury and temperature of 90°F.
C. Raise temperature gradually,
about 10°F. per hour, until 230°F.
is reached at which point the
vacuum is released and the heat

shut off. The cement is fluid at this temperature and may be cast into small sticks for future use. The cement is more flexible than De Khotinsky cement and may be used where the rubber content of De Khotinsky cement is objectionable.

Universal Cement

½ second Pyroxylin	3½
Acetone	3½
Ethyl Acetate	2
Rezyl 12	2¼
Toluol	4
Paraplex 5B	½
Amyl Acetate	1
Aluminum Paste	¾

Quick-Setting Waterproof Cement

Sodium Silicate	50
Kaolin	40
Zinc Oxide	9½

Pettman Cement
Formula No. 1

Iron Oxide	50±3.0
Alcohol	20±2.0
Pine Tar	12±1.5
Shellac, Type D	18±1.5

No. 2

Iron Oxide	33±2
Alcohol	19±2.0
Pine Tar	17±2.0
Rosin	30–2.0
Ethyl Cellulose	1–0.5

Lacquer Cement

This lacquer cement is a strong, waterproof adhesive. On porous surfaces a thin coating should be applied and allowed to dry. A thick coating is then applied and the parts pressed firmly together. On non-porous surfaces such as chinaware a moderately thin coat-

ing is applied and the parts pressed together at once. This cement may be put up in collapsible tubes. It makes a very useful household cement.

Film Scrap	12.5
Raw Castor Oil	1.0
Dibutyl Phthalate	2.5
Ester Gum	2.0
Denatured Alcohol	25.0
Ethyl Acetate	16.0
Butyl Acetate	16.0
Toluol	25.0

Weigh out film scrap, ester gum, castor oil, and dibutyl phthalate and charge into a slow speed mixer. Add the solvents, soak for 2 to 4 hours, and mix until uniform.

Asphalt Emulsion as Millboard Adhesive

1 Asphalt	100
2 Tall Oil Soap (50% Water)	30
3 Kaolin	30

Mix 2 and 3 with heating and then add molten 1 slowly with stirring. Before use add hot water in quantity desired. One ton of board needs enough of above emulsion to give 100 kg. asphalt.

Adhesive for Sound-Deadening Pads
(Waterproof Asphalt Emulsion)
U. S. Patent 2,333,779

Asphalt	50–60
Bentonite	2–3
Oxalic Acid	0.02
Kerosene	3–10
Water	To make 100

Warm the asphalt in kerosene until dissolved. Put bentonite and oxalic acid in water and heat to a boil. Mix both solutions vigorously and run through a colloid mill if necessary.

Pressure-Sensitive Adhesive
U. S. Patent 2,285,570

Polystyrene	40–15
Triphenyl Phosphate	60–85

Pressure-Sensitive Adhesive Tape
British Patent 559,271

Porous crepe paper is impregnated with a 10% solution of polymerized methyl methacrylate in acetone, squeezed, dried and coated with:

Latex Crepe	100
Zinc Oxide	100
Hydrogenated Rosin	60
Betanaphthol	1
Heptane or other solvent	450

Masking Tape

Rubber (Neutral Latex 70%)	100
Castor Oil	50
Rosin	5

The rosin is heated in the castor oil, until it is dissolved. The solution is then added to the latex.

Removing Surgical Adhesive Tape

Soften up adhesive by applying cotton soaked in ether. The ether softens up the adhesive and allows the tape to peel off without any difficulty. Keep away from open flames.

Adhesive Tape
Formula No. 1
British Patent 545,713

An adhesive tape, for medical use, capable of being sterilized by subjecting it to steam at 240°F. (heat causing no damage to tape) can be made by coating one side of a fabric with the following composition:

Ethyl Cellulose	5.4
Titanium Dioxide	35.2
Synthetic Resin Plasticizer	8.8
Toluene	23.8
Ethyl Acetate	22.3
Ethyl Alcohol	4.5

The volatile solvents are removed by heating. The other side of the tape is then coated with a pressure-sensitive rubber adhesive mass.

No. 2
British Patent 481,593

Copal Resin	80
Castor Oil	20

Melt together and stir until uniform.

Surgical-Strapping Adhesive

Guiac	9
Myrrh	9
Mastic	18
Ethyl Cellulose	12
Castor Oil	1
Isopropyl Alcohol	50
Acetone	100

Waterproof Surgical Adhesive

Polyvinyl Butyral Resin	20 g.
Alcohol	120 cc.
Ether	20 cc.
Castor Oil	10 cc.

Grafting Wax
Formula No. 1

Rosin	80
Rosin Oil	5
Petrolatum (Yellow)	15

No. 2

Rosin	50
Wool Fat	20
Ceresin (58–60°C.)	20

Beeswax   5
Rosin Oil   10

Wax Wood Filler
Paraffin Wax   50
Montan Wax   10
Rosin   40

The wax and rosin mixture is colored either with mineral colors or coal tar dyes.

Wood Filler
Beeswax   10
Rosin   10
Sawdust   To suit

Melt the beeswax and rosin and stir in enough sawdust to obtain a hard mass on cooling.

Wax Furniture Filler
Japan Wax   30
Paraffin Wax   10
Wool Fat (Neutral)   10
Rosin   40

The waxes are simply melted together.

Putties
Metal to Glass—Metal to Wood
Metal to China
Metal to China
Shellac   1
Pumice (Powdered)   1

Fuse, stir and use while hot.

Metal to Horn**
Guttapercha   6
Coal Tar Pitch   4

Fuse, stir and use while hot.

Rubber to Glass—Rubber to Stone
Rubber to Metal
(a) Guttapercha   10
     Carbon Disulfide   40

** Horn refers to shell, ivory, horn, bones.

Dissolve while cold.
(b) Shellac   14
     Venetian Turpentine   1
     Alcohol   35

Dissolve b and mix with a.

Metal to Linoleum
Coumarone   20
Coal Tar Pitch   3

Fuse and mix with 2 naphtha solvent, use while hot.

Metal to Stone
Stir cold sodium silicate (38° Bé) with whiting to a thick dough; use immediately.

Brass to Marble
Sodium Oxide   7.5
Water   40.0
Rosin   22.5

Boil and mix with 35 gypsum. Use when warm; hardens after 40 minutes.

Metal to China
Guttapercha   2
Beeswax   1
Sealing Wax   3

Fuse the ingredients and warm the joints before applying the putty.

Glass to Glass
Rubber   1
Mastic   16
Chloroform   70

Dissolve by shaking.

Glass to Wood
Whiting   75
Linseed Oil   9

Knead to a tough dough.

Marine Glue
Pitch   3
Shellac   2

Crude Rubber 1
Melt and stir.

Putty for Knife Handles (and other parts which are subjected to heavy strain):
Mix powdered litharge with syrupy glycerin and apply. This compound hardens very rapidly.

Metal to Celluloid

Rosin 33
Camphor 2
Alcohol 65
Brush joints with the above solution and press them together.

Stone to Stone

Kieselguhr 1
Litharge 3
Slaked Lime 2
Mix and stir with linseed oil varnish to tough dough.

Stone to Horn

Shellac 5
Guttapercha 4
Fuse and use like sealing wax.

Stone to Hard Rubber

Guttapercha 2
Rosin 5
Fuse and mix with 2 pine tar.

China to China

Shellac 15
Mastic 5
Venetian Turpentine 1
Alcohol 60
Dissolve with gentle heat.

White Lead Putty

White Lead-Linseed Oil
Paste 50
Fine Bolted Whiting 50

Mix the two ingredients together, kneading the mixture with the hands, and then beat it with a wooden mallet until a smooth uniform product is obtained.
This is an excellent glazing putty. It may be used to repair surface defects and does not discolor subsequent paint coats.

Putty
U. S. Patent 2,346,408

Reclaimed Rubber 10
Whiting 75
Linseed Oil 5
Gasoline 10

Preventing Putty Sticking to Hands

1% ricinoleic acid is added to linseed oil, cooked to 400° C. and cooled.

Alberene Stone Filler or Cement

Litharge 94
Ultramarine Blue 5
Carbon Black 1
Pass through 60 mesh sieve and make into putty with
Glycerol 85
Water 15
After 24 hours, smooth patches with wet abrasive stone.

Stoneware Cement

Paraffin Wax 30
Rosin 20
White Sand 25
Emery (Grainy) 15
Tripoli 15
This cement is applied hot.

China, Stoneware Adhesive

Lead 55.5
Tin 27.8
Bismuth 16.7

Heat both surfaces and apply as molten fluid.

Refractory Cement
U. S. Patent 2,259,844
Calcium Carbonate	22.0
Magnesium Carbonate	14.0
Calcium Metasilicate	6.7
Magnesium Silicate	6.6
Silicon Dioxide, Free	20.9

Metal Filler and Calking Compound

This filler is used on iron castings and metal structures to fill holes and cracks. The material is hard, adhesive, elastic, and non-shrinking if properly used. Apply it in layers about one-eighth inch thick in very deep holes. When thoroughly dry it may be cut down level with the surface with abrasive wheels or sandpaper.

400 Mesh Crystalline Silica	10.0
Diatomaceous Earth	10.0
Talc	15.0
10-Gallon Varnish (50% Solids)	30.0
Dipentene	3.3
V M & P Naphtha	31.7

Put all liquids in a pony mixer and work in the silica and diatomaceous earth. Then work in enough talc to form a heavy paste.

Calking Compound
Formula No. 1
British Patent 550,003
Paraffin Wax	33½
Tallow	15
Rosin	50
Asbestos Fibers	1½
Caustic Soda	Small amount

No. 2
U. S. Patent 2,345,598
Asbestine or Blown Asphalt	2
Asbestos Fibers	1
Castor Oil	3
Soybean Oil	1

Heat at 140–200°F. and mix until uniform. This can be extruded to form continuous, flexible cords or ribbons.

Boat Calking Wax
Formula No. 1
Beeswax (Yellow)	20
Tallow (Beef)	25
Lard	40

For technical purposes such as the above, one may substitute B.Z. Wax A or other synthetic waxes for beeswax.

No. 2
Beeswax	4
White Lead Paste (in Linseed Oil)	5

Heat-Sealable Label Base
U. S. Patent 2,348,688
Aluminum Stearate	4–10
Ester Gum	3–10
Microcrystalline Wax, To make	100

Heat with good mixing until viscosity is somewhat reduced.

Conductive Adhesive for Rectifiers
British Patent 554,972
Ground Graphite	30 g.
Chlorinated Rubber Varnish	100 cc.

Denture Adhesive
Tapioca Dextrin (Soluble, Pure)	20.0
Casein (Finely Powdered)	10.0

Borax 0.1
Mix thoroughly.

Preparation of Surfaces for
Cementing Leather with
Gutta-Percha

An improvement in the adhering of gutta-percha solutions to leather can be obtained (1) by roughening the leather with emery cloth and (2) by treating the roughened surface with solutions of polar substances, such as rosin, butyric acid and stearic acid.

Leak-Sealing Composition
U. S. Patent 2,315,321

Tetrasodium pyrophosphate or trisodium phosphate is first dissolved in the proper volume of water, and rosin or a synthetic resin, together with a suitable amount of glue, corn starch, wheat flour, gelatin, gum arabic, gum tragacanth, etc., is added. The mixture is then heated to a temperature sufficient to cause the glue and the rosin or synthetic resin to be melted; then with continuous agitation the organic amine or ammonia is added, followed after a few minutes by the filler. At this point sodium silicate may be added. The pH of the final composition is adjusted to a value less than 10 and greater than 5.

Formula No. 1

Water	700.0
Tetrasodium pyro-	
phosphate	10.0
Glue	40.0
Rosin	40.0
Monoethanolamine	7.5
Asbestos Fiber	40.0

Sodium Silicate	25.0
Orthophosphoric Acid	8.5

No. 2

Water	700.0
Tetrasodium Pyro-	
phosphate	10.0
Glue	40.0
Synthetic Resin	40.0
Monoethanolamine	5.5
Asbestos Fiber	40.0
Sodium Silicate	25.0
Orthophosphoric Acid	8.5

Elastic Refrigerator Seal
U. S. Patent 2,149,975

Polymerized Chloro-	
prene	100–38.2
Lead Monoxide	59–19.1
Zinc Oxide	2–0.8
Abietic Acid	2.5–0.9
Carbon Black	100–38.2
Mineral Oil	5–1.9
Sulfur	1

Oil-Well Water-Sealing Cement
U. S. Patent 2,320,633

Oil-Well Cement	70
Calcium Oxide	15
Beidellite	10
Iron Oxide	5

Bonding Aluminum to Steel
British Patent 544,888; 545,023

(A) A 1:2 mixture of aluminum powder and ferric oxide or 2:9 mixture of aluminum powder and cupric oxide is placed between the surfaces to be bonded, the two parts are held together under high pressure, and the temperature is raised until an exothermic reaction occurs which bonds the parts together.

(B) The surface of the steel is coated with copper or a copper alloy and the surface of the aluminum with zinc or a zinc alloy, the copper and zinc surfaces are placed together, and bonding is effected by applying pressure and heating until the zinc and copper form an alloy.

Chapter III

FLAVORS AND BEVERAGES

Pure Fruit Concentrates

In the process of manufacturing pure fruit concentrates, equipment plays a very important part. Fresh or frozen fruit can be utilized for this purpose. The fruit is crushed by regular crushers, then the juice is expressed from the crushed fruit by means of a hydraulic press. The liquid is pumped into vertical cylinder tanks to which alcohol is added. This process precipitates pectin from the fruit. After 48 hours of constant agitation in closed tanks, by means of a propeller, the product is pumped through a filter press (preferably made of bronze), having at least 18 plates, and then pumped into the steam jacketed distilling apparatus. The latter should be provided with fractionating column, steam, adequate condensation equipment, and a receiver. The concentration should be carried on as slowly as possible, under low temperature. It should never exceed 65°C. in order not to lose the volatile constituents of the products subjected to concentration. The first fraction of distillation containing a mixture of alcohol and aromatic principles, of the fruit, is carefully taken off and kept in a cool place. The second fraction is taken off and pumped into a special tank. There it is washed with a volatile solvent extracting the remaining odoriferous components of the fruit. This fraction consists of 80% of the total amount of fruit juice employed in concentration. The condensed fruit juice after cooling is pumped into the tank where it is mixed with the first fraction of the distillate. The volatile solvent extract with the odoriferous components after being placed in a cylindrical container where it is agitated for 48 hours, is then pumped into a steam jacketed still with a high column for the recovery of the solvent. After the solvent is distilled off under ordinary atmospheric pressure, the residue is placed in a small vacuum still to eliminate all traces of the volatile solvent in the product. The residue containing the valuable aroma of fruit extract is used in making reinforced fruit flavors or fruit and other natural flavors. These flavors consist of fruit base and are reinforced with other natural flavorings. Raspberry, pineapple, grape, currant, and apple are fine intensifiers of other natural flavors. They are blended in various proportions for different flavors. Concrete examples are given of the four most popular flavors.

Strawberry with Other Natural
Flavors

True Fruit Strawberry Flavor	55
Volatile Extraction Base	1

True Fruit Raspberry
Flavor   10
True Fruit Currant
Flavor   10
True Fruit Grape Flavor   2
True Fruit Pineapple
Flavor   10
Cognac Oil (10% Solu-
tion in Alcohol)   2
Otto Rose (10% Solu-
tion in Alcohol)   2
Jasmin Absolute (5%
Solution in Alcohol)   1
St. John's Bread Extract   2
Vanilla Extract   3
Tonka Extract   3

Raspberry with Other Natural
Flavors
True Fruit Raspberry
Flavor   55
Volatile Extraction Base   1
True Fruit Currant
Flavor   12
True Fruit Grape Flavor   12
True Fruit Pineapple
Flavor   5
Orris Concrete 10X (2%
Solution in Alcohol)   1
Jasmin Absolute (5%
Solution in Alcohol)   1
Otto Rose (10%
Solution in Alcohol)   2
Resinol Foenugreek (10%
Solution in Alcohol)   3
Orange Oil, Calif. (10%
Solution in Alcohol)   3
Vanilla Extract   5

Grape with Other Natural Flavors
True Fruit Grape Flavor   55
Volatile Extraction Base   1
True Fruit Raspberry
Flavor   10
True Fruit Currant
Flavor   10

Cognac Oil (5%
Solution in Alcohol)   10
Vanilla Extract   7
Distilled Vinegar   4
Orange Oil, Calif. (10%
Solution in Alcohol)   1
Otto Rose (10%
Solution in Alcohol)   1
St. John's Bread Extract   2

Cherry with Other Natural Flavors
True Fruit Cherry Flavor   55
Volatile Extraction Base   1
True Fruit Raspberry
Flavor   10
True Fruit Currant Flavor 10
True Fruit Grape Flavor   5
Bitter Almond Oil (10%
Solution in Alcohol)   10
Cinnamon Oil, Ceylon
(10% Solution in
Alcohol)   1
Clove Buds Oil (10%
Solution in Alcohol)   1
Orange Oil, Calif. (10%
Solution in Alcohol)   1
St. John's Bread Extract   2
Vanilla Extract   2
Tonka Extract   2

Nectars
Orange Nectar
Orange Oil   8.5 lb.
Acacia, Powdered   4.0 lb.
Water   7.0 pt.
Emulsify and add to the follow-
ing:
Tartaric Acid   150 lb.
Sugar   200 lb.
Orangeade Color   2 gal.
Water   To make 100 gal.

Lime Nectar
Lime Oil   2.5 lb.
Acacia, Powdered   1.25 lb.
Water   2.5 pt.

Emulsify and add to the following:

Tartaric Acid	150 lb.
Sugar	200 lb.
Light Green S.F.	
Yellowish Certi-	
fied Color	100 g.
Naphthol Yellow	
Certified Color	140 g.
Water To make 100 gal.	

Imitation Nectars

Grape

Methyl Authranilate	75	lb.
Ethyl Butyrate	10	lb.
Methyl Salicylate	5	lb.
Ethyl Acetate	10	lb.

Grape Flavor

(above)	1.25	gal.
Gum Acacia	4.00	lb.
Water	1.00	gal.

Emulsify above and add:

Citric Acid	150	lb.
Sugar	200	lb.
Amaranth F. D. &		
C. No. 2	18	oz.
Brilliant Blue F. C. F.;		
F. D. & C. Blue		
No. 1	50	g.
Water To make 100 gal.		

Cherry

Benzaldehyde	30	lb.
Vanillin	4	lb.
Acetic Ether	6	lb.

Cherry Flavor

(above)	12	pt.
Tartaric Acid	150	lb.
Sugar	200	lb.
Amaranth Cert. Color	20	oz.
Fast Blue F.C.F.		
Cert. Color	20	g.
Water To make 100 gal.		

Strawberry

Vanillin	3	lb.
Orisol, Alpha	2	lb.
Amyl Acetate	11	lb.
Acetic Ether	12	lb.
Butyric Ether	9	lb.
Formic Ether	3	lb.

Strawberry Flavor

(above)	6	pt.
Tartaric Acid	150	lb.
Sugar	200	lb.
Amaranth Cert. Color	20	oz.
Fast Blue F.C.F.		
Cert. Color	16	g.
Water To make 100 lb.		

Raspberry

Ethyl Formate	3	lb.
Ethyl Acetate	13	lb.
Ethyl Butyrate	9	lb.
Amyl Acetate	6	lb.
Amyl Butyrate	3	lb.
Orisol, Alpha	3	lb.
Benzyl Butyrate	3	lb.

Raspberry Flavor

(above)	14	pt.
Tartaric Acid	150	lb.
Sugar	200	lb.
Amaranth Cert. Color	20	oz.
Fast Blue F.C.F.		
Cert. Color	10	g.
Water To make 100 gal.		

Imitation Walnut Extract

Black Walnut Essence		
(40% Alcohol)	5	fl. oz.
Alcohol	1.5	pt.
Caramel	0.5	oz.
Water To make 1.0 gal.		

Let stand overnight and filter.

Imitation Pineapple Extract

Pineapple Flavor,		
Imitation	6.5	fl. oz.

| Alcohol | 64.0 fl. oz. |
| Water | To make 1.0 gal. |

Almond Extract

Oil of Bitter	
Almonds	7.5 oz.
Alcohol	2.5 gal.
Water	To make 5.0 gal.

Spearmint Extract

Spearmint Oil	4	fl. oz.
Alcohol	7¼	pt.
Water	To make 1	gal.

Lemon Extract

Lemon Oil	3.1
Alcohol	55.0
Water	4.3

Imitation Spices

Imitation Allspice

Eugenol U.S.P.	74
Bay Oil U.S.P.	11
Eucalyptol U.S.P.	4
Methyl Eugenol	11

Thoroughly mix this imitation oil with an equal weight of tannic acid contained in about 48 times its weight of a prepared cereal base.

Imitation Allspice Oil	2
Tannic Acid	2
Prepared Cereal Base	96

Imitation Nutmeg

Prepare the imitation oil by mixing the following:

Rectified Oil of	
Turpentine	28.0
Linalool, Bois de Rose	13.0
Thymol	33.0
Geraniol, Pure	9.0
Safrol	5.5
Eugenol	5.5
Terpineol, Extra	3.0
Isoeugenol	3.0

Now mix:

Imitation Nutmeg Oil	
(above)	2.5
Prepared Cereal Base	97.5

Imitation Sage Oil

Thujone—Concentrated	
Thujone Fraction 80%	0.7
Prepared Cereal Base	99.3

Concentrated Root Beer Extract

Sassafras Oil,	
Natural	45 fl. oz.
Wintergreen Oil,	
Artificial	30 fl. oz.
Anise Oil	2 fl. oz.
Cinnamon Oil	4 fl. oz.
Coumarin	¾ fl. oz.
Vanillin	10 oz.
Caramel	20 gal.
Alcohol	½ gal.
Water	To make 50 gal.

Quick Dissolving Beverage Syrup
U. S. Patent 2,151,499

Sucrose	593 g.
Glucose, Anhydrous	593 g.
Water	814 cc.
Sodium Bi-Carbonate	20 g.
Sodium Carbonate	20 g.
Sodium Phosphate	20 g.
Sodium Hydroxide	4–5 g.

Sugarless Syrup Substitutes
Formula No. 1

Gum Tragacanth	2 g.
Glycerin	Sufficient
Water	To make 100 cc.

No. 2

Karaya	1
Glycerin	Sufficient
Water	To make 100

In both of these formulas, saccharin (0.1%) is used as the sweetening agent, and methyl-para-hydroxy benzoate (0.1%)

added as a preservative. Glycerin is used to increase miscibility and to form a cream. In making syrup substitutes of the above type, the mucilages are brought to the boiling point, allowed to simmer for half an hour and then strained through flannel.

Hop Extract Emulsion
U. S. Patent 2,248,153

Extract pulverized hops with acetone, in presence of active carbon (¼ ounce of carbon per pound of extract), set aside for 2–4 hours, filter, under pressure, the solution containing the same proportion of hop oil, lupulin, tannin, and hop-seed oil as is present in fresh hops, add malt syrup (equal parts of extract and syrup), and beat the mixture to form an emulsion.

Emulsifier and Stabilizer for Flavors
U. S. Patent 2,165,828

Wheat Flour	50
Soya Flour	50
Papain	0.5–10

Hydrolyze flours with papain before use.

Rye Bread Flavor

Caraway Seed	801.25
Wheat Flour	183.75
Ammonium Chloride	10.60
Magnesium Sulfate	5.00

Marmalade Flavoring from Orange Peels

Orange Peel and Pulp	6
Water	15
Corn Syrup	10
Standardized Invert Sugar	10

This formula is designed to use the orange peel and pulp remaining after the juice has been extracted at soda fountains or restaurants. The product can be used for flavoring rolled and cast cream centers, nougat, fudge, hard-candy centers, jellies and other confectionery products.

Dice or grind the peel, add the water and allow to boil until soft, which requires about one-half hour. Add the corn syrup and standardized invert sugar. Cook the batch slowly to 222°F. Transfer at once to a container, place the cover in position and allow to set for 24 hours.

If corn syrup is limited, use 25 lb. of the corn syrup and 75 lb. of standardized invert sugar, cooking the batch slowly to 225°F.

Chocolate Syrup

Water	2	qt.
100 Grade Exchange Citrus Pectin (1-RS/100)	3	oz.
Chocolate or Cocoa	8	oz.
Granulated Sugar	5¾	lb.
Vanilla Extract	As desired	

Heat the water to about 180°F. Then thoroughly mix the pectin with about 1 pound of the sugar. Add this pectin-sugar mixture to the water as the latter is being stirred vigorously. Continue the stirring, heat to boiling, and then add the chocolate or cocoa mixed with the remainder of the sugar.

The batch is again heated to boiling and then stirred with a mechanical mixer until cool.

The syrup is then packaged. A subsequent sterilization process is always necessary.

Lemon Oil Emulsion
(Containing about 15% Oil
by Volume)

Cold Pressed Oil of
Lemon   74.0 ml.
Exchange Citrus
Pectin (1-RS/100)
(about 8 level tea-
spoons)   28.4 g.
Water   393.3 ml.
Glycerin   9.9 ml.

Method—See Orange Oil Emulsion (following formula).

Orange Oil Emulsion
(Containing about 15% Oil
by Volume)

Cold Pressed Oil of
Orange   74.0 ml.
Exchange Citrus Pec-
tin (1-RS/100)   28.4 g.
Water   393.3 ml.
Glycerin   9.9 ml.

The most satisfactory apparatus for making this emulsion in small quantities is the ordinary malted milk mixer. Any violent means of agitation is suitable.

Place in the container of a malted milk mixer the cold pressed oil of orange and add to this the 100 grade exchange citrus pectin. This mixture is then stirred violently to obtain uniform suspension of the pectin in the oil and during the stirring, the water is mixed with the glycerin and added quickly. The mixture is then stirred for five minutes, allowed to stand for five minutes, and then again stirred for five more minutes. The emulsion is allowed to stand for a few minutes to permit air to come to the surface. It is then bottled, care being taken to fill the bottles full. The

bottles should be tightly corked and kept in a cool place away from light.

This emulsion will mix very satisfactorily with a sugar syrup causing the oil to be uniformly distributed. The oil will not separate from the syrup under ordinary storage conditions. The emulsion is equally satisfactory when used in bakery goods or candy.

Upon standing the emulsion may "cream" slightly. This is not detrimental and the emulsion may be easily restored by shaking.

Orangeade Syrup
Formula No. 1

Concentrated Orange
Juice   12   fl. oz.
#21 Pulpy Orange
Juice   19   fl. oz.
Frozen Orange Juice
(defrosted) or
Freshly Extracted
Orange Juice   42½ fl. oz.
Granulated Sugar 5¼ lb.
50% Citric Acid
Solution   2   fl. oz.
15% Orange Oil-
Pectin Emulsion 1½ fl. oz.
Certified Orange
Color   As desired
Benzoate of Soda
U.S.P.   3   g.

The above formula should produce 1 gallon (128 fl. oz.) of finished orangeade syrup. Add water, if necessary, to make 128 fl. oz. total.

Mix the frozen orange juice (defrosted) or the freshly extracted orange juice with the pulpy orange juice, add the benzoate of soda, and then the sugar. Stir the mixture until the sugar is practically dissolved. Do not heat to hasten solution of the sugar. Now add the concentrated orange juice, citric acid solution, emulsion, and color. Stir the ingredients for a few minutes to insure uniformity.

The finished orangeade syrup is diluted with 5 parts of ice water to produce an orangeade drink which contains about 18% orange juice by volume.

No. 2

Concentrated
Orange Juice 11 fl. oz.
Concentrated
Lemon Juice 5½ fl. oz.
Pulpy Orange
Juice 19 fl. oz.
Water 40 fl. oz.
Granulated Sugar 5¼ lb.
15% Orange Oil-
Pectin Emulsion 1¼ fl. oz.
Certified Orange
Color As desired
Benzoate of Soda
U.S.P. 2.8 g.

The above formula will produce 1 gallon (128 fl. oz.) of the finished orangeade syrup.

Mix the water and pulpy orange juice and then add the benzoate of soda. Now add the sugar and stir the mixture. Do not heat to hasten solution of the sugar. After the sugar is practically dissolved, add the concentrated orange and lemon juices, emulsion, and color. Stir the ingredients for a few minutes to obtain uniformity and the syrup is finished.

The finished orangeade syrup is diluted with 4 parts of ice water to produce an orangeade drink

containing about 17% orange and lemon juice (by volume).

Orange Juice Syrup

Concentrated
Orange Juice 12 fl. oz.
Concentrated
Lemon Juice 3 fl. oz.
Orange Juice-
Pulpy Type 19 fl. oz.
Water 42½ fl. oz.
Granulated Sugar 5¼ lb.
15% Orange Oil-
Pectin Emulsion 1¼ fl. oz.
Certified Orange
Color (F.D. &
C. Yellow #6) As desired
Benzoate of Soda
U.S.P. 3 g.

If necessary, add water to make a total of 128 fluid ounces of the finished syrup.

Dissolve the benzoate of soda in the water and then add the orange juice-pulpy type. Stir well, add the sugar, and continue the stirring until the sugar is dissolved. (Do not heat to hasten solution of the sugar.) Add the concentrated orange juice, concentrated lemon juice, orange oil-pectin emulsion, and certified orange color. Mix thoroughly.

1 fluid ounce of the finished orange juice sirup is diluted with 5 fluid ounces of ice water to produce the finished drink that will contain not less than 15% fruit juice (orange and lemon) by volume.

Ice Cream or Soda Fountain Crushed Fruits

Water-Packed Fruit
or Fresh Fruit 13 lb.

100 Grade Exchange
Citrus Pectin   5   oz.
Granulated Cane or
Beet Sugar   40   lb.
Powdered Benzoate of
Soda U.S.P.   1.2 oz.
Certified Color and
Flavor   As desired
Yield about 53 lbs., 24 kilos
Place the fruit in the kettle.
With constant stirring, heat until
the fruit is warm (180° F. or 82°
C.). Thoroughly mix the pectin
with about 8 times its weight of
granulated sugar. Add this pectin-
sugar mixture to the hot fruit,
stirring until it is all dissolved.
Continue to stir and heat to boiling. Add the remainder of the
sugar and heat to boiling again.
Boil vigorously for a moment,
then add the benzoate of soda
which has been dissolved previously in a little water. Add certified color and flavor. Fill hot
into jars, seal immediately, and
store in a cool place.

This crushed fruit is used at
the soda fountain by diluting with
the desired amount of sugar
syrup.

Prepared Lemon Beverage
Sugar   64 lb.
Salt   2 oz.
Powdered Lemon
Juice   2 lb. 8 oz.
Citric Acid   1 lb. 8 oz.
California Lemon
Oil   2 oz.
Lemon Coloring   As desired
Place 48 lb. of sugar, the salt,
powdered lemon juice and citric
acid (or substitute) in a mixing
bowl and blend thoroughly. Blend
the lemon oil and coloring with
16 lb. of sugar and add this to
the first mixture. Additional sugar
may be used if desired. Put the
entire mixture through a fine
sieve.
To serve, make up in the proportion of 1 lb. of mixture to 1
gal. or more of water.

CHAPTER IV

COSMETICS AND DRUG PRODUCTS

Cleansing Cream
Formula No. 1

Paraffin Oil	350
Spermaceti	130
Diglycol Stearate	100
Propylene Glycol	40
Distilled Water	380

No. 2

Paraffin Oil	310
Beeswax	20
Spermaceti	110
Triethanolamine Stearate	80
Glycerin	16
Perfume	4
Distilled Water	360

No. 3

Spermaceti	125
Diethylene Glycol Stearate	100
Paraffin Oil	335
Glycerin	40
Distilled Water	380

No. 4

Stearin	150
Lanolin	100
Beeswax, White	80
Paraffin Oil	165
Propylene Glycol	80
Triethanolamine	19
Distilled Water	475

No. 5

Cholesterin	5
Spermaceti	35
Beeswax, White	25
Triethanolamine Stearate	100
Paraffin Oil	390

Hexadecyl Alcohol	100
Glycerin	20
Perfume	5
Distilled Water	360

No. 6

Hydrogenated Vegetable Oil (Having Consistency of Lard)	2
Snow White Petrolatum	1

Mix together in the cold and perfume to suit.

No. 7

Yellow Beeswax	30.4
Sesame Oil	130.0
Mineral Oil	24.0
Stearic Acid	12.1
Spermaceti	15.0
Chlorothymol	0.3
Water	104.0
Triethanolamine	4.1
Carbitol	20.3

No. 8

Emulsifying Base	7
Technical White Mineral Oil	20
Refined Petroleum Jelly	5
Hard Paraffin Wax	½
Water	To make 100

Liquid Skin Cleanser

Stearic Acid	47.50
Lanolin	5.94
Sesame Oil	123.50
Mineral Oil	123.50
Chlorothymol	0.36
Distilled Water	376.00
Triethanolamine	15.60

Cosmetic Emulsifying Base
British Patent 554,859

Purified Kerosene	5
Purified Hydrous Wool Fat	1
Commercial Wool Wax	1

In preparing the emulsifying solution, the kerosene is heated to about 90–100°C. and the purified hydrous wool fat and commercial wool wax stirred in and dissolved.

Thus a relatively non-oily cream or emulsion can easily be prepared by mixing the required quantity of the emulsifying base with water. For example, 1 to 2 per cent of the emulsifying base may be mixed with water; the water being added slowly to the emulsifying liquid with continuous stirring. At the beginning of the mixing the water is added very gradually until the brownish tint of the liquid changes to the white color of the finished product, after which the water may be added more rapidly with continued mixing.

Other typical examples of cosmetics manufacture are given as follows:

Toilet Cream

Emulsifying Base	4
Liquid Paraffin	4
Petroleum Jelly	1
Paraffin Wax	½
Castor Oil	½
Olive Oil	½
Water	To make 100

Ointment Base

Cetyl Alcohol	5
Lanolin, Anhydrous	5
Paraffin Wax	15
Petrolatum, White	50
Mineral Oil, Medicinal	30

Warm slightly and mix. This yields a white creamy ointment when water is worked in.

Washable Ointment Base
Formula No. 1

Cetyl Alcohol	15
Propylene Glycol	10
White Wax	1
Sodium Lauryl Sulfate	2
Water	72

Melt the cetyl alcohol and white wax in the propylene glycol on a water bath and heat to about 65°C. Dissolve the sodium lauryl sulfate in the water and heat on the water bath to about 65°C. Slowly add the oil phase to the well-stirred water phase and continue stirring on the water bath for about 10 minutes. Remove from bath and continue stirring to the point of congealing.

No. 2
Cream

Sodium Lauryl Sulfonate	50
Cetyl Alcohol	125
Triethanolamine	125
White Petrolatum	250
Cocoa Butter	100
Distilled Water	800

Emulsifier

Sodium Lauryl Sulfonate	5.0
Cetyl Alcohol	12.5
Triethanolamine	5.0
Stearic Acid	11.0
Cocoa Butter	10.0
Distilled Water	125.0

Cream: Melt sodium lauryl sulfonate and cetyl alcohol on water bath, heat for ½ hour with constant stirring. Add triethanolamine and stir, then add petrolatum and stir for 15 minutes. Add water in small portions with constant stirring until homogeneous. Add cocoa butter and stir until it liquefies and is thoroughly mixed. Remove from water bath and immerse in cold water. Stir vigorously until smooth, white cream. The faster the mixture is stirred the smoother and whiter the cream. The cream will absorb up to 60 per cent of water and retain a firm consistency.

Emulsifier: Melt sodium lauryl sulfonate, cetyl alcohol and cocoa butter on water bath. Add stearic acid and stir until liquefied. Mix triethanolamine and water in graduate and add to liquefied mixture in small quantities with constant stirring. When thoroughly mixed, remove from water bath, immerse in cold water and stir until mixture congeals.

Sulfathiazole-Urea Cream

Sulfathiazole	5.0
Urea	10.0
Chlorbutanol	0.1
Distilled water	25.0
Washable Ointment Base	To make 100.0

The urea is dissolved in the water and filtered. The sulfathiazole and chlorbutanol are triturated in dry mortar and enough of urea solution added in small portions to make a paste. Add about 30 g. of non-greasy cream and triturate until smooth. Add balance of urea solution, triturate and add enough ointment base to make 100 g. Triturate to smooth cream. Do not use metal utensils in preparing or storing.

This preparation is used in burns, as postoperative dressing, and in necrotic areas and streptococcic and staphylococcic infections. It may be made with 10 per cent of sulfathiazole, with or without the urea. Sulfanilamide or sulfadiazine may be used in place of sulfathiazole.

Hormone Creams
Formula No. 1

Beeswax, U.S.P.	10
Spermaceti	3
Cocoa Butter	5
Diglycol Laurate	2
Lanolin	6
Olive Oil	28
Mineral Oil	30
Borax	½
Water	15
Hormone Oil	½

No. 2

Beeswax, U.S.P.	4
Cocoa Butter	4
Lanolin	10
Olive Oil	40
Borax	1
Water	36
Hormone Oil	½
Stearic Acid	2
Diglycol Stearate	2

No. 3

Beeswax, U.S.P.	4
Diglycol Laurate	4
Lanolin	4
Borax	4
Water	31
Hormone Oil	½
Stearic Acid	1
Diglycol Stearate	1
Almond Oil	50

Camouflage Face Cream
Formula No. 1

Hydrogenated Castor Oil	25

No. 612 Insect
Repellent   30
*Cold Mixture   38
Phenyl Mercuric
Benzoate   0.004
Beeswax, Yellow 4
Lanolin, Anhydrous 3
        No. 2
Hydrogenated Castor
Oil   25
No. 612 Insect
Repellent   25
*Cold Mixture   38
Dimethylphthalate 5
Phenyl Mercuric
Benzoate   0.004
Beeswax, Yellow 4
Lanolin, Anhydrous 3

Acid Cream

Cetyl Alcohol Flakes	15.0
Spermaceti	5.0
"Duponol" C or	
Diglycol Laurate	1.0
Lactic Acid (85%)	1.2
Water	72.8
Glycerin	5.0

Four-Purpose Cream

Beeswax	8.0
Triethanolamine	0.4
Paraffin Wax	7.0
Cetyl Alcohol Flakes	11.0
"Duponol" C or	
Diglycol Laurate	1.0
Mineral Oil	26.0
Water	46.6

Concentrated Day Cream

Stearic Acid	180
Spermaceti	55

---

* Color mixtures shall be blended from five pigments: yellow iron oxide, titanium dioxide, chromium oxide, carbon black and umber. The pigments shall have a neutral reaction and conform to toxicity requirements.

Cetyl Alcohol	45
Potassium Hydroxide	
(50° Bé)	20
Triethanolamine	5
Glycerin	150
Water	525

Preservative and perfume oil to be added in the usual proportions.

The stearin and spermaceti are melted at approximately 85°C. and then added to the remainder (except for the cetyl alcohol), which has also been heated to a temperature of 85°. Continue to heat for approximately 15 minutes while stirring, then continue stirring until cool. When the temperature is down to approximately 50°C., add the cetyl alcohol and finally the preservative and perfume oil.

Glycerin Face Gel

While preparations of this type may not be as generally in demand as others more commonly used, they may nevertheless be of interest, especially because of their good emollient effect.

Gelatin	20
Glycerin	44
Water	540
Perfume	To suit

The gelatin is first dissolved in somewhat more than half the quantity of water and during cooling the glycerin, perfume oil, and the rest of the water are added. The gelatin used for this purpose should be of highest quality, white, and free from odor. Heating the gelatin solution insures stability but despite this fact, the addition of a preservative would be recommended. In place of gelatin, tragacanth or methyl cellulose could, of course, be used.

Cosmetic Base Cream
Formula No. 1
A

Cetyl Alcohol	10
*Sealex Base	10
Rose Oil (Mineral Oil)	20
Dodecaethylene Glycol Monostearate	10

No. 2
B

Water	70
Span #85	5
Tween #85	5

Heat A and B each to 80°C. and add together.

No. 3

Stearic Acid	37.20	oz.
Sesame Oil	4.13	oz.
Cocoanut Oil	4.13	oz.
Water	111.00	oz.
Sodium Benzoate	83.00	gr.
Triethanolamine	2.01	oz.
Carbitol	16.50	oz.

Tissue Cream
A

Sesame Oil	12	oz.
Mineral Oil	108	oz.
Lanolin	32	oz.
Beeswax	16	oz.
Spermaceti	14	oz.
Stearic Acid	8	oz.

B

Distilled Water	84	fl. oz.
Borax	5.44	fl. oz.

Heat A to 180°F. and stir. Heat B to 180°F., stir and add A. Beat 3–4 minutes cool with occasional stirring and add odor and color at about 115°F. and pour immediately.

* Sealex Base is:

Rose Oil	17½
Diglycol Stearate	10
Diglycol Laurate	2½

Modern Tissue Cream
In kettle 1 place:

Parachol (Absorption Base)	12 lb.
Mineral Oil	3 lb.
Lanolin	2 lb.
Beeswax, Sun-Bleached	1 lb.
Cetyl Alcohol	1 lb.

Heat gently to 150°F.
In kettle 2 place:

Water	16 qt.

Heat gently to 150°F. Add the water to kettle 1 and stir gently. Let cool, stirring occasionally till creamy. Add 10 to 12 drams perfume. Mix well and fill jars with a spatula when cold.

Cold Cream
Formula No. 1

Beeswax	6.5
Cetyl Alcohol Flakes	4.5
"Duponol" C or Diglycol Laurate	0.5
Borax	0.325
Mineral Oil	41.925
Water	46.0
Perfume	0.25

No. 2
In kettle 1 place:

Mineral Oil	1 qt.
Parachol	2 oz.
Lanolin	1 oz.
White Beeswax	7 oz.

Heat gently to 150°F.
In kettle 2 place:

Water	1 pt.
Borax	½ oz.

Heat gently to 150°F. Pour slowly the solution from kettle 2 into the oils in kettle 1 stirring slowly till well mixed. When cooled down to about 120°F. add 1 to 2 drams perfume as desired. Stir and fill into jars.

Suntan Oil
Formula No. 1

Lauryl Alcohol	20
Highly Refined Mineral Oil	80

Add some dark brown oil-soluble dye to make color of oil definitely brown; add some perfume.

No. 2

Sun Screen	5
Peanut Oil	30
Coconut Oil	5
Sweet Almond Oil	60
Perfume and Color	To suit

Suntan Lotion
Formula No. 1

Sun Screen	5
Diethylene Glycol or Glycerin	5.5
Tragacanth, Flaked (or Powdered Tragacanth 2.5)	2.0
Alcohol	20.0
Water	67.5
Perfume and Color	To suit

No. 2

Sun Screen	6
Glycerin	60
Water	30
Perfume and Color	To suit

No. 3

Æsculin or Other Sun Screen	5
Gelatin	4
Glycerin	40
Rose Water	60
Fluid Extract of Tormentil Root (Polentilla tormentilla)	40

Add a suitable preservative and perfume and color to suit.

No. 4

Æsculin or other Sun Screen	4
Glycerin	120

Starch	10
Water	10

Add a suitable preservative and color and perfume to suit.

The action of æsculin in these formulas may be increased by the addition of potash (approximately 10 drops of a 10% solution). A lotion containing æsculin is generally more satisfactory than one containing quinine bisulfate since æsculin is non-irritating, whereas extensive use of quinine bisulfate might cause a slight irritation.

The perfume used in these preparations must be stable in water, acids, and alkalis and be non-irritating. Very light and particularly eau de cologne types are recommended.

Sunburn Ointment

Peanut Oil	60
Spermaceti	8
Beeswax	7
Heat and mix, then add:	
Calcium Acetate	5
Water	20

Suntan Cream
Formula No. 1

Sun Screen (such as Quinine Bisulfate)	5
Lanolin Base	20
Cocoa Butter	4
Peanut Oil	15
Lanolin	2
Cetyl Alcohol	2
Alcohol	10
Water	42
Perfume and Color	To suit

Dissolve the quinine bisulfate in the alcohol and melt the oils and fats at as low a temperature as possible. After gradual cooling add the water-alcohol mixture to the fatty remainder.

No. 2

Quinine Sulfate or	
Quinine Hydrochloride	25
Stearin	100
Sodium Carbonate	15
Paraffin Oil	15
Alcohol	100
Distilled Water	745

The quinine sulfate and a suitable perfume are dissolved in the alcohol and added as the last ingredient to the otherwise completed cream.

Anti-Sunburn Ointment

Melt up lanolin and a good grade of vanishing cream with p-aminobenzoic acid in the following proportions, and then beat the mixture well, as it cools, in order to cream as much as possible:

p-Aminobenzoic Acid	5
Lanolin	47.5
Vanishing Cream	47.5

Apply freely to the area to be protected from sun-burn or to the sun-burned area.

Milky Suntan Oil

Lanolin Anhydrous	10
Sesame Oil	5
Water White Mineral Oil	35
Burrow's Solution	5
Water	44.98
Perfume	0.02

Dissolve Burrow's solution in water and mix this well into lanolin; add oil mixture under constant stirring. Keep on stirring for at least ½ hour; add perfume last. This milky product protects against sunburn by oiling and cooling the skin. It settles out,

after long standing only, but can be brought back into emulsion by shaking. If complete protection is required against intense sun rays in high altitude, small amounts of quinine should be added.

Milk of Almond (Synthetic)
Formula No. 1

Paraffin Oil (Viscosity 30–32)	350
Triethanolamine Stearate	80
Beeswax	20
Perfume	10
Distilled Water	500

No. 2

Sweet Almond Oil	50
Pulverized Almond Oil Soap	15
Spermaceti	10
Beeswax	10
Diethylene Glycol	25
Alcohol 90%	50
Distilled Water	840

Face Lotion

Glycerin	50	cc.
Alcohol	325	cc.
Water	625	cc.
Menthol	0.6	g.
Boric Acid	4.5	g.
Coloring (Dye)	As needed	
Perfume	As needed	

Hand Lotions
Formula No. 1*

Stearic Acid	5.0
Triethanolamine	0.5
Cetyl Alcohol	0.5
Alcohol	5.0
Sodium Alginate	0.3
Sodium Benzoate	0.002

* Distilled water to make 100 cc. and perfume as desired, throughout.

No. 2*

Stearic Acid	4.0
Potassium Hydroxide	0.2
Cetyl Alcohol	0.5
Glycerin	2.0
Alcohol	5.0
Sodium Alginate	0.3
Sodium Benzoate	0.002

No. 3*

Stearic Acid	5.0
Triethanolamine	0.25
Glycerin	6.0
Cetyl Alcohol	0.5
Alcohol	10.0

No. 4*

Stearic Acid	3.0
Triethanolamine	0.3
Glycerin	2.0
Cetyl Alcohol	0.3

No. 5*

Stearic Acid	2.5
Triethanolamine	0.25
Glycerin	2.0
Cetyl Alcohol	0.25

The lotions are all made following more or less the same procedure. After dissolving the preservative in the warm distilled water, the sodium alginate is is slowly added, stirring until dissolved. The triethanolamine or potassium hydroxide is then added, stirring constantly, and heated to 85°C. on the water bath. The stearic acid and cetyl alcohol are melted at a temperature under 85°C. and added to the alginate mixture under continuous stirring until cooled to room temperature. The perfume, dissolved in alcohol, is added slowly, and the mixture well stirred.

No. 6

Bay Rum	80

* Distilled water to make 100 cc. and perfume as desired, throughout.

Glycerol	20
Camphor	To saturate

No. 7

Stearic Acid	4.0
Almond Oil	9.0
Spermaceti	3.59
Chlorothymol	0.13
Water	109.0
Triethanolamine	1.0
Honey	3.59

No. 8

This hand lotion is a stable, creamy emulsion which is very effective for relieving dryness of the skin.

Trihydroxyethanolamine Stearate	18	oz.
Light Mineral Oil	19	fl. oz.
Glycerin	32½	fl. oz.
Alcohol	19	fl. oz.
Water	288	fl. oz.

Yield 2⅝ gal.

Heat water, triehydroxyethanolamine stearate, and glycerin to boiling. Add stearic acid and mineral oil melted together. Mix well and stir in alcohol. When cool, perfume and coloring may be added if desired.

No. 9

In kettle 1 place:

Diglycol Stearate S	7	oz.
Triethanolamine	1¼	oz.
Water	3½	qt.

Heat to boiling for 5 min. then stir and cool to 140°F.

In kettle 2 place:

Castor Oil	2½	oz.
Mineral Oil	1¼	oz.
Cetyl Alcohol	½	oz.
Lanolin	½	oz.
Moldex	1	g.

Heat to 140°F. Pour 2 slowly into 1 and stir till cool and smooth. The water may be increased to 5 quarts if you wish a less heavy lotion, which dries more quickly.

No. 10

Bay Rum	1
Rose Water	1
Spirits of Camphor	1
Glycerol	1

Vanishing Cream, Hand Cream, Foundation Cream, etc.
Formula No. 1

In pan 1 place:

Propylene Glycol Monostearate	3½ oz.
Triethanolamine	1 oz.
Water	2¼ qt.

Heat up to boiling for 5 minutes. Stir and cool until down to 140°F.

In pan 2 place:

Triple Pressed Stearic Acid	14 oz.
Mineral Oil	2 oz.
Lanolin	1 oz.

Heat to 140°F.

Pour 2 into 1 slowly stirring until cold and creamy.

Fill cold into jars.

No. 2
Vanishing Cream

Stearic Acid	14.0
Triethanolamine	0.7
Cetyl Alcohol Flakes	3.0
"Duponol" C or Diglycol Laurate	0.5
Glycerin	5.0
Water	76.8

Alginate Hand Lotion Base
British Patent 555,940
Formula No. 1

Sodium Citrate	20
Water	1000
Glycerin	750
Calcium Alginate	30

Mix well, until dissolved. A gel forms on standing.

No. 2

Sodium Hexametaphosphate	1
Water	100
Calcium Alginate	2½

Invisible Glove
(Skin Protective)

Shellac	13
Isopropyl Alcohol	31
Linseed Oil	4
Titanium Oxide	12
Sodium Perborate	13
Talcum	20
Carbitol	3

Hand Cream
(Protective Against Alkali)

Lanolin, Refined	9.0
Stearic Acid	50.0
Triethanolamine	2.7
Carbitol	18.0
Boric Acid (Saturated Solution)	120

Lithographers' Protective Hand Wash

Carbolic Acid	10 drops
Glycerin	3 oz.
Ammonia	1 dram
Methylated spirits	1 oz.
Water	10 oz.

After dipping the hands into this lotion, they should be wiped dry on a paper towel, which should be discarded.

Hand Protective Cream
Formula No. 1

Beeswax	5.0
Glyceryl Monostearate	12.5
Hydrous Wool Fat	5.0
Sodium Silicate	5.0

Ammonium Hydroxide
(10% Solution)   0.5
Petrolatum   72.5

If desired this formula can be modified by the addition of five per cent, by weight, of latex to produce a rubbery film on the skin.

No. 2

Another in this series of compositions is one recommended where there is prolonged contact with water:

White Wax   10
Hydrous Wool Fat   5
Sulfonated Olive Oil   10
Petrolatum   75

No. 3

Glyceryl Monostearate   12.0
White Wax   12.0
Wool Fat   6.0
Cholestrol   1.0
Sodium Silicate   5.0
Ammonium Hydroxide
(10% Solution)   0.5
Water   63.5

No. 4

Ethyl Cellulose   5
Mastic   8
Castor Oil   1
Acetone   86

No. 5

Zinc Oxide   25
Kaolin   25
White Petrolatum   50

No. 6

Glycosterin
(Diglycol Stearate S)   70
Diglycol Laurate,
Synthetic   40

Melt and cool to 55°C. and add with constant stirring a mixture of:

Aquaresin   120
Cellosize W S   100
which has been heated to 55°C.

Stir slowly till cold and perfume as desired at 35°C.

No. 7
(Oil-Resistant)

Stearic Acid   2.5 oz.
White Beeswax   1.0 oz.
White Petroleum
Jelly   2.5 oz.
Mineral Oil   1.5 oz.

Melt these four ingredients and emulsify with 4 dr. of triethanolamine. Then add boiding water to make 24 oz. and mix in magnesium stearate 2 oz.

No. 8

Glyceryl Monostearate   20
Glycerin   5
Spermaceti   5
Zinc Oxide   10
Water   60

All of the components are boiled together, stirred to complete emulsification; the stirring continued until cold.

Nail Creams

The dryness and brittleness of finger nails caused by nail polish is generally considered to be due to the solvents which these materials contain such as, for instance, acetone, possibly in combination with other solvents which are equally good solvents of fats. The polish remover also consists largely of acetone. It can therefore easily be understood that upon repeated application and removal of nail polish, the natural oil as well as the cholesterin and lipoids, with which the nails are provided by nature to keep them supple, are removed and as a result the nail begins to be dry and brittle and apt to break.

It is because of this widespread

condition of the nails that nail creams have come to fill a definite need and now have their own place in modern cosmetics.

In extremely severe cases the nails should be bathed in a mixture containing 10 parts of alum dissolved in 90 parts rosewater or other suitably perfumed watery solution, after which lanolin cream is applied. Specially medicated creams of this type may be made as follows:

Formula No. 1

White Wax	15
White Ceresin	10
Almond Oil	30
Cream of Tartar	35
Citric Acid	4
Alum	6

No. 2

Almond Oil	25
Petroleum Jelly	15
White Wax	5
Anhydrous Lanolin	5
Water	75

No. 3
Non-Fatty Cream

Stearin	150
Cetyl Alcohol	20
Paraffin Oil	10
Glycerin	50
Triethanolamine	8
Borax	2
Distilled Water	260
Hamamelis Water 1:1	50

No. 4

Glyceryl Monostearate	160
Ceresin	40
Glycerin	40
Paraffin Oil	40
Distilled Water	720 ·

The glyceryl monostearate is melted and when liquid, the ceresin is added, followed by the paraffin oil. The water required has in the meantime been brought to a boiling point and is added, stirring vigorously, allowed to boil for a few minutes, and then allowed to cool while continuing to stir. The perfume compound is added when sufficiently cool.

No. 5
(Honey Gel)

Gelatin	10
Distilled Water	110
Glycerin	300
Honey	25
Distilled Water	50
Perfume	5

No. 6

Vitamin F (250 million Shepherd-Linn units per gram)	200 to 300
Paraffin Oil	700 to 800
Beeswax	500
Water	500
Triethanolamine	20

An addition of Vitamin F has also been recommended for nail lacquers themselves. It is soluble in acetone, amyl acetate, ethyl acetate, isopropyl alcohol, and other solvents used, and the suggested quantity is 2 to 3 million Vitamin F units per pound nail lacquer, which is considered a good preventative dose. Vitamin F can, of course, also be added to the polish remover.

No. 7

Beeswax, White	15
White Ceresin	10
Almond Oil	30
Cream of Tartar	35
Citric Acid	4
Alum	6

No. 8

Almond Oil	25

Petroleum Jelly	15
Beeswax, White	5
Anhydrous Lanolin	5
Water	75

Nail Polish
(Clear)

Rezyl 12, Solution E (50% in Toluol)	8
½ Sec. RS Nitrocellulose (35% Alcohol)	20
*Solvents	72

Non-Volatile Content—17.4%.
Viscosity — E (Gardner-Holdt System).
This polish may be colored by the addition of suitable oil-soluble aniline dyes. The polish adheres well, dries quickly, and has a pleasant odor.

Oily Nail Polish Remover

Butyl Acetate	24
Ethyl Acetate	24
Acetone	24
Alcohol	24
Dibutyl Phthalate	4

Antiseptic Dusting Powder
Formula No. 1

| Iodoform, Powdered | 5 |
| Boric Acid, Powdered | 15 |

Mix and pass through No. 60 sieve.

No. 2

Camphor, Powdered	12
Zinc Oxide	30
Starch	58

Mix and pass through No. 60 sieve.

*Solvents	
Acetone	10
Butyl Acetate	50
Butyl Alcohol	25
Alcohol	15

No. 3

Boric Acid	80
Tannic Acid	10
Alum	10

Mix and pass through No. 60 sieve.

Astringent Dusting Powder

Boric Acid	80
Alum	10
Tannic Acid	10

Mix powdered ingredients and pass through a No. 60 sieve.

Douche Powder
Formula No. 1

| Alum, Finely Powdered | 100 |
| Zinc Sulfate, Finely Powdered | 100 |

Mix well and screen through No. 60 sieve. Use 1 teaspoonful per qt. of water.

No. 2

| Boric Acid, Powdered | 100 |
| Tannic Acid, Powdered | 100 |

Mix and use as above.

No. 3

1. Alum, Powdered	200 g.
2. Phenol, Liquefied	75 cc.
3. Boric Acid, Powdered	675 g.
4. Wintergreen Oil	40 cc.

Mix 1 and 3 then 4 and 2. Pass through a No. 40 sieve.

Face Powder
Formula No. 1

Osmo Kaolin	55
Zinc Oxide	5
Magnesium Stearate	7
Talc	26
Titanium Oxide	7

Tints as follows:
Synthetic Ochre, Scarlet, Geranium lake, etc., in pigments in

proper proportions for example Rachelle:

Synthetic Ochre YO 12.5
Brilliant Scarlet 3.2
Add to above batch.

No. 2

Talcum 38
Zinc Stearate 10
Zinc Oxide 13
Kaolin 39
Magnesium Carbonate 1

Solid Perfume

To produce a solid perfume which is to retain its shape and appearance for a reasonable length of time, acetanilid should be added and may be considered one of the most important ingredients, along with magnesium carbonate which acts as an absorbent, and a certain percentage of crystalline aromatics. To aid in removing the cones or sticks of solid perfume from the forms in which they have been allowed to cool, a small percentage of wax should be added, which at the same time gives a smooth lustrous surface. The selection of this wax is of importance. It has for instance been observed that Japan wax will produce a very thin frosted surface on the finished perfume after comparatively limited storage, while other waxes such as spermaceti and beeswax were found to be absolutely free of these defects and even upon prolonged storage no change was observed in the appearance of the outer surface. A suggested formula for a base for solid perfumes which has been tested in actual practice would for instance be the following:

Formula No. 1

Acetanilid 142
Magnesium Carbonate 15
Beeswax or Spermaceti 5
Musk Xylol 50
Heliocrete 8
Perfume Compound 18

The first three ingredients, acetanilid, magnesium carbonate and wax, are melted while stirring continuously, after which the solid aromatic chemicals, musk xylol and heliocrete, are added, which soon melt and combine with the mass. The perfume compound itself is added last and the mixture is then poured into forms. The mass will solidify very quickly. Cooling with water is not recommended, as the cones or sticks are removed while still warm. Remelting of the base and its use in the manufacture of further sticks is not recommended and, therefore, the batch to be made should be adjusted to exactly the quantity required for the forms to be filled, and so measured that the mixture can easily be poured throughout until all forms are filled, before it has the chance to solidify.

Experiments have been carried out which indicate that on a batch of 220 weight units, an addition of 18 units of perfume represents the maximum quantity possible. If this quantity is exceeded, the firmness of the sticks or cones is reduced, and a tendency to sweat will furthermore be noticed, resulting in a simultaneous loss of perfuming effect. In developing a perfume compound adjusted to these particular purposes, an appreciable quantity of benzyl alcohol, approximately one third, is necessary. It is likewise imperative that the other ingredients, no matter whether of natural origin or synthetically produced, be carefully examined as to their suitability and for their chemical reactions, etc., as well as for their physical properties, such as evaporation velocity. The correct dosage of suitably effective fixatives is very essential, as these, acting as stabilizers, are most important in developing and securing the harmonious effect of the perfume used, and particularly so in view of the fact that the perfume oil has to be added to the basic mass while hot. Although also good fixatives, the crystalline aromatics, musk xylol and heliocrete, are by no means sufficient as they will only increase the durability of the perfume's after-effect, and without the simultaneous use of odor stabilizers, the final perfume effect would be decidedly flat and insipid.

No. 2

Beeswax	2
Japan Wax	2
Diethyl Phthalate	4
Perfume Oil	2

A solid cube used as a sachet is most simply prepared by adding the essential oils in any mixture desired to melted paraffin in the proportion of one-half to one dram of the perfume to one ounce of paraffin.

Cream Cologne

Glycerin	15.0
Stearic Acid	1.5
Quince Seed	1.0
Potassium Hydroxide	0.2
Gum Karaya	0.3
Cetyl Alcohol	0.5
Lanolin	1.0
Preservative (Moldex)	0.1
Water	79.9
Perfume Oil	0.5

Eau de Cologne
Modified Scoville Formula

Bergamot Oil	120
Cold Pressed Lemon Oil	60
Cold Pressed Orange Oil	29
Petit-Grain Oil	7
Rosemary Oil	20
Lavender Oil	5
Sandalwood Oil	5
Orange Flower Water	1000
Methyl Anthranilate	2
Linalool	10
Geranyl Acetate	4
Linalyl Acetate	8
Citral	7
Neroli, Synth.	7
Tincture Siam Benzoin	80
Alcohol, to make	10,000

Dissolve oils in alcohol, add tincture Siam benzoin, and then orange flower water. Let stand a week or more, then filter.

Spanish "Paste" Leather Perfume

Powdered Ambergris	6	dr.
Powdered Benzoin	1½	oz.
Powdered Musk	6	dr.
Powdered Vanilla	6	dr.
Powdered Orris Root	6	dr.
Powdered Cinnamon	6	dr.
Bergamot Oil	1½	oz.
Rose Oil	6	dr.
Gum Acacia	1½	oz.
Glycerin	1½	oz.

Mix the whole, and add water drop by drop until a doughy mass is obtained. Divide into pieces of suitable size.

Mouth Wash
Formula No. 1

Ammonium Octyl Oxy-acetate (25% Aqueous Solution)	92.0
Peppermint Oil	5.0
Anise Oil	1.0
Clove Oil	0.5
o-Benzyl-p-Chlorophenol	1.5

No. 2

25 parts of cresoxy-acetic acid are dissolved in 50 parts of water. The solution is then neutralized with ammonium hydroxide and added with 8 parts of peppermint oil and a few drops of fennel oil.

No. 3

25 parts of naphtheneoxy-propionic acid are dissolved in 140 parts of water and neutralized with ammonium hydroxide. The mixture is then added to 12 parts of peppermint oil, 1.5 parts of o-benzyl-p-chlorophenol and some cochineal red.

Glycerin Thymol Mouth Wash

Borax	20	g.
Sodium Bicarbonate	10	g.
Sodium Benzoate	8	g.
Sodium Salicylate	5	g.
Menthol	1/3	g.
Terpineol	1/2	cc.
Methyl Salicylate	1/3	cc.
Thymol	1/2	g.
Eucalyptol	1 1/3	cc.
Alcohol	26	cc.
Glycerin	100	cc.
Bordeaux B (Dye)	1/4	g.
Talc, Purified	25	g.
Water	1000	cc.

Dissolve menthol, thymol, eucalyptol, methyl salicylate, dye and terpineol in alcohol. Rub this into talc. Make solution of salts in water and glycerin. To the latter add slowly with mixing the alcohol-talc suspension. Filter.

Germicidal Dental Mouth Rinse
Formula No. 1

Oxyquinoline Sulfate	1
Peppermint Oil	1
Alcohol	60
Distilled Water, to make	100

Mix and filter using, 2.5 parts of talc.

No. 2

Iodine	0.4
Potassium Iodide	0.8
Sodium Chloride	17.0
Distilled Water, to make	125.0

Gargle Powder

Alum, Powdered	100
Sodium Borate, Powdered	100
Potassium Chlorate	90

Mix and pass through No. 60 sieve a few times.

Use 4 g. per 120 cc. warm water.

Antiseptic Dentifrice

Sodium Oleate	6.0
Saccharin	0.2
Clove Oil	1.0
Cassia Oil	0.5
Peppermint Oil	0.5
Alcohol	50.0
Glycerin	25.0
Water	To make 100.0

Use 10 drops full strength on a dry brush as a dentifrice.

Pumice-Bentonite Dental Paste

Saccharin, Soluble	0.25
Bentonite	13.00
Cinnamon Oil	0.25
Peppermint Oil	1.50
Calcium Carbonate, U. S. P.	20.00

Sol. Amaranth (10%)   1.00
Pumice Flour   64.00
Mix, make a powder and place in No. 000 capsules.

Salt-Lime Dentifrice

Menthol
Saccharin   0.6
Calcium Oxide   12.0
Methyl Salicylate   2.0
Sodium Chloride, Fine   500.0

One-third teaspoonful in a glassful of water as a mouthwash-dentifrice.

Sodium Oleate Dentifrice

A sodium oleate dentifrice leaves a film of sodium oleate on tooth surface. The film is broken down by saliva and leaves a thin coating of oleic acid on teeth. Such a dentifrice is useful in tooth brush abrasion, when special protection is needed.

Dental Epithelial Solvent

Epithelial tissue sometimes interferes with pocket healing and a caustic solution is useful to dissolve it, permit its removal and stimulate connective tissue growth.

Sodium Sulfide   70 gr.
Sodium Carbonate,
    1 H$_2$O   20 gr.
Water   To make 1 oz.

Trench Mouth Treatment

Brilliant Green   1 g.
Crystal Violet   1 g.
Alcohol (50%)
            To make 100 cc.

This dye solution is applied following gross scaling and a thorough spraying with hydrogen peroxide. It is applied to the dried gums and teeth, and because of its low surface tension, penetrates evenly into the gingival crevices and interproximal spaces for a considerable distance.

Gingivitis Mouth Wash

Zinc Iodide   3 g.
Iodine   5 g.
Glycerin   60 cc.
Distilled Water
            To make 100 cc.

Toothache Drops

Benzyl Alcohol   60
Chloroform   30
Clove Oil   4
Creosote   4
Benzocaine   2

Dental (Dentin) Desensitizer

Sodium Fluoride   1
White Clay   1
Glycerin   1

Rub together to a fine paste.

In using the paste, a preliminary cleansing with cotton moistened with a 4 percent sodium fluoride solution, followed by isolation of the area with cotton rolls and subsequent drying is recommended. A small amount of the paste is then applied on the end of a small plastic instrument, the paste being rubbed vigorously on the sensitive surface of the tooth with special concentration on painful areas. The time necessary for disappearance of all sensation ranges from one to five minutes. Where the pain becomes acute, the paste is washed off with a warm spray and a second treatment given immediately, without difficulty. After disappearance of the pain as much of the paste is wiped off as possible; followed by washing with a spray bottle and thorough rinsing of the mouth by the patient.

Topical Dental Anesthetics
Formula No. 1

Ethyl Aminoben-zoate	6.00 g.
Alcohol	44.50 cc.
Cinnamon Oil	0.13 cc.
Liquor Amaranth	0.13 cc.
Distilled Water	
To make	60.00 cc.

No. 2

Ethyl Aminobenzoate	4
Eugenol	8
Benzyl Alcohol	
To make	30

No. 3

Benzocaine	10.0
Thymol	0.3
Pectin	0.4
Water To make	30.0

No. 4

Benzocaine	7.5
Peppermint Oil	6.0
Phenol Crystals	3.5
Ethylene Glycol	
To make	50.0

The benzocaine, peppermint, and phenol crystals are mixed in a flask and heated until the benzocaine dissolves, and sufficient ethylene glycol is added to make 50 cc.

Dental Cavity Cement Lining

This consists of two parts, a powder mixture and a liquid mixture, to be combined before use by mixing ten parts of the powder with one part of the liquid.

Powder:

Zinc Oxide	69.0
White Rosin	29.3
Zinc Stearate	1.0
Zinc Acetate	0.7

Liquid:

Eugenol	12.75
Olive Oil	2.25

Denture (False Teeth Plate) Adhesive
Formula No. 1

Powdered Traga-canth	22.5 g.
Powdered Gum Karaya	7.5 g.
Sassafras Oil	0.5 cc.

Gum acacia may be substituted for the tragacanth, and other flavoring agents may be employed if desired.

No. 2

Powdered Tragacanth	50
Corn Starch	50
Vanillin	½

Styptic Dental Cotton

Solution of Iron Chloride	80
Glycerin	16
Water	225
Purified Cotton	100

Mix the fluids and immerse the cotton, allowing it to remain for one hour. Remove the cotton and press free of excess fluid, spread in thin layers to dry in a warm place protected from dust and light. When dry, transfer to a well-closed amber glass container.

Dental Instrument Sterilizing Fluid

Formaldehydye	266 cc.
Sodium Borate	30 g.
Distilled Water	
To make	480 cc.

Dental Pulp Devitalizer

Paraformaldehyde	1.00
Procaine Base	0.30

Powdered Asbestos	0.50
Petrolatum	1.25
Carmine	0.02

Dental Antiseptic

Potassium Iodide	48 gr.
Iodine Crystals	84 gr.
Wintergreen Oil	24 min.
Glycerin	To make 8 oz.

Denture Cleaner
Formula No. 1

Trisodium Phosphate	400
Cinnamon Oil	1
Nonaethylene Glycol Laurate	1

No. 2

Calcium Carbonate, Precipitated	15
Triisopropylamine	50
Water	20

No. 3

Calcium Carbonate, Precipitated	75
Magnesium Carbonate, Heavy	20
Sodium Metasilicate, Powdered	14
Sodium Lauryl Sulfate	20

Hair "Tonic"
Formula No. 1

Beta Naphthol	1 oz.
Castor Oil	10 oz.

Dissolve in 1 gal. alcohol, color and perfume to suit.

No. 2

Water	3 qt.
Alcohol	1 qt.
Oxyquinolin Sulfate	1 oz.

To each 7 ounces of this solution float on top one ounce of light mineral oil or deodorized kerosene. Some like to color the oil layer yellow. Perfume to suit.

No. 3

Diglycol Laurate	50 cc.
Cantharidin	1 g.
Glycerin	150 g.
Coloring	As desired
Perfume	As desired
Alcohol	To make 1 l.

No. 4

Chloral Hydrate	1½ oz.
Resorcinol	¾ oz.
Beta Naphthol	¾ oz.
Glycerin	2 oz.
Water	2 qt.
Alcohol	2 qt.

Perfume and color to suit.

Hair Lotion

Dichlorethyl Ether	20
Propylene Glycol	20
Cholesterin	3
Tannin	2
Peruvian Balsam	50
Phosphoric Acid	1
Perfume	5
Isopropyl Alcohol	84
Alcohol	720
Distilled Water	95

Permanent Waving Compositions
U. S. Patent 2,310,687
Formula No. 1

Wheat Starch	10
Cetyl Alcohol	8
Lanolin	6
Sodium Sulfite	6
Water	70

Heat the water to about 70°C., and then add the mixture of cetyl alcohol and lanolin, with continuous stirring, until the mixture is homogeneous.

The wheat starch is then added in powdered form, very slowly, while stirring is continued, and the temperature is raised to 90°C.

At this point the heating is suspended, and the sulfite is added, continuing to stir. Stirring is continued until the temperature has fallen to less than 40°C. The result is a viscous fluid in paste form, homogeneous throughout, and stable.

No. 2

Wheat Starch	5
Cetyl Alcohol	9
Lanolin, Anhydrous	2
Sodium Sulfite	1
Water	83

No. 3

Gelatin	30
Stearyl Alcohol	10
Sulfated Alcohol	1
Sodium Sulfite	5
Water	54

No. 4

Wheat Starch	10
Cetyl Alcohol	4
Olive Oil	6
Sodium Sulfite	6
Water	74

Chemical Heating Powder
(For Permanent Hair Waving)
Formula No. 1
U. S. Patent 2,279,589

Aluminum Powder	30
Sodium Nitrate	20
Calcium Hydroxide	10
Soda Ash	10
Inert Filler (Infusorial Earth)	30

No. 2
Canadian Patent 417,499

Sodium Chlorate	5–18
Talcum Powder	7–18
Aluminum Powder	5–18
Copper Sulfate	2–18

This is moistened with water to start reaction producing heat.

Hair-Waving Self-Heating Composition
Canadian Patent 409,086

Potassium Permanganate	5–11
Sodium Sulfite	3–11
Feldspar	3–11
Glycerin	320–640
Potassium Sulfite	32
Ammonium Carbonate	32
Hydroquinone	1
Water	255–640

Cold Hair-Waving Solutions
A. Direct Application

Sodium or Ammonium Thioglycollate	4–8
"Sulfatate" (Wetting Agent)	0.1
Water	To make 100

B. Circulation Method

Ammonium or Morpholine Thioglycollate	4–8
"Sulfatate" (Wetting Agent)	0.2
Water	To make 100

Prior to use, hair must be washed free of oil, grease and soap residues.

These solutions must be used cautiously as they affect the skin of the head and hands and must be washed off well after the hair has been treated.

Wave Set (Hair-Waving) Fluid
Formula No. 1
100 Grade Exchange

Citrus Pectin	3½	g.
Powdered Gum Karaya	28	g.
Anhydrous Ethyl Alcohol	1	fl. oz.
Salicylic Acid	½	g.

Thoroughly mix the pectin, gum karaya, and salicylic acid. Add 1 fluid ounce anhydrous ethyl alcohol to the dry ingredients stirring thoroughly to obtain a good suspension. (Only the salicylic acid will dissolve in the alcohol.) Pack in a 2 fluid ounce bottle.

Shake the bottle thoroughly until the contents are uniform. Add the entire contents of the bottle to 16 fluid ounces (1 pint) of hot water (almost boiling). Stir thoroughly until a smooth suspension is obtained. Allow to stand until the gum karaya and pectin are completely dissolved. Use as a wave set.

The consistency may be changed to meet individual requirements by increasing or decreasing the amount of gum karaya in the above formula.

No. 2
British Patent 550,746

Sodium Sulfite	50
Castor Oil, Sulfonated	6
Ammonia	8
Glyceryl Monostearate	10
Water	246

Bandoline (Hair-Gloss)
100 Grade Exchange

Citrus Pectin	2½ oz.
Glycerol	6½ fl. oz.
Water	1 gal.
Benzaldehyde	⅓ fl. oz.
Perfume and Color	As desired
Salicylic Acid	4 g.
95% Ethyl Alcohol	10 cc.

A smooth paste is made by mixing the pectin with the glycerol. The water is heated to about 120°F. and added rapidly to the pectin-glycerol mixture. The liquid is then stirred vigorously with a mechanical mixer until the pectin is completely dissolved. Then the perfume and color are added.

Finally, 4 grams salicylic acid are dissolved in 10 cc. warm ethyl alcohol. This solution is added to the bandoline with stirring, and the whole batch thoroughly mixed.

The consistency of the bandoline may be regulated to meet individual requirements by increasing or decreasing the amount of pectin specified in the above formula.

Hair Pomade
Formula No. 1

Cholesterol	1
Cetyl Alcohol	3
Anhydrous Lanolin	5
Paraffin Wax	20
Liquid Petrolatum	30
White Petrolatum	50
Water	70

No. 2

Cholesterol	1
Cetyl Alcohol	3
Anhydrous Lanolin	5
Paraffin Wax	20
Liquid Petrolatum	27
White Petrolatum	50
Water	106

For each 1000 g. of pomade, 2.5 g. of the following perfume mixture is suggested:

Lavender Oil	2.2	cc.
Artificial Neroli Oil	2.0	drops
Bergamot Oil	10.0	drops
Rose Oil	4.0	drops
Jasmine Oil	5.0	drops
Rosemary Oil	10.0	drops
Essence of Ylang Ylang	15.0	drops
Coumarin	0.4	drops

No. 3

White Petrolatum	3 lb.
Lanolin	1 lb.
Tar Oil	3 oz.
Cade Oil	2 oz.
Resorcinol Monoacetate	2 oz.

Melt the lanolin and petrolatum together and mix in the other ingredients.

Wavy Hair Pomade

1. Snow White Petrolatum, Heavy	94
2. Beeswax, Bleached	5
3. Perfume	1

Warm 1 and 2 and mix, until uniform. Mix in 3 before mass sets.

Add oil soluble coloring if desired.

Hair Cream

Mineral Oil (High Viscosity)	48.0
Spermaceti	1.0
Glycerin	1.2
Quince Mucilage	0.5
Stearic Acid	0.5
Perfume Compound	0.6
Sodium Para Oxy Benzoate	0.2
Fatty Acid Sulfonate	2.0
Distilled Water	46.0

The quince mucilage is made from 3 parts quince seed dissolved in 97 parts distilled water. All fats and waxes, together with the sulfonated fatty alcohol, are melted at about 60°C. and then added to the aqueous balance of the mixture, which latter has also previously been heated to 60°C. The addition is made under constant stirring and when cold, the perfume is added.

Another formula is based on a solution of sodium alginate in water, 2.5 parts sodium alginate being dissolved in 100 parts distilled water, to which is added 1 part perfume compound dissolved in 5 parts alcohol. A small quantity of calcium citrate is then triturated in 90 parts distilled water and added to the sodium alginate solution. After two hours the mixture will thicken, the proportion between sodium alginate and calcium citrate being the determining factor as to the degree of viscosity.

Washing and Bathing Composition
British Patent 559,137

Water	1 gal.
Sulfonated Castor Oil	480 cc.
Oleic Acid	96 cc.
Sodium Ortho Silicate	10 oz.

For personal use, and particularly in the bath, about one tablespoonful is used to each gallon of water. When used in the washbasin, the hands should first be wetted with water and then a few drops of the detergent applied.

Soapless Shampoo
Formula No. 1

Epersol-Y (Hydrolyzed Protein)	100 cc.
Water	296 cc.
Sodium Benzoate	4 g.
Deodorant Oil B-7652 (Givaudan)	5 cc.

No. 2

Epersol-Y (Hydrolyzed Protein)	250
Sodium Benzoate	10
Calgon	25
Water	715

Water-Soluble Per-
fume Oil   To suit

No. 3
U. S. Patent 2,290,908

Terpeneless Bay Oil	0.03
Geraniol	0.03
Tertiary Amylphenol	0.07
Carvacrol	0.07
Borneol	0.04
Menthol	0.06
Alcohol	43.00
Sodium Bicarbonate	1.20
Boric Acid	0.10
Citric Acid	0.15
Water	57.00

Eyebrow Dressing
U. S. Patent 2,292,645

Green Soap, U.S.P.	100
Beeswax	48
Borax	3
Water	49

Bluing for White Hair

Aniline Violet	2
Sodium Sulfate (Anhydrous)	1
Sodium Carbonate (Anhydrous)	1

Finely powder all ingredients and mix well. To use, dissolve one dram in one pint of warm water.

"Crude" Oil for Shampoo

Gasoline	400
Kerosene	249
Fuel Oil	300
Cylinder Oil	150
Ethyl Mercaptan (Optional)	1

Head Lice Lotion

Phenyl "Cellosolve"	40
Ethyl Alcohol	30
Methyl Salicylate	5
Water	25

It should be applied so that the hair is thoroughly wet but it must not be permitted to get in the eyes. A single treatment is said to be 100 per cent effective.

Electric Shaving Aid

The preparation is intended to impart rigidity to the hairs prior to shaving with a blade or electric shaver:

Paraffin Wax	25.5
Petrolatum	10.0
Stearic Acid	30.0
Triethanolamine	5.0
Lanolin	3.0
Ichthyol	1.5
Castor Oil	20.0

The composition forms a solid stick and produces, when applied, a thin film on the skin.

Brushless Shaving Cream
Formula No. 1

Glycosterin	40 g.
Diglycol Laurate Synthetic	90 g.
Triple Pressed Stearic Acid	180 g.
Lanolin	30 g.
Cocoa Butter	20 g.
Cetyl Alcohol	20 g.
Moldex (Preservative)	1 g.

Melt these together and cool to 55°C.

In another pan mix:

Aquaresin	60 g.
Sulfatate 2%	100 cc.
Water	500 cc.

Heat to 55°C. and add to the melted waxes in the first pan slowly and with constant stirring. Continue till cool adding 10 grams perfume at 30°C.

This makes a very smooth cream which washes off the razor, is not greasy and spreads well. Any excess left on the skin after shaving should be rubbed in, not washed off, as it has a pronounced softening effect on the skin.

No. 2

In pan 1 place:

Propylene Glycol	
Monostearate	1 oz.
Triethanolamine	½ oz.
Glycerin	¾ oz.
Water	2 qt.

Heat until mixture boils. Cool to 150°F.

In pan 2 place:

Stearic Acid	7 oz.
Mineral Oil	1 oz.
Lanolin	½ oz.

Heat to 150°F. Then add slowly to pan 1. Stir gently until cool. Perfume with ¼ ounce perfume and add water-soluble color if desired. Add color to water in pan 1.

After-Shaving Lotion
Formula No. 1

Water	600 cc.
100 Grade Exchange	
Citrus Pectin	6 g.
Glycerol	25 cc.
95% Ethyl Alcohol	25 cc.
Menthol	¾ g.
Salicylic Acid	¼ g.
Benzaldehyde	1 cc.
Perfume and Color	As desired

A smooth paste is made by mixing the pectin with the glycerol. The water is heated to about 120°F. and added rapidly to the pectin-glycerol mixture. The liquid is then stirred vigorously with a mechanical mixer until the pectin is completely dissolved.

The menthol and salicylic acid are dissolved in 25 cc. warm ethyl alcohol. After the pectin solution has cooled somewhat, the alcoholic solution is added to it with thorough stirring to obtain uniform mixing of the whole batch. Finally, the benzaldehyde, perfume, and color are added.

The consistency may be regulated to meet individual requirements by increasing or decreasing the amount of pectin.

No. 2

Witch Hazel	15
Alcohol	10
Alum	½
Menthol	1/20
Boric Acid	1
Glycerin	5
Water	69

Throat and Nasal Spray

Dissolve three grams of sulfathiazole in eight grams of triethanolamine. Then dissolve this solution in ninety-one grams of water, and apply as a spray to the infected area.

Benzedrine Nasal Spray

Benzedrine	10
Petrolatum, Light Liquid	990

Nasal Spray

Menthol	10
Camphor	20
Eucalyptol	10
Mineral Oil, Light	
Refined	960

Nasal Inhalant

Menthol	11
Linaloe Oil	1
Lavender Oil	2
Pennyroyal Oil	1
Peppermint Oil	2
Thymol	3
Chloroform	80

Acetone may be substituted for the chloroform. This inhalant will quickly relieve nasal congestion and is used as follows: Place a few drops on a folded cloth and allow to evaporate for about 15 to 30 seconds. Now inhale gently, gradually deepening the inhalations.

Vapor Inhalant
Formula No. 1

Dwarf Pine Needle Oil	100 cc.
Compound Tincture of Benzoin	100 cc.

Mix.
Use 4 cc. per qt. of boiling water.

No. 2

Alcohol	200 cc.
Camphor	95 g.
Menthol	65 g.
Terpineol	5 g.
Cinnamon Oil	30 cc.
Eucalyptol	105 cc.

Use 4 cc. per qt. of boiling water.

Nasal Constrictor

Introduced into the sinuses by the displacement method, 0.2% solution of 2-amino-heptane sulfate is effective and well tolerated.

Isotonic solution of 2-amino-heptane sulfate may be prepared as follows:

2-Amino-Heptane Sulfate	2.0
Sodium Chloride	0.406
Distilled Water	
To make	100.0

Dilutions may be made as required with 0.9% sodium chloride solution. The pH of a 2% solution of 2-amino-heptane sulfate is 5.7.

Ephedrine Nose Drops

Ephedrine Sulfate	1.00 g.
Menthol	0.10 g.
Camphor	0.10 g.
Chlorobutanol	0.50 g.
Sodium Benzoate	0.50 g.
Eucalyptol	0.25 cc.
Acacia, Powdered	12.50 g.
Liquid Petrolatum, Heavy	50.00 cc.
Distilled Water	
To make	100.00 cc.

Dissolve the camphor, menthol, chlorobutanol, and eucalyptol in the mineral oil. Add this mixture a little at a time to the acacia in a mortar and triturate to form a smooth paste. Then add all at once 25 cc. of distilled water and triturate rapidly until emulsified. Dissolve the ephedrine sulfate and sodium benzoate in 10 cc. of distilled water and add gradually to the emulsion with constant stirring. Finally add distilled water to make a volume of 100 cc.

Isotonic Nose Drops

Ephedrine Hydrochloride	.300
Dextrose U.S.P.	.750
Sodium Chloride	.135
Chlorobutanol	.150
Menthol	.015
Water	30.0

Asthma Spray

Methyl Atropine Nitrate or Bromide	0.14
Papaverine	0.08
Sodium Nitrate	0.08
Adrenalin	0.05
Lactic Acid	2.50
Glycerin	10.00
Distilled Water	
To make	100.00

Believed to be more beneficial than plain solutions of adrenalin for the treatment of asthma and related conditions, this is given in the form of a nasal or oral spray three or more times per day as a prophylactic or during an asthmatic attack.

Asthma Nose Drops

Ephedrine Sulfate	0.50
Sodium Chloride	0.25
Dextrose	1.20
Chlorobutanol	0.25
Distilled Water	
To make	50.00

Asthma Smoke Powder
(Asthma Cigarettes)

Stramonium, Powdered	1
Potassium Nitrate	2

Hay-Fever Asthma Inhalants
Formula No. 1

Suprarenalin, Crystal	10.0	g.
Sodium Chloride	9.0	g.
Chlorobutanol	5.0	g.
Sodium Bisulfite	0.9	g.
Dilute Hydrochloric Acid (10% U.S.P.)	20.0	cc.
Glycerin	50.0	cc.
Distilled Water		
To make	1000.0	cc.

The glycerin in 850 cc. of water is heated to boiling to remove the dissolved air. Remove from source of heat and add the chlorobutanol and sodium chloride. Cool to room temperature and add the sodium bisulfite. Dissolve the suprarenalin crystals in the hydrochloric acid and add immediately to the above solution. Adjust the pH to a value of 3.0; using a few drops of dilute sodium hydroxide to adjust the value if necessary.

No. 2

Atropine Sulfate	0.10	g.
Sol. Adrenalin Hydrochloride (1:100)	12.50	cc.
Ephedrine Hydrochloride	0.50	g.
Hyoscine Hydrobromide	0.025	g.
Chlorbutol	0.25	g.
Glycerin	12.50	cc.
Distilled Water		
To make	100.00	cc.

This solution provides a preparation which is effective in the treatment of asthma by the spray method. The product is physiologically stable, has an acid reaction, and does not deteriorate, no discoloration developing during a period of three months when stored in well-filled dark containers.

Hay-Fever Treatment

Twelve hay-fever patients were cured of their symptoms by spraying nasally and dropping into the eyes a simple solution of sodium oleate dissolved in 10,000 parts of water. A momentary smarting was produced, but two treatments a day are said to have resulted in definite improvement within four days.

Non-Irritating Eye Drops

By the addition of sodium citrate, the pH of eye drops can be raised to 6.2 or higher without precipitation of zinc hydroxide, making the eye drops non-irritating. A suitable formula consists of the following: zinc salicylate 10, crystal sodium citrate 12, boric acid 20, borax 15, salt 3, parahydroxybenzoic acid ester 0.8, and water to make 100.

Hay-Fever Eye Wash

Sodium Bicarbonate	2
Water	98

Also efficacious as a nose wash.

Cake Mascara
Formula No. 1

Triethanolamine	
Stearate	40
Paraffin	30
B-Z Wax	12
Lanolin	5
Lampblack	13

No. 2

Glyceryl Monostearate	60
Paraffin	15
Carnauba Wax	7
Lanolin	8
Lampblack	10

Melt, mix, and mill the ingredients, then cast or extrude to the proper form.

Kohl (Eye-Brow Black)

Powdered Gum	
Tragacanth	12½
Alcohol	25
Rose Water	1000

Mix gum well with alcohol and stir in rose water, rapidly. Then grind in 10–100 finely divided lampblack.

Eye-Burn Treatment

For acid burns, a buffer two per cent solution of sodium bicarbonate should be instilled until a neutral reaction is obtained. For alkali burns the buffer solution should be:

Acetic Acid	2.5

Sodium Acetate	3.0
Sodium Chloride	4.5
Distilled Water	1000

Neutral Eye Wash

Zinc sulfate 0.25–0.5, sodium citrate 0.5, borax 1.2, sodium salicylate 0.1%. The pH is 7.2–7.4, and the freezing point is –0.8°C., corresponding to that of lacrimal fluid.

Treating Hives on Eyelids

Dab the hives with cotton wool moistened with this solution:

Sodium Thiosulfate	30
Rose Water	30
Distilled Water	250

Ointment Bases
Formula No. 1

Glyceryl Monostearate	12.0
Paraffin, Liquid	2.0
Glycerin	3.0
Spermaceti	5.0
Nipagin	0.1
Water To make	100.0

No. 2

Pectin	5
Zinc Oxide	15
Talc	15
Glycerin	15
Water To make	100

No. 3

Bentonite	25
Zinc Oxide	25
Cottonseed Oil	25
Hydrogenated Cottonseed Oil	25

Intimately mix the powders and the fat and fat-like products before incorporating and finely dispersing any powder.

No. 4

Stearic Acid	15
Triethanolamine	8

Methyl Cellulose

Mucilage (10%)   20

Water   To make 100

No. 5

Stearic Acid   5.0

Triethanolamine   2.5

Paraffin, Liquid, and

Methyl Cellulose

Mucilage (10%),

equal parts   100.0

Absorption Base
Formula No. 1

Cholesterol   3

Cholesterol Stearate   3

Lanolin, Anhydrous   25

Petrolatum, White   69

No. 2

Cholesterol   3

Cholesterol Stearate   3

Anhydrous Lanolin   25

White Petrolatum   69

Heat the petrolatum and lanolin to 150°C., add cholesterol and cholesterol stearate and stir until solution is effected and stir occasionally while cooling. This base is reported to emulsify readily with 300% and more of water, and will not show any separation of water when rubbed on the skin.

Healing Salve

Petrolatum   100

Phenol (88%)   0.5

Cajeput Oil   0.1

Hemlock Oil   0.1

Camphor Oil   0.2

Mix and apply as needed.

Stainless Ointment Base

Cetyl Alcohol   14

Beeswax   1

"Aquaresin G. M." or

Glycerin   9

"Sulfatate"   2

Water   74

Procedure: Melt the alcohol and wax, mix with the aquaresin, and heat to 65°. Dissolve the sulfatate in water and heat to 65°. Add the oil slowly to the water with continuous stirring until congealed.

Analgesic Ointment

Menthol   ¾ g.

Camphor   100 g.

Chloral Hydrate   100 g.

Terpineol   1 g.

Lanolin   800 g.

Mix first four materials until liquid then work in lanolin.

Cod-Liver Oil Ointments

Cod-liver oil ointment generally contains, in addition to cod-liver oil, zinc oxide, adeps lanae, and soft paraffin; other additions are caustic potash, calcium carbonate, kaolin, and balsam of Peru. In such preparations the vitamin A is rapidly destroyed. The addition of suitable antioxidants, hydroquinone (0.2 per cent) or triethanolamine (0.02 per cent) prevents spoilage. A suitable formula is as follows: kaolin, 100 g.; zinc oxide, 100 g.; cod-liver oil, 100 g.; soft paraffin (yellow), 700 g.; triethanolamine, 2.5 g. This shows a loss in activity of not more than 25 per cent after three months.

Lanolin Substitute
Formula No. 1

Refined Lanolin   80

Wool Fat, Crude   20

Wax, Yellow   50

Petrolatum, Yellow   1000

No. 2

Refined Sublan	90
Glyceryl Monostearate	10

Disinfectant
Formula No. 1

Eucalyptus Oil	100
W. Rosin	100
Caustic Soda Solution (25%)	50
Water	150
Methylated Spirit	20

The procedure is first to prepare the soap by dissolving the rosin in the caustic soda solution and then diluting it with half the quantity of water. The soap solution is allowed to cool a little and then the oil is gradually stirred in. The balance of water is added while the mass is still warm. After cooling, the methylated spirit is stirred in. The product is then ready for use by simple dilution with water from 10 to 100 times its volume.

No. 2

Eucalyptus Oil	250
Oleic Acid	105
Caustic Soda Solution (25%)	60
Water	300
Methylated Spirit	60

Germicidal Cream

A

White Petrolatum	12
Rose Oil (Mineral Oil)	9
Beeswax (White)	2.25

B

Water	6
Borax	1.5

C

Parachlormetacresol	1.5

Heat A and B to 80°C. and then add together. When the emulsion is formed, add C, also at 80°C. Stir constantly until cold.

Germicidal Bombs (Aerosols)
Formula No. 1

Propylene Glycol	5
Ethanol	20
Dichlorodifluoromethane	75

No. 2

Hexylresorcinol	0.15
Olive Oil	9.85
Dichlorodifluoromethane	90.00

Germicide
U. S. Patent 2,347,012

Formaldehyde	4
Alcohol	70
Water	10
Hexamethylenamine	0.1–0.5
Methanol	15

Antiseptic
U. S. Patent 2,291,735

Colloidal Silver Bromide	16
Acacia, Powdered	49
Sorbitol Syrup	35

Antiseptic Solution

Boric Acid	25.0	g.
Thymol	0.4	g.
Chlorthymol	0.5	g.
Menthol	0.1	g.
Eucalyptol	0.1	cc.
Methyl Salicylate	0.1	cc.
Thyme Oil	0.01	cc.
Alcohol	300.0	cc.
Water To make	1000.0	cc.

Disinfecting Shoes

There are at least two ways in which shoes may be adequately treated with formalin. The first way is to insert in the toe of each shoe a small piece of absorbent cotton moistened with half a teaspoonful of formalin, and then place the shoes overnight in an ordinary shoe box or airtight container of approximately the same size. The second way is to place one or two pairs of shoes overnight in a small airtight box which contains a saucer or jar holding a teaspoonful or two of formalin.

Silver Fluoride Soap
(Strong Antiseptic)

Sodium Laurylsulfonate	910
Silver Fluoride	10
Glycerin	30
Lanolin	25
Mineral Oil	25

Antiseptic Hand Soap

Water	1000
White Soap Chips	110
Naccanol NR	40
Glycerin	30
Cresol	10
Liquid Phenol	10

Aluminum Chloride Soap
(Anti-Perspiration Soap)

Triethanolamine Lauryl- sulfonate	955
Aluminum Chloride	30
Hydrochloric Acid (1/10 N)	15

Anti-Perspirant Liquid
Formula No. 1

12% solution of aluminum chloride. Apply to dry skin after cleansing. Wash off the solution before putting on clothing.

No. 2

Aluminum Chloride	130
Urea	30
Distilled Water	840

Dissolve aluminum chloride and urea in the water and filter. The use of a filter press is recommended, but filtration through paper with the aid of diatomateous earth (kieselgur) will do. The solution should not come in contact with metal, especially not with iron.

Under Arm Cream

Stearic Acid, Triple Pressed	150
Water	715
Triethanolamine	25
Alum, U.S.P., Powdered	110

To the stearic acid, melted in 500 parts of water add triethanolamine in 215 parts of water. Cool, add alum and grind until smooth, with pestle and mortar. Perfume to suit.

Anti-Perspirant and
Body Deodorant
U. S. Patent 2,350,047

Aluminum Sulfate	18
Zinc Oxide	3
Alcohol	10
Water	69

Control of Foot Perspiration

Use the following routine: twice daily antiseptic soaks, for which a warm solution of potassium permanganate (1:4,000) is the best preparation. The feet are soaked for ten to fifteen minutes and are allowed to dry thoroughly afterward. This should leave a brown tan on the skin; if not, it means that it is not being done correctly or that sweating is so profuse that it immediately washes the stain off the skin. The feet and socks are then well dusted with 3 per cent salicylic acid in talc. This controls the disease in the usual forms. The milder forms may be relieved by use of:

Methenamine	0.5
Tragacanth	0.5
Talc	25.0
Water	74.0

This makes a creamy lotion which is easy to spread over the feet and leaves a liberal residue of powder on drying. Action of the acid sweat on the methenamine liberates formaldehyde, which inhibits bacterial growth and hardens the skin, preventing maceration.

Deodorant Cream
Formula No. 1

Glyceryl Monostearate	15
Glycerin	5
Mineral Oil	2
Spermaceti	5
Titanium Dioxide	3
Moldex (Preservative)	½
Hexamethylenetetramine	3
Water	66

No. 2

Glycerin	3
Spermaceti	5
Titanium Oxide	3
Benzoic Acid	6
Salicylic Acid	½
Water	68
Acimul	15

No. 3

Glyceryl Monostearate	20
Glycerin	10
Methenamine	3
Water	67

No. 4
(Vanishing Cream Type)

a. Acid-Stable Glyceryl Monostearate	170
b. Spermaceti	90
c. Glycerin	39
d. Distilled Water	380
e. Preservative (Moldex)	1
f. Aluminum Sulfate	150
g. Distilled Water	150
h. Titanium Dioxide	20

Dissolve (f) in (g), then stir (h) into the solution.

Heat (c) + (d) + (e) in the mixing kettle to 90°C.

Heat (a) + (b) also to 90°C. and introduce into the water phase with fairly vigorous agitation. Stir until the emulsion is well formed and almost cool, then add gradually the mixture of (f) + (g) + (h).

No. 5
(Absorption Base Type)

a. Absorption Base (Parachol)	240
b. Aluminum Sulfate	150
c. Distilled Water	600
d. Titanium Dioxide	10

Dissolve (b) in (c) add (d). Add this mixture gradually to (a) while stirring, with both phases at 50°C. Stir until 25°C. is reached, pass through an ointment mill.

This type of cream can be adjusted for consistency by addition of waxes like spermaceti or ozokerite. The great advantage of an absorption base cream is that it will not dry out and harden. But its formulation and handling have to be carefully worked out to give a product which holds up satisfactorily.

Deodorant-Fungicide
U. S. Patent 2,314,125

Cadmium Chloride	1
Sodium Dioctylsulfosuccinate	1

Alcohol (30%) 98

Liniment

Chloral Hydrate	50 g.
Menthol	40 g.
Diglycol Laurate	10 cc.
Camphor Oil	50 cc.
Alcohol	850 cc.

Bay Rum

Citric Acid	5.0 g.
Myrcia Oil	2.0 cc.
Myristica Oil	0.1 cc.
Orange Oil	1.0 cc.
Alcohol	500.0 cc.
Glycerin	30.0 cc.
Purified Talc	20.0 g.
Water To make	1000.0 cc.

Dissolve the oils in the alcohol, add the glycerin and citric acid, and then add sufficient distilled water to make the volume measure 1000 cc. Add the purified talc. Filter, returning first portions until the liquid filters clear.

Menthol Liniment

Menthol	100 g.
Chloroform	125 cc.
Diglycol Laurate	50 cc.
Peanut Oil	450 cc.

Athlete's Rub (Liniment)

Methyl Salicylate	1 oz.
Tincture of Green Soap	1 oz.
Fluid Extract of Arnica	1 oz.
Witch Hazel	½ gal.
Water To make	1 gal.

Isopropyl Rubbing Alcohol

Isopropyl alcohol is suitable for purposes of rubbing compounds, the following formula having been used for several years by the University Hospital, of the University of Michigan, as a back lotion:

Isopropyl Alcohol	25.0
Glycerin	10.0
Acetic Acid (4%)	2.5
Water To make	100.0

Coloring and perfume may be added.

Foot Bath

Menthol	5
Alum	10
Boric Acid	20
Magnesium Sulfate	30
Water	500

Antiseptic Military Foot Powder
Formula No. 1

Zinc Peroxide	10
Tannic Acid	5
Boric Acid	20
Bentonite	10
Talc	55

No. 2

Thymol 1 per cent, boric acid 10 per cent, zinc oxide 20 per cent, talc 69 per cent. Use of foot powders containing as much as 3 per cent salicylic acid is contraindicated for soldiers because the salicylic acid adds to the macerating effect of the perspiration and causes a denudation of epithelium.

Athlete's Foot Powder
Formula No. 1

Salicylic Acid	5
Alum	4
Zinc Oxide	4
Talc	45
Starch	30
Lycopodium	15
Tannic Acid	6
Bergamot Oil, Synthetic	¼
Lemon Oil	¼

No. 2

Salicylic Acid	350
Precipitated Sulfur	300
Petrolatum	2000
Cornstarch	200–250 g.

The salicyclic acid and the sulfur are mixed well with the petrolatum, either yellow or white. To the resultant mixture a sufficient quantity of cornstarch is added.

Soaking Bath for Athlete's Foot

Zinc Chloride	5 g.
Sodium Chloride	20 g.
Sodium Nitrate	5 g.
Sodium Silico-Fluoride	1 g.
Boric Acid	5 g.
Warm Water	2 gal.

Mix all the above in lukewarm water, and soak affected feet two or three times per week, not any oftener, in this solution.

Athlete's Foot Remedy

Glycerin Monostearate	10.0
Glycerin	20.0
Bentonite	2.0
Water	67.0
Monochlormercuricarvacrol	1.5
Salicylic Acid	3.0
Benzoic Acid	6.0

Callous Softener and Remover

Castor Oil	65 cc.
Paraffin Wax	60 g.
White Soap	10 g.
Sodium Thiosulfate	5 g.

Melt the wax and soap together, then add the castor oil, and the thiosulfate, which has been pulverized to a fine powder. Pour out and let cool.

Apply a thin sliver or a small amount of the material to a piece of cloth and bandage overnight. Then soak feet in hot water, and loosen the callous. Stubborn callouses may require a second or third treatment.

Corn Remover

Pulverized Calomel	1
Pure Pork Lard	25

Mix the calomel with the lard thoroughly and bandage this material to the affected corn, using the proper amount for the size of the corn. On arising, soak foot thoroughly, and corn should come off easily, if the water is not too cool. A stubborn corn may require a second or third application.

Corn-Collodion
Formula No. 1

Salicylic Acid	8
Cannabis Indica Extract	1
Flexible Collodion	60

This preparation softens and breaks up the corn, which may then be picked away gradually or carefully abraded with pumicestone. It is suggested that, after removal of the corn, the skin of the foot should be hardened by daily bathing for some time in salt water or spirit.

No. 2

Salicylic Acid	1 oz.
Chloretone	2 dr.
Chlorophyll	20 gr.
Flexible Collodion	10 oz.

No. 3

A more modern, effective corn remedy consists of 20 parts each of 25 per cent 40-second cotton in ethyl acetate, specially denatured alcohol No. 1, and ethyl acetate; 10 parts sulfuric acid; 12 parts salicylic acid; 1 part each of chlorbutanol, castor oil and gum camphor and 15 parts toluol to make 100. This solution may be tinted an amber or green, and coloring is best accomplished with an oil-soluble color.

No. 4

Collodion	180
Salicylic Acid	25
Turpentine	10
Chloroform	15

Callous Skin Remover

Diethylene Glycol Laurate	10
Salicylic Acid	10
Sesame Oil	27
Hydrogenated Castor Oil	13

Antacid Mixture
Formula No. 1

Magnesium Sulfate	125 g.
Magnesium Carbonate	65 g.
Aromatic Spirits of Ammonia	60 cc.
Distilled Water	
To make	1000 cc.

Dissolve the magnesium sulfate in 500 cc. of distilled water. Triturate the magnesium carbonate with this solution until a smooth mixture is obtained. Add the aromatic spirits of ammonia and finally sufficient distilled water to make the product measure 1000 cc. Average dose—Metric, 4 cc.—Apothecaries, 1 fluid drachm.

No. 2
(Colloidal Aluminum Hydroxide)
U. S. Patent 2,166,868

Ammonium Alum	100 g.
Gum Acacia	⅞ g.
Water	1 l.

Add ammonia until in slight excess. Filter and wash precipitate. Dry at 40°C. It gels readily on addition of water.

Anti-Acid Powder

Tribasic Calcium Phosphate	350 g.
Magnesium Trisilicate	320 g.
Tribasic Magnesium Phosphate	340 g.
Peppermint Oil	1 cc.

Dose 1 teaspoonful.

"Acetic" Acid Tablets

70 parts of sodium acetate (powdered) are added with 3.15 parts of granulated sugar, dried at 25–35°C. and subsequently mixed with 26.85 parts of dried citric acid. The mixture is then formed into tablets. In using, the tablets are dissolved in a quantity of water sufficient to produce a 4 to 5% acid.

Medicinal Barium Sulfate Suspension

350 grains of methyl cellulose is placed in a wide-mouthed bottle and sixteen fluid ounces of water added. This is well shaken, left overnight, and finally stirred well the next morning. Two and a half pounds of barium sulfate is then rubbed down in a mortar with water until a smooth cream is obtained and to this the whole of the mucilage is added with further trituration. Suitable flavoring is then added and the whole made up to four pints with water. A smooth thickish white cream results and the solid remains well suspended.

Although methylcellulose is not subject to fungal attack it has no antiseptic properties and thus there is a possibility of a musty taste developing on long standing due to molds growing in the water or attacking the flavoring or sweetening agents added. This may be overcome by adding twenty grains of sodium benzoate to each four pints of the cream, a much smaller amount than is required in starch paste.

Acne Lotions
Formula No. 1

Calamine	4
Zinc Oxide	8
Phenol	2
Glycerin	8
Spirit of Camphor	4
Distilled Water	
To make	240

For patients who have a dark complexion and for those whose skin is quite oily, a more "drying" lotion is prescribed:

No. 2

Calamine	2
Zinc Oxide	4
Phenol	1
Spirit of Camphor	8
Precipitated Sulfur	8
Alcohol	120
Distilled Water	
To make	240

Chloracne Treatment

Sulfonated Castor Oil	98
Dupanol WA (Wetting Agent)	2

Waterproofing Skin Cream

Beeswax	100
Lanolin, Hydrous	50
Diglycol Laurate	100
Petrolatum	To suit

Melt and mix until uniform. Pour into jars while warm.

Skin Treatment Cream
(For Taking up Formalin, Poison Ivy, etc.)

A

Rose Oil	17.5 lb.

Diglycol Stearate	10	lb.
Diglycol Laurate	2.5	lb.

B

Water	60	lb.

Heat each to 80°C. Add together. When nearly cold, add perfume to suit.

Stable Zinc Peroxide Ointment
Formula No. 1

Zinc Peroxide	10
Glyceryl Monostearate	3
Peanut Oil	To make 100

Such suspensions showed no loss of oxygen after a month of storage under ordinary conditions, but after this time there is a progressive loss of oxygen; about 10 per cent in three or four months.

No. 2

Zinc Peroxide	10
Wool Fat, Anhydrous	3
Glyceryl Monostearate	3
Liquid Petrolatum	
To make	100

If desired, chemically compatible drugs can be incorporated in these suspensions. For example, five per cent of scarlet red is often added to speed epithelialization.

Coal-Tar Skin Treatments
Formula No. 1

Coal Tar	50
Bentonite	50

Agitate before application to skin or scalp.

No. 2

Coal Tar	50
Bentonite	25
Triethanolamine	25

Agitate before application to skin or scalp.

Ulcer Ointment

"Carbowax"	120.0
Nonaethylene Glycol	80.0

Zinc Oxide,
Activated   10.0
Cetyl Pyridinium
Chloride   0.2

Dermatitis Lotions
Formula No. 1

Sodium Benzoate   0.1
Spermaceti   10.0
Sodium Lauryl Sulfate   1.0
Stearyl Alcohol   10.0
Cetyl Alcohol   3.0
Glycerin   10.0
Water   65.0

No. 2

Sodium Benzoate   0.1
Cetyl Alcohol   20.0
Sodium Lauryl Sulfate   1.0
Glycerin   10.0
Water   64.0

Both of these emulsion bases are prepared as follows: Heat the water, glycerin and sodium lauryl sulfate. Melt the lipoid ingredients separately. Mix the two thoroughly and continue mixing until the mixture is cool.

Cable Rash Ointment
(Halowax Dermatitis)

White Corn Meal   50
Butyl Stearate   22½
Sulfonated Castor Oil
(75%)   22½
Oleic Acid   5

Mix until uniform. Apply to hands and wipe off completely, after working it into the skin very thoroughly.

Mucilage of Quince Seed

Quince Seed   100 g.
Moldex (Preservative)   50 g.
Water   5000 cc.
Macerate the seeds for ½ hr.;

mix well and filter through muslin.

Sulfathiazole Suspension
(For Application to Skin)

Bentonite   10
Sulfathiazole, Powdered   5
Water   To make 100

Medicines and Salves for Chromic Acid and Chromate Skin Injuries

The following simple preparations have proven very satisfactory:

Nasal Ointment

Boric Acid   60 gr.
Menthol   10 gr.
Petrolatum, White   480 gr.

Nasal Spray

Ephedrine containing preparations, or 2 per cent Antiseptic Dye.

Antiseptic Dye

Gentian Violet, Acriflavine or Mercurochrome (2 to 5 per cent).

Antiseptic Wash

1–1000 solution of Bichloride of Mercury for washing.
1–2000 solution of Bichloride of Mercury for wet dressing.

Eye Wash

Saturated solution of boric acid or 3–5 per cent solution borax or 3–5 per cent solution bicarbonate of soda.

Eye Drops

Castor Oil, 5 to 15 per cent solution of Argyrol (freshly made, not over two weeks old).

Eye Ointment

1 per cent mercuric oxide, yellow petrolatum base. Any good analgesic ointment suitable for the eye.

Skin Ointment

Mix up 2 oz. lanolin, 1 oz. castor oil, 1 oz. zinc oxide and stir in

20 cc. of 5 per cent solution of wetting agent (such as aerosol).

Skin Salve

Zinc Oxide—petrolatum base.

Zinc Oxide—lanolin base.

Zinc Oxide — petrolatum base, containing methyl salicylate.

Zinc Oxide — petrolatum base, containing phenol and oil of eucalyptus.

Accelerator

Thymol iodide powder.

Note: Never use tincture of iodine on chrome ulcers.

Dermatitis Salve for Metol Poisoning

Many organic developers cause dermatitis. Prevention is the best cure. Wash developer from hands with dilute acetic acid (1%). The following salve has been found helpful for dermatitis resulting from metol.

Ichthyol	10 g.
Lanolin	40 g.
Petrolatum	30 g.
Powdered Boric Acid	25 g.

Metol Poisoning Salve

Ichthyol	10
Lanolin	40
Petrolatum	30
Boric Acid, Finely Ground	25

Poison Ivy Treatment
U. S. Patent 2,322,565

Urea	16
Casein	7½
Water	16

Poison Ivy and Oak Remedies

There seem to be many remedies for the relief of poison ivy, poison oak and poison sumach, but it appears that each is selective in its action on different persons. One solution may work wonders with one person, and be entirely negative with another. The following "cures" have been tried, and while some people report a positive result, still others say that the results are negative.

(1) Yellow laundry soap and water.

(2) Saturated solution of trisodium phosphate in water.

(3) Lead acetate solution in 95% alcohol.

(4) Potassium permanganate solution, 5% in water.

(5) Pine oil with salt.

(6) Black gunpowder and sour cream.

(7) Corrosive sublimate solution, 1–4% solution in water and alcohol. (POISON).

(8) White iodine.

(9) Household ammonia.

(10) Homemade potash soap and water.

(11) Mercurochrome solution.

(12) Tincture of Merthiolate solution.

(13) Lime water, saturated solution.

(14) Zinc oxide ointment.

The following remedies seem to cure a great many more cases of poisoning, than the above, hence may be relied upon to a greater degree:

(15) Calamine lotion with phenol.

(16) Infusion of Jewel Weed (Impatiens pallida) or Snap Weed, Wild Lady Slipper, Touch-Me-Not, or Silver Weed in alcohol or water.

(17) Lotion composed of the following:

Zinc Oxide	5	g.
Calcium Oxide	3	g.
Glycerin	2	g.
Phenol (88%)	10	drops
Alcohol (95%)	100	cc.
Bismuth Sub-nitrate	0.5	g.
Water	50	cc.

Scabies Preparations
Formula No. 1

Sodium Benzoate	0.1
Spermaceti	10.0
Sodium Lauryl Sulfate	1.0
Stearyl Alcohol	10.0
Cetyl Alcohol	3.0

No. 2

Precipitated Sulfur	2
Balsam of Peru	8
Castor Oil	4
Petrolatum	To make 30

No. 3

Styrax	50	g.
Alcohol	25	cc.
Linseed Oil	25	cc.

No. 4

Benzyl Benzoate	25
Soft Soap	35
Alcohol	40

No. 5

Tetraethylthiuram Monosulfide	25
Polyglycerol Monoricinoleate	10
Industrial Methylated Spirits	65

No. 6

Soft Soap	1	oz.
Derris Powder	4	oz.
Water	1	gal.

No. 7

Derris Powder	4	oz.
Soap Flakes	18	oz.
Water	1	gal.

Antipruritic Lotion

Menthol	1
Methyl Salicylate	2
Chloral Hydrate	12
Zinc Oxide	12
Magnesium Carbonate	6
Camphor Water	24
Alcohol	24
Witch Hazel Extract	To make 100

Itch Remedy

N-Butyl-p-Aminobenzoate	100	g.
Benzyl Alcohol	170	cc.
Anhydrous Lanolin	20	cc.
Cornstarch	640	g.
Sodium Lauryl Sulfonate	64	g.

N-butyl-p-aminobenzoate is dissolved in the benzyl alcohol, which is first warmed, making an approximately saturated solution. Melted lanolin is added, and the mixture kept warm and stirred until as much of the lanolin as will dissolve is in solution. While the liquid is warm, it is added slowly, a little at a time, to a thorough mixture of the cornstarch and sodium lauryl sulfonate, and the whole is kneaded carefully to distribute the liquid evenly throughout the mass. Additional benzyl alcohol, about a tenth the amount already used, is added as before to give the material desired consistency. Final preparation should be a doughy, non-greasy, cake-like material that can be packed in ointment jars (preferably non-metallic) or other suitable containers.

This material is rubbed onto the skin, previously moistened as desired, until it forms a moderately thick layer over the affected region. No attempt should be made to rub the preparation into the skin. It will dry gradually to a powder, which should be left on the skin undisturbed. Beneficial results are reported from the use of this product in treating mosquito bites, fungus and poison ivy infections, yellow jacket stings and other miscellaneous skin infections accompanied by extreme itching.

Grain Itch Lotion
(Mite)

Phenol	1 cc.
Glycerin	4 cc.
Lime Water	15 cc.
Zinc Oxide	30 g.
Rose Water	
To make 120 cc.	

Chlorophyll Impetigo Ointment

Chlorophyll (Oil-soluble)	3
Nupercaine Base	1
Petrolatum	30
Lanolin To make	100

Impetigo and Wound Powder
Formula No. 1

Sulfanilamide	4
Sulfathiazole	2
Zinc Peroxide	1 to 2

This combination finds very wide use in a large variety of conditions such as: abrasions, lacerations, avulsed wounds, puncture or stab wounds, exploratory wounds, operative wounds, hand infections, human bites, abscess cavities, carbuncles, burns, leg ulcers and miscellaneous disorders including vaginitis, cervicitis, otitis media, suppurative sinusitis and certain skin infections (e.g. pyodermia, impetigo, furunculosis, etc.).

No. 2

Copper Sulfate	3 gr.
Zinc Sulfate	2 gr.
Precipitated Sulfur	5 gr.
Sulfathiazole	
10 tablets powdered	
Zinc Oxide	½ oz.
Starch	½ oz.
Soft Paraffin Wax	1 oz.

Sulfa-Zinc Peroxide Ointments
Formula No. 1

Sulfanilamide	10.0 g.
Sulfathiazole	5.0 g.
Zinc Peroxide	2.5 g.
Lanolin, or	
Monostearin	100.0 g.

No. 2

Sulfanilamide	4.0 g.
Sulfathiazole	2.0 g.
Zinc Peroxide	1.0 g.
Gentian Violet	1.0 g.
Acriflavine	0.1 g.
Monostearin Base	
(With wetting agent) To make	100.0 cc.

Sulfa-Drug Surgical Film

Sulfadiazine or	
Sulfanilamide	3.0
Methylcellulose	2.5
Triethanolamine	3.0
Sorbitol	0.5
Acetone or Alcohol	50.0

This is sprayed on a smooth, horizontal glass surface with a pressure gun and allowed to dry. The films so produced are stable, and can be sterilized by dry heat.

Wound and Burn Film Treatment

This solution contains 3 per cent sulfadiazine or sulfanilamide,

2.5 per cent methylcellulose, 3 per cent triethanolamine and 0.5 per cent sorbitol, with 50 per cent alcohol or acetone added. This emulsion is sprayed on a smooth horizontal surface with a pressure gun or paint spray apparatus and allowed to dry. The sheets can be made in any desired size to fit the immediate need.

Burn Ointment

Tannic Acid	5
Sodium Sulfadiazine	2
Petroleum Jelly	100

Mix and apply as needed.

Tannic Acid Burn Emulsion

Tragacanth	1– 1.5
Tannic Acid	1– 5
Polyisobutylene	5–10
White Oil	30–40
Benzocaine	1– 2
Glycerin	5–10
Water	40–50

Burn Cream

Cetyl Trimethyl Ammonium Bromide	1.0
Sulfanilamide	3.0
Castor Oil	25.0
Beeswax	1.8
Wool Fat	1.8
Cetyl Alcohol	5.0
Glycerin	10.0
Water	52.4

Melt the castor oil, beeswax and wool fat and cetyl alcohol at as low a temperature as possible. Dissolve the CTAB* in the water with the aid of heat and mix with the oil, etc., at about 60°C. and stir till set. The sulfanilamide is then rubbed

* CTAB = cetyl trimethyl ammonium bromide.

up with the glycerin and incorporated, with thorough mixing in the cream.

The cream should be applied with a knife-blade or spoon previously sterilized by dipping for two minutes in boiling water or passing through a flame. The burn should not be washed before the cream is applied, nor the blisters be snipped. It should not be left on the burn for more than two days, because prolonged application involves a slight risk of inducing a sensitization dermatitis.

Mustard-Gas Ointment

Benzyl Alcohol	50
Stearic Acid	30
Glycerin	10
Ethyl Alcohol	8
Pontocaine	1
Menthol	1

Anaerobic Wound Infection Germicide

Proflavine	1
Sulfathiazole	99

Aseptic Wax

Beeswax, U.S.P.	875 g.
Almond Oil	125 cc.
Salicylic Acid	10 g.

Melt wax with oil, stir, strain through muslin and with continued moderate heating, for a half hour, add salicylic acid, with stirring. Pour into jars, while fluid.

Baby Skin Oil

White Mineral Oil, U.S.P.	95 cc.
Anhydrous Lanolin	5 g.
Antiseptic Oil B-5671 (Givaudan)	3 cc.

Dissolve the lanolin in the mineral oil at 110°F. Cool to room temperature and filter. Add the deodorant and antiseptic oil, shake well, and allow to stand for three days, and filter again.

Insect-Repellent Cream

A

Glycosterin	40
Diglycol Laurate, Synthetic	90
Triple Pressed Stearic Acid	180
Cocoa Butter	20
Lanolin	30
Cetyl Alcohol	20
Ethyl Hexanediol	60

B

Aquaresin	60
2% Sulfatate Solution	100
Water	440

Melt A and cool to 60°C. Heat B to 60°C. and add to A with constant stirring. Stir slowly till cold.

To repel the attacks of mosquitos, black flies or other pests, cover the exposed areas with a layer of the cream, which will not stain the clothes and can be left on the skin but washes off readily if so desired.

Insect Repellent
Formula No. 1

Coumarin	10
Calcium Chloride	10
Alcohol	80

No. 2

Clove Oil	5
Coumarin	3
Strontium Bromide	8
Calcium Chloride	6
Alcohol (85%)	78

No. 3

Petrolatum	8
Citronella Oil	2
Spirits of Camphor	1
Cedar Oil	1

Apply to face and hands, renewing every 1–2 hr.

No. 4
U. S. Patent 2,356,801

Dimethyl Phthalate	1
2-Ethyl-1,3, Hexanediol	1
n-Butylmesityloxide Oxalate	1

Treatment of Chigger Bites

In the treatment of chigger bites, there are three objectives: (1) the destruction or removal of all remaining parasites, both free and attached; (2) the relief of the severe itching by palliative measures; (3) the treatment or prevention of secondary infection. The mites, if any remain, are most readily removed by an application of benzine, kerosene or copper compound, followed by bathing for a half hour with plenty of soap lather. This should be followed by thorough rinsing with fresh water and patting dry rather than rubbing with a towel. Since active mites may remain in infested clothing, it is advisable that all articles be boiled or sent to a dry cleaner.

Brief application of rubbing alcohol to the affected areas, followed immediately by a mild antiseptic, antipruritic ointment is satisfactory. A clean and generally effective application has been that of boric acid ointment U.S.P., to which may be added from 1 to 2 per cent of phenol, the strength being in inverse proportion to the area of the skin to be covered, and 0.2 per cent of menthol. This ointment should be applied sparingly at least three times a day and also

used as needed to relieve itching;
it is to be rubbed in gently, and
the remainder is wiped off with
cotton. A little plain talc may
then be dusted over the surface.
These applications are made after
the daily bath and at least two
other times daily. Scratching
must be avoided. Secondary infection, when it occurs, should be
treated as required.

Infestation may be prevented by
the use of protective clothing and
by parasitocidal applications. Sulfur has long been recommended as
a repellent. One investigator
dusted it uniformly over the skin
or into the clothing with a shaker,
and found it effective. It should
be dusted freely on the legs and
ankles and inside the hose and
trousers. As an added precaution
one should bathe as promptly as
possible after exposure to chiggers
in order to remove the somewhat
irritating sulfur and to destroy
any surviving mites.

Cedar Chest Compound

Cedarwood (Fine chips)	80
Naphthalene	9
Paradichlorbenzene	8
Cedarwood Oil	2
Spike Lavender Oil	1

Mix the two oils with the cedar-
wood and add the remaining two
ingredients. Mix well. Dispense in
sealed packages. To be sprinkled
over clothes, or liberally distributed throughout a closet.

Hemorrhoidal Suppository
Formula No. 1

Bismuth Oxyiodide	1.0
Basic Bismuth Gallate	1.0
Zinc Oxide	1.0

Resorcin	0.1
Balsam of Peru	0.5
Cocoa Butter	26.4

No. 2

Ichtyol	4.00
Tannic Acid	4.00
Extract of Stramonium	0.25
Cocoa Butter	24.00

Vaginitis Suppository

Sulfanilamide	1.0 g.
Sulfathiazole	1.0 g.
Zinc Peroxide	0.5 g.
Sodium Tetradecyl Sulfate	0.2 cc.
Cocoa Butter To make	4.0 g.

"Sulfa" Vaginal Suppository

Sulfanilamide	1.0
Sulfathiazole	1.0
Zinc Peroxide	0.5
Sodium Tetradecyl Sulfate	0.2
Cocoa Butter To make	4.0

Rectal Suppository

Chloral Hydrate	1
Lactose	1
Cocoa Butter	2

Cocoa-Butter Suppository

Cocoa Butter	20
Glycerin	20
Lanolin	½

Warm and mix well until ready
to cast.

Suppository Mold Lubricant

Castor Oil	1
Hard Soap	2
Alcohol	18
Water	2

Sulfa Drug Compositions
Foot Fungus Ointment

Sulfathiazole	6.0

Cod Liver Oil   60.0
Beeswax   40.0

Melt the white wax at a high temperature, add the cod liver oil and stir quickly. Continue the heating for at least an hour and a half. After the mixture has been allowed to cool, the sulfathiazole is added.

Vaginal Ointment

Sulfathiazole Powder   4
Mercurous Chloride, Mild   8
Wool Fat, Anhydrous   8
Petrolatum, White   8

Washable Jelly
Formula No. 1

Pectin, N.F. VII   5.0
Tannic Acid   10.0
Glycerin   12.0
Sulfadiazine   5.0
Methyl Parahydroxy-
  benzoate   0.2
Sodium Sulfite   0.2
Ringer's Solution*   67.6

Mix well the pectin, glycerin and sulfadiazine to a smooth paste. Dissolve the sodium sulfite, methyl parahydroxybenzoate and the tannic acid in boiling Ringer's solution and add to the pectin paste, stirring well until it cools down to room temperature.

No. 2

Sulfathiazole   5.0
Pectin   10.0
Benzoic Acid   0.2
Ringer's Solution   84.8

*Ringer's solution, U.S.P. XII:
Sodium Chloride   8.60 g.
Potassium Chloride   0.30 g.
Calcium Chloride   0.33 g.
Distilled water   1000 cc.

No. 3

Sulfadiazine   5.0

Tragacanth   1.0
Glycerin   83.0
Preservative (Methyl
  Parahydroxybenzoate) 0.2
Distilled Water
          To make 100.0

Use sufficient distilled water to make a smooth paste of sulfadiazine. To the balance of the distilled water, add the glycerin and the preservative and bring to a boil. Then discontinue the heat and add the tragacanth, allowing the mixture to stand for 24 hours. Strain the tragacanth mucilage (i.e., through muslin) if necessary. Combine the sulfadiazine paste and the tragacanth mucilage.

Ointment Emulsion
Formula No. 1

Sodium Sulfathiazole   5
Paracho (Absorption
  Base)   95

No. 2

Sodium Sulfathiazole   5
Petrolatum   0.5
Hydrous Wool Fat   0.5
Wool Fat   20
White Wax   10
White Petrolatum   65

No. 3

Five per cent sulfathiazole or sodium sulfathiazole in:

Cod Liver Oil   8.75
Concentrated Cod Liver
  Oil   0.15
Hydrous Wool Fat   5
Petrolatum   To make 30

No. 4

Sulfanilamide   30
Oleic Acid   10
Lanolin   60

No. 5

Sulfanilamide   30.0

Wool Fat	46.5
Water	23.5

No. 6

Hexadecyl Alcohol	6.1
Octodecyl Alcohol	6.1
Sodium Lauryl Sulfate	1.4
White Petrolatum	13.6
Liquid Petrolatum	20.5
Sulfadiazine	5.0
Water	47.3

Melt the alcohols together over a water bath at 65°C., and add the sodium lauryl sulfate and stir well. Add the white petrolatum and the liquid petrolatum, continuing the heat until the mixture is melted. Add the sulfadiazine and stir until the mixture cools to room temperature. Add the water slowly with constant stirring.

No. 7

Ringer's Solution	58.0
White Petrolatum	20.0
Glycerin	11.8
Sulfadiazine	5.0
Pectin	5.0
Methyl Parahydroxy- benzoate	0.2

Wet the pectin and the methyl parahydroxybenzoate with the glycerin, add the Ringer's solution previously heated to boiling, and stir well to make a smooth paste. Keep the mixture on a water bath and add the sulfadiazine and the white petrolatum. Remove from water bath and continue stirring until cool.

No. 8

Sulfathiazole	5
Carbitol	10
Stearic Acid	20
Peanut Oil	4
Potassium Hydroxide	1
Water	60

No. 9

Sulfathiazole	5
Sodium Lauryl Sulfate	1
Stearyl Alcohol	10
Cetyl Alcohol	3
Spermaceti	10
Glycerin	10
Water	61

No. 10

Sulfathiazole	1.0
Sodium Lauryl Sulfate	0.5
Cetyl Alcohol	8.0
Petrolatum, White	20.0
Water	To make 100

No. 11

Sulfathiazole (Finely powdered)	5
Triethanolamine	2
Water	24
Beeswax	5
Liquid Petrolatum	64

No. 12

Sulfanilamide Powder	30.0
Cod Liver Oil	50.0
Oleic Acid	3.0
Beeswax	2.5
Triethanolamine	1.0
Water	To make 100.0

No. 13

Sulfanilamide	30
Wool Fat	30
Stearic Acid	5
Triethanolamine	2
Water	33

No. 14

Sulfanilamide	30.0
Liquid Paraffin	28.5
Cetyl Alcohol	1.4
Stearic Acid	2.0
Triethanolamine	1.0
Arachis Oil	1.5
Water	35.6

No. 15

Sulfanilamide	30.0
Beeswax	0.7
Cod Liver Oil	49.0

Oleic Acid	3.6
Sodium Hydroxide	0.5
Water	16.2

No. 16

Sulfathiazole	2.5
Chlorobutanol	0.1
Urea	10.0
Distilled Water	25.0
Vanishing Cream	
Base To make	100.0

Triturate the sulfathiazole with the chlorobutanol and the distilled water in small portion (in the ointment dissolve the urea in the distilled water) to make a smooth paste. Then add the vanishing cream base, triturate well, add the balance of the water and the proper base.

No. 17

Sulfanilamide	10.0
Allantoin	2.0
Chlorobutanol	0.5
Greaseless Ointment	
Base	100

No. 18

Cetyl Trimethyl Ammonium Bromide	1.0
Sulfanilamide	3.0
Castor Oil	25.0
Beeswax	1.8
Wool Fat	1.8
Cetyl Alcohol	5.0
Glycerin	10.0
Water	52.4

Mix the castor oil, beeswax, wool fat and cetyl alcohol together at as low a temperature as possible. Dissolve the cetyl trimethyl ammonium bromide in the water with the aid of heat; mix the oil, etc., at about 60°C., and stir till set. The sulfanilamide is then rubbed up with the glycerin, and the two incorporated in the cream and thoroughly mixed.

No. 19

Sulfanilamide	10
Glycerin	10
Castor Oil	25
Lanette Wax SX	10
Water	45

Heat the castor oil to 70°C. and add the Lanette wax SX. When the wax is completely melted, add the water (previously heated to 65°C.), with gentle stirring to avoid incorporation of air. Heat the whole to 65°C. for two hours, to kill off non-sporing pathogens. Intimately mix the sulfanilamide powder with the glycerin. Heat to 60°C. and add slowly to the cream, allowing the mixture to cool before use.

In combining sulfanilamide or sulfathiazole with the three bases below, sift the drug and make a paste with an equal amount of boiling water.

No. 20

Bentonite	25
Zinc Oxide	25
Cottonseed Oil	25
Hydrogenated Cottonseed Oil	25

Mix until smooth, then add the sulfathiazole.

No. 21

Bentonite	33
Triethanolamine	5
Water To make	100

Incorporate the bentonite into a mixture of triethanolamine and water, then add the sulfathiazole.

No. 22

Bentonite	25
Liquid Petrolatum (Low viscosity)	25
Petrolatum	25
Water To make	100

Thoroughly mix the petrolatums, add the water and incorporate the bentonite. Then add the sulfathiazole.

No. 23

Bentonite	10
Sulfathiazole	5
Water	To make 100

The sulfathiazole is agitated with the water before incorporation with the bentonite.

No. 24
(Paste)

Copper Sulfate	3 gr.
Zinc Sulfate	2 gr.
Precipitated Sulfur	5 gr.
Sulfathiazole	10 tablets, powdered
Zinc Oxide	½ oz.
Starch Powdered	½ oz.
Soft Paraffin	1 oz.

Medicinal-Vanishing Creams

The following vanishing creams can be used as bases for sulfa drug ointments.

Formula No. 1

Glyceryl Monostearate	12
Beeswax	2
Glycerin	6
Glyceryl Laurate	4
Water	To make 100

Heat together until solids are melted, and stir to form a cream.

No. 2

Liquid Petrolatum (Low Viscosity)	15
Stearic Acid	5
Lanolin	5
Glyceryl Monostearate	12
Petrolatum	20
Water	To make 100

No. 3

Liquid Petrolatum (Low Viscosity)	20
Glyceryl Monostearate	12
Petrolatum	20
Water	To make 100

The ingredients are placed into one container and heated with constant stirring until a homogeneous mixture results. When this cools, the sulfathiazole is added.

No. 4

Cetyl Alcohol	10
Glycerin	10
Sodium Lauryl Sulfate	1
Sulfadiazine	5
Water	74

No. 5

Stearic Acid	1 oz.
Glycerin	1 oz.
Distilled Water	4 oz. 6 dr.
Sodium Borate	8 gr.
Potassium Carbonate	16 gr.
Perfume, As desired	45 min.

Melt the stearic acid on a water bath and heat to 85°C. Dissolve the alkali in the water, add the glycerin, warm the solution to 85°C., and gradually pour the warmed solution into the stearic acid, stirring briskly. Continue to stir actively and at the same temperature to assure complete saponification and absence of free alkali. Then remove the heat, and continue to stir until cold, at the same time incorporating the perfume. The cream should afterward be beaten for several hours, preferably by mechanical means.

Ophthalmic-Ointment Base
Formula No. 1
(Absorption Base)

10% cholesterin (cholesterol) is dissolved, by stirring, in a soft-grade, white petrolatum which has been heated to 130°C. Stir until cool. Small portions of light liquid petrolatum may be used to soften the white petrolatum to any desired degree. In the preparation of the ointment, a solution of the drug is made in a quantity of water equal to at least 20% of the finished ointment. The aqueous solution is stirred thoroughly with the cholesterinated base. As the stirring continues, the ointment base begins to absorb water and as this occurs, the base loses its translucency and becomes opaque.

No. 2

| Sodium Alginate | 4 |
| Boiling Water | 75 |

Emulsify and strain, stir until cool.

Add:

Anhydrous Wool Fat	16
Petrolatum, White	78
Sodium Chloride	1

(Dissolve the sodium chloride in 4 parts of water.)

No. 3

Sodium Lauryl Sulfate	0.5
Cetyl Alcohol	8.0
Theobroma Oil, U.S.P.	6.0
White Petrolatum	20.0
Water	65.0

Use a water bath to melt the first four ingredients, and then add the water, stirring constantly.

No. 4

Liquid Petrolatum	35
Spermaceti	13
Glycostearin	10
Water	38

Heat the first three ingredients to 140°F., add to the water heated to the same temperature, and stir until cool.

Sugarless-Pharmaceutical Syrups
Formula No. 1

| Gum Tragacanth | 2 |

| Glycerin | A sufficient quantity |
| Water | To make 100 |

No. 2

Karaya Gum	1
Glycerin	A sufficient quantity
Water	To make 100

In both these preparations, saccharin (0.1%) is used as a sweetening agent and methyl parahydroxybenzoate (0.1%) as a preservative. The glycerin increases miscibility and forms a cream. In making these syrup substitutes, the mucilages should be brought to the boiling point, allowed to simmer for a half hour and then filtered through flannel.

Drug-Penetrating Aid

Aerosol MA (Sodium Dihexyl Sulfosuccinate)	1
Xylene	1
Antipyrine	1
Propylene Glycol	4

Anthelmintic Tablet
U. S. Patent 2,282,290

Phenothiazine	80
Starch	8
Sodium Bicarbonate	5
Tartaric Acid	4

A wetting agent, e.g., sodium choleate, is added to ensure complete disintegration, and a laxative, e.g., phenolphthalein (1 part).

Sedative

Sodium Bicarbonate	32½	g.
Potassium Iodide	40½	g.
Potassium Bromide	81	g.
Ammonium Bromide	81	g.

Tincture of
Nux Vomica   46   cc.
Compound Tincture
of Gentian   60   cc.
Water, Distilled
To make 1000   cc.
Dissolve the salts in half the water. Mix in the tinctures, then the balance of the water.

Non-Toxic Bitter-Tasting Product

Sucrose octa-acetate has an extremely bitter taste. Small amounts may be used for medicinal bitters or stomachics, dissolved in alcohol.

To utilize this glycerin-containing preparation in an elixir of B vitamins, the combination given below is suggested:

Vitamin B Elixir

Thiamine Hydro-
chloride   0.125 g.
Riboflavin   0.042 g.
Pyridoxine Hydro-
chloride   0.038 g.
Calcium Panto-
thenate   0.038 g.
Nicotinic Acid   0.375 g.
* Elixir Raspberry
To make 1000.0   cc.

Each 8 cc. of this product yields 1 mg. of vitamin B_1; $\frac{1}{3}$ mg. of riboflavin; 0.3 mg. each of pyridoxine and pantothenate, and 3 mg. of nicotinic acid.

---

* Elixir of Raspberry:
Benzoic Acid   1 g.
Tincture Cardamom Com-
pound   5 cc.
Alcohol   150 cc.
Glycerin   110 cc.
Syrup of Raspberry, N.F.   160 cc.
Distilled Water To make 1000 cc.

Pharmaceutical-Flavoring
Vehicles

Formula No. 1

Wild Cherry Bark   150 g.
Glycerin   150 cc.
Granulated Sugar   675 g.
Alcohol   20 cc.
Filtered Percolate
To make 1000 cc.

Macerate the bark with a sufficient quantity of distilled water for one hour. Run out 400 cc. of percolate using additional water as menstruum as needed. Filter percolate perfectly clear. Add sugar and dissolve by agitation, then add the glycerin, alcohol and sufficient percolate to make 1000 cc. and strain.

No. 2

Benzoic Acid   1 g.
Tincture Cardamom
Comp.   5 cc.
Alcohol   150 cc.
Glycerin   110 cc.
Syrup of Rasp-
berry, N.F.   160 cc.
Distilled Water, to
make   1000 cc.

No. 3

Liver Extract Powder   22.5
Benzoic Acid   0.5
Syrup of Orange   80.0
Glycerin   55.0
Tincture Cardamom
Comp.   10.0
Sherry Wine To make 500.0

Mix, macerate for 24 hours with occasional shaking and filter. Then add vitamins.

Surgical Lubricating Jelly

Tragacanth   35.0 g.
Boric Acid   15.0 g.
Benzoic Acid   0.5 g.
Formaldehyde   3.0 cc.

Methyl Salicylate	0.4 cc.
Alcohol	120.0 cc.
Water To make	1000.0 cc.

Flexible Adhesive Surgical Dressing

Polyvinylbutyral Resin	20 g.
Alcohol	120 cc.
Ether	20 cc.
Castor Oil	10 cc.

This is painted on the usual dressing.

Surgical-Cast Plaster
British Patent 545,539

Plaster of Paris, Dry Powdered	80–90
Kaolin, or Fuller's Earth	10–20

The above ingredients are formed into a cast by the aid of water and may be carried on a surgical bandage.

Capsule-Filling Hints

Use crystals where it is possible, such as for aspirin and phenacetin, when they must be triturated in a mortar. When powders fluff around in the mortar and become electrified, such as cinchophen and amidopyrine, use ten or twelve drops of pure alcohol and the mixing will be accomplished in one-third of the usual time. This method may be used with nearly everything but glandular products.

Pharmaceutical-Tablet Coating

A

Sugar, Powdered	450
Glucose	150
Gelatin	40
Water, Distilled	360

B

Talcum	150
Potato Starch	500
Sugar, Powdered	350

First coat with A and then tumble with B. A pure food color may be added (in solution) to B if desired.

Enteric-Pill Coating

Shellac	25
Castor Oil	5
Alcohol	25

Experiments indicate that the kind of coating and its thickness influence the degree of resistance to gastric juices. The coating should be 26 to 39 microns in thickness.

Pill Excipient

Glucose	60
Dextrin	20
Wheat Starch	20

Liquid-Petrolatum Emulsion
Formula No. 1

Liquid Petrolatum	500	cc.
Mannitan Mono-stearate	0.5	g.
Syrup, Sugar	100	cc.
Vanillin	.040	g.
Alcohol	60	cc.
Distilled Water To make	1000	cc.

Dissolve the mannitan monostearate in the oil by warming and add the other ingredients, emulsification takes place immediately, no special care is required to insure a permanent, stable emulsion.

No. 2

Liquid Petrolatum	50	cc.
Pectin	1	g.
Syrup	10	cc.

Vanillin	0.004	g.
Alcohol	6	cc.
Distilled Water	34	cc.

No. 3

Methyl Cellulose	25
Water	350

Dissolve and then add slowly with good stirring:

Mineral Oil, Medicinal	500
Citric Acid	2
Saccharin, Soluble	1/10
Phenolphthalein	7½
Lemon Oil	1½

Vegetable Physic
(Laxative)

Prunes, De-pitted	60
Dates, De-pitted	60
Raisins, Seedless	56
Figs	60
Senna, Powdered	15

Pass twice through a fine meat grinder and pack in closed jars.

Cod-Liver Oil, Orange Juice, Malt Extract, Emulsion
Formula No. 1

Cod-Liver Oil	25	g.
Pure Concentrated Orange Juice	15	g.
Exchange Citrus Pectin, 100 Grade	3¼	g.
Glycerin	10	g.
Malt Extract	20	g.
Distilled Water	26¾	g.

Mix the glycerin and the pectin producing a smooth paste. Add the water, then the orange concentrate, stirring thoroughly during the addition. Now add the malt extract and any other desired ingredients. Finally emulsify the oil into the mixture by adding it slowly in small amounts. Stir thoroughly.

Remove entrapped air by vacuum deaeration or light centrifu-

gal treatment, mix to insure uniformity, and place into containers.

No. 2

Cod-Liver Oil	40	g.
Pure Concentrated Orange Juice	20	g.
Exchange Citrus Pectin, 100 Grade	1¾	g.
Glycerin	6	g.
Malt Extract	25	g.
Distilled Water	7¼	g.

Substitutes for Glycerin in Cosmetics

Suitable substitutes for glycerin are, first of all, materials which are chemically related to glycerin, such as glycol and its derivatives. Diethylene glycol monethyl ether (Carbitol) may be mentioned as an example, 2 to 15% of which has been used satisfactorily in creams. As a matter of fact it is also used abroad in the manufacture of toothpastes.

A very good glycerin substitute for use in toothpowders, incidentally, is silica gel, which is believed to be very beneficial to the gums. To prevent the drying out and hardening of toothpastes, glucose is almost as effective as glycerin. It is somewhat more viscous than glycerin and is quite inexpensive. Another suitable preparation is a mixture as follows:

Formula No. 1

Magnesium Carbonate	5
Calcium Carbonate	45
A Solution of Starch	18
in Water	38

to which preparation additional ingredients may be added, including essential oils.

Free cholic acid 0.5, together

with triethanolamine 1.5, and a little water is a type of combination showing great emollient and penetrative properties, which are of particular importance in shaving creams. Preparations of this type are offered to the trade under the names Curacit Soda and Curacit Triethanolamine, both of which may be considered satisfactory substitutes for glycerin.

Other combinations which also offer very interesting possibilities as glycerin substitutes and which also excel because of their emollient properties, their penetration and hygroscopic quality, are concentrated solutions of calcium lactate and sodium lactate, as have been known in Europe under the names of Perka-glycerin and Per-glycerin. These combinations, however, may irritate the skin if used in too large a proportion. Not more than 3% should be added to soap and even less to creams.

Finally, butyl stearate might be mentioned, which also has a place as a glycerin substitute.

No. 2

Magnesium Chloride	33.3
Urea	34.5
Water	32.2

Depilatory

Strontium Sulfate	60
Wheat Starch	15
Zinc Oxide	15
Lithium Carbonate	8
Menthol	2
Water	To make a paste.

Apply thin layer and leave on skin for 10–20 min. Wash off with a dilute solution of borax, alcohol and water.

Cosmetic Stocking
Formula No. 1

Precipitated Chalk	10.0
Talc	5.0
Titanium Dioxide	3.0
Bentonite	2.0
Alcohol	8.0
Aquaresin	3.0
Sulfatate	0.5
Methyl Cellulose	0.5
Water	68.0
Dye and Pigment	To suit

No. 2

Spermaceti	3
Mineral Oil	1
Trigamine Stearate	21
Glycerin	5
Zinc Stearate	8
Titanium Dioxide	3
Pigment (Ochre)	3
Water	56

Feminine-Hygiene Preparations

The following acid vaginal jelly, adjusted to pH 4, five cc. of which is introduced by means of a vaginal applicator before retiring:

Lactose	6.0
Citric Acid	0.275
Tragacanth	1.73
Irish Moss	1.75
Glycerin	22.58
Boric Acid	2.0
Water	To make 100.0

Another preparation, a teaspoonful in two quarts of warm water as a douche before retiring, provides both local and emotional relief in cases of non-infective leukorrhea, and generally controls the discharge. A suggested prescription consists of:

Menthol	1.5 dr.
Camphor	1.5 dr.

Sodium Bicarbonate	3.0 dr.
Sodium Borate	3.0 oz.

Embalming Fluid
U. S. Patent 2,318,319

Aluminum Nitrate	8.3
Methyl Formate	20.0
Ethyl Alcohol	20.0
Acetamide	0.8
Water	50.9

Chapped Lip Stick

Camphor	12
Spermaceti	92
Peanut Oil	40
Diglycol Laurate	40
Beeswax, White	16–20

Melt and mix at low temperature. Cool and pour into molds at lowest possible temperature. Hardness can be varied by increasing or decreasing amount of beeswax.

Mustache Wax

Propylene Glycol	100
Hard Soap	100
White Beeswax	250
Trigamine	10
Gum Arabic	50
Water	480

Perfume and Color,* to suit.

Bubble Bath
Formula No. 1

Sodium Lauryl Sulfate	400
Soluble Starch	50
Sodium Cholate	20
Sodium Bicarbonate	260
Tartaric Acid	180
Adipic Acid	50
Pine Needle Oil	20
Fluorescein	1/10

* For black use 10% powdered lamp black. For brown 6–9% powdered sienna.

No. 2

Sodium Lauryl Sulfate	130
Sodium Bicarbonate	450
Tartaric Acid	350
Sodium Cholate	10
Soluble Starch	50
Pine Needle Oil	10

No. 3

Epersol-Y (Hydrolyzed Protein)	150 cc.
Water	150 cc.
Sodium Benzoate	3 g.
Water-Soluble Perfume Oil	5 cc.

Dissolve the sodium benzoate in the water, then add the Epersol and finally the water-soluble perfume oil.

Air-Odor Neutralizer
(Wick Type)
Formula No. 1

Siberian Fir Needle Oil	40
Wintergreen Oil	5
Clove Oil	5
Spearmint Oil	5
Sassafras Oil	5

This mixture of oils is then combined with the emulsifying agent to form the following concentrate:

The Above Oil Mixture	50
Sulfonated Oil (Emulsifying Agent)	50

This concentrate may be stored until needed. To use it, it is diluted in the following proportions:

Concentrate	40
Distilled Water	200

No. 2
U. S. Patent 2,326,672

Chlorophyll	3 oz.
Water	½ gal.
Alcohol	½ gal.

Stir together and add slowly while mixing vigorously Formaldehyde (40% Solution) 16 oz.

Air Deodorant

Naphthalene	730
Camphor, Synthetic	70
Hexachloroethane	20
Paradichlorobenzene	100
Bornyl Acetate	50
Eucalyptol	30

Melt naphthalene on the water bath, add camphor, and later hexachloroethane and paradichlorobenzene. When dissolved, cool to 80°C., add bornyl acetate and eucalyptol.

Air Sterilization
U. S. Patent 2,344,536

A method of sterilizing air comprises contacting a minor portion of the air to be sterilized with an extended surface of triethylene glycol at a temperature of from 150°F. to 250°F. and mixing this minor portion of the air with the remaining portion of the air to be sterilized.

Telephone-Mouthpiece Antiseptic

Completely Denatured	
Alcohol	100
Liquid Phenol	1
Pine Needle Oil	10

Let stand 48 hours and filter.

CHAPTER V

EMULSIONS AND COLLOIDS

Paraffin Wax Emulsion
Formula No. 1
Paraffin Wax	90
Stearic Acid	10
Triethanolamine	5
Water	120

No. 2
Paraffin Wax	90
Stearic Acid	10
Water	125
Morpholine	3

Ceresin Wax Emulsion
Ceresin Wax	90
Stearic Acid	10
Triethanolamine	5
Water	200

Ouricury Wax Emulsion
Ouricury Wax	90
Stearic Acid	10
Triethanolamine	5
Water	200

Cetyl Alcohol Emulsion
Cetyl Alcohol	90
Stearic Acid	10
Triethanolamine	5
Water	700

Dichloroethyl Ether Emulsion Spray
(Cabbage-Maggot Spray)
Dichloroethyl Ether	20 cc.
Tergitol F	0.4 cc.
Water	To make 1 gal.

Apply 8½ oz. per linear foot of cabbage rows.

Benzyl Benzoate Emulsion
Benzyl Benzoate	250 cc.
Triethanolamine	10 g.
Stearic Acid	50 g.
Distilled Water	To make 1000 cc.

Dissolve the stearic acid in the benzyl benzoate with the aid of gentle heat; add the solution to a mixture of the triethanolamine and 500 cc. of water, previously warmed to the same temperature. Shake until emulsified and then add sufficient water to make the required volume.

Neat's Foot Oil Emulsion
Neat's Foot Oil	90
Stearic Acid	10
Triethanolamine	5
Water	120

Castor Oil Emulsion
Castor Oil	300 g.
Alcohol	100 cc.
Water	570 cc.
Sodium Benzoate	1 g.
Soybean Lecithin	50 g.

Suspend the melted lecithin in the warm alcohol mixing thoroughly. Now add the castor oil and mix again. Dissolve the sodium benzoate in the hot water, and add this solution to the first mixture. To form the emulsion rapidly agitate in a mechanical mixer for 10 to 20 minutes. A suitable flavor, such as peppermint oil, may be used to cover up the flavor of the castor oil.

The above emulsion may be diluted, is permanent, and is more palatable than plain castor oil.

Cod-Liver Oil Emulsion

Cod-Liver Oil	300 g.
Alcohol	60 cc.
Glycerol	50 g.
Water	500 cc.
Gum Tragacanth	½ g.
Sodium Bicarbonate	1 g.
Sodium Benzoate	1 g.
Soybean Lecithin	100 g.

Suspend the melted lecithin in the warm alcohol and glycerol mixture, and mix in the warm cod-liver oil. Dissolve the gum tragacanth, sodium benzoate, and sodium bicarbonate in the hot water, and add this solution to the first mixture. Now agitate in a mechanical mixer for 10 to 20 minutes until a permanent creamy emulsion results.

The above emulsion may be diluted, is permanent, and is easier to take than plain cod-liver oil.

Peppermint Oil Emulsion
Formula No. 1

Peppermint Oil	10
Tincture of Quillaja	5
Glycerin	10
Water	To make 100

No. 2

Peppermint Oil	10
Tragacanth	1
Glycerin	10
Water	To make 100

Polyvinyl Chloride Polymerization Emulsion

Emulsions of polyvinyl chloride are obtained when 100 g. of vinyl chloride are mixed with 0.25–1.0 g. of hydrogen peroxide and 2–3 g. of acetaldehyde in water containing an emulsifier. Instead of acetaldehyde acetic or formic acid at pH 2.8 can be used.

Cumar Emulsion
Formula No. 1

Cumar V2½	20
Cumar RS	10
Toluol	10

Warm together until dissolved. Cool to 95–100°C. and add following previously made solution heated to 70°C.:

Ammonium Linoleate S	10½
Water	24½

Stir until temperature falls to 35°C.

No. 2

1

Cumar V2½	20
Cumar QS	10
Naphtha	10

2

Ammonium Linoleate S	10
Water	50

Mix and warm No. 1 until uniform and then add warm solution of No. 2 slowly with good mixing.

Polyisobutylene Emulsion
U. S. Patent 2,330,504

Polyisobutylene Polymer (M.W. 50,000)	665
Casein	54
Borax	8
Water	650

Knead polymer in a Banbury mixer for five minutes and add borax. Knead together for ten minutes. Then add, in small portions, casein suspended in warm water (after suspension has stood for a half hour). Do not add more casein suspension until previous amount has been absorbed.

Pentaerythritol Abietate Emulsion
U. S. Patent 2,333,887

Pentaerythritol Abietate	50.0
Oleic Acid	20.0
Triethanolamine	4.2
Mineral Spirits	15.0
Water	500.0

"Staybelite" Ester Emulsions

Emulsions of Staybelite esters may be prepared by two general methods as illustrated in the following table. Method A requires the use of a colloid mill and only a minimum of emulsifier. Method B yields spontaneous emulsions on rapid agitation, but requires the use of large amounts of a soap such as potassium oleate. Staybelite Esters No. 10, 1, and 2 are introduced as 80% solvent solutions; Staybelite Ester No. 3 is so fluid it may be emulsified without the aid of solvent.

Method A—Blend the solvent solution of the Staybelite ester (in the case of Staybelite Ester No. 3, where no solvent is used, warm to 90°C. before blending) with the water containing the Duponol ME and sulfated castor oil, then pass through colloid mill.

Method B—Stir the oleic acid into the solvent solution of the Staybelite ester (or into the resin itself, when Staybelite Ester No. 3 is used). Dissolve all the potassium hydroxide and sulfated castor oil in half the water. Add the resin-oleic acid mixture to this solution with vigorous agitation to produce a spontaneous emulsion. Stir vigorously for 10–15 minutes, dilute with remaining water, stir 30–40 minutes, filter, and the emulsion is ready for use.

Formulation of Staybelite Ester Emulsions

Material	Parts by Weight Method A		Method B	
Staybelite Ester No. 10, 1, or 2 (80% solution*)	100.0	....	47.5	....
Staybelite Ester No. 3 (Solids basis)	....	80.0	....	42.0
Duponol ME	0.8	2.0	....	....
Potassium Hydroxide	....	....	0.3	0.3
Oleic Acid	....	....	1.1	1.2
Sulfated Castor Oil (75%)	0.8	2.0	0.9	1.0
Water	138.4	136.0	50.2	55.5
% Solids	37	38	40	45

* In Varsol, HF naphtha, toluene, xylene, tollac, Solvesso No. 2, Solvent 50D, etc.

Colloidal Sulfur

Powdered Sulfur	100
Gum Acacia	1
Water	56

Mix in ball mill for 20 hrs. Remove and dilute with 3 volumes of water. Agitate well; allow to settle; siphon off upper layer of colloidal sulfur.

Oleoresin Capsicum Emulsion

Oleoresin Capsicum	30 g.
Water	963 g.
Sodium Bicarbonate	2 g.
Soybean Lecithin	5 g.

Melt together the oleoresin and the lecithin, using a moderate temperature to prevent escape of volatile oils. Dissolve the sodium bicarbonate in the hot water and add to the first mixture. Agitate in a mechanical mixer for 10 to 20 minutes until a uniform emulsion is formed.

This emulsion is convenient to use in flavoring spiced meats such as sausages and the like.

Sulfur Emulsion
U. S. Patent 2,343,860

Sulfur	3.5–6.3%
Kerosene	5%
Asphalt	< 0.6%
Modinal (Emulsifier)	1–2%
Water	87%

Cottonseed Pitch Emulsion

Cottonseed Oil Pitch	90
Triethanolamine	5
Stearic Acid	10
Water	200

Asphalt Emulsion
U. S. Patent 2,157,698

Asphalt is dispersed in 50% of water in the presence of (0.6%) sodium phosphate, and to this are added 12.5% of a 4% aqueous solution of a weak organic acid, e.g., tannic, and then 7.5% of a 3.5% aqueous solution of boric acid and 8.75% of a 3.5% solution of oxalic acid, citric or tartaric acid.

Emulsions, Miscellaneous

Please refer to "emulsions" in index. There you will find many examples of agricultural, paint and other types of emulsions.

Aqueous Colloidal Graphite

Graphite (300 Mesh)	13
Diglycol Stearate	8–12
Water	200

Warm to 60°C. and mix well until cool.

Breaking Emulsions
Petroleum Demulsifier
Formula No. 1

Diglycol Laurate	78
Cresylic Acid	12
Sodium Silicate	9
Soda Ash	1

No. 2
U. S. Patent 2,287,567

Naphthalene	1¼ lb.
Phenol	2 lb.
Stove Oil	1 gal.

Flotation Reagent for Oxide Ores
U. S. Patent 2,164,063

Oleic Acid	45.1
Kerosene	40.6
Soda Ash	5.5
Sodium Silicate	8.8

Mix to uniform jelly and add to water.

Fluorspar Flotation Frother
U. S. Patent 2,168,762

| Oleic Acid | 1.0 lb. |
| Quebracho Extract | 0.9 lb. |

Use the above amount per ton of ore or tailings.

CHAPTER VI

FARM AND GARDEN SPECIALTIES

Soilless-Growth Plant Foods
(Hydroponic Plant Nutrients)
Formula No. 1
For 500 Gallon Tank. Use Distilled Water and C. P. Chemicals

Solution A
Dissolve in 10 gallons water.

Potassium Nitrate	3 lb.
Calcium Nitrate	2 lb.
Magnesium Sulfate	1 lb.
Ammonium Nitrate	½ lb.
Potassium Acid Phosphate	½ lb.

Solution B
Dissolve in 1 gallon water.

Manganese Nitrate	10 oz.
Boric Acid	8 oz.
Zinc Sulfate	4 oz.
Copper Sulfate	4 oz.
Potassium Permanganate	4 oz.
Nickel Sulfate	4 oz.
Potassium Iodide	2 oz.
Sodium Chloride	1 oz.
Cobalt Nitrate	4 oz.
Lithium Chloride	2 oz.
Tin Chloride	6 oz.
Cadmium Bromide	2 oz.
Aluminum Sulfate	4 oz.
Molybdic Acid	4 oz.

Solution C
Dissolve each separately in 1 pint of water. Keep in separate containers. (Each used separately.)

Slaked Lime	1 lb.
Sulfuric Acid	5 oz.
Iron Chloride	8 oz.

No. 2
For 25 Gallon Tank of Tap Water for Use with Technical Grade Chemicals

Solution A

	Ounces	Grams
Potassium Nitrate	2.4	68.04
Calcium Nitrate	1.6	45.36
Magnesium Sulfate	0.8	22.68
Potassium Acid Phosphate	0.8	22.68

Dissolve each of the above in one pint of water and keep in separate bottles.

Solution B

	Ounces	Grams
Manganese Nitrate	0.190	5.40
Zinc Sulfate	0.015	0.42
Copper Sulfate	0.005	0.14
Potassium Iodide	0.040	1.12
Boric Acid	0.2	5.6

Dissolve each component of B, except boric acid, in a quart of water. Use 50 cc. for 25 gallons solution, starting third week, and then use once every two weeks. Dissolve boric acid in one quart water but keep in a separate bottle, and use only when needed, as shown in diagnostic chart in a following chapter, usually about once each 3 weeks.

Solution C

	Ounces	Grams
Ferric Nitrate	0.3	8.5

	Ounces	Grams
Citric Acid	0.9	25.5
Calcium		
Hydroxide	0.8	22.68

Pour the above 2 liters of solution together, stirring vigorously.

Test for neutrality. If the liquid still is alkaline add very small amount of calcium chloride solution to bring to the neutral point.

Let this settle for two days. Pour off (decant) 1.3 liters of clear solution (throw away) by siphoning off into a graduate. This leaves 0.7 liters. Add water to make 1 liter.

Note. Use exactly calibrated graduates and distilled water for this special solution.

No. 3
(Buffered Formula)

Use 5 one-quart bottles, one bottle for each numbered solution.

1. Potassium Nitrate   4 g. in 1 l. water
2. Ammonium Nitrate   6 g. in 1 l. water
3. Magnesium Sulfate, Heptahydrate   10 g. in 1 l. water
4. Ferric Citrate (Colloidal)   0.40 g. in 1 l. water
5. Potassium Phosphate   13.7 g. ⎰(Specially prepared
    Cuprous Chloride   10.75 g. ⎱colloidal complex)

For practical purposes a quart may be considered as equal to a liter. To prepare the culture solution for the plant, make up as follows:

Measure the total capacity of all the plant containers (liquid level) in liters (or in quarts). Into a container large enough to hold total solution put in the following for each liter of finished product:

50 cc. of each of stock solutions No. 1, 2, 3, 4, 5	250 cc.
750 cc. of water	750 cc.
	————
	1,000 cc.

The ammonium nitrate, colloidal ferric citrate, and colloidal tricalcium potassium phosphatochloride complex will maintain a fairly uniform pH for at least one week. It is best to stir the solution daily at the beginning.

Stock solution No. 5 must be specially prepared.

Note: Make a little extra calcium chloride solution.

Dissolve 13.7 g. potassium phosphate in 1 l. water.

Dissolve 10.75 g. calcium chloride in 1 l. water.

  (Calcium chloride must be anhydrous salt).

Dissolve ferric nitrate in 1 quart of water. Use 100 cc. weekly or more often to keep leaves green.

Dissolve citric acid in 1 pint water. Should be added to keep pH near 6 for most plants.

Calcium hydroxide is dissolved in a quart of water, and used only in rare cases, when solution is too acid for certain plants.

No. 4
(For water of large cities of Eastern U. S. and for soft water)

For 25 Gallon Tank
Solution A

1. Potassium Nitrate   3.0 oz.

2. Calcium Nitrate   2.5 oz.
3. Magnesium Sulfate 0.8 oz.
4. Calcium Acid
      Phosphate   0.6 oz.

Dissolve each of the above in 1 pint of water and keep in separate bottles.

Solution B
Same as in Formula No. 2.
Solution C
Same as in Formula No. 2.

Directions for use are same as Formula No. 2.

Formula No. 2 was made for use with hard water that contains a lot of calcium and magnesium salt. The eastern cities contain less minerals in the water supply than in the west and, therefore, require a slightly different formula, as above.

No. 5
Mix the following ingredients:

Calcium Dihydrogen		
Phosphate	10	lb.
Copper Sulfate	1/3	oz.
Ferrous Sulfate	1	lb.
Magnesium Sulfate	10	lb.
Manganese Sulfate	2½	oz.
Zinc Sulfate	1	oz.

To this mixture add slowly, with adequate stirring:

Sulfuric Acid   5.4 lb.

Then add and mix thoroughly:

Calcium Nitrate	7	lb.
Potassium Nitrate	40	lb.
Sodium Metaborate	2	oz.

After thorough mixing, dissolve one-fifth pound of the mixture in 25 gallons of water.

Traces of arsenic, lead, selenium, titanium, and tellurium must be avoided.

No. 6
For use in hydroponics or for watering and nourishing potted plants.

Solution A

Ammonium Nitrate	11.0	g.
Calcium Sulfate	0.8	g.
Magnesium Sulfate	4.0	g.
Dipotassium		
Phosphate	2.8	g.
Water	9.0	l.

Solution B

Boric Acid	1.6	g.
Manganese Sulfate	1.6	g.
Zinc Acetate	1.6	g.
Copper Sulfate	0.35	g.
Water	1.0	l.

Solution C

Ferric Ammonium		
Sulfate	0.8	g.
Water	0.5	l.

For use add 4 cc. of B and 40 cc. of C to 9.0 l. of A.

Tomato-Plant Nutrient
(For use in soilless culture of tomato plants when employing sub-irrigation methods.)

Sodium Nitrate	1.80	g.
Magnesium Sulfate	4.20	g.
Potassium Chloride	1.20	g.
Ferric Chloride	0.01	g.
Mono Calcium		
Phosphate	1.70	g.

Dissolve each in one pint of water, then combine in the order listed above. Make the total volume up to one gallon.

Plant-Growth Regulator
Formula No. 1

Indoleacetic Acid	0.3	mg.
Lanolin	5.0	g.
Soap Flakes	5.0	g.
Agar	0.2	g.
Water	100.0	cc.

No. 2

Thiamin Hydrochloride   1
Water   1000
Use 1 drop per gal. of water.

Root-Growth Stimulant
Formula No. 1
U. S. Patent 2,168,550

Indoleacetic Acid   1–32
Alcohol (50%)   1000
Dilute with 100 times water before use.

No. 2

Indolebutyric Acid   12
Talc   88
Moisten ends of cutting and dip into above.

Granulated Fertilizer
Formula No. 1
British Patent 550,782

Urea   97–93
Water   1–3
Starch   2–4
Heat in a revolving horizontal cylinder for about 3 minutes at 70°C. and dry to 0.5% of water to yield 3–35–mesh granules. 0–20% of insoluble mineral powder (e.g., calcium carbonate, dolomite, phosphate rock) and about 1% of cacao shell, castor bean or peanut hulls may be added.

No. 2
U. S. Patent 2,136,069

Calcium carbonate (4) is stirred into a molten 9:1 ammonium nitrate-water mixture (5) at 100–105°C., and the mixture is granulated by spraying it from a horizontal rotating disc into a conical hopper from which it falls into a drying kiln, where it is mixed with more calcium carbonate to coat the granules and prevent them from sticking together.

Conserving Agent for Plants

Add to 1000 cc. of a saturated arsenic trioxide solution of 7 cc. of a 3% hydrochloric acid solution, 2 cc. of a 1% copper sulfate solution and a few drops of formaldehyde. Place plant in this solution and cover airtight.

Cut-Flower Preservative
U. S. Patent 2,317,631

A composition adapted to be added to and dissolved in water in which the stems of cut flowers are placed for causing the continuous growth and development of cut flowers and prolonging their span of life, consists of the following materials in powdered form in substantially the following proportions by weight: 23 to 43 pounds of hydrazine sulfate, 42 to 82 pounds of a compound selected from the class consisting of manganese sulfate and iron oxide, 3 to 5 pounds of calcium hypochlorite, 23 to 43 pounds of a compound selected from the class consisting of aluminum sulfate and soda alum and 3,125 to 5,125 pounds of sugar.

To regulate pH of Soil for Creeping Bent Lawns: Use 6 lb. of aluminum sulfate per 100 square feet of lawn.

This amount lowers the pH 1 to 1.5.

It is best to apply in early spring or late fall using a grass seeding machine.

Preventing Dropping of Apples

Pre-harvest dropping of apples can be prevented by spraying the tree with 7 or 8 gal. of solution containing 5–10 parts per million of methyl alpha-naphthoic acid. The spray requires 2–3 days to take effect and is effective for 10–15 days; therefore, the time of application has to be carefully chosen. The action is to inhibit the formation of abscission layers in the stem of the fruit.

Protecting Potatoes against Decay
U. S. Patent 2,348,946

A method of protecting potatoes against decay comprises treating the surfaces of the potatoes with sodium hypochlorite solution to dissolve deleterious formations on the skin, then treating the surfaces of the skin with a dilute solution of ammonium hydroxide to soften the skins, then drying the potatoes to harden the skins.

Grafting-Wax Salve

Rosin	80
Rosin Oil	5
Petrolatum (Yellow)	15

Grafting Wax

Rosin	50
Wool Fat	20
Ceresin (58–60°C.)	15
Beeswax	5
Rosin Oil	10

Tree Gall Treatment

Expose the gall, clean with a brush, and paint the entire surface with one of two mixtures. The first contains 1 part sodium dinitrocresol and four parts of methanol; the second, 100 parts by volume of methanol, 15 parts glacial acetic acid, and 12 parts by weight of crystal iodine.

General Insecticides
Mosquito Larvicide
Formula No. 1

Fuel Oil # 2	6 gal.
Sulfated Sperm Oil	1 qt.
Water	30–40 gal.

The above will treat effectively 1 acre of water surface.

No. 2

Diesel Fuel Oil	1
Sulfated Sperm Oil	4
Water	95

Mix and spray about 4 gal./acre.

No. 3

2–5% carbon tetrachloride in kerosene, or a 5% aqueous solution of a cresote disinfectant to be used in out-of-the-way places such as cellars and barns. Another suggestion is 2% citronella oil mixed with 1% spirit of camphor, 1% cedar wood oil, and 8% petroleum jelly. This is to be used as a repellent. A larvicide may be made from 1 part of crude oil and 4 parts of kerosene mixed with 0.1–0.2% castor oil, and, if desired, 1% cresol. To disinfect sewage-disposal plants and other large operations the following formula is suggested:

Light Fuel Oil	95
Coconut-Oil Potash Liquid	
Soap	5
Water	45
Conc. Pyrethrum Extract	5

This mixture should be diluted 10 times before being used.

Insecticidal Aerosol

Pyrethrum Extract,	
Purified	5
Sesame Oil, Refined	2
Dichlorodifluormethane	93

The above solution is made at about 90 lb. pressure per sq. in.

and packed in pressure cylinders.

D.D.T. Fly Spray

Dissolve ten grams of bis-(p-chlorophenyl)-1,1,1-trichloroethane (D.D.T.) in ninety grams of any fly spray or in ethylene chloride or any other relatively non-toxic solvent. Apply this as a spray to screen doors, and walls if desired, to keep flies away. One spraying will usually be effective for about three months.

D.D.T. Insect Powder

Thoroughly mix one part of bis-(p-chlorophenyl)-1,1,1-trichloroethane (D.D.T.) with nine parts of an inert filler, such as talc or other fine powder. Dust the area to be protected with this powder. It is effective against almost every variety of pest, but it should be used with caution as it is slightly toxic to the higher forms of life also. It should not be used on areas visited by honey bees, as they too are very sensitive to it.

Vermin Exterminator
U. S. Patent 2,344,105

A mixture of 40 parts by weight of methyl bromide with 60 parts by weight of trichloracetonitrile, whereby 100 parts by weight of this liquid mixture, for instance, may be absorbed in 100 to 120 parts by weight of kieselguhr or 40 to 50 parts by weight of woodpulp.

Silver-Fish Poison

| Arsenic Oxide | 1 |
| Flour | 9 |

Mix well and place on small cards. Place cards wherever silverfish roam.

Roach Powder
Formula No. 1

Powdered Pyrethrum	25
Sodium Fluoride	25
Ordinary Talc	50
Color	To suit

No. 2

Powdered Pyrethrum	15–20
Sodium Fluoride	25
Starch Pyrophyllite	
	To make 100

Sowbug Poison Bait
Formula No. 1

Paris Green	1
Bran	10
Fish Meal	10

No. 2

| Paris Green | 1 |
| Dry Dog Food | 20 |

Dry Insecticide

Calcium		
Carbonate	212	lb.
Talc	178	lb.
Bentonite	10	lb.
Lime, Air-Slaked	150	lb.
Crude		
Naphthalene	360	lb.
Cresylic Acid	4	lb.
Metallic Brown		
Oxide	75	lb.
Light Cresote Oil	1½	gal.

Rotenone-Derris Substitutes

As a substitute for Rotenone-Derris powdered sabadilla or yam bean seeds are used.

Disinfectant Spray
(Concentrate)

| Mineral Oil | 77 |
| Cresylic Acid | 3 |

Diglycol Laurate   20
Dilute with 100 times volume of water.

Pine Oil Disinfectant
Formula No. 1

Pine Oil	80.0
Linseed Soap (15% Anhydrous)	20.0

No. 2

Pine Oil	66.6
Rosin	26.5
Caustic Soda Solution 35° Bé	6.9

No. 3

Pine Oil	74.0
Rosin Soap	8.0
Linseed Soap	8.0
Water	10.0

Formula No. 1 is a simple solution that should have a phenol coefficient of about 5.5. Formula No. 2 results in a simple solution of rosin soap in pine oil. The amount of alkali required depends upon the type and acid value of the rosin used. A phenol coefficient of 3.5 to 4.0 should be obtained. Rosin does noot emulsify oils of high alcohol content as readily as vegetable oil soaps and will not hold as much pine oil in stable emulsion. This may be overcome by the addition of two to five per cent of a low titer soap such as linseed, or possibly, sulfonated castor oil. On the other hand, pine oil is more readily emulsified with a rosin soap than with a vegetable soap. Formula No. 3 represents a mixture of rosin and vegetable oil soaps. Such mixtures give better emulsions than either of the soaps alone and usually yield a more satisfactory phenol coefficient. The coefficient of this formula is between 5.3 and 5.8.

Agricultural Insecticides
Cutworm Control

Bran	50 lb.
Sodium Arsenite	1 qt.
Water	4 gal.

Use about 20 lb. of this per acre.

Lawn Grub and Japanese Beetle Larvae Control

Resin Fish-Oil Soap	1
Water	3
Carbon Disulfide	10

Place the soap and water in a bottle and shake until the solution is uniform. Then add the carbon disulfide and shake for 1 or 2 minutes or until a creamy emulsion has formed.

To treat the soil, stir 4 teaspoonfuls of this emulsion into a gallon of water and apply with a sprinkling can at the rate of 3 pints per square foot. Carefully measure the area to be treated and apply the emulsion uniformly without excess in any part, otherwise injury to grass roots or other plants will result.

If the lawn is kept moist for several days prior to the application of the insecticide, the grubs will tend to feed near the surface, where they can be reached by the emulsion.

Subterranean Grass Caterpillar Poison

Paris Green	2
Bran	50

with or without Molasses
Apply 16 lb. per acre.

Slug Insecticide

Metaldehyde	1	oz.

Bran or Bread
    Crumbs   3½ lb.

To prepare the bait, mix the metaldehyde and the bran or bread crumbs and store the bait in a jar or other container until needed. When ready to use the material, place a portion of it in a pan and add water slowly, while stirring it, until the bait is moistened, yet remains crumbly when a handful is squeezed together.

At dusk this moistened bait is scattered over the beds that are infested with slugs. The treatment may be repeated in 2 or 3 weeks if the snails reappear, or if their injury and slimy trails are discovered.

After having eaten baits containing metaldehyde, the slugs are stupefied and finally die. Where the vegetation is dense and humidity is high, the affected snails are killed more slowly or they may recover. Under such conditions the following bait containing calcium arsenate is more effective and kills more rapidly than baits containing only metaldehyde:

Calcium Arsenate	1 oz.
Metaldehyde	½ oz.
Bran	1 lb.
Molasses	2 tsp.
Water	1 pt.

In situations where no domestic or farm animals have access to the bait, apply it in piles of about a tablespoonful each, spaced about 2 feet apart; otherwise it should be scattered in the infested area. Unless washed away by watering or rains it remains effective for some time, and baiting two or three times during the year gives adequate control.

Tomato Moth Poison

Molasses (Cane) or	
Malt Extract	100
Ale	400
Sodium Fluoride	5

Tomato Pinworm Control

Powdered Cryolite	70
Talc	30

Four applications are needed at rate of 20–25 lb. per acre at 10 day intervals. Begin when fruit is about inch in diameter.

Boxwood Leaf Miner Spray

Nicotine Sulfate	1½	pt.
Molasses	12	gal.
Water	88	gal.

The material is applied to both surfaces of the leaves as a fine spray. It kills adults as they emerge from the leaves and entangles others in the sticky deposit.

Potato Leafhopper Dust

Pyrethrum Powder	
(1.3% Pyrethrins)	4
Sulfur	96

Cherry Fruit Fly Control
Formula No. 1

Lead Arsenate, Acid	2½	lb.
Water	100	gal.
Agricultural Spray		
Spreader	5	lb.

No. 2

Rotenone (4%		
Extract)	3	lb.
Molasses	2½	gal.
Water	100	gal.

No. 3

Acid Lead Arsenate	10
Dusting Sulfur	90

Parasiticide Spray
British Patent 552,879

Mannitan Laurate	1.0
Water	100.0
Pyrethrum Extract	0.1

Control of Pea Aphids
Formula No. 1

Ground derris root of rotenone content of 1%; a wetting or spreading agent (such as peanut or soya oil) 2%; and an inert carrier. A combination of Celite and talc makes a good carrier.

No. 2

Ground derris root of rotenone content of 1%; a wetting or spreading agent 1%; terpene ether 4%; and an inert carrier.

Codling Moth Spray

Phenothiazine, Micronized	1.8	lb.
Monoethanolamine Oleate	0.5	lb.
Stove Oil	0.25	gal.
Water	100	gal.
Casein	0.5	lb.
Hydrated Lime	0.4	oz.

Codling Moth Ovicide and Larvicide

Xanthone	2	lb.
Zinc Sulfate	4.5	oz.
Sodium Oleate	1.7	lb.
Water	100	gal.

Apply early in season.

Water Hyacinth Spray
U. S. Patent 2,248,159

A spray composition, harmless to animal life but destructive to water plant growth, comprises calcium arsenate containing total arsenic trioxide 42–57% (water-soluble portion 2–6) and metallic arsenic 31.5–43 (water-soluble 1.5–4%), in a water carrier (1 pound of solids, 15 gallons of water).

Rotenone Spray

Effective against red spiders, thrips (except the gladiolus thrips) on certain flowering plants, the cyclamen mite on chrysanthemums, aphids, cucumber beetles, tarnished plant bugs, certain species of leaf rollers, and leaf tiers.

Rotenone-Containing Root Powder (4% Rotenone)	1 lb.
Pyrethrum Extract (Alcoholic Extract, containing 2% of Pyrethrins)	2 qt.
Sulfonated Castor Oil	1 qt.
Water	50 gal.

In preparing this spray, add the sulfonated castor oil to the water. Next add a small quantity of this oil-and-water mixture to the derris or cube powder to make a uniform paste. Then stir the paste slowly into the remainder of the oil-and-water mixture. Finally add the pyrethrum extract to this mixture in case it is intended for the control of thrips or the cyclamen mite. For either red spiders or whiteflies, the pyrethrum may be omitted. A proprietary spreader-sticker, such as sodium oleyl sulfate plus synthetic resinous base, may be substituted for the sulfonated castor oil in the above formula, since the oil may at times injure the petals of open flowers and also the foliage of some plants. This material is used at the rate of ¾ teaspoonful per gallon, or

1½ pints per 100 gallons of spray mixture.

Horticultural Spray Base
U. S. Patent 2,327,152
Blown Rapeseed Oil   0.25–3%
Sodium Lauryl
  Sulfate   0.5–3%
Ammonia   0.1–1%
Mineral Oil To make 100%

Thrip, Aphid, Mite and Mealy Bug Control

Immersion of plants, corms, or bulbs in heated water, maintained at a constant temperature ranging from 110° to as high as 120°F. for the period of treatment, is a method used in the elimination of a number of pests, including the gladiolus thrips, aphids, and mealy bugs on gladiolus corms, the larvae of bulb flies and mites in narcissus and other bulbs, and the cyclamen mite in crowns and distorted growths of some ornamental plants.

The cyclamen mite and broad mite, 15 minutes at 110°F., except 20 minutes for large clumps of delphinium or gerbera and for trays of loosely placed strawberry plants.

Bulb mites on tuberoses, narcissus, and other bulbs, 1 hour at 110°F.

Bulb flies in narcissus and amaryllis, 1½ hours at 111°F.

The grape mealy bug on gladiolus corms, 30 minutes at 116°F.

The gladiolus thrips on gladiolus corms, 30 minutes at 112°F.

The boxwood leaf miner on boxwood, 5 minutes at 120°F. during late fall and early spring.

Gladiolus Thrip Spray
Formula No. 1
Sodium Antimony
  Lactophenate   8 lb.
Brown Sugar   4 lb.
Water   To make 100 gal.
No. 2
Paris Green   0.4 lb.
Corn Syrup   3 gal.
Water   To make 100 gal.

Thrip Spray

A spray solution for control of the gladiolus thrips on gladiolus, the flower thrips on roses, the orchid thrips on orchids, the chrysanthemum thrips, the banded greenhouse thrips, and the onion thrips on various ornamentals is made up as follows:
Tartar Emetic   2 lb.
Brown Sugar   4 lb.
Water   100 gal.

Dissolve each in a small quantity of water, then dilute to the quantity desired. Dissolving the tartar emetic may be hastened by using hot water. After this spray solution has been made up, no agitation is required to maintain a uniform spray.

Apply this spray as a fine mist to infested foliage of gladiolus or other plants or to flowers of roses or other plants when infested. The spray should cover the plant parts as tiny droplets. Do not apply so much spray that these droplets will unite and run off. Applications are made weekly, and if rain falls within 24 hours after the spray has been applied the treatment is repeated.

Paris Green Sprays

Paris green is highly toxic to

most insects but is also toxic to many plants and is rarely used on fruit trees. It is used as a spray, constant agitation is necessary to keep it in suspension. One of its chief uses as a spray is for the control of the Colorado potato beetle, combined with bordeaux mixture.

Paris Green   2 lb.
Hydrated Lime   8 lb.
Water   100 gal.

The lime is added to the spray to combine with the soluble arsenic in the Paris green and thus reduce plant injury.

As a dust it is mixed with a carrier such as talc or lime and is used for the control of cabbage "worms" before the cabbage head begins to form. It is also used on tobacco in some areas to control hornworms and flea beetles.

Spray for Gladiolus Thrips
Paris Green   4 lb.
Brown Sugar   66 lb.
Water   100 gal.

To obtain best results with this spray, use a nozzle that produces a fine mist and apply only enough to form small droplets on the foliage. If more is applied, the droplets coalesce and run to the base of the plant, causing waste and plant injury.

Ornamental Plant Insecticide
Spray
For red spiders:
Derris or Cube Powder
   (4% Rotenone)   1 tbs.
White Oil Emulsion
   (83% Oil)   4 tsp.
Water   1 gal.

For mealy bugs and scale insects:
Nicotine Sulfate Solution
   (40% Nicotine)   1½ tsp.
White Oil Emulsion 3   tbs.
Water   1 gal.

For newly hatched scale insects on hardy shrubs and also against lacebugs:
White Oil Emulsion
   (83% Oil)   ½ pt.
Soap Flakes   1½ cups
Nicotine Sulfate
   Solution   4 tsp.
Water   3¼ gal.

Another spray that may be used against lacebugs on such shrubs as azalea or rhododendron is:
Derris or Cube Powder
   (4% Rotenone)   5 tbs.
White Oil Emulsion
   (83% Oil)   ½ cup
Water   3 gal.

Some ornamental plants, including sweet peas, ferns, and orchids, are injured by oil sprays. Other plants may be injured where the spray collects in cavities or leaf axils; as the water evaporates, excess oil is left at these points. Palms and other plants having cavities in which spray material collects should be syringed with water or laid on their side after being sprayed with oils. Certain pyramidal junipers and spruces may also be injured by oil sprays. It is advisable to wash or syringe the more tender plants with water an hour or so after applying the spray.

Shade Tree Insecticides
Formula No. 1
Spray
   Soybean Flour   1 lb.

Paraffin Oil (100	
Sec.)	1 gal.
Water	1 gal.
Dinitro-o-cyclohexyl-	
cresol	0.5%

No. 2

Lead Arsenate	4 lb.
Soybean Flour	¼ lb.
Water	100 gal.

No. 3

Fixed Nicotine	2 lb.
Paraffin Oil	
(Summer)	½ gal.
Water	100 gal.
Soybean Flour	¼ lb.

No. 4

Dust

Dusting Sulfur	75
Hydrated Lime	15
Lead Arsenate	10

General Agricultural Spray

Water	2	gal.
Nicotine Sulfate		
(40%)	3	tsp.
Bordeaux Mixture	2	oz.
Paris Green	0.5	oz.
Lead Arsenate	1	oz.
Gum Arabic	0.1	oz.
Nekal NS (25%)	4	tsp.

Mix very well by vigorous shaking.

Kerosene Spray Emulsion

Kerosene emulsion is an effective contact insecticide. If not properly prepared, kerosene emulsion may cause injury to succulent plants such as coleus, ferns, heliotrope, begonia, and crucifers, although it can be safely used on chrysanthemums, crotons, palms, and rubber plants.

Other, hardier plants are not injured by even a 10-per cent emul-sion. Diluted to 5 per cent, this emulsion is effective against mealy bugs, rose midge larvae in the soil, immature scales, and red spiders, while a 1 per cent emulsion can be used successfully against aphids, thrips, and ants in the soil. Kerosene emulsion should be applied preferably late in the afternoon and the plants thoroughly syringed with water the next morning before sun-up. Soil overrun with ants may be freed of these pests without injury to the plants by drenching the infested areas with a 1-per cent emulsion.

A stock emulsion of kerosene is prepared according to the following formula:

Kerosene	2 gal.
Fish-Oil Soap or	
Laundry Soap	½ lb.
Water	1 gal.

If hard bar soap is used, first cut the soap into chips and then dissolve it in hot water, and while it is still hot add the kerosene very slowly, stirring constantly. The mixture should be pumped through a bucket pump back into the container for several minutes, or until a creamy emulsion has formed. Small quantities may be made with an egg beater.

The stock emulsion may be kept until needed in a tightly stoppered bottle or fruit jar. However, it will deteriorate with age and the kerosene will collect at the top of the mixture. This is the case with some of the commercial emulsions which have been prepared for some time. This free oil is the cause of much injury when applied to plants. The emulsion may be reclaimed by reheating and agitating, with or without the addition of soap.

Agricultural Spray Spreader
Skim Milk   2 qt.
Hydrated Lime   2 oz.

Dry Agricultural Spray
Canadian Patent 412,635
Dry Bentonite   100
Dry Clay   50
Tobacco Dust, Flue
    Cured   50
Nicotine Sulfate (40%)   100
Light Engine Oil   13

Bordeaux Mixture
Copper Sulfate   8 lb.
Fresh Hydrated
    Lime   12 lb.
Water   100 gal.

When small quantities are needed for use with compressed-air or knapsack sprayers, the total quantity of water to be used is divided and placed in two pails. The powdered copper sulfate is dissolved in one pail, and the lime mixed with the water in the other. Then the copper sulfate solution and the lime-water mixture are poured together and thoroughly mixed. The mixture is then poured through a strainer into the sprayer. If copper sulfate crystals or lumps are used, they should preferably be dissolved in a quantity of hot water representing one-half the total volume desired. If hot water is not available, place the copper sulfate crystals in a cloth sack and suspend this in the vessel containing cold water in such a way that the bottom of the sack is just below the surface of the water. Complete solution should occur in 1 to 2 hours. The suspension of lime in the other half of the water is then added, as the mixture is being agitated, and the whole poured through a strainer into the sprayer.

Nicotine Bentonite Insecticide
Bentonite has the property of combining with nicotine to form a compound more resistant to weathering than other nicotine preparations, hence it has been used extensively on apples during the past few years as a substitute for lead arsenate in the control of the codling moth, and to some extent on grapes for the control of the grape berry moth.

Nicotine when combined with bentonite may persist on fruit for 2 or 3 months, though in amounts too small to be toxic to man.

Tank mixtures are the cheapest and most effective but are more adhesive and may leave visible bentonite residues at harvest.

An effective tank-mixed spray for use on apples during the cover spray period consists of:

Nicotine Sulfate
    (40%)   1 pt.
Wyoming Bentonite   5 lb.
Crude Raw Soybean
    Oil   1 qt.
Water   100 gal.

Place about one-third of the water in the tank, add the nicotine sulfate, then add the bentonite slowly with strong agitation, followed by the soybean oil and the remainder of the water. Continue the agitation while spraying.

Proprietary nicotine bentonites are obtainable on the market.

Nicotine Decoctions
(Home-Made)

For many years gardeners have used tobacco decoctions prepared in different ways. The most common method is to soak tobacco stems or high-grade tobacco refuse for 24 hours, stir occasionally, and use the liquid. It requires 1 pound of stems for each gallon of water to make a satisfactory spray. If high-grade refuse is used, less is required—in some instances only one-fifth to one-tenth as much refuse as stems.

Nicotine Dust

Nicotine dust may be prepared with an ordinary flour sifter, using 1 pound of hydrated lime and 1 to 1½ ounces of 40 per cent nicotine sulfate solution. Or, place a quart of fresh hydrated lime in a container which can be tightly closed. Then add a handful of small stones or marbles, pour in 1 fluid ounce of nicotine sulfate, close the lid, and shake well for several minutes. To prepare larger quantities, roll the ingredients together in a drum or keg for at least 20 minutes with a peck of stones the size of goose eggs. Until used, the nicotine dust must be preserved in tight metal or glass containers, as it loses its strength very rapidly when exposed to the air.

Nicotine dust is used against aphids, such as occur on pea, cabbage, melon, turnip, and other plants, as well as against the striped cucumber beetle, adults of the greenhouse leaf tier, adults of the boxwood leaf miner, and the orchid fly.

For greater effectiveness, nicotine dusts should be applied to dry foliage when the temperature is above 65°F. and the air is still. Apply thoroughly to reach all insects present and repeat the treatment if control is not complete. For the boxwood leaf miner, make applications daily throughout the period of adult emergence.

Tobacco Blue Mold Control

Ferric Dimethyldiothiocarbamate	2
Lime	2
Water	1000

Spray plants twice weekly.

Rust-Fungus Disinfectant Dust
German Patent 715,636

Anhydrous Copper Sulfate	1
Powdered Lime	99

Disinfecting Seeds
U. S. Patent 2,309,289

Seeds are coated with a fungicidal organic mercury compound dissolved in an oily vehicle. For example 2.3–3.0 g. of methyl mercury bromide is incorporated in a vehicle consisting of equal parts of benzene and fuel oil. For disinfecting one kilogram of rye, wheat or barley in a mixing drum 1.7–2.0 g. of above mixture is used; for one kilogram of oats 3.0 g., and for one kilogram of beet seed 5.0–6.0 g.

Control of Potato Ring Rot and Scab

The spread of ring rot during the cutting of seed tubers is largely prevented by dipping the cutting knife in a solution containing:

Iodine	38 g.

Potassium Iodide 76 g.
Glycerin 1 pt.
Water 2 gal.

Scab (Rhizoctonia) organisms in cut pieces are killed by immersion in 1% iodine solution. The tubers are uninjured.

Disinfectant and Insecticide for
Poultry Houses
Orthohydroxy-
diphenyl 10 g.
Sufonated Mineral Oil 80 cc.
Pine Oil 10 cc.
Water To make 1000 cc.

Tetrachloroethylene Anthelmintic
Emulsion
Water (Heated to
70°C.) 500 cc.
Caustic Soda 6 g.
Casein 40 g.
Heat to 85°C. and rapidly stir in
Rosin, Ground 40 g.
Stir for 15–20 min. at 85°C. and then mix in
Water To make 800 cc.

Then add slowly with efficient high speed mixing
Tetrachlorethylene 2400 g.

Phenothiazine Suspension
U. S. Patent 2,294,888
Phenothiazine 100
Bentonite 8¼
Water To disperse Bentonite
Sodium Dioctylsulfo-
succinate 1
Water 250

Sheep Anthelmintic
Phenothiazine 1
Granular Salt 9

Sheep Tick Dip
Formula No. 1
Cubé (5% Rotenone) 10 lb.
Wettable Sulfur 100 lb.
Water 1000 gal.
No. 2
Nicotine Sulfate 10 lb.
Wettable Sulfur 100 lb.
Water 1000 gal.

Cattle Louse Dust
Formula No. 1
Derris (5% Rotenone) 1
Wettable Sulfur 10
No. 2
Powdered Sabadilla Seed 1
Wettable Sulfur 10
No. 3
Yam Bean Seed, Powdered 1
Wettable Sulfur 10
No. 4
Complete control of both chewing and sucking lice which infest cattle may be had by dusting the cattle occasionally with a mixture containing:
Phenothiazine 1
Sodium Fluosilicate 2
White Flour 1
Mix thoroughly and dust well with a good duster.

Cattle Grub Control
1:1 mixtures of wettable sulfur (325-mesh) and ground cubé or derris powder containing 5% of rotenone (200-mesh or finer) gives control of grubs on cattle when gently rubbed into their coats.

Control of Hide Beetles
Hides are dipped in 0.5% sodium silicofluoride solution for ¼ hr., before or after curing.

Cattle Horn Fly Spray

Nicotine	½
Lactic Acid	1
Water	98½

Treating Fungous Diseases in Pet Fish

Transfer the fish to a 2% solution of potassium permanganate for 15 to 30 minutes, then return to the aquarium. A 0.05% solution of phenyl mercuric nitrate may be used in the same way, and is highly effective.

Weed Killers
Formula No. 1
U. S. Patent 2,344,063

1 part of 2,6-dichloro-4-nitrophenol, 3 parts of urea and 3 parts of water are heated while stirring until solution takes place. The compound crystallizing on cooling, is water soluble in concentrations of 1-2%, as it is used for spraying.

No. 2

Spray ground early in spring with a 4% copper chloride solution. Especially suitable for controlling weeds in cereal grains.

No. 3
Honey Suckle Weed Killer

Ammonium Sulfamate	1 lb.
Wetting Agent (Sulfo-Turks)	1 lb.
Water	200 gal.

Poison-Ivy Killer

Ammonium Sulfamate	15
Water	75

Spray on poison ivy foliage but keep it away from valued trees or shrubs.

Mole Cricket Poison Bait
Formula No. 1

Wheat Bran	100 lb.
Sodium Fluosilicate	8 lb.
Water	To moisten

Apply at the rate of 20 pounds per acre.

No. 2

Fresh Earthworms	50 oz.
Strychnine Sulfate	1 oz.

Mix and use two inch lengths of worms. It must be used within six hours of preparation.

Solution for Cleaning Eggs
Formula No. 1
Canadian Patent 411,792

Eggs are immersed for 5–10 seconds in an aqueous solution containing:

Hydrochloric Acid	40
Sodium Dichromate	20
Wetting Agent	Small Amount

No. 2
U. S. Patent 2,287,147

Acetic Acid	9–15%
Alcohol	0.9%
Methyl Salicylate	0.1%
Water	To make 100%

Brush or rub eggs with above.

Feather and Animal Hair Depilatory
U. S. Patent 2,326,609

Cottonseed or Mineral Oil	5–25%
Polymerized Rosin	To make 100%

Apply warm and allow to cool: then strip off.

Poultry Mashes

Formula No.	1	2	3	4	5	6	7	8	9	10	11	12	13	14	15	16	17	18
	All Mash						Regular Laying Mash (To be fed with additional grain)											
Ground Yellow Corn	42	25	41½	37.1	20	30	200	22	225	200	30	200	400	29	350	150	100	
Ground Oats	12		10	10	15			13	100		15	200	350	10	100	25	100	19
Ground Barley		30				11½												
Ground Wheat		20				10												
Wheat Bran	15	8	7	6	15	16	100	15	150	200	15	200	400	5	150	100	100	10
Wheat Middlings or Shorts	15		20	20	20		100	15	200	200	15	200	400	20	150	100	100	15
Meat Scraps	8		4	2	16	7½	100	5	150	115	7½	75	300	10	75	75	75	10
Fish Meal		6		2		7½		5			2½	75			50			
Dry Skim Milk	3	2½	3	5	5	5	25	5	50	35	2½	50	100	6	50	25	25	10
Alfalfa Meal		5	5			7½		8						5	75	25	25	
Alfalfa Leaf Meal	3			7	6				100		5							15
Soybean Oil Meal			6	2½				5			5			9				5
Linseed Oil Meal				2														4
Ground Oyster Shell or Limestone	1½	1½	1	3.3	2	2		2			2½	20	40	1	10			5.6
Steamed Bone Meal		1	1	1.0		1		2						1				2.4
Salt	½	½	½	0.7	1	1	5	1	5	7	½		10	1	5	5	5	1.2
Fish Oil or Cod Liver Oil 85 D to 100 D		½	3.1	1.4		1		2	20		2	20	40	1	20			2.8

Poultry Mashes (*continued*)

Formula No.	19	20	21	22	23	24	25	26	27	28	29	30	31	32	33	34
	Regular Type Egg or Laying Mash (To be fed with grain or scratch feed)										All Mash Laying Mash					
Ground Yellow Corn	500	400	450	650	350	400	450	450	600	650	600	600	850	650	700	600
Ground Whole Oats	250	300	300		200	250	300	200	200	200	400	200		200	300	200
Ground Barley					200			100				200		200		100
Wheat Bran	100	200	200	100	200	200	100	100	100	200	200	100	200	200	100	200
Wheat Middlings or Shorts	500	500	500	600	500	500	600	600	500	400	450	600	600	400	600	600
*Fine Ground Alfalfa	100	100	100	100	100	100	100	100	100	100	50	50	50	100	50	50
Meat Scraps (50% Protein)	100	150	150	150	200	250	150	300	150	250	100	150	50	200	100	50
Fish Meal (60% Protein)	50	100	50	50	150		50	50	100	100	50	50	50		100	50
Dry Skim Milk	100	100	100	100	100	100	100	100	100	100	50	50	50	50	50	50
Soybean Oil Meal	200	100	100	200		150	100		100		50		150			100
Ground Oyster Shell or Limestone	60	40	50	60	40	50	50	40	40	40	40	40	40	20	20	20
Steamed Bone Meal	30	20		20					20		20		10			20
Reinforced Vitamin A & D oil	8	8	8	8	8	8	8	8	8	8	4	4	4	4	4	4
Salt	20	20	20	20	20	20	20	20	20	20	10	10	10	10	10	10
Total	2016	2038	2028	2058	2068	2028	2028	2068	2038	2068	2024	2054	2064	2034	2034	2054
% Protein (not less than)	19	20	19	19	20	20	19	20	20	20	16	16	16	16	16	16
% Fat (not less than)	4	4	4	4	4	4	4	4	4	4	4	4	4	4	4	4
% Fiber (not more than)	7	7	7½	6½	7	7	7½	7	6½	6½	7	6½	5½	7	7	7

* Alfalfa leaf meal may be used in place of fine ground alfalfa meal. When so used the fiber content may be reduced ½% for the regular type, and ¼% for all the mash feeds. Manganese sulfate may be added at the rate of 8 ounces per ton of regular mash, or 4 ounces per ton of all mash feed. It may help to improve shell texture.

Dehusking Carob Beans
U. S. Patent 2,326,868

Carob Beans	100
Sulfuric Acid (95%)	80

Stir the whole for 40–50 min., cool and keep temperature at about 50°C. Allow to stand for two hours. Remove husks by washing with

Water	75

Stir and drain off liquid. Repeat this washing until acid free and then dry in warm air.

Artificial Honeycomb
U. S. Patent 2,331,231

Hydrogenated Castor Oil	30–50
Beeswax	70–50

Broiler Feed

Bran	75
Dried Greens	10
Soya Bean Meal	10
Fine Sand	5

Chick Starting and Growing Mash
Formula No. 1

Yellow Corn Meal	50
Bran	15
Wheat Middlings	15
Dry Skim Milk	15
Steamed Bone Meal	3
Fine Salt	1
Cod Liver Oil (When chicks are indoors)	1

No. 2

Yellow Corn Meal	50
Bran	15
Wheat Middlings	15
Dry Skim Milk	10
Meat and Bone Meal	5
Steamed Bone Meal	2
Fine Salt	1
Cod Liver Oil (When chicks are indoors)	1

No. 3

Yellow Corn Meal	37
Ground Oats without Hulls (Wheat middlings may be substituted)	25
Ground Wheat or Middlings	15
Bran	10
Meat and Bone	5
Dry Skim Milk	5
Ground Oyster Shell or Bone Meal	2
Fine Salt	1
Cod Liver Oil (When chicks are indoors)	1

Poultry Feed
All Mash Ration Indoor

Ground Yellow Corn	64.5
Soybean Oil Meal	20.0
Alfalfa Leaf Meal (Dehyd.)	8.0
Steamed Bone Meal	4.5
Ground Limestone	1.0
*Mineral Mixture	0.5
85-D Oil	1.5

Grain and Mash

Grain: Whole yellow corn—12 pounds daily per 100 Plymouth Rocks.

Ground Yellow Corn	38
Soybean Oil Meal	40
Alfalfa Leaf Meal	10
Steamed Bone Meal	9
Ground Limestone	2
Mineral Mixture	1

Broiler All-Mash

Meat and Bone Scraps	35
Ground Yellow Corn	722
Soybean Oil Meal	165
Alfalfa Leaf Meal	33
Ground Limestone	7

* Salt and MnSO₄.

Steamed Bone Meal	23
Mineral Mix (Salt and MnSO$_4$)	5
85-D Oil	10

Breeding Mash
California

Ground Grain and Grain Products	69½
Dry Skim Milk	5
Fish Meal	12
Alfalfa Leaf Meal	8
Bone Meal	2
Ground Limestone or Oyster Shell	2
Cod Liver Oil	½
Salt	1

Michigan

Ground Yellow Corn	20
Ground Oats	18
Wheat Bran	20
Wheat Middlings	12
Dry Skim Milk	10
Meat Scraps	10
Alfalfa Leaf Meal	5
Bone Meal	2
Cod Liver Oil	2
Salt	1

Ohio

Ground Yellow Corn	20
Ground Oats	20
Wheat Bran	9
Wheat Middlings	20
Dry Skim Milk	5
Meat Scraps	20
Alfalfa Leaf Meal	5
Cod Liver Oil	2
Salt	1

A. D. M. I. No. 1

Ground Yellow Corn	20
Ground Oats	15
Wheat Bran	10
Wheat Middlings	25
Dry Skim Milk	10
Meat Scraps	5
Fish Meal	5

Soy Bean Oil Meal	5
Alfalfa Leaf Meal	5
Bone Meal	1
Ground Limestone or Oyster Shell	2
Cod Liver Oil	1½
Salt	½

Chick Starters
Illinois

Ground Yellow Corn	40
Fine Ground Whole Oats	8
Wheat Bran	10
Wheat Middlings	15
Dry Skim Milk	10
Meat Scraps	10
Fine Ground Alfalfa	5
*Cod Liver Oil	1
Salt	1

Kentucky

Ground Yellow Corn	70
Wheat Middlings	25
Dry Skim Milk	10
Meat Scraps	7½
Soy Bean Oil Meal	5
Alfalfa Leaf Meal	2½
*Cod Liver Oil	1
Salt	1

New England States

Ground Yellow Corn	200
Fine Ground Whole Oats, Feeding Oatmeal or Rolled Oats	100
Wheat Bran	100
Wheat Middlings	100
Dry Skim Milk	50
Meat Scraps	50
Fish Meal	25
Alfalfa Leaf Meal	25
Ground Limestone or Oyster Shell Meal	15
*Cod Liver Oil	7
Salt	5

Pennsylvania

Ground Yellow Corn	350

Fine Ground Whole Oats	100
Wheat Bran	150
Wheat Middlings	150
Dry Skim Milk	75
Meat Scraps	50
Fish Meal	50
Alfalfa Leaf Meal	75
Ground Limestone or Oyster Shell Meal	10
*Cod Liver Oil	10
Salt	5

U. S. D. A. No. 1

Ground Yellow Corn	25
Fine Ground Whole Oats	10
Wheat Bran	10
Wheat Middlings	10
Dry Skim Milk	10
Meat Scraps	10
Soy Bean Oil Meal	10
Corn Gluten Meal	5
Alfalfa Leaf Meal	5
Oyster Shell Meal	2½
*Cod Liver Oil	2
Salt	½

Wisconsin No. 2

Ground Yellow Corn	45
Wheat Bran	15
Wheat Middlings	15
Dry Skim Milk	8
Meat Scraps	8
Alfalfa Leaf Meal	5
Ground Limestone or Oyster Shell Meal	3
*Cod Liver Oil	1
Salt	½

A. D. M. I. No. 2

Ground Yellow Corn	35
Fine Ground Whole Oats	10
Wheat Bran	10
Wheat Middlings	20
Dry Skim Milk	10
Meat Scraps	5
Fish Meal	2½
Alfalfa Leaf Meal	5
Ground Limestone or Oyster Shell Meal	2
*Cod Liver Oil	1
Salt	½

A. D. M. I. No. 5

Ground Yellow Corn	30
Fine Ground Whole Oats	15
Wheat Bran	10
Wheat Middlings	20
Dry Skim Milk	7½
Meat Scraps	2½
Fish Meal	2½
Soy Bean Oil Meal	2½
Corn Gluten Meal	2½
Alfalfa Leaf Meal or Fine Ground Alfalfa	5
Ground Limestone or Oyster Shell Meal	2½
*Cod Liver Oil	1
Salt	½

Turkey Starters Formula No. 1

Ground Yellow Corn	300
Fine Ground Oats	100
Wheat Bran	200
Wheat Middlings or Shorts	300
Dry Skim Milk	250
Meat Scraps (50% Protein)	200
Fish Meal (60% Protein)	100
Alfalfa Leaf Meal	200
Corn Gluten Meal or Soybean Oil Meal	300
Ground Oyster Shell or Limestone	20
Salt	10

* Based on 85 AOAC units of vitamin D. Concentrated oils may be used in proportion of potency. Sardine oil and other oils of similar potency may be used.

Cod Liver Oil, Sardine	
Oil (or Equivalent)	40
No. 2	
Ground Yellow Corn	450
Fine Ground Oats	200
Wheat Bran	100
Wheat Middlings or	
Shorts	200
Dry Skim Milk	200
Meat Scraps (50%	
Protein)	300
Fish Meal (60%	
Protein)	100
Alfalfa Leaf Meal	200
Corn Gluten Meal or	
Soybean Oil Meal	200
Ground Oyster Shell	
or Limestone	20
Salt	10
Cod Liver Oil, Sardine	
Oil (or Equivalent)	40
No. 3	
Ground Yellow Corn	400
Fine Ground Oats	200
Wheat Bran	200
Wheat Middlings or	
Shorts	200
Dry Skim Milk	200
Meat Scraps (50%	
Protein)	200
Fish Meal (60%	
Protein)	100
Fine Ground Alfalfa	
Meal	200
Corn Gluten Meal or	
Soybean Oil Meal	250
Ground Oyster Shell or	
Limestone	40
Salt	10
Cod Liver Oil, Sardine	
Oil (or Equivalent)	40

If concentrated cod liver oil or similar anti-rachitic oil is used, the amount should be at least twice as much as is recommended for chicks.

Turkey-Growing Mashes
Formula No. 1

Ground Yellow Corn	400
Fine Ground Oats	200
Wheat Bran	200
Wheat Middlings or	
Shorts	400
Dry Skim Milk	200
Meat Scraps (50%	
Protein)	100
Fish Meal (60%	
Protein)	100
Alfalfa Leaf Meal	200
Corn Gluten Meal or	
Soybean Oil Meal	200
Ground Limestone or	
Oyster Shell	40
Salt	20
No. 2	
Ground Yellow Corn	400
Fine Ground Oats	200
Wheat Bran	200
Wheat Middlings or	
Shorts	400
Dry Skim Milk	150
Meat Scraps (50%	
Protein)	100
Fish Meal (60%	
Protein)	100
Fine Ground Alfalfa	
Meal	200
Corn Gluten Meal or	
Soybean Oil Meal	200
Ground Limestone or	
Oyster Shell	40
Salt	20
No. 3	
Ground Yellow Corn	500
Wheat Bran	250
Wheat Middlings or	
Shorts	500
Dry Skim Milk	100
Meat Scraps (50%	
Protein)	200
Fine Ground Alfalfa	
Meal	200

Corn Gluten Meal or	
Soybean Oil Meal	200
Ground Limestone or	
Oyster Shell	40
Salt	20

Turkey Mashes
Nevada Starting Mash

Ground Yellow Corn	10
Wheat Middlings	10
Dry Skim Milk	10
Meat Scraps or	
Fish Meal	20
Corn Gluten Meal or	
Soy Bean Oil Meal	15
Ground Barley	15
Ground Wheat	15
Alfalfa Leaf Meal	5

Pennsylvania Growing

Ground Yellow Corn	270
Fine Ground Oats	100
Wheat Bran	150
Wheat Middlings	150
Alfalfa Meal	50
Dry Skim Milk	40
Meat Scraps	70
Corn Gluten Meal or	
Soy Bean Oil Meal	140
Ground Oyster Shell	
or Limestone	20
Salt	10

U. S. D. A. Starting

Ground Yellow Corn	17
Fine Ground Oats	12
Wheat Bran	12
Wheat Middlings	12
Dry Skim Milk	17
Meat Scraps	13
Fish Meal	8
Alfalfa Leaf Meal	6
Cod Liver Oil	2
Salt	1

A. D. M. I. Starting

Ground Yellow Corn	20
Fine Ground Oats	10
Wheat Bran	10

Wheat Middlings	10
Alfalfa Meal	10
Dry Skim Milk	10
Meat Scraps	10
Fish Meal	5
Corn Gluten Meal or	
Soy Bean Oil Meal	12½
Cod Liver Oil	2
Ground Oyster Shell	
or Limestone	2
Salt	1

A. D. M. I. Growing

Ground Yellow Corn	25
Wheat Bran	12½
Wheat Middlings	25
Alfalfa Meal	10
Dry Skim Milk	5
Meat Scraps	10
Corn Gluten Meal or	
Soy Bean Oil Meal	10
Ground Oyster Shell	
or Limestone	2
Salt	1

Calf Meal (Feed)
Formula No. 1

Fine Ground Yellow	
Corn	600
Dry Skim Milk or	
Sweet Cream Butter-	
milk	400
Flour Middlings or	
White Shorts	400
Ground Oat Groats	300
Blood Flour or Meal	200
Linseed Oil Meal	100
Steamed Bone Meal	40
Salt	20
Vitamin A and D	
(1000 A, 400 D)	5

Guaranteed analysis: Protein, 24%; fat, 3%; fiber, 3½%.

Some commercial calf meals are flavored with a natural or artificial flavoring material. It may also be colored with a suitable coloring material. Iron oxide is often used for this purpose. The following formula contains the coloring and flavoring materials in the amounts often used.

No. 2

Dry Skim Milk or Sweet Cream Buttermilk	500
Ground Yellow Corn	200
Oat Groats or Rolled Oat Groats	100
Linseed Oil Meal	200
Soybean Oil Meal	100
Blood Flour or Blood Meal	100
Second Clear Flour or Red Dog Flour	400
Flour Middlings	350
Iron Oxide (Venetian Red)	10
Steamed Bone Meal	15
Ground Limestone or Oyster Shell	10
Iodized Salt	10
Vitamin A and D Oil (1000 A, 400 D)	5
Anisol (Anise Oil Substitute)	5/8

Guaranteed analysis: Protein, 25.0%; fat, 3.5%; fiber, 3.5%.

No. 3

Finely Ground Corn	50
Linseed Oil Meal	15
Finely Ground Rolled Oats	15
Dried Blood Flour	10
Dry Skim Milk	10
Salt	1/2

Feed wet by mixing with 6–8 times weight of warm water.

No. 4

Dry Buttermilk or Dry Skim Milk	25
Rolled or Ground Barley	25
Rolled or Ground Oats	25
Wheat Bran	17
Linseed Oil Meal	5
Steamed Bone Meal	2
Iodized Salt	1

Dry Calf Feed

After the calf is six to eight weeks old mix one-half pound of dry milk per day (per calf) with about as much of the following grain mixture as the calf will clean up readily:

Formula No. 1

Ground Grain (Corn and Oats)	600
Wheat Bran	200
Linseed or Soybean Oil Meal	200

No. 2

Rolled or Ground Barley	25
Rolled or Ground Oats	25
Wheat Bran	15
Dry Buttermilk or Dry Skim Milk	35
Steamed Bone Meal	2
Iodized Salt	1

No. 3

Ground Yellow Corn	30
Ground Oats (or Rolled Oats)	30
Skim Milk Powder	20
Wheat Bran	10
Linseed Oil Meal	10

No. 4

Dry Skim Milk	250
Ground Barley	200
Ground Oats	200
Wheat Bran	150
Blood Meal	100
Linseed Oil Meal	70
Sterilized Bone Flour	20
Salt	10

Feed whole milk until two weeks old, then gradually change to remixed dry skim milk until five weeks old. Then gradually discontinue liquid feed. Keep grain mixture, good quality alfalfa hay and water available. Offer grain feed when calves are one week old.

Reinforced Calf-Starter Mixture

Ground Yellow Corn	32.25
Rolled Oats	28.00
Wheat Bran	10.00
Linseed Oil Meal	5.00
White Fish Meal	3.00
Dry Skim Milk	20.00
Salt	0.50
Ground Limestone	0.50
Steamed Bone Meal	0.50
Reinforced Cod Liver Oil	0.25

Whole milk is fed up to a maximum of 10 pounds per day during the third week. This is reduced until at the end of the seventh week no more liquid milk is fed.

Calf starter is offered at the beginning of the third week, twice a day, until a maximum of 5 pounds per day is consumed.

Hay is fed after the calves are four weeks old, giving all they want.

Preserving Green Feed (Fodder)
German Patent 728,563

Add 50–200 g. bleaching powder per 100 kg. green feed, mixing well.

Cattle Salt Iodine Blocks
U. S. Patent 2,170,611

Potassium Iodide	0.02%
Corn Syrup	1–2%
Salt	To make 100%

Chapter VII

FOOD PRODUCTS

Canned Fruit Salad

The fruit is prepared separately, then mixed in the proportions of 50% peaches, 30% pears, 10% pineapple, 8% grapes and 2% cherries.

Mix until uniform and then place in the can so that the fruit represents 60% of the net contents.

A syrup is then prepared containing:

Sugar	40.00
Dispersible Pectin-320 (400°F.)	1.12
Sodium Citrate (Buffer)	0.20
Tricalcium Phosphate	0.25
Citric Acid	0.50
Water	57.93

Add the pectin to the rapidly agitated water and stir until thoroughly dispersed, then add sugar, buffer and citric acid and bring the mixture to a boil while stirring. Add the tricalcium phosphate dispersed in about 10 parts of water. Bring to a boil and shut off steam. A good grade of 150-mesh TCP is essential. Hold at 180°F. and run into a filler for syruping fruit-filled cans.

Proceed in the regular manner to fill the containers with syrup and process. Cool the cans in an agitating cooler until the center temperature is reduced to 100°F. Dry the cans and stack.

Let the cans remain undisturbed for at least 24 hours to jell properly.

Samples of jellied canned fruit salad prepared in the above manner have been held in storage for 18 months at normal temperatures and for 9 months at 100°F. without any visible change in flavor or texture. No deleterious action on the can has been observed throughout the period of these tests.

Precautions in Canning

In the canning of fruit salad, and more particularly fruit cocktail, it is customary to place the fruit in the can in layers and then cover with syrup. If this preparation were jellied it would show a stratified condition, peaches in one layer, pears in a second, and so forth. To avoid this, it is necessary that the fruit go into the can as a uniformly distributed mixture. The mechanics of such an operation is up to the individual canner or the particular plant in question, but should not present any serious operating difficulties.

Syruping—The syrup is prepared in the usual manner with the desired amount of sugar. The pectin-320 buffer salt and tricalcium phosphate are dissolved in the syrup. This causes no difficulty if properly handled. Because the increased viscosity makes the syrup difficult to handle in the ordinary gravity syrupers, vacuum syrupers are suggested, although not essential.

Cooling and Drying—In order

to get a uniformly dispersed mix, it is necessary to bring about the jelling in a rotating cooler. The cooler should be of sufficient length to reduce the center temperature of the can to 100°F. or less. Because of the low cooling temperature, the stacked cans may be likely to rust. It would appear advisable to include a small tunnel dryer in the tail-off line to insure control of this condition.

It should be kept in mind that pectin-320 is a specialty product, and as such, variations from a proved formula, such as that described, should not be made without some expert advice or some experimental tests to act as a guide.

With the obvious modifications, the above technic can also be applied to vegetable salads, or any other combination of solids dispersed in a jelly base.

In passing, it should be noted that the example covers a 40% "choice" sugar syrup salad. This sugar concentration is not essential for proper jelling. 15 or 25% sugar syrups would jell equally well.

Canned Jellied Consommé

Because of special properties and high jelly melting point, pectin-320 is especially suited for use in jellied consommés, madrilenes, and so forth. In contrast to gelatin consommés which require chilling to assure jell formation, pectin-320 consommés remain jelled at room temperature, but can be refrigerated without adversely affecting the tenderness of the jell. In addition, when heated to the normal serving temperature of soups, the pectin jell melts to contribute a desirable "body" to the hot soup.

Because of the absence of solid matter in which to disperse the pectin, it is essential that dispersible pectin be used in this type of product. The operations are simple and readily adapted to normal plant routine. A typical example of the preparation of a canned jellied consommé follows:

To each 90 lb. of prepared soup stock, add 0.7 lb. of standard dispersible pectin-320. This is stirred into the soup stock rapidly and the mixture brought to a boil.

To ten pounds of soup stock, add 0.14 lb. of monocalcium phosphate and stir until the phosphate is thoroughly wetted and well dispersed. The whole is then brought to a boil. If desired, the calcium slurry can be prepared from water rather than from soup stock.

The calcium mix is added to the pectin soup mix while stirring, and the combined mixture brought to a boil. Consommé is then filled into cans and the cans are processed, cooled, stacked and labeled by the customary procedures.

The two important points in the preparation of such consommés are the use of the dispersible form of pectin-320, added to the soup stock rapidly while stirring, and having the calcium slurry well dispersed and above 175°F. before it is added to the pectin-containing base. With these two modifications in mind, the remainder of the canning operation is routine.

The texture of the consommé may be varied by regulating the amount of pectin used.

Canned Tomato Aspic

Cold Tomato Juice	100.000
Dispersible Pectin-320 (250°)	0.835
Salt	1.500
Citric Acid	0.334
Tricalcium Phosphate	0.251
Spices, to meet desired flavor requirement.	

Add pectin-320 to rapidly churning juice, then heat to a boil while agitating, and hold at the boil three to five minutes. Add salt, citric acid and spices or spice oils, then tricalcium phosphate dispersed in about one pint of water. Continue boiling for three to five minutes while agitating. Viscolize (if desired), fill into containers and process as usual. Cool and stack, leaving undisturbed for 24 hours.

New Fruit Desserts

For best results, the formula for the new dessert must be varied slightly for different fruits, but one of the following typical formulas will usually prove satisfactory:

1. The following formula is used with unsugared fruit puree with high acid and low pectin content—such as raspberries, boysenberries, loganberries, youngberries, Stanta Rosa plums, strawberries, and similar fruits. To make approximately 100 gallons of mix, combine the following ingredients:

Puree	640 lb.
Sucrose	265 lb.
Gelatin (275 Bloom)	5 lb., 13 oz.
Water	60 lb.

(Mix the gelatin and water, sterilize, and add.)

In general, with these highly acid fruits no citric acid need be added. An exception is strawberries, which can be improved occasionally by the addition of not over 0.2 per cent (1 lb., 14 oz.) of citric acid to the mix. The soluble solids content, including the sugar in the fruit, should be about 37 to 38 per cent. Sugar should be added in approximately the ratio of 1 part for every 2.4 parts of fruit puree. High-conversion corn sirup can be substituted for one-third of the sugar, in which case slightly less fruit is necessary to make 100 gallons of mix. The corn syrup must be substituted for the sugar in a 3-to-2 ratio in order to maintain the same sweetness. With the syrup the formula becomes:

Puree	610 lb.
Sucrose	170 lb.
High-Conversion Corn Syrup	125 lb.
Gelatin (275 Bloom)	5 lb., 13 oz.
Water	60 lb.

(Mix the gelatin and water, sterilize, and add.)

2. The following formula is used for fruits with low acid and high pectin content, such as unsugared apricots, cantaloupe, pears, or other naturally sweet fruits. To make approximately 100 gallons of mix, combine the following:

Puree	680 lb.
Sucrose	225 lb.
Gelatin (275 Bloom)	5 lb., 13 oz.

Water   60 lb.
Citric Acid   1 lb., 14 oz.

A soluble solids content of 34 to 35 per cent, including the natural sugar of the fruit, is sufficient because of the lower acid content of these fruits. A 3-to-1 ratio of fruit to sugar is usually satisfactory, though even less sugar can sometimes be used.

In the preparation of a mix, the puree, sugar, and citric acid (if used) are mixed together until well dissolved. Keeping the puree and the mix cool and avoidance of excessive mixing tend to preserve the ascorbic acid of the fruit. The gelatin is mixed with ten times its weight of water, and is heated to 170–180°F. to dissolve and sterilize it. The mix itself is not pasteurized. During the addition of the gelatin solution the mix is stirred.

Syrup for Pickling Fruits

Various fruits, particularly figs and peaches, can be packed in a heavy vinegar syrup as fruit pickles. The fruit is prepared as for canning and cooked in syrup to about 60% sugar.

Dry Ginger Root   12 oz.
Whole Cloves   14 oz.
Stick Cinnamon   18 oz.
Cider Vinegar (40
    grain)   3 gal.
Sugar   125 lb.
Water   20 gal.

Place spices in cheesecloth bags. Cook fruit in syrup, then allow to stand overnight.

Remove spices. Pack in vacuum sealed jars and pasteurize 30 minutes at 175 to 180°F.

Preventing Discoloration of Cut Fruit
U. S. Patent 2,298,933

Sodium Thio-
    sulfate   0.01–0.5
Sodium Sulfite   0.001–0.05
Water   To make 100

Immerse in above, drain and dry.

Grape Shipment Protection

The sawdust or wood wool used in packing table grapes is sprayed with 20 cc. (per box) of:

Sodium Bisulfite   20
Water   100

Fruit Protective Coatings (Emulsions)
U. S. Patent 2,333,887
Formula No. 1

Beeswax   2.55
Aeresol O.T.   0.25
Trigamine Stearate   0.70
Water   96.50

No. 2

Carnauba Wax   10
Trigamine Stearate   3
Water   260

No. 3

Candelilla Wax   13.5
Oleic Acid   2.9
Morpholine   2.6
Water   81.0

No. 4

Manila Resin D.B.B.   50.0
Oleic Acid   20.0
Morpholine   18.3
Water   500.0

No. 5

Ester Gum   95.0
Morpholine   28.5
Water   1000.0

These are emulsions that are applied to citrus fruits to prevent spoilage.

Non-Discoloring Peeled Potatoes
U. S. Patent 2,241,436

To prevent the discoloration of peeled potatoes they are immersed in 2% aqueous sodium pyrosulfite for about 10 minutes.

Vegetable Preservation
British Patent 550,076

Addition of 0.22% of sodium sulfite to the water reduces the losses of color and of ascorbic acid and the development of abnormal flavors both during blanching or cooking and during any subsequent treatment such as drying, freezing, or canning, to preserve the vegetable.

Pea-Soup Cubes

Pea Flour	80
Corn Flour	40
Potato Flour	40
Onion Powder	12
Salt	18
Gelatin Powdered	12
White Pepper	$\frac{1}{2}$
Beef Extract	12

Place all the ingredients, except the beef extract, in a steam-heated kettle, fitted with a stirrer. Heat and stir continuously.

Dissolve the beef extract in 5 parts boiling water and add very slowly, while continuing to heat and stir, until dry.

Press into cubes and wrap.

Bleaching Nut Shells
U. S. Patent 2,155,923

The shells are soaked in dilute, non-toxic, alkaline solution (aqueous sodium hypochlorite), drained, treated with hydrogen peroxide solution, drained (hydrogen peroxide = 5–25 vol.; 10–15 vol.), and dried.

Removal of Filbert Nut Skins
U. S. Patent 2,273,183

The kernels are soaked in aqueous 3% solution of sodium hydroxide, sodium carbonate, sodium acid carbonate, or borax for 1–4 minutes, and after rinsing in 2–4% hydrochloric acid, acetic acid, or citric acid, the skins are pushed off by water jets.

Removing Nut Skins
U. S. Patent 2,156,406

Treat with a solution of:

Water	60.00
Sodium Hypochlorite	50.00
Caustic Soda	2.80
Trisodium Phosphate	0.60
Sodium Carbonate	0.60
Soap	0.02

Wash thoroughly and then dry.

Orange Marmalade
(Using California Valencia Oranges)

Water	$\frac{1}{2}$ gal.
Orange Juice	4 gal.
Lemon Juice	2 gal.
Orange Peel (Sliced Very Thin), Cooked	8 lb.
Lemon Peel (Sliced Very Thin), Cooked	4 lb.
100 Grade Exchange Citrus Pectin, Rapid Set, #436	10–12 oz.
Granulated Cane or Beet Sugar	75 lb.
50% Citric Acid Solution	6 fl. oz.

Cook to 223°F. at sea level or

11°F. above the boiling point of water at your factory.

Yields about 115 pounds of finished marmalade.

Prepare 4¼ gallons of orange juice and 2 gallons of lemon juice from the fresh fruit. Mix the juices, place in a large jar or crock, and allow to settle. Siphon off the top juice and strain through fine-weave cloth. The juice may be filtered if desired. Use a total of 6 gallons of juice for the marmalade.

Slice the orange and lemon peel very thin, put in a kettle, cover with water, and cook until soft. Strain off the liquid from the peel. Use 12 pounds of this cooked peel for the marmalade.

Put the 6 gallons of juice and ½ gallon of water in a kettle, and heat hot (180°F.). Thoroughly mix the pectin with about 10 times its weight of dry granulated sugar (taken from the total amount required for the batch), and add this pectin-sugar mixture with stirring. Continue to stir and heat to boiling. Boil vigorously for ½ minute.

Now add the remainder of the sugar and cook to 223°F. (or 11°F. above the boiling point of water at your factory). Turn off the steam, add the 50% citric acid solution, stirring thoroughly, and fill into containers hot.

When using California navel oranges, less pectin is required to produce the same consistency.

Strawberry Jam or Preserves

Water	20 lb.
Strawberries	82 lb.

100 Grade Exchange

Citrus Pectin	6 to 8 oz.
Sugar	100 lb.
50% Citric Acid	
Solution	11 fl. oz.

Cook to 223°F. (106.1°C.) at sea level or 11°F. (6.1°C.) above the boiling point of water at your factory.

Yields about 151 pounds of the finished jam or preserves at 68% soluble solids.

Put the water in a kettle and heat hot (almost boiling). Thoroughly mix the pectin with about 8 times its weight of granulated sugar, and add this pectin-sugar mixture to the water as the latter is being stirred vigorously with a paddle. Continue to stir and heat to boiling.

Now add the strawberries and cook slowly for a few minutes if necessary to soften the fruit. Add the remainder of the granulated sugar and cook as quickly as possible to the final temperature.

Citric Acid	1 lb.
Hot Water	16 fl. oz.
Dissolve.	

Remove the batch from the kettle and place in a shallow pan to cool at the same time adding the 50% citric acid solution being sure that it is thoroughly mixed with the batch.

Stir the batch occasionally as it is cooling and when it thickens sufficiently to hold the fruit and berries in suspension, fill it into containers. If the jam is filled at a temperature below 185°F. (85°C.), a subsequent sterilization process is necessary.

This formula for strawberry jam was developed for a package

not exceeding about 2 pounds net (0.91 kilos). When larger containers are used, more pectin should be added and the batch should be cooked from 1 to 3°F. higher. This applies also to the formula for jelly which follows.

Grape Jelly

Grape Juice	100 lb.
100 Grade Exchange Citrus Pec-tin	10 to 12 oz.
Sugar	100 lb.
50% Citric Acid Solution	12 fl. oz.

Cook to 220°F. (104.4°C.) at sea level.

Put the grape juice in the kettle and heat hot (almost boiling). Thoroughly mix the pectin with about 8 times its weight of granulated sugar and add this pectin-sugar mixture to the grape juice as it is being stirred vigorously with a paddle. Continue to stir and heat to boiling. Boil vigorously for a few minutes.

Now add the remainder of the sugar and cook as quickly as possible to the proper temperature. Add the citric acid solution being sure that it is mixed thoroughly with the batch. Fill into containers immediately.

If the jelly is at a temperature lower than 185°F. (85°C.) at the time of filling, a subsequent sterilization process is necessary.

This formula for grape jelly was developed for a package not exceeding about 2 pounds net. When larger containers are used, more pectin should be added and the batch should be cooked from 1 to 3°F. higher. This applies

also to the formula for jelly which follows.

Wine Jelly

Pectin Solution:

100 Grade Exchange Citrus Pectin Slow Set #451	11.0	oz.
Granulated Cane or Beet Sugar	1.5	lb.
Water	9 lb. 6	oz.
Sweet Wine	4.75	gal.
Granulated Cane or Beet Sugar	59	lb.
*Standard Citric Acid Solution	4.5	fl. oz.

First prepare the pectin solution by thoroughly mixing the pectin with the sugar and stirring it into the boiling or nearly boiling water. Shut off heat and stir until the pectin has all dissolved. Draw off and set aside until needed.

Add the wine and the sugar (larger amount shown in the table) to the kettle. Warm and stir until the sugar has dissolved. Add the pectin solution, stir well and heat to boiling. Shut off heat and stir in the citric acid solution. Allow to stand until the bubbles have risen to the surface (not over 15 minutes), skim, and package.

Cocoa Jelly

Water	2½	gal.
100 Grade Exchange Citrus Pectin, Slow Set, #451	12	oz.
Glucose (43° Bé.)	20	lb.
Cane or Beet Sugar	20	lb.
Natural Process Cocoa	4	lb.
Salt	¾ to 1	oz.

* 1 pound crystals per pint of water.

15% Phosphoric
Acid Solution   8½ fl. oz.
*Flavor

(1) Put 2½ gallons of water in a kettle and heat hot (170°F.). (Open fire or steam-jacketed kettle may be used.)

(2) Thoroughly mix 12 ounces of pectin with about 8 pounds of granulated sugar.

(3) Add the pectin-sugar mixture to the warm water as it is being stirred with a paddle. Continue to stir and heat to boiling. Boil vigorously for a moment.

(4) Add the glucose, salt, and the 15% phosphoric acid solution. Heat to boiling.

(5) Mix the natural process cocoa and the remaining cane or beet sugar. Add to the kettle and stir until all lumps have dissolved. Boil 16°F. above the boiling point of water at your factory. (This is 228°F. at sea level.)

(6) Add the vanilla extract or the citrus oil. Cast into starch molds.

Pectin Solution
100 Grade Exchange
Citrus Pectin for
Confectioners No.
451   18   oz.

------

* Flavor with 12 fluid ounces vanilla extract or ½ to ¾ fluid ounces exchange cold pressed oil of lemon or oil of orange.

It may be possible to use less than 8.5 fluid ounces of the acid solution with some cocoa. It is well to use the minimum amount needed to set the candy to satisfactory firmness.

Natural process cocoa is specified because Dutch process cocoa will not give satisfactory results.

Granulated Cane or
Beet Sugar   8   lb.
Water   1.5 gal.
                  or 12.4 lb.
Citric Acid U.S.P.   3   oz.
Acetate of Soda
U.S.P.   2   oz.

1. Thoroughly mix the 18 oz. pectin with the 8 lb. granulated sugar.

2. Put 1½ gal. of water in kettle and heat just to boiling. Add the pectin-sugar mixture to the water as it is being stirred with a paddle. Break up all lumps.

3. Stir in the citric acid and acetate of soda. Keep pectin solution hot and use portions in candy making as described below.

Pectin Candy Formula
Granulated Cane or
Beet Sugar   12   lb.
Glucose (43° Bé.)   20   lb.
Nulomoline   5½   lb.
Pectin Solution   11   lb.

1. Heat glucose and Nulomoline until quite fluid, then add the granulated sugar. Heat to boiling and stir to dissolve the sugar. As soon as sugar is all dissolved, turn off heat.

2. Add the hot pectin solution carefully to avoid boiling over. Heat to boiling and cook rapidly to 231°F. (19°F. above the boiling point of water at your factory).

3. Color and flavor. Cast into small starch molds. Place filled starch trays in hot-room and allow to remain 6 hours at 140°F. or for a longer time at a lower temperature. This formula will produce about 45 pounds of candy.

Jelly Candy

Tart and Moderately Firm Fruit
Flavors, Cast or Slab

Water	2½ gal.	
100 Grade Exchange Citrus Pectin for Confectioners No. 451	12	oz.
Acetate of Soda U.S.P.	2	oz.
Citric Acid U.S.P.	3½	oz.
Glucose (43° Bé.)	20	lb.
Granulated Cane or Beet Sugar	20	lb.
Color and Flavor	As desired	

1. Thoroughly mix 12 ounces of pectin with about 8 pounds of the granulated sugar.

2. Put 2½ gallons of water in kettle and start heating.

3. Add the pectin-sugar mixture to the water as it is being stirred with a paddle. Continue to stir and heat to boiling. Boil vigorously for a moment.

4. Add the acetate of soda, dissolved in a little water, and the 20 pounds of glucose. Dissolve the citric acid in a small amount of warm water and add half of it to the kettle. Heat to boiling again.

5. Add the remainder of the granulated sugar (12 pounds) and cook rapidly to 229°F. (17°F. above the boiling point of water at your factory).

6. Add the remaining citric acid solution and the desired amount of color and flavor. Cast into starch or onto a slab. This formula will produce about 47 pounds of candy.

This batch can be cooked to 226°F. (14°F. above the boiling point of water at your factory) to produce a more tender piece. This is possible where storage and shipping conditions do not produce sweating.

Mildly Tart and Moderately Firm
Pieces, Cast or Slab

Water	2½ gal.	
100 Grade Exchange Citrus Pectin for Confectioners No. 451	12	oz.
Acetate of Soda U.S.P.	1	oz.
Citric Acid U.S.P. (Crystals or Powdered)	2	oz.
Glucose (43° Bé.)	20	lb.
Granulated Cane or Beet Sugar	20	lb.
Color and Flavor	As desired	

1. Thoroughly mix 12 ounces of pectin with about 8 pounds of the granulated sugar.

2. Put 2½ gallons of water in kettle and start heating.

3. Add the pectin-sugar mixture to the water as it is being stirred with a paddle. Continue to stir and heat to boiling. Boil vigorously for a moment.

4. Add the acetate of soda, dissolved in a little water, and the 20 pounds of glucose. Dissolve the citric acid in a small amount of warm water and add half of the solution to the kettle. Heat to boiling again.

5. Add the remainder of the granulated sugar (12 pounds) and cook rapidly to 229°F. (17°F. above the boiling point of water at your factory).

6. Add the remaining citric acid solution and the desired amount of color and flavor. Cast into starch or onto a slab. This

formula will produce about 47 pounds of candy.

This batch can be cooked to 226°F. (14°F. above the boiling point of water at your factory) to produce a more tender piece. This is possible where storage and shipping conditions do not produce sweating.

Novel Pieces with Chocolate
Flavor, Cast or Slab

Water	2½	gal.
100 Grade Exchange Citrus Pectin for Confectioners No. 451	12	oz.
15% Phosphoric Acid Solution	8½	fl. oz.
Natural Process Cocoa	4	lb.
Glucose (43° Bé.)	20	lb.
Granulated Cane or Beet Sugar	20	lb.
Salt	¾ to 1	oz.
Flavor	12	fl. oz.

(Vanilla extract, or ½–¾ fl. oz. Cold Pressed Oil of Lemon, or Oil of Orange.)

1. Thoroughly mix 12 ounces pectin with about 8 pounds of the granulated sugar.

2. Put 2½ gallons of water in kettle and start heating.

3. Add the pectin-sugar mixture to the water as it is being stirred with a paddle. Continue to stir and heat to boiling. Boil vigorously for a moment.

4. Add the 20 pounds of glucose, the salt, and the 8½ fl. oz. of phosphoric acid. Heat to boiling.

5. Mix the natural process cocoa and the remaining granulated sugar. Add to the kettle and stir until all lumps have dissolved. Cook rapidly to 228°F.

6. Add the vanilla extract or the orange or lemon oil. Cast into starch molds or onto a slab. This formula will produce about 50 pounds of candy.

The 15% phosphoric acid solution is made by dissolving 1 pint of 85% syrupy phosphoric acid in 1 gallon of water.

It may be possible with some cocoas to use a little less than 8½ fl. oz. of 15% phosphoric acid solution. Use the minimum amount necessary to give the candy good firmness.

Dutch process cocoa does not give satisfactory results.

High-Grade Licorice Flavored
Piece, Cast or Slab

Water	2½	Gal.
100 Grade Exchange Citrus Pectin for Confectioners No. 451	15	oz.
Acetate of Soda U.S.P.	1	oz.
Citric Acid U.S.P.	2½	oz.
Glucose (43° Bé.)	20	lb.
Granulated Cane or Beet Sugar	20	lb.
Solid Licorice Extract	10	oz.
Oil of Anise	1	tsp.
Caramel Color	2	lb.
Black Color	As desired	

1. Thoroughly mix 15 ounces pectin with about 8 pounds of the granulated sugar.

2. Put 2½ gallons of water in kettle and start heating.

3. Add the pectin-sugar mixture to the water as it is being stirred with a paddle. Continue to stir and

heat to boiling. Boil vigorously for a moment.

4. Combine the acetate of soda and the citric acid. Dissolve in a small amount of hot water.

5. Add the acetate of soda-citric acid solution to the kettle and then the 20 pounds of glucose. Heat to boiling again.

6. Add the remainder of the granulated sugar (12 pounds) and heat to boiling.

7. Dissolve the 10 ounces of solid licorice extract in a small amount of hot water. Add the solution and the caramel color late in the boiling process.

8. Cook rapidly to 228°F. Add the black color and oil of anise. Cast into starch or onto a slab. This formula will produce about 48 pounds of candy.

Licorice of various types such as powder, flake, solid and syrup, differ in their effects upon the setting of the candy. It may be necessary to use a little more or less acetate of soda than is specified when using other than the solid extract.

Moderately Firm Pieces Using Orange Pulp and Juice, Cast or Slab

Water	2	gal.
100 Grade Exchange Citrus Pectin for Confectioners No. 451	12	oz.
Acetate of Soda U.S.P.	2	oz.
Citric Acid U.S.P.	3½	oz.
Glucose (43° Bé.)	20	lb.
Granulated Cane or Beet Sugar	20	lb.

Ground Oranges (Use the whole fruit)   5   lb.

1. Thoroughly mix 12 ounces pectin with about 8 pounds of the granulated sugar. Cut 5 pounds of whole oranges into pieces, remove all seeds, and grind in a suitable food chopper.

2. Put the water in kettle and start heating.

3. Add the pectin-sugar mixture to the water as it is being stirred with a paddle. Continue to stir and heat to boiling. Boil vigorously for a moment.

4. Add the acetate of soda, dissolved in a little hot water, and the 20 pounds of glucose. Dissolve the citric acid in a small amount of warm water and add half of it to the kettle. Heat to boiling again.

5. Add the remainder of granulated sugar (12 pounds), the ground fruit, and cook rapidly to 229°F.

6. Add the remaining citric acid solution. Color and flavor if desired. Cast into starch molds or onto a slab. This formula produces about 50 pounds of candy.

If desired, the ground fruit can be added toward the end of the boiling process.

Moderately Firm Pieces Using Grapefruit Pulp and Juice, Cast or Slab

Water	2	gal.
100 Grade Exchange Citrus Pectin for Confectioners No. 451	12	oz.
Acetate of Soda U.S.P.	2¾	oz.
Citric Acid U.S.P.	3½	oz.

Glucose (43° Bé.)   20   lb.
Granulated Cane or
  Beet Sugar   20   lb.
Ground Grapefruit
  (Use the whole
  fruit)   6   lb.

1. Thoroughly mix 12 ounces pectin with about 8 pounds of the granulated sugar. Cut 6 pounds of whole grapefruit into pieces, remove all seeds, and grind in a suitable food chopper.

2. Put the water in kettle and start heating.

3. Add the pectin-sugar mixture to the water as it is being stirred with a paddle. Continue to stir and heat to boiling. Boil vigorously for a moment.

4. Add the acetate of soda, dissolved in a little hot water, and the 20 pounds of glucose. Dissolve the citric acid in a small amount of warm water and add half of it to the kettle. Heat to boiling again.

5. Add the remainder of granulated sugar (12 pounds), the ground fruit, and cook rapidly to 229°F.

6. Add the remaining citric acid solution. Color and flavor if desired. Cast into starch molds or onto a slab. This formula produces about 50 pounds of candy.

If desired, the ground fruit can be added toward the end of the boiling process.

Moderately Firm Pieces Using Fig
        Pulp, Cast or Slab
Water   2½ gal.
100 Grade Exchange
  Citrus Pectin for
  Confectioners No.
  451   12   oz.

Acetate of Soda
  U.S.P.   1   oz.
Citric Acid U.S.P.   3½   oz.
Glucose (43° Bé.)   20   lb.
Granulated Cane or
  Beet Sugar   20   lb.
Ground Figs
  (Dried White
  Figs)   7   lb.

1. Thoroughly mix 12 ounces pectin with about 8 pounds of the granulated sugar. Grind up 7 pounds of dried figs or use 7 pounds of fig paste.

2. Put the water in kettle and start heating.

3. Add the pectin-sugar mixture to the water as it is being stirred with a paddle. Continue to stir and heat to boiling. Boil vigorously for a moment.

4. Add the acetate of soda, dissolved in a little hot water, and the 20 pounds of glucose. Dissolve the citric acid in a small amount of warm water and add half of it to the kettle. Heat to boiling again.

5. Add the remainder of granulated sugar (12 pounds), the ground fruit, and cook rapidly to 229°F.

6. Add the remaining citric acid solution. Color and flavor if desired. Cast into starch molds or onto a slab. This formula produces about 50 pounds of candy.

If desired, the ground fruit can be added toward the end of the boiling process. The addition of a small amount of cold pressed oil of lemon will enhance the flavor of these fig candies.

Slab-Cast Apple Confection
Water   2   gal.
100 Grade Exchange

Citrus Pectin for
Confectioners No.
451   10   oz.
Acetate of Soda
U.S.P.   2½   oz.
Citric Acid U.S.P. 3½   oz.
Glucose (43° Bé.)   20   lb.
Granulated Cane or
Beet Sugar   20   lb.
Peeled, Cored, and
Cooked Apples   9   lb.
Walnut Meats,
Chopped   5   lb.
Apple Flavoring   As desired

1. Peel, core and slice enough ripe apples to give 9 pounds of slices. Cook until soft with a little water (2 to 4 pounds). Mash or sieve to produce a uniform pulp.

2. Thoroughly mix 10 ounces pectin with about 6 pounds of the granulated sugar.

3. Put the water in the kettle and start heating.

4. Add the pectin-sugar mixture to the water as it is being stirred with a paddle. Continue to stir and heat to boiling. Boil vigorously for a moment.

5. Add the acetate of soda, dissolved in a little hot water, and the 20 pounds of glucose. Dissolve the citric acid in a small amount of warm water and add half of it to the kettle. Heat to boiling again.

6. Add the remainder of the granulated sugar (14 pounds), the cooked apples, and cook rapidly to 228°F.

7. Add the remaining citric acid solution, the apple flavor, and the chopped walnut meats. Pour onto a slab. When cool, cut into squares and finish by rolling them in a mixture of equal parts of starch and powdered sugar. This

formula produces about 53 pounds of candy.

If desired, the cooked apples can be added toward the end of the boiling process.

Old Fashioned Apple Leather
(Candy)

This material is made in the South, and often takes the place of an appetizer, or a between-meal candy-like snack. Essentially, it is a dried applesauce confection and is prepared as follows:

A thick applesauce is prepared by peeling, coring, and slicing a quart of apples. They are placed in a sauce pan with two cups of water, and boiled until the apples are all disintegrated. This thick pasty-like mass is now passed through a sieve, and a cup of sugar is added. The sauce is now allowed to cool, and several buttered pie pans are fixed for the drying operation.

Now the cooled sauce is spread on the pie tins about ¼ inch in thickness, and allowed to dry in the sun, or in a very low temperature oven, not over 300°F. until all the moisture has been driven off, down to about 20%, or when the material turns a light brown.

Next, it is removed from the drying pans sponged over with apple brandy, or vanilla extract, then rolled into small cigar-shaped rolls and placed in a stone jar, with a tight lid, to mellow down. Usually this mellowing operation takes about a week.

Turkish Paste
Granulated Sugar   55 lb.

Standardized Invert

Sugar	45 lb.
Cream of Tartar	3 oz.
Cold Water	110 lb.

Thick or Heavy

Boiling Starch	11 lb.

Orange, Lemon,
Peppermint (or
other flavors with
coloring to
match)   To suit

The granulated sugar, invert sugar and cream of tartar are placed into a kettle along with approximately 30 pounds of cold water. The batch is then brought to the boiling point.

Meanwhile the remaining 80 pounds of water is mixed with the corn starch and this is added to the boiling batch, a gallon at a time, care being taken to keep the batch boiling while adding the starch suspended in the cold water. After all of the starch has been added, the batch is allowed to slowly cook for approximately five minutes, then it may be cooked more rapidly to a heavy jelly sheet. At this point, the heat is turned off, the flavoring and coloring added and mixed thoroughly. The batch is poured at once into heavy wooden trays which have first been lined with heavy wrapping paper. The surface of the Turkish paste is immediately sprinkled with a combination of 8 pounds of finely powdered sugar mixed with 2 pounds of pre-dried corn starch. The batch is allowed to set for a minimum of three days after which the sheet of jelly is removed from the tray, the wrapping paper is removed by first moistening with water and the sheet of jelly is then dusted well with the powdered sugar—starch combination. The Turkish paste is then cut into squares or oblongs and rolled in the dusting medium and allowed to set until the surface is partly dry before packing.

Fondant Icing Sugar
U. S. Patent 2,299,287

Sucrose, passing through a 300 mesh screen, is placed in a mixing machine. Into this is then injected and mixed a portion of partially or wholly inverted sugar. The finished product should contain not less than 70% sucrose nor more than 30% invert sugar. This powder, on being mixed with a correct proportion of water, produces a fondant icing of excellent quality.

Butterscotch Sauce

Egg Yolks	0.85
Butter	2.5
Corn Syrup	5.0
Water	2.5
Brown Sugar	5.6

Beat egg yolks slightly, add remaining ingredients and cook in steam-jacketed kettle, stirring frequently until a thick syrup forms.

Starch-Albumen Marshmellow

Granulated Sugar	25	lb.
Modified Starch		
(50–60 fluidity)	7	lb.
Cold Water	56	lb.
Cream of Tartar	2	oz.
Corn Syrup	50	lb.
Standardized Invert		
Sugar	25	lb.
Albumen—Dissolved		
in	1¼	lb.

Water	2½ lb.
Flavoring and	
Coloring	As desired

The sugar, corn starch and all of the cold water are placed into a kettle. The batch is mixed well and gradually brought to the boiling point, then the cream of tartar, corn syrup and standardized invert are added, the batch being continually stirred until the temperature of the batch registers 222°F. The cooked batch is immediately transferred to a marshmallow beater, the beater started and the albumen solution gradually added and the batch beaten until it attains a volume of approximately 5 pounds to the gallon. Flavoring and coloring are added and the batch cast into starch impressions which have been heated to approximately 110–115°F. Dry corn starch is sprinkled over the surface and the cast marshmallows allowed to remain in the starch for from one to three days, or, if a drying room is available, the starch trays containing the marshmallows may be transferred to a drying room tempered to 115°F. and allowed to remain for a minimum of ten hours and as long as twenty-four hours.

The marshmallows are then removed from the starch and coated with chocolate or caramel or hot butterscotch or they may be packed in a mixture of 2 parts of powdered corn starch to 8 pounds of powdered sugar.

Reducing Viscosity of Chocolate

Minimum viscosity of chocolate is reached at 89.6°F. by the addition of 0.1% triethanolamine.

Glycerin-Gelatin Bases

Glycerin-gelatin combinations for the bases for a wide variety of candies including gum drops, jujubes, pastilles, throat lozenges, cough drops, "tickle stoppers" and the like.

Sometimes vegetable gums are included in these glycerin-gelatin bases, as in the following example:

Granulated Edible	
Gelatin	20
Glycerin	30
Sugar	15
Tartaric Acid	3
Mucilage of Acacia	
	To make 100

The glycerin, sugar and acid are mixed with the mucilage, and the gelatin stirred in. After about three minutes, heat on a water bath until the gelatin is dissolved, usually within 5 minutes.

Heat-Resistant Candy and Cake
Cocoa Coating
(Dark Vanilla)

Hydrogenated Hard	
Butter (98°)	30 lb.
Sugar	48 lb.
Cocoa (Under 10%	
Fat)	22 lb.
Vanillin	½ oz.
Ethyl Vanillin	¼ oz.
Chocolate Flavor	½ oz.
Salt	1 oz.
Soybean Lecithin	5 oz.

The sugar, cocoa, flavors, and salt are mixed with about ⅔ of the melted hard butter in a dough-mixer. The mixture is then refined in a chocolate refiner and placed in a chocolate kettle. The rest of the melted hard butter is then added, and the lecithin dissolved in an equal weight of the melted hard butter. The coating is now mixed until blended for 1 to 3 hours at a temperature between 100 and 120°F.

Heat-Resistant Candy and Cake Cocoa Coating
(Light Colored Milk)

Hydrogenated Hard Butter (98°)	30	lb.
Sugar	52	lb.
Cocoa (Under 10% Fat)	6	lb.
Buttermilk Powder	12	lb.
Vanillin	½	oz.
Ethyl Vanillin	¼	oz.
Chocolate Flavor	½	oz.
Milk Flavor	1½	oz.
Salt	1	oz.
Soybean Lecithin	5	oz.

The sugar, buttermilk powder, cocoa, flavors, and salt are mixed with about ⅔ of the melted butter in a dough-mixer. The mixture is then refined in a chocolate refiner and placed in a chocolate kettle. The rest of the melted hard butter is then added and the lecithin dissolved in an equal weight of melted hard butter. The coating is now mixed until blended for 1 to 3 hours at a temperature between 100 and 110°F.

These chocolate coatings contain no added cocoa butter, but have the advantage that they do not readily melt or turn white. They are widely used to coat candy and biscuits, and particularly 5 cent bars. The candy so coated can withstand summer or tropical temperatures without "blooming" or turning white.

Bakers' Icings, Glazes, and Washes

Citrus pectin produces a very superior glaze at low cost for all types of bakery goods. It is convenient to use in the largest or smallest bake shops for icings, glazes, and washes. For this purpose, it is most conveniently used in syrup form, the syrup being prepared in the following manner:

Pectin Syrup

Water	30	lb.
100 Grade Exchange Citrus Pectin	8	oz.
Glucose (43° Bé.)	20	lb.
Granulated Sugar	50	lb.

Put the water in a kettle and heat hot, almost boiling. Thoroughly mix the 8 ounces of pectin with about 4 pounds of the granulated sugar and add this pectin-sugar mixture to the water as it is being stirred vigorously with a baker's wire whip. Continue to stir and heat to boiling. Now add the 20 pounds of glucose and the remainder of the sugar (46 pounds). Heat to boiling again and boil for about 1 minute.

This is the stock syrup which may be stored and used as needed in the bakery.

Pectin syrup is used hot for glazing and in boiled icings. In those icings where volume is desired, such as in fat icings, pectin syrup should be used cold.

When used as a glaze for fruit tarts, open-face fruit pies, fruit cakes, etc., pectin syrup is acidified with a small quantity of 50% citric acid solution. It is used in this manner also as a wash for sweet rolls, coffee cakes, and other raised goods where a bright, lustrous finish is desired.

Coffee Cakes, Danish Pastry, Sweet Rolls, etc.

To every gallon (128 fl. oz.) of pectin syrup, add from a pint (16 fl. oz.) to a quart (32 fl. oz.) of hot water and mix together thoroughly. Apply this wash to the hot, sweet goods immediately or as soon as possible after they are taken from the oven. In this way the wash should be dry about the time the baked goods are cool. They are then ready for distribution or application of the icing.

Icing

Use the fondant type icing, given later, thinned-down to the desired consistency.

Glazed Topping for Nut Butter Coffee Cake
(To be applied to dough before proofing)

Exchange Citrus Pectin Syrup	3	lb.
Butter and Shortening	1½	lb.
Salt	½	oz.
Vanilla	As desired	
50% Citric Acid Solution	1	tsp.
Ground or Sliced Nuts	1	lb.

In all of these recipes a teaspoon is considered equivalent in volume to 5 ml.

Cream light and spread on coffee cake dough. This topping will bake with a beautiful luster and will not sink into the cake. Other spices may be added if desired.

Fruit and Berry Glazes

In order that the pectin syrup may set firmly to a beautiful lasting glaze, the addition of just the right amount of 50% citric acid solution is necessary. This knowledge of the correct amount is quickly acquired.

For berries and cherries, use 1 teaspoon (5 ml.) 50% citric acid solution to 3 pints (48 fl. oz.) of pectin syrup which should be at a temperature of about 190°F. (88°C.). Stir in well and start glazing either with a brush or a spoon. Stir the syrup gently each time you dip into it. This keeps it from setting and becoming too thick.

For fruits that are inclined to be moist, such as bananas, oranges, peaches, apples, and pears, 1 teaspoon (5 ml.) of 50% citric acid solution to each pint of pectin syrup is recommended. Stir well, and then apply in the same manner. Stir gently every time you dip into it.

Pectin syrup may be colored and flavored in any manner desired. Always use good flavors and certified colors. Add the desired amount of flavor and color to each batch before stirring in the 50% citric acid solution.

When working on moist fruits, permit the jelly to thicken a little before application. This will prevent it from running over the sides. Be careful not to stir the jelly too vigorously as this will tend to incorporate air bubbles.

Strawberry, and Other Short Cakes

Upon one layer of regular short

cake, arrange the strawberries (or any other kind of fruit or berry that you desire to use). Now, color and flavor the amount of pectin syrup necessary to cover all the cakes that you are going to make. Cover the berries with pectin syrup to the desired thickness using a spoon. It is important to gently stir the syrup every time you dip into it. The layer of jelly may look a little lumpy, but when you put on the top layer, cover this with whipped cream and when the whole cake is cut, the cut sides will look as smooth as if it were one solid piece.

Another method is to cover the bottom layer of the cake with the pectin syrup prepared as directed above and then to arrange the fruit or berries in the syrup.

Pecan Bun Glaze

A beautiful, durable glaze can be obtained either by washing the buns with pectin syrup containing citric acid solution as soon as they are taken from the oven, or, another successful method is to add about ½ pound of pectin syrup to every 5 to 8 pounds of regular paste with which the pans are lined before putting in the dough. To prevent the drying-out of soft cakes when cut, wash the sides with pectin syrup containing citric acid solution. For this purpose use ¼ teaspoonful (1¼ ml.) 50% citric acid solution to each pint (16 fl. oz.) of pectin syrup.

Piping Jelly

Pectin syrup with citric acid solution added may be used for decorating tarts, cookies, cakes, etc.

First, color and flavor the required amount of pectin syrup and after the 50% citric acid solution is added, it may be tubed or piped. Beautiful results are obtained in this manner.

Icings
Cream Icing

Icing Sugar	10	lb.
Shortening and/or Butter	2	lb.
Pectin Syrup	1½	lb.
Whole Eggs or Egg Whites	½	lb.
Cold Water	1	lb.
Milk Powder (Optional)	½	lb.
Salt	1	oz.
50% Citric Acid Solution	5	cc.
Vanilla	As desired	
Water (See directions below)	If needed	

Cream about one-quarter of the sugar, all the shortening, salt, and milk powder until light; then add slowly and cream in the eggs, then the pectin syrup containing the citric acid solution, and finally the flavor, balance of sugar, and whatever water is needed to give the desired consistency and lightness.

The consistency and texture of this icing can be changed either by adding to or decreasing the water content. More shortening, milk powder, and eggs may be used to give a richer, lighter texture.

These icings may be kept indefinitely under a moistened cloth or in a covered container.

Chocolate Icing

To the above may be added approximately 2 pounds of cocoa powder and whatever additional water is needed to give you the desired flavor, color, and consistency.

Icings with Fruits, Nut, etc.

Delicious varieties of icings can be made by the addition of ground nuts, fruits, jams, and fresh berries to the above. When using any fruits or berries that contain excessive amounts of moisture, add approximately 1 pound of icing sugar to every pound of fruit to take up the additional moisture, or you may reduce the water content of the original formula.

Boiled Chocolate Icing
Formula No. 1

Granulated Sugar	7½	lb.
Glucose (43° Bé.)	4	lb.
Water	4	lb.
Gelatin	1 to 1½	lb.
Icing Sugar	7½	lb.
Cocoa Powder	4	lb.
Pectin Syrup	2	lb.
50% Citric Acid Solution	½	fl. oz.

Heat granulated sugar, glucose, and 2 pounds of water to 240°F. (at sea level). Thoroughly dissolve the gelatin in the other 2 pounds of water, and together with the icing sugar and cocoa powder start to whip, so that this is about three-quarters up when the hot syrup is poured in slowly. Then add the pectin syrup containing the citric acid solution and whip until full volume is obtained.

This icing is particularly adaptable for angel food cakes and all other types where a permanent high gloss is desired.

No. 2

Granulated Sugar	8	lb.
Glucose (43° Bé.)	1¾	lb.
Water	3½	lb.
Pectin Syrup	1¾	lb.
50% Citric Acid Solution	5	cc.
Cocoa Powder	2¼	lb.
Shortening	1¾	lb.
Salt	½	oz.
Icing Sugar	5	oz.
Vanilla	As desired	

Place the pectin syrup, cocoa powder, shortening, and salt in the bowl and cream together. Cook the granulated sugar and water, and when it has reached 235°F. pour it into the bowl, together with the icing sugar, vanilla, and citric acid solution. Cream for 5 minutes and apply it while warm.

Pineapple Fluff or Fruit Icing
(Very Light)

Crushed Pineapple (Including Juice)	8	lb.
Egg Whites	3½	lb.
Pectin Syrup	16	lb.
Granulated Sugar	36	lb.
50% Citric Acid Solution	2	fl. oz.

Place all of the ingredients in the mixing bowl and whip until stiff.

Other delicious varieties may be made with the addition of various fresh and canned fruits and berries. For quicker drying, simply add a little more icing sugar. After icing, the cakes may be sprinkled with ground nuts, if desired.

Fondant Type Icing

This is a stock icing and may be used in many different ways. It is extremely popular for use on coffee cakes, sweet rolls, angel food, and Danish pastry; also as an icing and a doughnut glaze. It will not chip, crumble, or crystallize, and will remain lustrous for a long time.

The uses of this stock icing are regulated by the addition of water; and it may be colored, flavored, and mixed with fruit and nuts as desired.

XXXX Sugar	13	lb.
Pectin Syrup	1½	lb.
Glucose (43° Bé.)	½	lb.
Water (Hot or Cold)	1½	lb.
50% Citric Acid Solution	7½	cc.
Salt	As desired	
Vanilla	As desired	

Place all of the above ingredients with the exception of the citric acid solution and the flavor into the mixing bowl and cream until perfectly smooth. Now add the flavor and citric acid solution. Continue to stir for 2 or 3 minutes.

White Icing

Granulated Sugar	12	lb.
Shortening	4	lb.
Butter	2½	lb.
Water	3¾	lb.
Pectin Syrup	5	lb.
Icing Sugar	35	lb.
50% Citric Acid Solution	1	fl. oz.
Vanilla	As desired	

Heat the granulated sugar, shortening, butter, water, and pectin syrup to about 200°F. Then pour into the mixing bowl together with the icing sugar, citric acid solution, and flavor. Cream it slowly for about ten minutes.

Other varieties of many of the above icings may be made with the addition of maple flavor, walnut flavor, chocolate, cocoa, orange, and lemon oils, or extracts, etc.

Coffee Cake Icing

Icing Sugar	9	lb.
Pectin Syrup	1	lb.
Egg Whites	3	oz.
Water	1¼	lb.
Flavor	As desired	

Place pectin syrup, egg whites, and ½ pound of water in mixing bowl and whip until light. Add half the sugar and water, and when that is thoroughly mixed in, add the balance of sugar, water, and flavor. Cream until light. This icing will remain lustrous and soft for a long time. It will blend beautifully with chocolate, cocoa, fruits, and nuts for other delicious varieties.

Marshmallow Icing
Formula No. 1

XXXX Icing Sugar	10	lb.
Glucose (43° Bé.)	4	lb.
Pectin Syrup	4	lb.
Hot Water	4¼	pt.
Gelatin	6	oz.
Salt	¼	oz.
Vanilla	As desired	

Completely dissolve the gelatin in hot water. Place all of the ingredients, excepting the vanilla, in the bowl and whip briskly until light and fluffy. 10 to 12 minutes will be sufficient under normal conditions. Just before it is finished, pour in the vanilla. This marshmallow icing will set firm, but will be short and tender and very light. It is very good for fillers as well as for toppings.

A delicious chocolate marshmallow icing may be made by adding to the above when it is about three-quarters up, about 1½ pounds of cocoa powder dissolved in a little warm water.

No. 2

Granulated Sugar	7
Glucose (43° Bé.)	2½
Water	3
XXXX Icing Sugar	2
Egg Whites	3½
Pectin Syrup	3
Vanilla	As desired
Salt	As desired

Boil the granulated sugar and water to 240°F. (or 28°F. above the boiling point of water at your factory). When this just begins to boil, start in whipping the pectin syrup, glucose, XXXX sugar, egg whites, and salt, so that it will be about three-quarters up when you start slowly pouring in the hot syrup. While whipping at high speed, just before finishing, add the vanilla. Cocoa powder dissolved in a little warm water may be added to the above to give a delicious chocolate marshmallow icing.

No. 3

Granulated Sugar	10
Glucose (43° Bé.)	5½
Water	3½
Gelatin	2
Hot Water	½
Pectin Syrup	5
XXXX Sugar	10
Egg Whites	4
Salt	As desired
Vanilla	As desired

Heat the granulated sugar, glucose, and water to 240°F. At the same time, thoroughly dissolve the gelatin in the hot water. Place this, together with the pectin syrup, XXXX sugar, egg whites, and salt in the bowl and whip so that it will be about three-quarters up when the syrup has reached 240°F. Slowly add the hot syrup to the batch with vigorous stirring. Just before it is finished, pour in the vanilla. This marshmallow is particularly adaptable for fillings and toppings.

No. 4

*Gelatin	10 oz.
Water, Cold	2 lb.
Flavor and Color	As desired
†Syrup	31 lb.

Soak the gelatin in the cold water, then heat until fluid but not over 140°F. Add the fluid gelatin to the special syrup and beat to marshmallow consistency. Just before the beating is finished, incorporate the flavoring and color.

† Syrup

The syrup must be varied to take care of summer and winter conditions, so two types of formula are given here:

	Summer	Winter
Water	36	33
Granulated Sugar	90	63
Standardized Invert		
Sugar	90	120
Corn Syrup	60	60

Dissolve the sugar in the water by heating. Simply using hot water will be sufficient for the winter formula, but the increased sugar in the summer formula will require heating to 150°F. Add the invert sugar and corn syrup to the warm dissolved sugar. Stir until thoroughly mixed. This syrup

---

* Increase to 11 oz. in summer.

may be made in any quantity desired and used as needed.

No. 5

Cane Sugar Granulated	22
Dextrose, Anhydrous	22
Corn Syrup	32
Gelatin (150 Bloom)	2
Liquid Egg White	1
Water	21
*Vanillin	To flavor

Soak the gelatin in about half the water. Dissolve the sugar in the remaining water, then add corn syrup and heat to about 230°F. Cool to about 180°F. Add the wetted gelatin mixed with the egg white, by agitating thoroughly. Whip until the desired marshmallow consistency is obtained. Flavor during whipping.

Peach Fluff Icing

Granulated Sugar	4 lb. 8 oz.
Clear Syrup	7 lb.
Water	2 lb.
Egg Whites	3 lb.
Cream of Tartar	½ oz.
Salt	1 oz.
Peach Base	6 lb. 8 oz.

Boil together the granulated sugar, syrup and water to 238°F. Beat the egg whites, cream of tartar and salt to a wet peak and add slowly to the boiled syrup. Add the peach base slowly. Beat to a dry peak.

Peach Base

Peaches, Canned or Fresh	6 lb.
Clear Syrup	3 lb.
Water or Peach Juice	1 lb.
Lemon Juice	¼ oz.

* Use about 2 oz. for 100 lb. of mix.

Cornstarch	3 oz.
Egg Color	$\frac{1}{16}$ oz.

Bring the fruit and clear syrup to a boil, add the other ingredients and cook until clear.

Malted Chocolate Icing

Icing Sugar	7½ lb.
Glucose	1½ lb.
Hot Milk	1½ lb.
Powdered Malt	3 oz.
Melted Shortening	1½ lb.
Cocoa Powder	3 lb.
*Egg Whip	1½ lb.

Place sugar, glucose, milk, and powdered malt in machine and mix together thoroughly. Add the melted shortening, then the cocoa powder, and finally the egg whip.

This icing will be light in texture and can be used the following day by stirring it up thoroughly.

*Egg Whip

Pectin Syrup	16 lb.
Egg Whites	15 lb.
Orange Flower Water	5 cc.

Place ingredients in bowl of beater and whip vigorously until stiff.

This is particularly adaptable for lightening all types of icings.

Whipped Cream

Fresh Milk	2
Heavy Cream	4
Icing Sugar	1
Egg Whip	2

Whip up the cream, milk, and icing sugar until it has attained its maximum volume. Then fold in the egg whip by hand. This whipped cream will retain its body for a much longer period than ordinary whipped cream during all kinds of weather.

For economy, the use of milk is suggested above. It may be omitted if a richer whipped cream is desired.

Buttercream Filling

Shortening and		
Butter	4½	lb.
Icing Sugar	9	lb.
Pectin Syrup	2¼	lb.
Whole Eggs	½	lb.
Skim Milk Powder	6	oz.
50% Citric Acid		
Solution	5	cc.
Salt	As desired	
Vanilla	As desired	

First cream light the shortening and butter, then add slowly one-half of the icing sugar, which should be sifted. Slowly add the balance of the icing sugar, milk powder, and salt which have been sifted together. Cream in the eggs, pectin syrup, citric acid solution, and vanilla.

Cookie Flat Icings

Pectin Syrup	1¾	lb.
50% Citric Acid		
Solution	7½	cc.
Gelatin	2	oz.
Water	2¾	lb.
Granulated Sugar	7¾	lb.
Icing Sugar	¾	lb.
Salt	As desired	
Vanilla	As desired	

Cook granulated sugar, salt, and 2 pounds of the water to 230°F. (or 18°F. above the boiling point of water at your factory). Dissolve the gelatin in the balance of the water. Now place the pectin syrup, citric acid solution, and the dissolved gelatin into the mixing bowl together with the boiled syrup and, after mixing together thoroughly, permit this to cool to about 115°F. Then add icing sugar and flavor and beat at high speed for approximately 5 minutes. The icing should be applied to the cookies while it is warm. This icing will blend very nicely with all flavors.

Buttercream Icing

Fondant Type Icing	6	lb.
Water	2¾	lb.
Butter	6	lb.
Shortening	12	lb.
Salt	2	oz.
Icing Sugar	16	lb.
Milk Powder	4	lb.
Pectin Syrup	2	lb.
50% Citric Acid		
Solution	10	cc.
Egg Whites	8	oz.
Vanilla	As desired	

Cream fondant, butter, shortening, salt, and one-third of the sugar until light. Then in the following order, cream in the egg whites, pectin syrup, balance of the sugar, water, milk powder, citric acid solution, and flavor.

"Buttercream"

Shortening or		
Sweet Butter	3½	lb.
Icing Sugar	5	lb.
Pectin Syrup	½	lb.
50% Citric Acid		
Solution	¼	tsp.
Whole Eggs or		
Egg Whites	8	oz.
Vanilla	As desired	
Salt	As desired	

Cream light the icing sugar and shortening; then add slowly the eggs, pectin syrup, citric acid solution, salt, and vanilla.

Doughnut Icing and Glaze

For French doughnuts, use the

fondant type icing, thinned down with water to the proper consistency. For ordinary doughnuts, do the same but use more water.

Home-Made Boiled Icing

Pectin Syrup	3
Egg Whites	6
Granulated Sugar	18
Glucose	2
Water	4½
Salt	As desired
Vanilla	As desired

Cook sugar, glucose, water, and salt to 240°F. While cooking, beat together the egg whites and pectin syrup so they will be about three-quarters up when the syrup has reached 240°F.; then pour in the boiling syrup slowly and beat until stiff. Add the vanilla. This icing is very suitable for fillers or toppings.

If a quicker-drying icing is desired, the addition of 1 to 2 pounds of finely-sifted sugar is recommended to be added toward the finish of the process. Use while warm.

Chocolate Fudge

Butter or Shortening	6	oz.
Chocolate	1½	lb.
Granulated Sugar	1½	lb.
Glucose	4	oz.
Icing Sugar	6½	lb.
Pectin Syrup	1½	lb.
Egg Whites	3	oz.
Fresh Milk	1½	lb.
50% Citric Acid Solution	2½	cc.
Vanilla	As desired	
Salt	As desired	

Bring to the boiling point, the butter, granulated sugar, chocolate, milk, pectin syrup, citric acid solution and salt; then cool this to about 100°F. and add balance of the ingredients; cream slowly until smooth.

Biscuit Coatings

Lemon Coating

98°F. M.P. Special Hard Vegetable Butter	35	lb.
Dry Skim Milk	11	lb.
4X Powdered Sugar	54	lb.
Oil Soluble Yellow Color	2	oz.
Terpeneless Lemon Oil	2	oz.

Malt Milk Coating

98°F. M.P. Special Hard Vegetable Butter	35	lb.
Dry Skim Milk	20	lb.
4X Powdered Sugar	44	lb.
*Spray Malt	1	lb.
Fine Salt	3	oz.
Vanillin	½	oz.

Orange Coating

98°F. M.P. Special Hard Vegetable Butter	35	lb.
Dry Skim Milk	11	lb.
4X Powdered Sugar	54	lb.
Oil Soluble Orange Color	3	oz.
Terpeneless Orange Oil	2	oz.

Dark Chocolate Coating

98°F. M.P. Special Hard Vegetable Butter	35	lb.

* More or less. If more, reduce sugar content to equal amount of additional spray malt.

*Cocoa Powder	18	lb.
Dry Skim Milk	5	lb.
4X Powdered Sugar	42	lb.
Fine Salt	3	oz.
Vanillin	1	oz.

White Coating
98°F. M.P. Special

Hard Vegetable Butter	35	lb.
Dry Skim Milk	20	lb.
4X Powdered Sugar	45	lb.
Fine Salt	3	oz.
Vanillin	½	oz.

Custard Icing

Butter	20	lb.
Emulsifying Shortening	20	lb.
Whole Milk Powder	12	lb.
4X Sugar	30	lb.
Clear Syrup	15	lb.
Whole Eggs	8	lb.
Vanilla	4	oz.
Boiled Icing	35	lb.

Cream the butter, shortening, milk powder, sugar and clear syrup together until light. Add the eggs slowly, then the flavoring. Finally add the boiled icing.

This icing can be varied by adding 1 lb. 12 oz. of crushed pineapple or chopped glacé cherries.

Boiled Icing or Base

Egg Whites	8	lb.
Salt	2¼	oz.
Cream of Tartar	4	oz.
Clear Syrup	60	lb.
Gelatin, 200 Bloom	8	oz.
Water	1 lb. 8	oz.
Vanilla	3	oz.

Beat the egg whites, salt and cream of tartar to a wet peak. Cook

* Not over 10% cocoa butter fat content.

the clear syrup to 230°F. Add the hot syrup slowly to the beaten egg whites. Dissolve the gelatin in the hot water, add the vanilla, then add this mixture to the one previously prepared. Beat to a dry peak.

This can be used as an icing or as a stock or base to extend other icings.

Orange Crumb Topping

Cake Flour	1	lb.
Brown Sugar	1	lb.
Cinnamon	¼	oz.
Salt	¼	oz.
Orange Gratings	¾	oz.
Orange Juice	3	oz.
Shortening	5	oz.

Mix together all the dry ingredients, add the roange juice and the melted shortening. Mix until crumbly.

Special Icing

*Powdered Sugar	9 lb. 6	oz.
Shortening	2 lb. 8	oz.
Special Stock (See below)	3 lb. 2	oz.
Salt	1	oz.
Flavor	To taste	
Cake Flour	1 lb. 8	oz.

Cream one-third of the sugar (3 lb. 2 oz.) with the shortening until light. Add the special stock alternately with the balance of the sugar, salt and flavoring. Then add the cake flour. Whip until light.

Special Stock

Granulated Sugar	2 lb. 8	oz.
Water	6 lb. 4	oz.
Cornstarch	8	oz.

* Corn sugar may be used to the extent of 25 percent replacement of powdered sugar.

Bring sugar to boil with 5 lb. of water.

Dissolve the cornstarch in 1 lb. 4 oz. of water, add to the sugar mixture and cook until clear. Use cold.

This icing will carry fruit. For a chocolate icing use 1 lb. of cocoa or chocolate and reduce the powdered sugar by a corresponding amount. Add the cocoa or chocolate when creaming the sugar and shortening.

French Custard Cream Filling

Liquid Milk	4 lb.
Granulated Sugar	1 lb. 6 oz.
Emulsifying Shortening	3 lb. 2 oz.
Egg Yolk	12 oz.
Butter	10 oz.
Cornstarch	5 oz.
Salt	¼ oz.
Vanilla	1 oz.

Bring to a boil 3 lb. 12 oz. of liquid milk, 12 oz. of sugar, 2 oz. of shortening and 2 oz. of butter. Mix egg yolk, the remainder of the granulated sugar, the cornstarch, salt and the remainder of the milk, and add this mixture gradually to the boiling milk solution, stirring constantly with a wire whip until thick. Remove from fire, stir in the vanilla and set aside to cool.

Mix the shortening and butter in the machine bowl until smooth, and at 70 to 75°F. Add the cool, cooked mixture gradually and mix at medium speed until light and smooth.

Be sure that the cooked cream is cold before it is added to the shortening. This filling must be kept covered and under refrigeration when not in use.

Lemon Cream Pie
Formula No. 1

Granulated Sugar	1 lb.
Cornstarch	6 oz.
Egg Yolks	8 oz.

No. 2

Water	4 lb.
Granulated Sugar	1 lb.
Salt	½ oz.
Lemon Juice	10 oz.
Butter	4 oz.

Dry mix the first portion of sugar and cornstarch. Add the egg yolks and mix until smooth. Heat the water, the second portion of sugar, the salt, and the lemon juice until the solution reaches its boiling point. Pour a cup of the hot solution into the starch mixture and mix until smooth. Pour the starch mixture into the boiling solution, while stirring vigorously. As soon as the mixture begins to thicken, remove from the fire. Add the butter and stir until it is melted. Fill into prebaked shells at once. Cool.

Top with either meringue or whipped cream. Do not overcook.

Bakers' Flavored Pectin Jelly Powder

Many bakers prefer to make their own jellies and for this purpose a flavored pectin jelly powder can be produced with exchange citrus pectin such that 1 pound of the preparation will be sufficient for 30 pounds of the finished jelly. The formula and directions are given below. For this purpose the rapid set pectin is recommended.

100 Grade Exchange Citrus Pectin	4
Powdered Tartaric Acid	1

Artificial Flavor	As desired
Certified Color	As desired
Powdered Refined	
Corn Sugar	As desired

Mix the certified color and flavor with a part of the powdered corn sugar; then mix thoroughly with the pectin, tartaric acid, and the remainder of the corn sugar.

Put 1½ gallons of water in a kettle and heat hot (190°F. or 88°C.).

Add 16 ounces of the flavored pectin jelly powder to the hot water as it is being stirred vigorously with a wire whip. Continue to stir and heat to boiling, then boil vigorously for ½ minute.

Now add 18 pounds of granulated sugar, stir with the wire whip until the sugar is all dissolved, remove from the fire quickly, and pour at once into #10 tins or a 30-pound pail. This will make approximately 30 pounds of finished jelly.

Do not disturb the jelly while it is cooling.

Note: If the jelly batch made according to the above formula sets too rapidly, reduce the quantity of tartaric acid to ¾ ounce or lower. If a more rapid setting jelly is desired, increase the amount of tartaric acid to 1½ ounces. Jellies that have set too quickly have a "mushy" or pudding type consistency. They contain small pieces of very weak jelly usually surrounded by much syrup.

Powdered Pie Fillings

Kiln Dried Wheat		
Flour	20	lb.
Kiln Dried Corn		
Starch	30	lb.

Salt	2	lb.
100 Grade Exchange		
Citrus Pectin	3½	lb.
Dried Egg Yolk		
(Powdered)	12	lb.
Pure Concentrated		
Lemon Juice	2	gal.
Sugar (Cane or		
Beet)	155	lb.
15% Lemon		
Oil-Pectin		
Emulsion	As desired	
Certified Color	As desired	

Thoroughly mix the flour, corn starch, salt, pectin, and dried egg yolk. As these dry ingredients are being stirred, gradually add the lemon juice in small amounts so as to avoid the accumulation of large lumps in the batch. Continue to stir until the batch is uniform. Then add the granulated sugar, lemon oil emulsion, and certified color. Stir the mixture until it is uniform and dry.

Stir 1 pound of the powder into 1 quart of cold water and heat in a double boiler until it thickens. Additional sugar and thickener may be added if desired.

Pectin Solutions

The preparation of pectin solutions with exchange citrus pectin may be accomplished by any of the four more common methods described below. The consistency may be readily changed as desired by varying the amount of pectin used. Each formula makes about 1 gallon (128 fl. oz.) of the finished solution.

Formula No. 1

Used where the presence of a little glycerin in the solution is not objectionable.

100 Grade Ex-
    change Citrus
    Pectin   8 oz.
Glycerin   20 fl. oz.
Warm Water
    (180°F.)   108 fl. oz.

Thoroughly mix the pectin with
the glycerin by the aid of a spatula
and allow a few minutes for soaking. Then mix again to insure
absence of lumps. While the
pectin-glycerin mixture is rapidly
stirred, preferably with a mechanical agitator, the warm water is
added and the stirring is continued until dispersion is complete,
that is, until no particles are visible.

Note: Corn syrup may be used
in place of glycerin if desired.

No. 2

Used where the presence of a
little alcohol in the solution is not
objectionable.

100 Grade Ex-
    change Citrus
Pectin   8 oz.
190 Proof Ethyl
    Alcohol   20 fl. oz.
Warm Water
    (180°F.)   108 fl. oz.

Thoroughly mix the pectin with
the alcohol and then add all at
once to the warm water which is
being violently stirred. Continue
the stirring until dispersion is
complete, that is, until no particles
are visible.

No. 3

Used where glycerin or corn
syrup would be objectionable, but
sucrose (cane or beet sugar) would
not.

100 Grade Ex-
    change Citrus
    Pectin   8 oz.

Granulated Sugar   16 oz.
Lukewarm Water
    (about 100°F.)   128 fl. oz.

Thoroughly mix the dry pectin
and sugar, and then add the mixture all at once to the water in a
bucket or open pan, taking care
that the mixture is totally sub-
mersed. Then immediately stir
rapidly, preferably with a mechanical agitator, until dispersion
is complete, that is, until no particles are visible. Heat to 180°F.
with continuous stirring. Corn
sugar may be substituted for cane
or beet sugar if desired, and the
stirring of the solution vigorously
by hand, using a baker's wire
whip, will often suffice.

No. 4

Used where pectin alone is desired in solution without any other
ingredients (except for a small
amount of refined corn sugar
added to the pectin at the time of
manufacture for standardization
purposes).

100 Grade Ex-
    change Citrus
    Pectin   8 oz.
Cold Water
    (70°F.)   128 fl. oz,

Stir the water with a high speed
mechanical agitator so that a cone
or funnel is formed in the water
about the agitator shaft. Slowly
sift the pectin into the water at
this point. When all the pectin
has been added, taking care not to
add the pectin too rapidly, continue the stirring for a few minutes. Then heat to 180°F. while
stirring by hand.

Note: The above procedures, if
followed exactly, yield homogeneous liquid pectins. Occasionally a

few lumps may be formed if any part of the liquid is not sufficiently agitated during dispersion, therefore, it is advisable to strain the liquid preparation through a screen or cloth to remove any of those particles. A very thick and viscous pectin solution may sometimes be referred to as a "mucilage" or "slime".

Lemon Cheese
(Lemon Curd)

Cane or Beet Sugar	24	oz.
Glucose (43° Bé.)		
or Corn Syrup	30	oz.
Pure Concentrated		
Lemon Juice	2¾	oz.
Oleomargarine	2	oz.
Potato Starch	3¼	oz.
Whole Eggs	1¼	oz.
100 Grade Exchange		
Citrus Pectin	¼	oz.
Salt	As desired	
Cold Pressed		
Lemon Oil	As desired	
Certified Food		
Color	As desired	
Water	16	fl. oz.

(1) Thoroughly mix the pectin with about 3 ounces of granulated sugar.

(2) Heat the water hot (almost boiling), then add the pectin-sugar mixture, and stir until completely dissolved. Allow to cool.

(3) Add the pectin solution slowly to the starch, stirring thoroughly to obtain a uniform suspension. The mixture is then heated and the remainder of the sugar, the glucose, oleomargarine, salt, and whipped eggs are added with stirring. The mixture is boiled for about 3 minutes.

(4) Water is then added to make up for the evaporation and to regulate consistency of the final product.

(5) Allow the batch to cool and when a temperature of 185°F. is reached, the lemon juice, color, and cold pressed oil of lemon are added, stirring to obtain thorough mixing throughout the batch.

(6) The product is then filled into containers.

Meringue

Cold Water	2	pt.
Powdered Egg		
Albumen	1	oz.
100 Grade Exchange		
Citrus Pectin		
#436	½	oz.
Granulated Sugar	2¾	lb.
Citric Acid U.S.P.	3	g.
Flavor	If desired	

(1) Thoroughly mix the pectin and albumen with about 8 ounces of granulated sugar. Add this mixture to the water as the latter is being stirred vigorously with a beater or baker's wire whip.

(2) Continue to stir or beat until the pectin and albumen are completely dissolved, then add the remainder of the sugar, the citric acid, and flavor.

(3) Stir until the sugar is completely dissolved and use as desired.

Custard-Pie Base

This is an excellent base for any rich flavor such as caramel, maple, butterscotch, banana, vanilla, pistachio, etc. Shredded cocoanut added to the batch makes a cocoanut pie filling.

Powdered Skim Milk	30
Powdered Whole Milk	10

100 Grade Exchange
  Citrus Pectin
  (1-RS/100)   5
Salt   3
Dried Egg Yolk
  (Powdered)   15
Cornstarch   40
Sugar (Cane or Beet)   124
Certified Color   If desired
Flavor   As desired

Mix the dry ingredients together being sure to produce a uniform mixture. Now add the flavor gradually in very small amounts as the dry mixture is being stirred. Continue to stir until the flavor is uniformly mixed throughout the batch. The product is then ready for packaging.

Stir 1 pound of the powder into 1 quart of cold water and heat in a double boiler until it thickens. Additional sugar may be added, if desired.

Chocolate-Pie Filler

Dark Cocoa   30 lb.
Powdered Skim
  Milk   30 lb.
Powdered Whole
  Milk   10 lb.
100 Grade Exchange
  Citrus Pectin
  (1-RS/100)   5 lb.
Salt   2 lb.
Corn Starch   40 lb.
Sugar (Cane or
  Beet)   110 lb.
* Vanillin-Cou-
  marin Solution   2 fl. oz.

Mix the dry ingredients together being sure to produce a uniform

------

* Vanillin-Coumarin Solution is prepared by dissolving 2 ounces vanillin and ½ ounce courmarin in 1 pint of 95% alcohol.

mixture. Now add the vanillin-coumarin solution gradually in very small amounts as the dry mixture is being stirred. Continue to stir until the flavor is uniformly mixed throughout the batch. The product is then ready for packaging.

Stir 1 pound of the powder into 1 quart of cold water and heat in a double boiler until it thickens. Additional sugar and thickener may be added if desired.

Fruit-Flavor Pie Filler

This is an excellent base for any fruit flavor such as Orange, Lemon, Strawberry, Cherry, Pineapple, etc.

Kiln Dried Wheat
  Flour   20   lb.
Kiln Dried Corn
  Starch   30   lb.
Salt   2   lb.
100 Grade Exchange
  Citrus Pectin
  (1-RS/100)   3½   lb.
Dried Egg Yolk
  (Powdered)   12   lb.
Citric Acid U.S.P.
  (Powdered)   5   lb.
Sugar (Cane or
  Beet)   151   lb.
Pure Concentrated
  Lemon Juice   2   qt.
Certified Color   As desired
Flavor   As desired

Thoroughly mix the flour, cornstarch, salt, pectin, and dried egg yolk. As these dry ingredients are being stirred, gradually add the 2 quarts of lemon juice so as to avoid the accumulation of large lumps in the batch. Continue to stir until the batch is uniform. Then add the granulated sugar, citric acid, certified color, and flavor. Stir the mixture until it is uniform and dry.

Stir 1 pound of the powder into 1 quart of cold water and heat in a double boiler until it thickens. Fresh or canned fruit, additional sugar, and thickener may be added if desired.

Spreading Jelly for Bakers
U. S. Patent 2,059,541

Water	67	lb.
100 Grade Exchange Citrus Pectin, Slow Set, #460	20	oz.
Cane or Beet Sugar	100	lb.
Anhydrous Disodium Phosphate	1¾	oz.
Color and Flavor	If desired	
15% Phosphoric Acid Solution	12	fl. oz.

Cook to 220–221°F. at sea level.

(1) Put water in kettle and heat hot (190°F.).

(2) Thoroughly mix the pectin with about 10 pounds of granulated sugar and add this pectin-sugar mixture to the hot water all at once as the latter is being stirred with a paddle. Continue to stir and heat to boiling. Boil vigorously for a minute.

(3) Add the remainder of the sugar and bring to a boil again.

(4) At this point add the anhydrous disodium phosphate dissolved in hot water. (This brings the pH value to about 5.5.) Boil the batch for 8 minutes. If 70–72% soluble solids are not produced by that time, add half the phosphoric acid and continue boiling. Add the other half of the phosphoric acid just before pouring.

Note: After the first trial, the amount of water can be adjusted to produce 70–72% soluble solids at the end of 8 minutes of boiling.

The degree of salviness in the final jelly can be varied by changing the amount of anhydrous dibasic sodium phosphate, by changing the boiling time, or by a combination of the two. The more of the anhydrous dibasic sodium phosphate used, the softer the jelly; or the longer the boiling time the softer the jelly. The conditions given in this formula should be followed the first time, then if a softer consistency is desired, try boiling 10 or 12 minutes. For a firmer consistency boil only 6 minutes.

The purpose of adding half of the setting acid at the end of 8 minutes of boiling if 70–72% soluble solids are not produced by that time is to slow up the action of the anhydrous dibasic sodium phosphate.

The setting time of the batch will be long enough to allow all of the setting acid to be added to the kettle just before pouring into containers. If desired, from 2½ to 3 fluid ounces of 15% phosphoric acid can be added to each 30-pound container instead of adding the acid to the kettle.

Lemon Pie Filler

Kiln Dried Wheat Flour	20	lb.
Kiln Dried Corn Starch	30	lb.
Salt	2	lb.

100 Grade Exchange Citrus Pectin (1-RS/100)	3½	lb.
Dried Egg Yolk (Powdered)	12	lb.
Pure Concentrated Lemon Juice	2	gal.
Sugar (Cane or Beet)	155	lb.
15% Lemon Oil-Pectin Emulsion	As desired	
Certified Color	As desired	

Thoroughly mix the flour, corn starch, salt, pectin, and dried egg yolk. As these dry ingredients are being stirred, gradually add the 2 gallons lemon juice in small amounts so as to avoid the accumulation of large lumps in the batch. Continue to stir until the batch is uniform. Then add the granulated sugar, lemon oil emulsion, and certified color. Stir the mixture until it is uniform and dry.

Stir 1 pound of the powder into 1 quart of cold water and heat in a double boiler until it thickens. Additional sugar and thickener may be added if desired.

Fig Slab Jellies

Water	4½	gal.
100 Grade Exchange Rapid Set Citrus Pectin (1-RS/100)	1½	lb.
Dried Figs (Ground) or Fig Paste	30	lb.
Glucose (43° Bé.)	30	lb.
Granulated Sugar	30	lb.
Cold Pressed Oil of Lemon	¾	fl. oz.
Tartaric Acid	3½	oz.

(1) Put the water in the kettle and heat hot (190°F.). (Open fire or steam-jacketed kettle may be used.)

(2) Thoroughly mix the pectin with about 10 pounds of granulated sugar and add this pectin-sugar mixture to the hot water as it is being stirred vigorously with a paddle or baker's wire whip. Continue to stir and heat to boiling. Boil vigorously for a moment.

(3) Add the dried figs (ground) or fig paste, heat to boiling, then add the glucose. Heat to boiling again.

(4) Now add the remainder of the granulated sugar and cook as rapidly as possible to 222°F. or 10°F. above the boiling point of water at your factory. (This temperature corresponds to 75% soluble solids.)

(5) Turn off the steam or remove from the fire and add the cold pressed oil of lemon. Stir thoroughly.

(6) Finally add the tartaric acid dissolved in a small amount of hot water, stirring the batch thoroughly to obtain a uniform mixture. Pour on the slab as quickly as possible and allow to congeal.

After the batch has "set", cut into desired shapes with a knife. The finished pieces may be sanded, crystallized, iced, or coated with chocolate.

Firm Jelly for Bakers

Fruit Juice	26	lb.
Water	38	lb.
100 Grade Exchange Rapid Set Citrus Pectin	27	oz.

Granulated Cane or
   Beet Sugar   105 lb.
Color and Flavor   If desired
50% Citric Acid
   Solution   3 fl. oz.
    (in each 30 lb. container)

Cooking Temperature: Cook to 222°F. at sea level.

(1) Put the fruit juice and water in a kettle and heat hot (190°F.).

(2) Thoroughly mix the pectin with about 10 pounds of granulated sugar and add this pectin-sugar mixture to the hot water all at once as the latter is being stirred. Continue to stir and heat to boiling. Boil vigorously for a minute.

(3) Now add the remainder of the sugar and cook as quickly as possible to 222°F., or 10°F. above the boiling point of water at your factory. Add color and flavor if desired.

(4) Put 3 fl. oz. of 50% citric acid solution in each 30-lb. container and then fill quickly with the hot jelly batch. Do not stir. Skim momentarily and allow the jelly to stand until it is "set" or congealed.

If the finished jelly is too firm, reduce the quantity of pectin somewhat. If the jelly "sets" or congeals too quickly, reduce the quantity of acid solution used in each 30-pound container.

Orange-Jelly Paste

Granulated Sugar	20 lb.
Corn Syrup	55 lb.
Standardized In- vert Sugar	25 lb.
Corn Starch, 50-60 Fluidity	12 lb.

Water	130 lb.
Cream of Tartar	3 oz.
Orange-Peel Marmalade	15 lb.
Orange Color	As desired

Place all of the above materials, except the orange-peel marmalade and color, in a kettle, mix well; and gradually bring to the boiling point. Cook slowly to a heavy jelly.

Add the orange-peel marmalade and color and mix thoroughly.

Pour the batch into trays lined with heavy wrapping paper, sifting a mixture of 8 parts of cornstarch and 2 parts of powdered sugar over the surface. Allow the jelly to set for 3 days or longer, then cut into bars, small squares or oblongs and roll in the powdered sugar and cornstarch mixture. Allow to set several hours before packing or coating.

The orange-peel marmalade is made by boiling together (to 222°F.) of 6 lb. of orange peel and pulp (ground), 15 lb. of water and 10 lb. each of corn syrup and standardized invert sugar. After cooling, the mixture can be ground finer for use in candies.

Yeast-Raised Doughnuts with
Potatoes

Potatoes, Boiled and Mashed		2	lb.
Sugar		1	lb.
Shortening	1 lb.	4	oz.
Salt		3	oz.
Whole Eggs		12	oz.
Vanilla		⅛	oz.
Lemon Flavor		⅛	oz.
Whole Milk	3 lb.	4	oz.
Yeast		8	oz.
Bread Flour		3½	lb.

Cake Flour 4 lb.

Cream until light the potatoes, sugar, shortening, salt, eggs and flavoring, with 1 lb. of cake flour added. Dissolve the yeast in the milk and add to this mixture, then add the flour. Mix to a medium dough but do not overmix. The dough temperature should be 86°F.

Give 15 minutes' time, then roll out to the desired thickness for cutting. Cut doughnuts to weigh approximately 1¼ oz. Place on a cloth and proof until very light. Fry in fat at a temperature of 375 to 385°F.

Non-Staling Baked Products
U. S. Patent 2,285,065

Add 0.5–4% sorbitol to dough. Ferment and bake for 30 minutes at 199°C.

Sugar Wafers

Sugar	29 lb.	8 oz.
Cake Oleo- margarine	18 lb.	12 oz.
Shortening	18 lb.	8 oz.
Salt		10 oz.
Bicarbonate of Ammonia		1½ oz.
Soda		1¾ oz.
Whole Eggs	6 lb.	10 oz.
Vanilla Extract		6 oz.
Cake Flour	56 lb.	8 oz.

Cream together the sugar, shortening, oleomargarine, soda, ammonia, vanilla and salt until light. Add the eggs about one-quarter at a time. Add the flour and fold in.

Run on a wire-cut cookie machine, using a 5-hole die. Scale 5 oz. to the dozen, making about 400 dozen from this amount.

Bake at about 400°F.

Filler for Sugar Wafers

The general practice today in making wafer fillers is to use 2 to 2¼ lb. of dry materials to 1 lb. of plastic vegetable butter for use during the summer months. For winter, this ratio is commonly changed to 1¾ to 2 lb. of dry materials to 1 lb. of plastic vegetable butter. The dry materials referred to consist of powdered sugar, dry skim milk and cocoa powder. The cocoa powder is used when a chocolate filler is desired.

A representative white filler for summer use may be made as follows:

92°F. Plastic Vegetable Butter	32 lb.
Powdered Sugar 2X or 4X	30 lb.
Dry Skim Milk	12 lb.
Fine Salt	1 oz.
Tartaric Acid	¼ oz.
Powdered Sugar 2X or 4X	25–30 lb.

Flavor delicately with vanillin crystals.

Place plastic butter, salt, dry skim milk and 30 lb. powdered sugar in mixer and mix into a smooth paste. Add vanillin (also oil soluble color if any is being used). Sprinkle in the tartaric acid. Add last portion of sugar to the mix, one scoop at a time, allowing each portion to mix in thoroughly before adding another portion.

The sugar is specified as 2X or 4X. As previously stated, the granulation of the sugar will affect volume weight of the filler. The 2X sugar will produce a fluffier, lighter filler than will the 4X sugar.

Sugar-Wafer Shell

Soft Wheat Flour		
(Medium Strength)	45	lb.
Dry Skim Milk	4½	lb.
Salt	4	oz.
Sodium Bicarbonate	2	oz.
Ammonia	2	oz.
Water	90	lb.
Corn Oil	8	oz.

Mix for a few minutes, then add—

Yellow Color	3	oz.
Carmine Red	3	oz.

Flavor as desired with orange oil. Mix about 30 minutes.

To change this formula so as to use whole dry milk and egg yolks, use the following:

Soft Wheat Flour		
(Medium Strength)	45	lb.
Whole Dry Milk	2	lb.
*Dried Egg Yolk	1	lb.
Sodium Bicarbonate	5	oz.
Ammonia	5	oz.
Water	74	lb.

Mix for a few minutes and then add—

Yellow Color	3	oz.
Carmine Red Color	3	oz.

Mix about 30 minutes.

For lemon shell use lemon oil and increase amount of yellow color to about 8 ounces.

It will be observed that the formula containing whole milk powder and dry egg yolk will cost considerably more than the one containing the dry skim milk.

Caramel Wafer Shell

Soft Wheat Flour	45	lb.
Dry Skim Milk	4	lb.
Cocoa Powder	1½	lb.

* Dissolved in about 2 gal. or 16 lb. water, and allowed to stand until egg yolk is thoroughly reconstituted.

Chocolate Brown

Vegetable Color	1	lb.
Sodium Bicarbonate	2	oz.
Ammonia	2	oz.
Water	85	lb.

Frankfurter Roll

Short Patent Flour	
(No Mixed Flour to	
Be Used)	100
Water	66
Yeast (Best Grade)	2
Yeast Food	¼
Granulated Sugar	12½
Salt	1
Milk Solids, Dry	6
Vegetable Shortening	9

1 dozen rolls bake to weigh 14 oz.

Prepared Pancake Flour

Sodium Bicarbonate	2.0
Mono-Calcium	
Phosphate	2.5
Salt	2.0
Sugar	10.0
Powdered Whole	
Egg	0.0 to 3.0
Shortening	10.0 to 15.0
Dry Skim Milk	7.0 or more
Flour	100.0

(Better quality pancakes are obtained by folding beaten whole egg into the batter that is prepared with water and this dry ingredient mixture.)

Prepared Biscuit Flours

The two formulas quoted below differ only in the baking acids used and have been found to be very desirable for prepared biscuit flours. In these formulas the amounts of ingredients are ex-expressed as parts per 100 parts flour.

Formula No. 1

Sodium Bicarbonate	1.5
Mono-Calcium Phosphate	2.0
Salt	2.0
Shortening	15.0
Dry Skim Milk	3.0–6.0

No. 2

Sodium Bicarbonate	1.5
Mono-Calcium Phosphate	0.625
Sodium Acid Pyrophosphate	1.75
Salt	2.0
Shortening	15.0
Dry Skim Milk	3.0–6.0

Baking Powder
U. S. Patent 2,138,029

Sodium Acid Carbonate	70.00
Calcium Acid Tetraphosphate	106.25
Starch	73.75

Prepared Waffle Flours
*Formula No. 1

Sodium Bicarbonate	1.5
Mono-Calcium Phosphate	1.875
Salt	2.0
Powdered Whole Egg	3.0
Sugar	12.5
Shortening	20.0
Dry Skim Milk	12.5 or more

No. 2

Sodium Bicarbonate	1.5
Mono-Calcium Phosphate	1.187
Sodium Acid Pyrophosphate	1.125
Salt	2.0
Powdered Whole Egg	3.0
Sugar	12.5
Shortening	20.0
Dry Skim Milk	12.5 or more

These formulae are to be used with 100 parts of flour.

Bleaching and Maturing Flour
U. S. Patent 2,158,588

To 90 parts of a dry mixture of calcium orthophosphate and benzoyl peroxide (85% passing 300-mesh) are added 6 parts by weight of potassium bromate; 0.5 oz. of the dry mixture is added to 196 pounds of wheat flour in a power feeding machine.

Dry Lecithin Shortening

Flour	90
Soybean Oil	5
Soybean Lecithin	5

The lecithin is dissolved in the oil by heating to 130 to 165°F. and this hot solution is sprayed in the flour agitated in a suitable mixer until uniform. The resulting mixture is a dry powder which blends readily in the dough.

The emulsifying power of the lecithin induces better mixing of the ingredients in bread dough or cake batter. The shortening effect of the fat used is increased, enabling one to cut it down at least 20%. Between 1 and 5 lb. of this dry shortening is used per 100 lb. of flour.

Malto-Dextrin
U. S. Patent 2,155,374

Maize (corn) syrup is evaporated to dryness (<2% of water) in vacuum (28 in.) at 215°C., allowed to solidify into sheets in dry air. These are broken up and fed

---

* The quality of waffles resulting from the use of these preparations will be improved by folding a beaten whole egg into the batter.

by gravity to a granulating mill and suitable containers in a closed system to which only dry air is admitted.

Substitute for Cocoanut Oil in Chocolate Icing

Results indicate that 8—15% of low melting point fats and 10—25% of high melting fats can be used successfully as substitutes for cocoanut oil. The best product is made by using a combination of 5% added hydrogenated soybean oil and 5% soybean oil. A coating satisfactory for emergency use is made using 10% cocoa, 40% powdered sugar, 35% hydrogenated soybean oil and 15% soybean oil.

Liquid Red Food Color

Amaranth F.D.C.

Red No. 2	6¼	lb.
Glycerin	10	gal.
Water	To make 50	gal.

Liquid Green Food Color

Light Green S.F.
  Yellowish F.D.&C.

Green No. 2	6¼	lb.
Glycerin	10	gal.
Water	To make 50	gal.

Liquid Yellow Food Color

Naphthol Yellow S,
  F.O.&C. Yellow

No. 1	6¼	lb.
Glycerin	10	gal.
Water	To make 50	gal.

Soft Curd Milk

U. S. Patent 2,353,946

The process of preparing a soft curd milk having a curd tension not exceeding about 30 grams, as measured by the Hill Curdometer, consists in adding to the whole milk a small amount of pancreatic enzyme of the order of about 1 lb. of enzyme to about 5,000 to 40,-000 of milk, depending upon the initial curd hardness in the raw milk and the activity of the enzyme, allowing the enzyme to act at a temperature between 32°F. and 70°F. for a period not exceeding 24 hours, then raising the temperature to 105°F. and 160°F. and holding the same at a controlled temperature in last range for about fifteen minutes to retard the activity of the enzymes, and further heating to a higher temperature to completely inactivate them before substantial hydrolysis or digestion of the casein has occurred.

Artificial Cream

U. S. Patent 2,142,650

A cream which keeps well at low temperatures is produced by heating a mixture of milk and cream to about 75°C. adding gelatin, 9.5% by volume of the mixture dissolved in an equal quantity of boiling water, and agitating at about 75°C. to disperse the gelatin. The mixture is rapidly cooled and agitated for 8 hours at 10°C.

Substitute for Coffee Cream

Formula No. 1

Milk	4
Condensed Milk	1
Light Cream (18%)	1

No. 2

Milk	3
Condensed Milk	1

No. 3

Milk	4
Condensed Milk	1
Evaporated Milk	1

NUMBER OF POUNDS OF MILK TO ADD TO A CERTAIN NUMBER OF POUNDS OF CREAM IN ORDER TO OBTAIN 40 QUARTS OF 19 PER CENT CREAM

(40-quart can of 19 per cent cream weighs about 84 lbs.)

Fat content of cream used for standardizing

Fat content of milk used for standardizing	20 *	20 **	22 *	22 **	24 *	24 **	26 *	26 **	28 *	28 **	*
0	4.20	79.80	11.46	72.54	17.50	66.50	22.61	61.39	27.00	57.00	30.80
3	4.95	79.05	13.26	70.74	20.00	64.00	25.56	58.44	30.24	53.76	34.22
3.2	5.00	79.00	13.41	70.59	20.19	63.81	25.79	58.21	30.48	53.52	34.47
3.4	5.06	78.94	13.55	70.45	20.39	63.61	26.01	57.99	30.73	53.27	34.73
3.6	5.12	78.88	13.70	70.30	20.59	63.41	26.25	57.75	30.98	53.02	35.00
3.8	5.18	78.82	13.85	70.15	20.79	63.21	26.48	57.52	31.24	52.76	35.26
4.0	5.25	78.75	14.00	70.00	21.00	63.00	26.72	57.28	31.50	52.50	35.53
4.2	5.31	78.69	14.16	69.84	21.21	62.79	26.97	57.03	31.76	52.24	35.81
4.4	5.38	78.62	14.32	69.68	21.43	62.57	27.22	56.78	32.03	51.97	36.09
4.6	5.45	78.55	14.48	69.52	21.65	62.35	27.47	56.53	32.30	51.70	36.37
4.8	5.52	78.48	14.65	69.35	21.87	62.13	27.73	56.27	32.58	51.42	36.66
5.0	5.60	78.40	14.82	69.18	22.10	61.90	28.00	56.00	32.86	51.14	36.96

Fat content of milk used for standardizing	30 **	32 *	32 **	34 *	34 **	36 *	36 **	38 *	38 **	40 *	40 **
0	53.20	34.12	49.88	37.05	46.95	39.66	44.34	41.99	42.01	44.10	39.90
3	49.78	37.65	46.35	40.64	43.36	43.27	40.73	45.60	38.40	47.67	36.33
3.2	49.53	37.91	46.09	40.90	43.10	43.53	40.47	45.86	38.14	47.93	36.07
3.4	49.27	38.18	45.82	41.17	42.83	43.80	40.20	46.12	37.88	48.19	35.81
3.6	49.00	38.45	45.55	41.44	42.56	44.07	39.93	46.39	37.61	48.46	35.54
3.8	48.74	38.72	45.28	41.72	42.28	44.34	39.66	46.66	37.34	48.72	35.28
4.0	48.47	39.00	45.00	42.00	42.00	44.62	39.38	46.94	37.06	48.99	35.01
4.2	48.19	39.28	44.78	42.28	41.72	44.90	39.10	47.21	36.79	49.27	34.73
4.4	47.91	39.56	44.44	42.56	41.44	45.18	38.82	47.50	36.50	49.55	34.45
4.6	47.63	39.85	44.15	42.85	41.15	45.47	38.53	47.78	36.22	49.83	34.17
4.8	47.34	40.14	43.86	43.15	40.85	45.76	38.24	48.07	35.93	50.11	33.89
5.0	47.04	40.44	43.56	43.44	40.56	46.06	37.94	48.36	35.64	50.40	33.60

* Pounds of milk of the specific fat content to mix with the ** pounds of cream of the specific fat content.

Cream Whipping Aids

Throughout the years research and experience have established many practices which aid in cream whipping. Heavy creams whip better than light creams. Aging the cold cream, whipping while cold, and precooling the beater and dish all help to improve whipping. It has also been established that a turbine whipper does a better job of whipping cream than a dover type beater. The amount of cream whipped at one time should not be too large for the whipper, as too much cream greatly increases the time required to whip it.

Perhaps the first important discovery in the addition of substances to cream to improve whipping was the use of "viscogen." "Viscogen" is sucrate of lime prepared from ordinary sugar and lime. Its original use was to restore the normal clumping of the fat in cream which had been previously impaired by pasteurization at high temperatures. Although "viscogen" is an aid to cream whipping, it is not sufficiently effective for 19 per cent cream.

The addition to light cream of lemon juice, citric acid, calcium salts, cream of tartar, etc., may improve cream whipping, but their effect is too small to be of value in light cream.

Recently there has been renewed interest in the heat treatment of cream to increase its viscosity. It is true that treatment does aid in whipping, but here again the effect is not important in light cream.

The fact is that there is not enough fat in 19 per cent cream to give a whipped cream of satisfactory stiffness, even if the fat is well clumped. Under these conditions it is practically essential to add some pure food material to the cream to supplement the action of the clumped milk fat.

The value of gelatin in preparing a whipped product has long been recognized. Gelatin has been used in very small amounts in commercially whipped cream to give additional stiffness to the whipped cream and to prevent drainage of skim milk out of the whipped cream prior to final consumption.

Light cream may be satisfactorily whipped in the following manner:

Cold Light Cream   1 cup
Sugar   2 tablespoonfuls
Gelatin   1 level teaspoonful
Vanilla Flavor   To suit

About 2 tablespoonfuls of cold water should be placed in a small pan and the gelatin sprinkled upon it. Let it soak for a few minutes, then heat to 140°F. to dissolve the gelatin. Place the cream in a dish and stir while the gelatin solution is added. Let stand in a refrigerator for 1 to 2 hours to allow the gelatin to set. Remove and whip. The sugar may be added before or after whipping.

This method gives a very satisfactory whip. The finished product has a glossy surface and definitely shows its gelatinized character. The aspect of the whipped cream can be altered somewhat by varying the amount of gelatin used.

The starches and gums are good water absorbents and emulsifying agents. The value of ordinary starch is questionable because the task of boiling the starch in cream is too difficult. Starches soluble in cold water obviate this problem. About 1 level tablespoonful of soluble starch per cup of cream gives a very good whip, but the whipped cream produces the sensation of flour or starch in the mouth. This sensation is quite objectionable and may be due principally to the size of the starch particles.

Various vegetable gums give the most satisfactory whipped cream. They closely resemble ordinary whipped cream in taste and stand up well without drainage. The gums are easily added to the cream and no aging is required. The gums are inexpensive and pure food products.

Trials with Karaya gum, gum acacia, gum tragacanth, and locust bean gum give quite comparable results.

Light cream may be whipped satisfactorily with the use of gums as follows:

Gum   1 level teaspoonful
Cold Light Cream   1 cup
Sugar   2 tablespoonfuls
Vanilla   Few drops

Place sugar in dry dish, add the gum, and mix until uniform and free from lumps. Pour cream into a bowl previously cooled by placing in refrigerator or rinsing with cold water. Add the sugar-gum mixture slowly while stirring the cream. The lumps need not be stirred out. Add vanilla. Whip at once.

If other conditions are favorable, light cream should be whipped in 2 or 3 minutes. The speed of whipping and the stiffness of the whipped cream can be varied by slight changes in the amount of gum.

12% Vanilla Ice Cream

Heavy Sweet Cream		
(40% Fat)	12½	qt.
Whole Fresh Milk		
(4% Fat)	30	qt.
Egg Yolk Powder	½	lb.
Skim Milk Powder		
(Spray Process)	3	lb.
Granulated Cane		
Sugar	15	lb.
Sodium Alginate	5	oz.
Vanilla Flavoring	10	fl. oz.

Combine the liquid material in pasteurizer and start heating. Mix sodium alginate, egg yolk powder, skim milk powder and granulated sugar in dry form and incorporate into the mix when the temperature reaches 150°F. Heat the entire ice cream mix to 155°F. and hold at this point for 30 minutes. Homo-

genize it and cool to about 50°F. Add vanilla flavoring shortly before freezing. Freeze and store in hardening room at about 15° be-below zero F.

12% Chocolate Ice Cream

Heavy Sweet Cream		
(40% Fat)	12	qt.
Whole Fresh Milk		
(4% Fat)	30	qt.
Frozen Egg Yolk	2	lb.
Cocoa Powder		
(Dutch Process)	4	lb.
Granulated Cane		
Sugar	17	lb.
Granulated Gelatin	4	oz.

Mix the cream and milk in pasteurizing tank. Apply the heat, then add frozen egg yolk. Combine the cocoa, sugar and gelatin by mixing thoroughly in dry form. Incorporate the dry mixture into the liquid part of the mix when the temperature reaches 120°F. Continue heating to 155°F. and hold the entire ice cream mix at this temperature for 30 minutes. Homogenize and cool to about 50°F. Age the cold mix in the storage tank for about 10 hours then freeze. Place in containers and store in hardening room at about 15° below zero F.

Water Ice

Sugar (Cane ¾;		
Corn ¼)	14	lb.
Stabilizer	3–3½	oz.
50% Citric Acid		
Solution or		
Equivalent	·3–4	oz.
*Avenex	¾	lb.

* Dissolved in water and heated to 145°F. and cooled.

Fruit, Color and Water
　　　　　To make 5 gal.
The 14 lb. sugar includes the sugar in the fruits added.

Tangerine Water Ice

Granulated Sugar	8	lb.
Corn Syrup Solids	4	lb.
Water (14 qt.)	28	lb.
*Tangerine Puree	10	lb.
Gelatin	1½	oz.
Pectin	1½	oz.
Oat Flour Stabilizer	2	oz.
Tangerine-Peel Extract	2	oz.
Tangerine Certified Color	3	oz.
Citric Acid (50% Solution)	6	oz.

Combine the sugar, corn syrup solids, gelatin, pectin and stabilizer by mixing them thoroughly until dry. Heat the water to 180°F. Sift the dry mixture into the hot water and stir vigorously until completely dissolved. Cool this mix to about 50°F.

Add the tangerine puree, flavor, color and acid. Place in freezer. Start whipping and freezing until an overrun of about 40 per cent is obtained. Remove from freezer into containers and place in hardening room at about 15° below zero F.

In case a milk sherbet is desired, replace some of the water, about 1-qt. with an equal volume of ice cream mix or condensed milk. This is best accomplished at the freezer before adding the acid, in order to prevent curdling.

Sherbet

Skim Milk	10–16 lb.

* Unsweetened cold pack.

Sugar (Cane ¾ ; Corn ¼)	14	lb.
*Avenex	½–¾	lb.
Stabilizer	3–3½	oz.
50% Citric Acid Solution or Equivalent	3–4	oz.

Fruit, Color and Water
　　　　　To make 5 gal.
The 14 lb. of sugar includes the sugar in the fruits added.

Stabilizer for Sherbets and Water Ices

Exchange Citrus Pectin (1-RS/100)	3.2	lb.
Locust Gum	2.8	lb.
Powdered Cerelose (Refined Corn Sugar)	1.0	lb.

Mix these ingredients thoroughly.

10 Gallon Batch of Sherbert Mix Pineapple

Crushed Pineapple in Juice	10½	lb.
Stabilizer (see formula above)	6	oz.
Cane or Beet Sugar	19	lb.
Cerelose (Refined Corn Sugar)	5½	lb.
50% Citric Acid Solution	13	fl. oz.
Ice Cream Mix	100	fl. oz.
Water	6	gal.

The 6 ounces of above stabilizer are thoroughly mixed with 1 pound of granulated sugar (taken from the quantity of sugar weighed out for the freezer batch). This mixture is added with stirring to 1½ gallons of hot water (180°F.). The solution is allowed

* Dissolved in water or skim milk and heated to 145°F. and cooled.

to stand with occasional stirring for 10 minutes, the remainder of the water (4½ gallons) is added, then the cane or beet sugar, cerelose, ice cream mix, crushed pineapple, and citric acid solution in the order given. Whip to 30–35% over-run.

Other fruits or juices may be substituted for the crushed pineapple in the above formula if desired. Also, certified food color and flavor may be used.

Ice-Cream Bar Coating
Formula No. 1

Cocoa	1
*Dry Milk Solids	2
Powdered Sugar	3
Butter	7

Mix the cocoa, milk solids and powdered sugar thoroughly and carefully to eliminate all lumps and insure a uniform mixture. If a darker color is desired, increase slightly the cocoa. Stir the butter and dry material together thoroughly. Raise the temperature after mixing to 103 to 105°F. It is then ready for dipping.

The ratio of butter to dry mixture will determine the thickness of the final coating.

Use care not to agitate the butter and dry material too much during the heating process, as this may result in "oiling off." The amount of moisture in the butter affects the body of the coating, and where the normal moisture content of the butter is reduced by evaporation a brittle coating is produced. A softer coating can be made by adding water to the butter.

Use fresh or very mildly salted

* Not over 1½ per cent fat.

butter. Make up only as used, to prevent the possibility of development of rancidity. Also, sugar crystals may become larger in unused butter and powder mixture.

No. 2

Coconut Oil (76°)	45 lb.
Peanut Oil	10 lb.
Sugar	35 lb.
Cocoa	10 lb.
Vanillin	1 oz.
Salt	1 oz.
Soybean Lecithin	5 oz.

The sugar, cocoa, vanillin, and salt are mixed in a dough-mixer with about half the quantity of the oils to form a stiff paste. It is then refined to the desired degree of fineness in a chocolate refiner. The refined paste is then placed in a chocolate kettle, the rest of the oils added, and finally the lecithin dissolved in an equal weight of the oil. The coating is now mixed for 1 to 3 hours until a uniform blending of the ingredients is obtained. The temperature during processing should be between 90 and 100°F.

No. 3

Coconut Oil (76°)	45 lb.
Peanut Oil	10 lb.
Sugar	30 lb.
Skim Milk Powder	10 lb.
Cocoa	5 lb.
Vanillin	½ oz.
Salt	1 oz.
Soybean Lecithin	5 oz.

The sugar, skim milk powder, cocoa, vanillin, and salt are mixed in a dough-mixer with about half the quantity of the oils to form a stiff paste. It is then refined to the desired degree of fineness in a chocolate refiner. The refined paste is then placed in a chocolate kettle, the rest of the oils added, and finally the lecithin dissolved in an equal weight of the oil. The coating is now mixed for 1 to 3 hours until a uniform blending of the ingredients is obtained. The temperature during processing should be between 90 and 100°F.

These coatings contain no added cocoa butter. They are widely used to coat ice cream bars. Their cost is low and the flavor excellent. Owing to the lower melting point of the cocoanut oil, they do not taste "waxy" on the ice cream, and because of the lower viscosity of the coconut oil and the presence of soybean lecithin greater coverage per pound of coating is obtained.

Reducing Churning Time in
Butter Production
U. S. Patent 2,311,598
0.5–1 oz. Graham's salt is added to 10 gal. cream, before churning.

Non-Melting Butter
(For Tropical Use)

Butter-Fat	60
Beef-Fat	20
Dried Milk	20

Melt together, mix well and cool. This will stand up to 105°F.

Fat-Free Butter Substitute
U. S. Patent 2,150,649

Ceresin	13
Cholesterol	2
Water	50–90
Mineral Oil, Medicinal	
To make	100

Non-Weeping Oleomargarine
U. S. Patent 2,156,036

Cottonseed Oil	80.00

Buttermilk	16.35

Churn the above at 30°F. and spray into water and ice, remove surface layer of fat and work latter till water content is about 12%. Mix with

Sodium Alginate	6.15
Sodium Chloride	3.00
Sodium Benzoate	0.50

Stabilized Homogenized
Butter Spread
Formula No. 1

Salted Butter	55.5	lb.
Skim Milk	42.2	lb.
Skim Milk Powder	2.0	lb.
Gelatin	0.3	lb.
Salt	5.0	oz.
Butter Color	75	ml.
Starter Distillate	25	ml.

No. 2

Salted Butter	43.0	lb.
*Cream, 19%	54.7	lb.
Skim Milk Powder	2.0	lb.
Gelatin	0.3	lb.
Salt	5.0	oz.
Butter Color	25	ml.
Starter Distillate	25	ml.

Add the gelatin, salt and skim milk powder to the mix at 90°F. Agitate and heat to 150°F. for 30 minutes. Homogenize at 3,500 lb. pressure. Cool, add starter distillate, package and allow to set in the refrigerator until firm.

No. 3

Cream, 46%	98.0	lb.
Skim Milk Powder	1.7	lb.
Gelatin	0.3	lb.
Salt	1.5	lb.
Butter Color	150	ml.
Starter Distillate	30	ml.

The product obtained by this method is very smooth and can be

---

* The equivalent of heavier cream and skim milk may be used.

spread easily as soon as it is removed from the refrigerator. It can be packaged hot, directly from the homogenizer, but better results are obtained when packaged after cooling.

Spreads of this type can be made in plants equipped with either a viscolizer or homogenizer.

The butter color can be omitted. The starter distillate is added to improve the butter flavor and aroma, and should be added to the finished product after cooling. There are various commercial products containing coloring and vitamins that may be added to improve the appearance and increase the nutritional value of such spreads.

Since the moisture content is much greater than that of butter, there is a tendency for a slight shrinkage to take place on storage, and properly waterproofed containers, such as the paraffined containers used for cottage cheese, are necessary. A slight syneresis may occur if the spread is not kept under refrigeration.

When properly refrigerated, and when made from high quality products, these spreads should keep well for two or three weeks.

Cultured Buttermilk

Solution:

(a) The buttermilk must contain 10 per cent of total solids, therefore, 500 pounds of buttermilk must contain (500×0.10) 50 pounds of total solids.

(b) Since dry milk solids contain approximately 96 per cent total solids it will be necessary to use $(50 \div 0.96)$ 52.08 pounds of dry milk solids and 447.92 pounds of water in the preparation of 500 pounds of cultured buttermilk.

1. Place water at about 70°F. in vat and add dry milk solids in two or three installments, mixing thoroughly to avoid lumping and to secure complete solution.

2. Heat to 180°F. for 30 minutes, cool to 68° to 70°F., and culture with 5 per cent starter.

3. Allow the cultured buttermilk to stand until an acidity of 0.95 to 1.00 per cent has developed.

4. Where possible, cool the curd to 50°F. prior to breaking. When this is impractical, satisfactory results can be secured by cooling the curd with slow, intermittent agitation in the vat. Satisfactory results have also been secured by breaking the curd in the vat, drawing it off into cans, and placing the cans in ice water or in a cold room with frequent agitation to reduce it to the desired temperature of 50°F. or lower.

5. In every case the curd of fermented reconstructed milk should be broken to a smooth consistency with an agitator operating at a speed which results in a minimum amount of air incorporation.

6. In so far as possible, avoid pumping.

Reducing the Tendency of Blowing in Cheese
U. S. Patent 2,291,632

Add 5.5 g. sodium bromate or iodate per 100 kg. of milk.

Non-Molding Cheese Coating
U. S. Patent 2,292,323

Stearic Acid   1

Propionic Acid 1–10
Paraffin Wax
To make 100

Cheese Coating Wax
U. S. Patent 2,299,951

Paraffin Wax 50
Chlororubber 5–15
Amorphous Petroleum
Wax To make 100

French Dressing

Vegetable Oil (Corn
or Cottonseed) 43.58
Water 18.11
Cider Vinegar
(50 Grain) 7.60
Tarragon Vinegar 7.60
*Lemon Juice 10.30
Dark Malt Syrup 1.48
Mixed Flavoring
Materials (see
table below) 9.89
Exchange Citrus
Pectin (1-RS/100) 1.44

Composition of Dry Mixed
Flavoring Material for
French Dressing

Sugar 14 lb. 7 oz.
Salt 5 lb.
Paprika 5 lb.
Garlic Salt 3 oz.
White Pepper 1¼ oz.
Onion Salt 2¾ oz.
Orange I (Color) 1¼ oz.
Mongolia, Egg Shade
(Color) ¾ oz.

The well mixed dry seasonings, together with the pectin, are stirred into the oil, and then while this mixture is being agitated rapidly, the previously mixed

* Made by mixing 1 volume of pure concentrated lemon juice, with 5 volumes of water.

watery substances (vinegars, lemon juice, malt and water) are rapidly poured in. Agitation is continued for five or ten minutes until the product is smooth. If the dressing is to be homogenized, the materials should be mixed as described above before being homogenized.

Mayonnaise
Formula No. 1

Salad Oil 20 lb.
Egg Yolk 2 lb.
100 Grade Exchange
Citrus Pectin 1½ oz.
Sugar 4¾ oz.
Salt 4¾ oz.
Mustard 2½ oz.
Pure Concentrated
Lemon Juice 3 fl. oz.
Water 14 fl. oz.
50 Grain Vinegar 20 fl. oz.

The pectin, sugar, salt, and mustard are blended together thoroughly and placed in the bowl of the mixer. About half of the water is then added to the dry ingredients, then the egg yolk, and the beater is then started at low speed. After the ingredients are thoroughly mixed, the mixer is shifted to high speed and the oil is added slowly. When the emulsion is well established, the oil can be added somewhat more rapidly. The remainder of the water should be added as needed to keep the emulsion from becoming too stiff. If the amount of water specified is not adequate for this purpose, the vinegar may be used. When all of the oil has been added (which should ordinarily take from 20 to 30 minutes), the pure concentrated lemon juice and the vinegar are added and the entire batch mixed very thoroughly. Allow the mayonnaise to stand for about a minute and then mix for about one minute more. It is then ready for packaging.

To produce a heavier mayonnaise, more pectin should be used. Less pectin makes a more fluid product.

No. 2

Salad Oil	20¼	lb.
Egg Yolk	2	lb.
Sugar	4¾	oz.
Salt	4¾	oz.
Mustard	2½	oz.
Pure Concentrated Lemon Juice	2	fl. oz.
Water	11	fl. oz.
50 Grain Vinegar	19	fl. oz.

Add the sugar, salt, mustard, egg yolks, and about half of the amount of water specified to the bowl of the mixer. Beat at low speed until the ingredients are thoroughly mixed. The beater is then shifted to high speed and the oil is added slowly at first until the emulsion is well established. Then the oil can be added more rapidly. If the mayonnaise becomes too heavy during the addition of the oil, add the remainder of the water and, if that is insufficient, use some of the vinegar. When all of the oil has been added (which ordinarily requires from 20 to 30 minutes), the pure concentrated lemon juice and the vinegar are added and the batch is mixed very thoroughly to obtain uniformity. The mayonnaise is then allowed to stand in the bowl of the mixer for a minute, then it is stirred for one minute more. It is then ready to be packaged into containers.

Spice for Ketchup

Ground Allspice	12½	oz.
Ground Cinnamon	12	oz.
Ground Cloves	1½	lb.
Ground Paprika	1½	lb.
Ground Black Pepper	6½	oz.
Cayenne Pepper	1	oz.

Curry Powder

Tumeric	60
Foenugreek	36
African Ginger	20
Coriander	24
Black Pepper	12
Cumin Seed	12
Cardamon (Whole)	6
Mace	6
Cayenne Pepper	12
Celery Seed	6
Caraway Seed	6

Celery Salt

Dried Kiln Salt	365
Ground Celery Seed	125
Calcium Phosphate	10

Red Pepper Substitute

Flour of Mustard Seed	5
Potato Flour	3
Roasted Chicory Flour	1
Oil Soluble Red Dyestuff	1
Oil with a Red Pepper Aroma (like a distillate of red peppers)	1

Mix together.

Capsicine-Free Red Pepper (Grade Fine-Sweet)

Very finely ground and sieved flour of Mustard Seed	100.0

Potato Starch	2.5
Oil Soluble Red Dye	
(dissolved in 6 parts	
of vegetable oil)	0.3
Sugar	2.0
Pepper Aroma Distillate	
(Oil distilled from red	
peppers)	0.5

New-England-Style Pressed Ham
Cured Pork Blade

Meat	80 lb.
Salt	10 oz.
Bull Meat	20 lb.
Dry Milk Solids	4 lb. 3 oz.
Cure for Sausage	½ pt.
Cinnamon	2 oz.

Cure blade meat at the rate of 2¾ lb. salt and 1 quart cure to 100 lb. meat. Cure 4 to 5 days, then cut into one inch squares.

Grind fresh bull meat through ⅜ inch plate. Put in silent cutter and add shaved ice (about 22 pounds required) salt, cure, dry milk solids, and cinnamon and chop fine. Put cured blade meat in mixer and add emulsion of bull meat and mix well.

Stuff into beef bungs or artificial casings. Smoke and cook in usual manner.

Pork Luncheon Loaf

Pork Blade Meat	100 lb.
Salt	3 lb.
Cure for Sausage	1 qt.
Dry Milk Solids	10 lb.
Water	10 lb.

Place blade meat in meat truck. Sprinkle salt and cure over meat, and mix well. Run meat through a 1 inch plate, spread in pans 4 to 6 inches deep, and cure for 3 days.

Then place in mixer. While mixing add cold water and dry milk solids, and mix well. Stuff into Frank, Hoy or similar cooking molds and cook for about 3 hours at 160°F. until inside temperature is 152–155°F.

Blood and Tongue Loaf

Pickled Shoulder		
Fat or Back Fat		
(Cubed)	40	lb.
Pickled Hog or		
Lamb Tongues	20	lb.
Pickled Hogskins	15	lb.
Blood	25	lb.
Dry Milk Solids	10	lb.
Fresh Onions		
(Ground with		
Hogskins)	3	lb.
Salt	1	lb.
Pepper	7	oz.
Marjoram	2½	oz.
Ground Cloves	1½	oz.

The pickled shoulder fat, tongues, and skins must be cooked, shoulder fat about 1 hour to make it easier for cubing and add the skins from the shoulder fat to hogskins. Cook tongues and skins until tender. Cut shoulder fat in about ½ inch cubes; cut tongues into four pieces, and grind skins through a fine plate.

When all meat is made ready, place in mixer or meat truck and add strained blood. Mix a little and add dry milk solids and spices. Mix well.

Then stuff by hand in artificial casing or beef bungs, taking care that the right amount of the blood gets into each piece otherwise there may be too much blood left over for the last few pieces.

Cook for 2½ to 3 hours at 175°–

180°F. depending on size of casing.

Whole tongues may be used, but if cut into pieces they are more evenly distributed.

Baked Hamburger Loaf

Boneless Chucks	75 lb.
Pork Cheek Meat	25 lb.
Dry Milk Solids	15 lb.
Fresh Onions	5 lb.
Cold Water	35 lb.
Salt	4 lb. 10 oz.
White Pepper	8 oz.
Ground Mustard	4 oz.
Cure for Sausage	1 qt.

Run beef, pork, and onions through a one-inch plate. Put all in mixer and add water, cure, dry milk solids, and spices and mix well.

Put in pans of desired size and bake at 225–250°F. It will take about 3 hours to bake a 5-pound loaf, the inside temperature should be 152°–155°F. before taken out of oven.

Stuffing in cellulose casing enhances the appearance and gives protection in shipping and handling.

Pro-Lac Sandwich Special

Lean Pork Trimmings	55 lb.
Fresh Beef Tongues	45 lb.
Dry Milk Solids	10 lb.
Salt	2¾ lb.
White Pepper	5 oz.
Cure for Sausage	1 qt.
Ground Sage	1 oz.
Water	10 lb.

Scald tongues in water hot enough to loosen skin—usually about 160°F. temperature is required, then chill quickly. Remove glands, bones, fat, palate and loosened skin, and grind through ⅛-inch plate.

Grind pork trimmings through a 3/16-inch plate.

Place all meat in mixer and sprinkle the cure solution evenly over meat; next add dry milk solids and spices. Mix well.

Cure this mixture in shallow pans for twenty-four hours, then mix again and stuff into square molds. ("Frank" or "Hoy" meat loaf molds are adaptable for this.) Cook at 160°F. until inside temperature reaches 155°F. Chill overnight and remove from molds; wrap or stuff into cellulose casings, as desired.

Hamburger Style Patties

Beef Trimmings (70% Lean)	100 lb.
Dry Milk Solids	5 lb.
Ice Cold Water	5 lb.
Salt	2 lb. 8 oz.

Run beef through one-inch plate. Place evenly on bench of meat truck. Sprinkle ice water, dry milk solids and salt over this and mix well. Then run all through a ⅛ inch plate. Make into patties or sell in bulk. Not over 5% dry milk solids should be added in hamburger, to avoid danger of scorching.

Roast-Beef Loaf

Fresh Beef	100 lb.
Worcestershire Sauce	3 oz.
Tomato Catsup	10 lb.
Salt	3 lb.
White Pepper	8 oz.
Ground Bay Leaves	2 oz.
Grated Onions	2 lb.
Dry Milk Solids	12 lb.

100 pounds fresh beef, either

boneless chucks and one-fourth plate meat, or something of similar construction. It must not be too fat.

Cut in about ½ inch pieces. Place in steam jacketed kettle with just enough water to cover meat. When this comes to boil, put in Worcestershire sauce, tomato catsup, salt, white pepper, ground bay leaves and the grated onions. Cook slowly until tender.

When tender, place cooked meat in mixer, add the dry milk solids, then add cooking broth. It will take about 45–50 lb. of this broth. Mix well.

Place in pans holding about 5 lb. Chill over night. It can be sold either wrapped in parchment paper or stuffed in cellulose casing.

Cooked-Ham Loaf

Hams that are otherwise hard to dispose of can be used to good advantage as follows:

Ham	100	lb.
Dry Milk Solids	5	lb.
Gelatin (#2 grade)	1	lb.
Brown Sugar	1½	lb.
Salt	1	lb.

Flavoring (Optional)

Bone and take skin off ham and cut up in pieces small enough to go through the mouth of grinder. Leave all the fat on. Run through ¾ inch plate. Mix dry milk solids, gelatin, brown sugar, and salt.

Put cut hams in mixer; while rotating add above mixture and mix well. If the ham, in the first place, is a little salty, which does happen occasionally, do not add salt. If any particular flavor, such as pineapple, maple, etc., is desired

add ½ oz. of the extract to the mixture. If a clove flavor is desired, use 1 oz. ground cloves.

Place in ham molds, Frank or Hoy loaf molds, according to the shape desired and cook as you would boiled ham or other cooked loaf. When cold and chilled, remove from molds and stuff in artificial casings.

If smoke flavor is desired, place hams 2 inches apart on screen shelves and give a dense cool smoke for 2 hours after removing from molds.

Corned-Beef Loaf

Boned Beef (Briskets, Plates, etc.)	100 lb.
Dry Milk Solids	15 lb.

Run beef through large lard cutting plate and mix well with cure made of

Salt	5 lb.
Cure for Sausage	1 qt.

Pack and cure 5 days in 38–40°F. cooler.

Place meat in nets for easier handling. Cook in steam jacketed kettle for 3½ hours at 165–170°F. Use enough water to fully cover meat. Put the following spices in a muslin bag for each 100 pounds of meat:

Black Pepper	8 oz.
Allspice	2 oz.
Cloves	2 oz.
Bay Leaves (Crushed and Broken)	3 oz.
Onion Powder	1 oz.
Garlic Powder	½ oz.

When cooking is finished, place meat in mixing truck and sift 15 lb. dry milk solids into this while stirring continuously. Add enough of the cooking water to give it the proper consistency. This mixture will take about 30–35 lb. of cooking water.

Put in pans of desired size and let set and cool overnight. It can then be wrapped in wax paper or stuffed in artificial casing. In the long run, the artificial package will prove more economical.

Cooked Corned-Beef and Spaghetti Loaf

Boned Beef (Briskets, Rumps, Plates, etc.)	100 lb.
Dry Milk Solids	20 lb.
Pimentos (Cut in ¾ inch pieces)	10 cans
Pistachio Nuts	5 lb.
Spaghetti	15 lb.
Gelatin	3 lb.
Salt	
Pepper	
Mustard	
Onion Powder	
Garlic Powder	

Run beef through large lard cutting plate and mix well with cure made of

Salt	5 lb.
Cure for Sausage	1 qt.

Pack and cure 5 days in 38–40°F. cooler. Place cured meat in steam jacketed kettle. Put in enough water to cover meat about 3 inches and cook for 3½ hours at 165–170°F.

Prepare spaghetti as follows: Put spaghetti in about 6–7 gallons of cold water. Add about 2½ lb. of salt and let this stand about 15 minutes to soak. Then heat up to 160°F. and cook about 15 minutes. Do not stir after cooking. Drain off the water and immediately immerse in cold water. Stir this slightly so that the spaghetti does not stick together. When cold, drain off the water and it is ready for use. This spaghetti should be used within a short time after it is finished, as it has a tendency to mat.

Prepare gelatin by putting about one gallon of cold water in a bucket and pour gelatin into the water. Do not stir. It will soak up the water and will not lump.

When meat has cooked 3 hours, sift the dry milk solids into kettle, stirring continuously while sifting. Then add pimentos, pistachio nuts, spices, spaghetti, and gelatin. At the end of 3½ hours, turn off steam and let it stand in kettle 30 minutes, stirring occasionally.

Place in suitable pans and cool overnight in cooler. Remove from pans and wrap or stuff in cellulose casings. If the loaves have a tendency to stick to the pans, dip pans in hot water momentarily.

Veal Loaf

Veal Trimmings	50 lb.
Beef Trimmings	30 lb.
Pork Trimmings	20 lb.
Dry Milk Solids	12 lb.
Onions	5 lb.
Parsley	3 bunches
Lemon Extract	2 oz.
White Pepper	6 oz.
Salt	3 lb.
Cure for Sausage	1 qt.

Grind and chop meat in usual way, adding ice, cure, dry milk solids, onions and spices, while chopping. When chopped almost fine enough, add parsley and run a few revolutions.

Put in greased baking pans and bake at 200–250°F. for 2½–3

hours. Inside temperature should be 152–155°F.

After loaf is baked it can be stuffed in artificial casings, which add to the appearance and keeping qualities.

Liver and Bacon Loaf

Fresh or Frozen Pork Livers	50 lb.
Veal Trimmings	10 lb.
Pork Cheek Meat	15 lb.
Regular Pork Trimmings	10 lb.
Skinned Bacon Ends or Heavily Smoked Belly Trimmings	15 lb.
Dry Milk Solids	10 lb.
Shaved Ice	8 lb.
Fresh Onions	3 lb.
Salt	2 lb. 8 oz.
Cure for Sausage	1 qt.
White Pepper	6 oz.
Ground Celery Seed	2 oz.
Mace	2 oz.
Sweet Marjoram	1½ oz.
Lemon Extract	1 oz.

If fresh livers are used, slash and wash them in several changes of cold water. If frozen livers are used, this procedure is not necessary. Grind livers, onions, veal, and pork through ⅛ inch plate.

Put ground liver in silent cutter and chop fine; then add veal, pork, dry milk solids, shaved ice and spices, chop until fine; add bacon ends or smoked belly trimmings and chop a few revolutions, just enough to have a dice or cube effect.

Put mixture in pans lined with sliced bacon and cover with bacon strips; place lid on pans and cook for 3–3½ hours at 165–170°F. The inside temperature

should be 152–155°F. Chill overnight. Stuff in cellulose casing or wrap in pliofilm wrappers.

Baked-Veal and Pork Loaf

Veal Trimmings	60 lb.
Reg. Pork Trimmings	40 lb.
Salt	3 lb.
Cure for Sausage	1 qt.
Dry Milk Solids	10 lb.
White Pepper	6 oz.
Cinnamon	2 oz.
Onion Powder	1 oz.
Garlic Powder	¼ oz.
Mapleine Flavor	2 oz.

Grind veal trimmings through ⅛ inch plate, and pork trimmings through ⅜ inch plate. Put veal trimmings, salt, and cure in silent cutter and chop almost fine. While chopping, add enough shaved ice to have emulsion of right consistency. Then add pork trimmings, dry milk solids and seasoning. Chop fine.

Put in pans that have been prepared with a thin film of grease inside, smooth top of meat with paddle dipped occasionally in thin sugar syrup.

Bake for 2½ to 3 hours at 220–250°F. Stuff in cellulose casing, which adds to appearance and facilitates handling.

Luncheon Meat Loaf
Formula No. 1

Pork Blade Meat (Cut into ½ inch squares)	100 lb.

Cure for 5 days with 2½ lb. salt and 1 qt. cure for sausage.

When cured place in mixer and sprinkle with 10 lb. dry milk solids while rotating. Mix well.

When mixed, place solidly in

square molds that are lined with good grade oiled paper and cook at 160°F. for 3–3½ hours, depending on the size of the mold. When taken out of cooker, press lid down one notch and place in cooler over night. Then remove and stuff in cellulose casing in usual manner.

No. 2

Pork Blade Meat (Cured)	80 lb.
Bull Meat Emulsion (Cured)	20 lb.
Dry Milk Solids	10 lb.
Moisture (Ice Water)	10 lb.
White Pepper	5 oz.
Paprika	3 oz.
Onion Powder	¾ oz.
Garlic Powder	¼ oz.

Take pork blade meat, mix salt and cure and run through ⅜ inch plate. Cure this for two days. Prepare bull meat in the usual manner.

Place cured pork in mixer, add bull meat emulsion, water, dry milk solids and spices and mix well. (Spices must be mixed well before adding to mixture.)

Stuff mixture tightly in 2¾x18 cellulose casings, hang in smoke house that has a temperature of 125–130°F. Raise temperature gradually so that at the end of 2½ hours, it will be 180°F. Then raise temperature to 195–200°F. Smoke six hours in all.

The inside temperature after:

2½ hours should be 125°F.
3½ hours should be 140°F.
4 hours should be 145°F.
5 hours should be 148°F.
5½ hours should be 152°F.
6 hours should be 154°F.

When taken out of smoke house, rinse with hot water and then give a cold shower for a few minutes. When finished in this manner the product will be delicious and keep well.

Utility Meat Loaf

This loaf is intended to make practical and profitable use of bruised or other disfigured hams or shoulders. Meat should not be more than 20–25% fat and must be cured.

Cured Pork Meat	100 lb.
White Pepper	6 oz.
Ground Sage	½ oz.
Ground Cloves	½ oz.
Mapleine Flavor Dissolved in 1 pint water	1 oz.
Dry Milk Solids	8 lb.
Cold Water	12 lb.

Run cured pork through ½ inch plate. Add water, dry milk solids, and seasoning; mix well.

Place meat in molds for cooking (Frank, Hoy, or similar molds). Cook for 3½ hours at 165°F. Inside temperature should be at least 152–155°F. After cooking, chill, and remove from containers. They can then be wrapped in pliofilm or stuffed in cellulose casing.

Loaf can be baked as follows:

Place meat in pans, smooth top with sugar syrup and bake at 200–225°F. Bring inside temperature to 154°F., then chill, wrap or stuff as above specified.

Defense Meat Loaf

Cured Pork Trimmings (75–80% Lean)	55 lb.
Cured Boneless Chucks	45 lb.
Dry Milk Solids	10 lb.
White Pepper	8 oz.

Pistachio Nuts	4 lb.
Pimentos	6 lb.
Mapleine Flavor (Dissolved in 1 pint water)	1 oz.
Salt	1 lb.

Grind pork through a ½ inch plate, and beef through an ⅛ inch plate. Place beef in silent cutter and chop fine. During chopping add dry milk solids, pepper, salt, and Mapleine solution; also put in enough shaved ice to make emulsion a little softer than for a frankfurter. After beef is cut fine, add pimentos and run just long enough to have about ½ inch pieces of pimentos showing throughout the mixture. Put ground pork in mixer and add beef emulsion and pistachio nuts, and mix well.

Put mixture in molds lined with cured hog caul fat. Cook at 160°F. for 3–4 hours depending upon the size of molds. Inside temperature must be 152–155°F. when taken out of cook. Chill and take out of container. Wrap in pliofilm or stuff in cellulose casing.

Luxury Meat Loaf

Beef Trimmings	45 lb.
Pork Blade Meat	35 lb.
Reg. Pork Trimmings (All fresh)	10 lb.
Back Fat (Cubed)	10 lb.
Dry Milk Solids	12 lb.
Pimentos	5 lb.
Cure for Sausage	1 qt.
Salt	3 lb.
Pistachio Nuts	2 lb.
Sage	1 oz.
White Pepper	6 oz.
Paprika	1 oz.
Mace	3 oz.

First chop beef almost fine, then blade meat, then pork. Add back fat and chop just enough to get cubed effect. After beef is put in chopper, add dry milk solids, cure and spices, adding ice during the chopping operation to make it of the right consistency. Add pimentos and run 1 or 2 revolutions.

When chopped fine enough, put it in pans that have been greased, then let them stand in sausage room temperature for two hours to complete cure. Then put in oven, bake at 225–250°F. for about three hours. Take inside temperature; it should be 152°–155°.

When this is reached, the loaves are done; take out and cool off.

If it is desired to stuff loaves in artificial casing, cool them off about 3–4 hours to let them set. Then stuff, dipping loaf in hot gelatin before stuffing.

If loaf is to be cooked, put meat in a container with a lid that can be pressed down. Cook for 2–3 hours, according to what size loaf is made. Have inside temperature 152–155°F. When done, cool off in cold water. If desired in casings, follow direction as for baked loaves.

Southern Meat Loaf

Veal Trimmings	40 lb.
Beef Trimmings	20 lb.
Pork Trimmings	20 lb.
Bull Meat	20 lb.
Dry Milk Solids	12 lb.
Salt	3 lb.
Cure for Sausage	1 qt.
Queen Olives	4 lb.
Pimentos	6 lb.
Red Pepper	5 oz.
Allspice	2 oz.

Ground Mustard	3 oz.
Cinnamon	2 oz.

Grind and chop the usual way, adding ice, cure, dry milk and seasoning while chopping. Then place in mixer and add whole olives and pimentos cut into suitable size. Mix well. Place in greased pans and bake at 200–250°F. for 2½–3 hours, depending on size of pan.

When cool, the loaves may be stuffed in cellulose casings.

Corsica-Style Meat Loaf

Fresh Veal Trimmings	50 lb.	
Fresh Hog Livers	15 lb.	
Fresh Beef Trimmings	30 lb.	
Fresh Pork Backfat	5 lb.	
Dry Milk Solids	8 lb.	
Salt	3 lb.	8 oz.
White Pepper	7 oz.	
Paprika	2 oz.	
Garlic Powder	1 oz.	
Mace	2 oz.	
Italian Cheese	4 oz.	
#3 Cans Tomato Puree	2	

Grind veal through a ⅛ inch plate; beef and livers through a ¼ inch plate.

Place veal in silent cutter. While chopping add shaved ice, dry milk solids, cure, and spices. When veal is nearly fine enough, put in beef, livers, and backfat, and chop just enough to have a cube effect from backfat. Keep this overnight to cure. Put in loaf containers, cook at 160°F. for 3 hours or more, depending on size of container. Inside temperature must be 152–155°F. Chill and stuff in cellulose casing or wrap in pliofilm.

Cumberland-Style Meat Loaf

Veal Trimmings	15 lb.
Beef Tripe	15 lb.
Beef Trimmings	30 lb.
Pork Trimmings	30 lb.
Pork Backfat	10 lb.
Dry Milk Solids	10 lb.
Salt	3 lb.
Cure for Sausage	1 qt.
White Pepper	7 oz.
Cinnamon	2 oz.
Onion Powder	2 oz.
Garlic Powder	¼ oz.

Grind beef tripe and veal trimmings through $\frac{1}{16}$ inch plate, then put in silent cutter and chop fine. During chopping add shaved ice, salt, cure, and dry milk solids. Add enough shaved ice to make the consistency a little softer than for frankfurters.

Grind beef trimmings through ¼ inch plate and pork trimmings through ½ inch plate. Cut backfat into ½ inch cubes, scald cubes before mixing with other ingredients.

When ingredients are prepared, place beef and pork trimmings in mixer, add the chopped emulsion, backfat, and spices, and mix well. Should the mixture be too stiff, add more shaved ice.

When mixture is well mixed, place in meat truck and let it cure overnight in cooler. Next day put in parchment lined pans, cover and cook at 160–165°F. for 2½ to 3 hours, then chill. Remove from pans and stuff in cellulose casings.

Meat, Pickle and Pimento Loaf

Beef Trimmings	35 lb.
Veal Trimmings	25 lb.
Pork Trimmings	25 lb.

Bull Meat	15 lb.
Dry Milk Solids	12 lb.
Sweet Pickles	5 lb.
Pimentos	6 lb.
Onions	5 lb.
White Pepper	6 oz.
Coriander	2 oz.
Marjoram	½ oz.
Salt	3 oz.
Cure for Sausage	1 qt.

Grind and chop in usual way, adding ice, dry milk solids, onions, cure and spices. Chop fine, then place in mixer. Add pickles and pimentos that have been cut in suitable pieces and mix well.

This can be baked or cooked. For baking, place in greased pans and bake at 200–250°F. for 2½–3 hours, depending on size of pans. For cooking, stuff in long Frank, Hoy or similar molds and cook at 160°F. for 3 hours. Then chill. After chilling stuff in cellulose casing.

Celery Meat Loaf

Fresh or Frozen	
Hog Liver	15 lb.
Beef Trimmings	45 lb.
Regular Pork Trimmings	40 lb.
Dry Milk Solids	12 lb.
Stalks of Celery	6 lb.
#3 Cans Pimentos	3
Fresh Onions	3 lb.
*Diced Cheese	8 lb.
Salt	3¾ lb.
Cure for Sausage	1 qt.
White Pepper	7 oz.
Paprika (Mix well	

* A cheese that does not melt or run at the temperatures required must be used. Cheese especially made for this purpose is recommended.

with other spices)	5 oz.
Marjoram	2 oz.

Place livers in silent cutter, chop a few revolutions, and put in beef that has been run through ⅜ inch plate.

Add salt, cure, onions that have been ground through ⅜ inch plate, and add shaved ice and dry milk solids. Chop to about the same fineness as bologna. Then put in pork and spices.

When chopped almost fine enough, add the celery and run a few revolutions. The pieces of celery should be noticeable, but not too large. Put in mixer and add diced cheese and chopped pimentos. Mix so that cheese and pimentos are evenly distributed.

Put in pans and bake about 3 hours at 250°F. which will give an inside temperature of 152–155°F. in a 5–6 pound loaf, then drain and cool.

The appearance will be materially enhanced if stuffed in amber colored artificial casings.

Dutch Meat Loaf

Pork Trimmings		
(About 60% Fat)	25	lb.
Pork Cheek Meat	40	lb.
Veal Trimmings	35	lb.
Dry Milk Solids	12	lb.
Shaved Ice	30	lb.
Salt	3 lb.	12 oz.
Cure for Sausage	1	qt.
Fresh Onions	4	lb.
White Pepper	8	oz.
Sage	2	oz.

Run veal trimmings, onion, pork cheek meat through ⅛ inch plate; pork trimmings through ¼ inch plate.

Put veal and pork cheeks in

silent cutter, add shaved ice, cure, salt and spices while chopping. Chop about 3 minutes.

Put all meat in mixer and mix about 4 minutes. Then put in pans to bake. The original Dutch Loaf was round, but now most are baked in regular loaf pans. Bake at 225–250°F. for about 3 hours; time depends upon size of loaf. Inside temperature of 152–155°F. is desirable.

Southern Peanut Meat Loaf

Boneless Beef Chucks (Cured)	60 lb.
Lean Pork Trimmings (Cured)	25 lb.
Back or Shoulder Fat (Diced) (Cured)	8 lb.
Cooked Tongues (Diced) (Cured)	7 lb.
Peanut Butter	4 lb.
Roasted Peanuts	4 lb.
Dry Milk Solids	15 lb.
White Pepper	8 oz.
Salt	8 oz.
Rubbed Sage	2 oz.
Marjoram	1 oz.
Shaved Ice	55 lb.

Run beef and pork through a ⅛ inch plate, then put beef in silent cutter. Put in some ice and chop 3 minutes, put in dry milk solids, peanut butter, pork, and seasoning, adding ice during operations. When chopped almost fine enough, put in tongues, shoulder fat and roasted peanuts. Chop just long enough to obtain a diced appearance of tongues and backfat. If it is desired to dice tongues and backfat by hand, a mixing machine must be used.

This loaf can be cooked or baked. To cook, put in Hoy, Frank, or ham molds and cook 2½–3 hours. If stuffed in cellulose casings, it will greatly improve appearance.

To bake, put in pans and bake 2½–3 hours (depending upon size of pans used) at 225–250°F. After chilling they can be stuffed in cellulose casings.

Delectable Meat Loaf

Cured Pork or Sheep Tongues		15 lb.
Cured Pork Snouts		10 lb.
Regular Pork Trimmings (Fresh)		30 lb.
Veal Trimmings (Fresh)		40 lb.
Cured Pork Ears		5 lb.
Dry Milk Solids		10 lb.
Shaved Ice		35 lb.
Salt	2 lb.	4 oz.
Cure		1 pt.
White Pepper		7 oz.
#3 Cans Pimentos		2

Chop veal and pork in silent cutter. While chopping add salt, cure, ice, dry milk solids and pepper. When chopped fine, add pimentos and chop just long enough to cut pimentos so they show plainly.

Cut tongues, snouts, and ears by hand or in head cheese cutter. Put all in mixer and mix until the cut pieces are evenly distributed. Put mixture in molds and cook at 160°F. for 3–4 hours, depending upon size mold used. Chill and remove from molds, stuff in cellulose casing or wrap in pliofilm wrappers.

Mosaic Liver Loaf

Fresh Hog Livers	30 lb.

Cured and Cooked Hog
Livers (Diced)	10 lb.
Veal Trimmings	10 lb.
Reg. Pork Trimmings	20 lb.
Pork Jowls	20 lb.
Backfat (Diced)	10 lb.
Dry Milk Solids	12 lb.
Pistachio Nuts	4 lb.
Stuffed Olives	10 lb.
Shaved Ice	15 lb.
Salt	3 lb. 8 oz.
Cure for Sausage	1 qt.
White Pepper	7 oz.
Marjoram	2 oz.

Run fresh livers and veal trimmings through ⅛ inch plate. Run pork trimmings and jowls through ⅜ inch plate. Put fresh livers in silent cutter and chop 3 minutes, then add veal trimmings, cure, salt, dry milk solids and part of shaved ice. Add pork trimmings, jowls, ice, and spices and chop fine.

Place mixture in mixer, then add cured livers that have been cut into about ½ inch cubes, backfat that is cut into small cubes and pistachio nuts. Just before mixing is completed, add stuffed olives. Be careful not to mix long as stuffed olives may be crushed.

Line container with layer of ¼ inch thick backfat, fill even with mixture, and cover with backfat. Put on lid and cook about 3 hours at 160–165°F. Then place in cooler. When cold remove from container and wrap in wax paper, or stuff into cellulose casings.

Barbecue-Style Pork Loaf

Pork Cheeks (or Lean Pork)	60 lb.
Pork Giblets (or Lean Pork)	10 lb.
Salt	1¾ lb.
White Pepper	6 oz.
#10 Cans Tomato Catsup	3
Dry Milk Solids	10 lb.

Grind pork through large plate (preferably 1½ inch). Place in steam jacketed kettle. Cover meat with water and cook slowly until tender. Empty catsup into a tub and sift dry milk solids and seasoning into catsup. Stir well to avoid lumping.

When meat is tender, place in meat truck. Pour the catsup mixture over meat and mix well, then put in parchment lined pans, and place in cooler to chill.

After chilling, remove from pans and stuff into cellulose casings.

Liver Loaf

Hog Liver	50 lb.
Veal Trimmings	20 lb.
Regular Pork Trimmings	30 lb.
Dry Milk Solids	12 lb.
Salt	2½ lb.
White Pepper	6 oz.
Marjoram	1½ oz.
Cardamom	1 oz.
Cure for Sausage	1 qt.

Have livers and meat real cold. No ice is used in this loaf. Grind and chop until fine, add dry milk solids and seasoning while chopping.

Place in containers that are lined with thin sheets of backfat or pork caul fat. After filling put on lid and cook at 160°F. for 2½–3 hours, depending on size of containers. After cooking, chill.

Translucent Liver Loaf

Fresh Hog Livers	35 lb.
Fresh Veal Trimmings	25 lb.
Fresh Reg. Pork Trimmings	20 lb.
Fresh Pork Cheek Meat	10 lb.
Fresh Pork Backfat (Cubed)	10 lb.
Dry Milk Solids	10 lb.
Onions (Fresh, Peeled)	4½ lb.
Salt	3 lb.
Cure for Sausage	1 qt.
White Pepper	6 oz.
Sage	1 oz.
Sweet Marjoram	2 oz.
Ginger	1½ oz.

Grind hog livers and onions through a $\frac{1}{16}$ inch plate, and veal and pork through ⅛ inch plate. Place livers in silent cutter, add salt and cure. Chop fine, then add veal, pork, dry milk solids and spices. Chop mixture fine. Place all in mixer and add cubed backfat, mix until cubes are evenly distributed.

The cube effect of backfat may also be obtained by placing backfat evenly in silent cutter, and letting the machine run a few revolutions, but the cube effect will not be as even.

Prepare pans for cooking by lining the desired size pans with parchment paper, then placing another liner of hog caul fat.

Fill pans nearly full of the mixture, cover first with caul fat then with paper. Put on lid.

Cook at 160–165° F. for 2 to 3 hours, depending upon size of pan. A 5 pound pan will take 3 hours to cook.

Chill over night. Remove from pan. It is suggested that they be stuffed in cellulose casings, for protection and ease of handling.

Minced Luncheon Loaf
Formula No. 1

Boneless Chucks	35 lb.
Bull Meat	35 lb.
Regular Pork Trimmings	20 lb.
Backfat (Diced)	10 lb.
Dry Milk Solids	5 lb.
Salt	3 lb.
White Pepper	9 oz.
Cardamom	3 oz.
Nutmeg	3 oz.
Cinnamon	3 oz.
Onion Powder	¾ oz.
Cure for Sausage	1 qt.

Run bull meat, beef chucks, and pork trimmings through a ⅜ inch plate, then put bull meat, beef cheeks and dry milk solids in silent cutter. Add shaved ice, cure and salt and chop almost fine. Then put in pork trimmings and spices. Keep adding ice until emulsion is of right consistency.

When the mixture is fine enough, put in backfat and run a few revolutions. (This will give the same effect as when backfat is run through a dicing machine without the extra labor cost.)

Stuff in beef bungs or cellulose casing of corresponding size. Put in smokehouse and smoke for 2–3 hours, depending upon the color desired. Then cook at 160°F. for 3–3½ hours. The inside temperature of 152–155°F. should be maintained for 30 minutes before taking out of cook. Then give cold shower, let dry and hang in cooler.

No. 2

Boneless Chucks	35	lb.
Beef Cheeks	20	lb.
Hog Stomachs	20	lb.
Beef Tripe	10	lb.
Regular Pork Trimmings	15	lb.
Dry Milk Solids	5	lb.
Salt	3	lb.
White Pepper	8½	oz.
Cardamom	2¾	oz.
Nutmeg	2¾	oz.
Cinnamon	2¾	oz.
Onion Powder	¾	oz.
Cure for Sausage	1	qt.

Run all meat through a ⅜ inch plate. Place beef cheeks, shaved ice, beef chucks, salt and cure in silent cutter, then add hog stomachs, beef tripe and dry milk solids, adding enough shaved ice to keep meat cold, then add pork and spices and chop until fine.

Stuff in beef bungs or cellulose casing of corresponding size. Put in smokehouse and smoke for 2 to 3 hours, depending upon the color desired. Then cook at 160°F. for 3 to 3½ hours. The inside temperature of 152–155°F. should be maintained for 30 minutes before taking out of cook, then give cold shower, let dry and hang in cooler.

Skim-Milk Meat Loaf

Pork, Blade Meat or Extra Lean Trimmings, Cured	85	lb.
Bull Meat or Boneless Chuck, Cured	15	lb.
Dry Milk Solids, not over 1½% Fat	10	lb.
Ice Water	27	lb.
White Pepper	6	oz.
Salt	12	oz.
Paprika	2½	oz.

Onion Powder	1	oz.
Garlic Powder	½	oz.

Grind the pork through a 1-in. plate. Grind the bull meat or boneless chuck through a ⅜-in. plate, then chop fine in a silent chopper. During chopping, add the dry milk solids, water, and the spices which have been mixed together thoroughly. If the spices are not well mixed before being added, the paprika will not be evenly distributed.

Put the ground pork and beef in a mixer and mix thoroughly.

When mixing is completed, stuff in "Frank," "Hoy," or similar molds. Cook in water at 160 to 165°F. Cook until the inside temperature reaches 155°F., which should be about 2 to 2½ hours.

When cooked, place in cooler and chill. Remove from molds and wrap or stuff into cellulose casings.

Tongue and Cheese Loaf

Pickled Pork or Lamb Tongues (Cut into 1 inch cubes)	40	lb.
Hog Skins	20	lb.
Backfat (Cut into ½ inch cubes)	40	lb.
Swiss Cheese (Cut into ½ inch cubes)	10	lb.
Dry Milk Solids	10	lb.
Salt	2½	lb.
White Pepper	7	oz.
Onion Powder	2	oz.
Sweet Marjoram	2	oz.
Ground Cloves	1	oz.

Cook tongues and hog skins until tender. Peel and cut tongues into 1 inch cubes. Cook hog skins until tender, then grind through $\frac{1}{16}$ inch plate.

Cook backfat about 1 hour at almost boiling temperature, then cut into ½ inch cubes. Cook all ingredients in same water. Use just enough water to cover well.

When everything is ready, place tongues and hog skins in mixer. Add about 20–25 pounds of meat broth, then sprinkle in dry milk solids, backfat, cheese, and spices and mix well. Place in parchment lined pans, put on lid and cook for 2 hours at 160–165°F. Chill overnight and stuff in cellulose casings.

Liver and Cheese Loaf

Fresh Pork Livers (Chopped Fine)	20 lb.
Veal Trimmings (Chopped Fine)	10 lb.
Reg. Pork Trimmings (Chopped Fine)	20 lb.
Beef Trimmings (Grind through ⅛ inch plate)	35 lb.
Pork Cheek Meat (Grind through ⅜ inch plate)	10 lb.
Dry Milk Solids	10 lb.
Salt	3 lb.
Cure for Sausage	1 qt.
Fresh Onions	3 lb.
White Pepper	7 oz.
Ground Caraway	1½ oz.
Garlic Powder	¼ oz.
Swiss Cheese (½ inch cubes)	10 lb.

Grind livers and onions through ¹⁄₁₆ inch plate. Put in silent cutter and chop fine. Grind veal and pork trimmings through ⅛ inch plate. Add to livers and chop fine. While chopping, add about half of salt and cure to mixture, also all of dry milk solids and spices.

Before grinding beef trimmings and pork cheek meat, sprinkle the rest of salt and cure over meat. Do not chop the beef trimmings and pork cheeks. When chopped and ground meats are ready, put all in mixer and mix well.

Put mixture in parchment lined pans, put on lid and cook for 3 hours. Chill overnight. Remove from pans and stuff in cellulose cases.

Meat, Macaroni and Cheese Loaf

Veal Trimmings	40 lb.
Beef Trimmings	40 lb.
Reg. Pork Trimmings	20 lb.
Dry Milk Solids	12 lb.
Macaroni	10 lb.
*Cheese (Diced ½ inch size)	15 lb.
Salt	3 lb.
Onions	3 lb.
White Pepper	8 oz.
Cure for Sausage	1 qt.

Chop beef, veal and pork. Add cure, ice, dry milk solids, onions and spice while chopping. When chopped fine, place in mixer. Add macaroni and cheese and mix until evenly distributed.

Macaroni should first be thoroughly cooked. When thoroughly mixed, place in greased pans and bake at 225–250°F. (for about three hours) until the inside temperature is 152–155°. The loaves are done when this is reached; then take out and cool off.

If it is desired to stuff loaves in artificial casing, cool them off about 3–4 hours to let them set. Then stuff, dipping loaf in hot gelatin before stuffing.

If loaf is to be cooked, put meat in Frank, Hoy or similar molds.

Cook for 2–3 hours, according to what size loaf is made. Have inside temperature 152–155°F. When done, cool off in cold water. If desired in casings, follow directions above for baked loaves.

Hamburger and Cheese Loaf

Beef Trimmings (30% fat)	100 lb.
Dry Milk Solids	10 lb.
*Swiss Cheese (Cut into ½ inch cubes)	20 lb.
Fresh Onions	6 lb.
Salt	3 lb.
Cure for Sausage	1 qt.
White Pepper	7 oz.

Grind beef trimmings and onions through ⅜ inch plate. Mix dry milk solids, salt, cure, and pepper with meat and grind again through same plate.

When grinding is completed, place meat in mixer; then add cheese and mix until evenly distributed.

Put mixture in greased baking pans, smooth top of pans with paddle occasionally dipped in a thin sugar syrup.

Bake at 225 to 250° F. for 3 hours if 5 pound pans are used. Inside temperature should be 152–155° F. Stuff in cellulose casings, which adds greatly to appearance and facilitates handling.

Glaze for Meat Loaves

Cane Sugar	2 lb.
Paprika	8 oz.
Glucose	8 oz.
Gelatin	1 lb.
Boiling Water	2½ lb.

* A cheese that does not melt or run at the temperature required for processing must be used. Cheese especially made for this purpose is recommended.

After a meat loaf mixture has been filled into well-greased loaf pans, it can be brushed lightly with the above mixture.

Barbecue Sauce for Pumping Barbecued Hams

Crushed Bay Leaves	1 oz.
Crushed Cloves	1 oz.
Ground Black Pepper	2 oz.
Ground Nutmeg	1 oz.
Ground Marjoram	1 oz.
Ground Mustard	2 oz.

Put all in muslin bag. Cook in one gallon of water for 1½ hours at 170–180°F.

Paprika	½ lb.
Brown Sugar	1½ lb.
Mapleine Flavor	2 oz.
Tomato Puree	1 gal.
Vinegar	1½ pt.

Take ½ pound paprika, add just enough boiling water to stir into a thin paste.

When spices have been cooked enough, remove bag of spices, squeeze dry, then add to the cooking water 1½ pounds of brown sugar. Stir until dissolved. Then add 1 gallon of tomato puree, 1½ pints of full strength vinegar, paprika paste, and 2 oz. Mapleine. Mix well.

Use this mixture for pumping. Pump in the same manner as a ham is pumped with pickle.

Boil hams at 170°F., 30 minutes for each pound of ham. After cooking place in shallow pans, sprinkle brown sugar over ham and place in oven, bake at 400°F. until it is nicely browned. Ham can be stuffed in cellulose casings which will give it an attractive appearance and will facilitate handling.

When a boneless ham is desired, pump the boned ham in such a way that the liquid will be evenly distributed, then sprinkle throughout the open cuts of the ham a few ounces of dry milk solids. Place in parchment paper lined containers and pour a few ounces of the sauce over ham and boil as any other boiled ham.

After ham is cooled and chilled, remove from container, place in pans and sprinkle brown sugar over it, place in 400°F. oven and bake until brown. It can be stuffed in cellulose casings, if desired.

Scrapple
Formula No. 1

Veal Trimmings	50 lb.
Lean Pork Trimmings	50 lb.
Corn Meal	30 lb.
Dry Milk Solids	10 lb.
Salt	6 lb.
White Pepper	7 oz.
Rubbed Sage	3 oz.
Nutmeg	2 oz.
Cooking Water	180 lb.

Select the meat and place each variety in a separate net, as the cooking time may vary and the net facilitates handling. Put enough water in kettle to cover meats well, cook all until tender.

Grind all meats through $\frac{1}{4}$ inch plate. Leave 180 lb. of cooking water (approximately $22\frac{1}{2}$ gallons) in kettle, start agitator or begin stirring, then sift dry milk solids and corn meal into cooking water. Cook this 45 minutes, then add ground meats and seasonings. Cook until the mixture is thick. (If stirred by hand, it must be stirred continuously to keep bottom and sides from sticking. If agitator is used, it must run all the time during cooking.) Then pour into pans of desired size and cool.

Care must be taken that the mixture is not too liquid when poured in pans, as if this happens, the slices will crumble when fried. A mixture that will just about run when poured into pans is the right consistency.

When set and cold take out of pans and wrap. Some manufacturers stuff loaves in artificial casings.

To prepare for eating, slice about $\frac{1}{2}-\frac{3}{4}$ inch thick and fry. This is a very wholesome and nourishing food. The use of dry milk solids increases the nutritive value of the scrapple to a considerable extent and adds much to the flavor and texture.

No. 2

Hog Cheek Meat	45 lb.
Hog Snouts	30 lb.
Hog Skins	25 lb.
Corn Meal	30 lb.
Dry Milk Solids	10 lb.
Salt	6 lb.
White Pepper	7 oz.
Rubbed Sage	3 oz.
Nutmeg	2 oz.

Select meats and place each in separate net, as the cooking time will vary with different kinds of meat; this also facilitates handling.

Place meats in a steam jacketed kettle and pour in enough water to cover meat nicely. Cook hog skins until very soft, other meats until tender. Cook slowly.

When cooked run hog skins through $\frac{1}{8}$ inch plate, other meat through $\frac{1}{4}$ inch plate. Leave about $22\frac{1}{2}$ gallons or approximately 180 pounds of cooking water in kettle; start agitator or begin stirring and sift the corn meal and dry milk solids not over 1½% fat into the kettle. Cook 45 minutes, then add meat and seasoning and cook until thick. Stir or agitate continuously.

When thick enough (which is when mixture is just liquid enough to run when poured), pour in pans of desired size and cool. When cold, take out of pans and wrap.

Other meats than those mentioned can be used; in fact, any odd pieces that may be on hand, leftovers from stuffer, also weasand meat, beef and hog tripe, and other offal that may be on hand can be worked off.

Head Cheese, Country Style

Cured Pork Snouts	40	lb.
Cured Pork Tongues	25	lb.
Cured Hogskins	15	lb.
Cured Hog Ears	20	lb.
Dry Milk Solids	12	lb.
Cooking Water	60	lb.
Fresh Onions	5	lb.
Salt	1½	lb.
Pepper	10½	oz.
Marjoram	3½	oz.

Cook all meats in separate nets until tender. Pork tongues will take the longest time, about three hours. Use just enough water to cover meats nicely.

After cooking, grind hogskins and onions through a fine plate and cut all other meat either by hand or with head cheese cutter.

Then place cooking water in mixer and add ground hogskins, snouts, tongues, ears, dry milk solids and spices. Mix well.

After mixing is completed, pour in sauce pans. 3 pound and 5 pound pans are mostly used. After filling, place in cooler.

The same formula may be used for stuffing in hog stomachs, beef bungs, or artificial casings. If hog stomachs are used, they should be cooked for 45 minutes at 160°F. after stuffing. If beef bungs are used, 15 minutes at 160°F. will be sufficient; if artificial casings are used it is only necessary to rinse them in hot water, to clean outside of casings.

Then place them in cooler. When they are set they are ready to market.

Breakfast Pork and Apple Patties

Lean Pork Trimmings		50	lb.
Pork Belly Trimmings		37	lb.
Dried Apples		8	lb.
Dry Milk Solids		5	lb.
Salt	2 lb.	12	oz.
White Pepper		5	oz.
Red Pepper		½	oz.
Sage		1	oz.
Ginger		1	oz.
Marjoram		½	oz.
Paprika		2	oz.

Soak dried apples for about 4 hours in cold water, pour in colander and drain. Spread pork evenly on bench and sprinkle dry milk solids and spices that have been mixed well over meat. Then lay soaked apples evenly over same. Mix well and grind through $\frac{3}{16}$ inch plate. Make into patties of suitable size.

Chili Con Carne

Boneless Chucks	45	lb.
Beef Trimmings	30	lb.
Beef Suet	25	lb.
Dry Milk Solids	8	lb.

Salt	2 lb. 12 oz.
Chili Pods	3 lb.
Chili Pepper (Ground)	6 oz.
Cloves (Ground)	1 oz.
Oregano	2 oz.
Onion Powder	2 oz.
Garlic Powder	½ oz.
Paprika	8 oz.

Grind beef suet through one inch plate, place this in steam jacket kettle and render until cracklings begin to float, during this process stir frequently.

Grind beef through a ¼ inch plate and put in kettle with rendered suet. Cook at 200°F. for one hour, stirring very frequently. (If there is a kettle with mechanical agitator available it would be best to use it.)

After cooking one hour add dry milk solids and spices, being careful to sprinkle dry milk solids evenly, stirring all the time (if dumped in there is a possibility of lumps appearing in the finished product).

Cook all 45 minutes more and turn off steam. Remove from kettle and put in meat truck. If fast cooling is desired, run truck in cooler. The mixture must be stirred frequently to keep meat from settling to bottom. When cooled to about 120°F. pour in pans holding one pound.

If it is desired to stuff chili in cellulose casing, put mixture into stuffer that has been flushed with hot water. Stuff into 1⅞x11 cellulose casings. This casing will hold one pound. Dip this package into hot water to rinse off any particles that may adhere, then hang up and place in cooler.

The addition of dry milk solids reduces the shrink, improves the flavor, and increases the nutritive value very materially.

Smoked Roast Chicken

Use any size fresh clean roasting chicken. Stuff with meat prepared as follows:

Lean Pork Trimmings	30	lb.
Veal Trimmings	50	lb.
Dry Milk Solids	20	lb.
Salt	2¾	lb.
Cure for Sausage	1½	pt.
White Pepper	6	oz.
Ground Ginger	2	oz.
Paprika	2	oz.
Ground or Rubbed Sage	4	oz.

Grind pork and veal through a $\frac{3}{16}$ inch plate. Put veal, cure, and salt in first, also shaved ice. Chop 3 minutes, then add dry milk solids, spices and pork. Add enough shaved ice to make mixture about consistency and fineness as for Bologna.

Sew up opening (where the craw has been removed) and stuff chicken with meat mixture very tightly. Then sew up other opening.

Lay chicken on screen shelves and put in smokehouse. Smoke in dense cool smoke for 2 hours, then bake 4 hours at 160–170°F. The inside temperature should be 152°F. before removing from oven.

To prepare for the table, place chicken in roasting pan, add about 2 inches of water in pan, cover, and heat about ¾ hour at roasting temperature. Serve while hot. The flavor is distinctive and will satisfy the most discriminating taste.

Nutrition Sausage

Boneless Beef		
Chucks	70	lb.
Pork Trimmings		
50/50	30	lb.
Dry Milk Solids	3½	lb.
Ice Water	4	lb.
Salt	2¾	lb.
Cure for Sausage	1	qt.
White Pepper	6	oz.
Mustard Seeds	3	oz.
Ground Mustard	4	oz.
Nutmeg	2	oz.
Onion Powder	1	oz.
Garlic Powder	½	oz.
Dried Ground		
Parsley	1	oz.
Rubbed Sage	1	oz.

Use fresh meat. Grind beef and pork through a one-inch plate, then place meat in mixer and add ice cold water in which cure has been dissolved. Then sprinkle dry milk solids and spices and mix well. Grind the whole mixture through a ¼ inch plate.

Stuff in cellulose casings and hang in 38–40°F. cooler for 24 hours. Then take out of cooler and let hang in sausage room temperature for 2 to 3 hours, give them a momentary water shower before hanging in smokehouse. Smoke 6 hours, holding temperature at 120–150°F. for 2 hours, at 150–170°F. for 2 hours and at 180° the last two hours. When chilled, the sausage is ready for sale, or can be kept several months.

Cervelate Sausage

Beef Trimmings	35	lb.
Pork Cheek Meat	35	lb.
Regular Pork Trim-		
mings	20	lb.
Lean Pork Trimmings	10	lb.

Dry Milk Solids	3 lb. 8	oz.
Ice Water	5	lb.
Salt	3	lb.
Cure for Sausage	1	qt.
White Pepper	6	oz.
Onion Powder	1	oz.
Garlic Powder	½	oz.

Grind beef and pork cheeks through ⅛ inch plate and regular and lean pork trimmings through ⅜ inch plate. Put all meat in mixer, add the cure, water, dry milk solids and spices and mix thoroughly.

Place on pan 4–6 inches deep and put in cooler at 38–40°F. Let stand for 24 hours, then stuff in wide beef middles or corresponding size of cellulose casing and hang in smokehouse. Smoke about 8 hours, the last 2 hours at 160–170°F. Then dip for a few seconds in hot water 180–190°F. Hang up to cool. The sausage should be chilled before shipping.

Marbled Sausage

Extra Lean Pork		
Trimmings	70	lb.
Regular Pork Trim-		
mings	10	lb.
Backfat (Diced)	10	lb.
Cured Pork Hearts	10	lb.
Dry Milk Solids	4	lb.
Pistachio Nuts (Peeled)	2	lb.
Pimento (#3 cans)		
Cut into ¾ in.		
Pieces	3	
Salt	2¾	lb.
Cure for Sausage Meat	1	qt.
White Pepper	6	oz.
Savory	1	oz.
Onion Powder	1	oz.
Mace	1	oz.
Cinnamon	1	oz.
Shaved Ice	17	lb.

Grind lean pork and regular pork trimmings through ⅛ inch plate. Scald and chill diced back fat. Cook pork hearts and dice them the same size as backfat, cutting out the tough sinews.

Chop lean pork about 2 minutes, adding salt, cure, ice and dry milk solids. Then add pork trimmings and spices and chop 2½ to 3 minutes more.

Put chopped meat in mixer and add hearts, backfat, pistachio nuts and cut pimentos. Mix well. Care must be taken not to mix too long as the pimentos will tear and smear. To avoid this, it is best to sprinkle pimentos slowly as the meat in mixer revolves.

Stuff in beef bungs or corresponding size of artificial casings. Smoke about 2 hours at 130–165°F. and cook at 160°F. until the inside temperature has reached 152–155°F. It will take about 3 hours if diameter of sausage is 4 inches.

Chill and cool.

Farmer Style Sausage

Boneless Chucks	50 lb.
50/50 Pork Trimmings	50 lb.
Dry Milk Solids	3 lb. 8 oz.
Salt	3 lb.
Cure for Sausage	1 qt.
White Pepper	6 oz.
Onion Powder	1 oz.
Garlic Powder	¼ oz.

Run beef and pork separately through a one-inch plate. Then put beef in mixer. While meat is revolving in mixer, add pork, water, cure, dry milk solids and seasoning. Mix about 3 minutes, then run all through a ³⁄₁₆ inch plate.

Stuff meat in 1⅞″ x 20 cellulose high stretch, or 2¼″ select beef middles. Hang sausage in cooler at 38–40°F. for 48 hours, the meat will be thoroughly cured by this time. Take out of cooler and let hang in sausage room temperature not less than 2 hours, then hang in smokehouse and smoke at 120–170°F. for 8 hours. The last 30 minutes the temperature should be held at 170°F.

After taking out of smokehouse, spray with very hot water or dip in water at 160°F. for one minute. This sausage can be marketed the next day or can be kept for several months. If kept for several months it should be stored in cooler at 55–60°F. and 70–75% humidity.

Summer Sausage

Boneless Chucks	65 lb.
Regular Pork Trimmings	35 lb.
Dry Milk Solids	3½ lb.
Cold Water	5 lb.
Salt	2 lb. 14 oz.
Cure for Sausage	1 qt.
White Pepper	6 oz.
Ground Mustard	2 oz.
Whole Mustard Seed	2 oz.
Onion Powder	2 oz.
Garlic Powder	½ oz.

Grind beef and pork through a 1 inch plate. Place meat in mixer. Sprinkle cure dissolved in water, dry milk solids, and seasoning (in the order named) over meat. Mix well. After mixing, grind all through a ³⁄₁₆ plate. Then stuff in beef middles or corresponding size of cellulose casings. Hang in

38–40°F. cooler for 24 hours to cure.

Before placing in smokehouse they should hang in sausage room temperature about 1½ to 2 hours. (If placed in hot smoke after coming from cooler a dark ring may develop around sausage.) Hang in smokehouse with temperature about 120°F. Gradually increase temperature so that at the end of 7 hours the temperature is 170°F. Maintain this temperature for 30 minutes. Rinse with hot water, or place under hot shower to wash off greasy film; then chill under cold shower; hang in sausage room temperature to dry. Then place in cooler.

Thueringer Style Sausage

Boneless Chucks	60 lb.
Pork Trimmings 50/50	40 lb.
Dry Milk Solids	4 lb.
Salt	3 lb.
Cure for Sausage	1 qt.
White Pepper	6 oz.
Ground Mustard	3 oz.
Whole Mustard Seed	2 oz.
Garlic Powder	¼ oz.

Grind beef through ⅜ inch plate. Grind pork through 1 inch plate. Place all meats together in mixer. While mixing, add the cure, sprinkle in dry milk solids and spices and mix well. After mixing run all through ⅛ inch plate.

Place in pans not over 8 inches deep and put in cooler for 24 hours at 38–40°F. Mix again and stuff in beef middles, sewed beef middles or artificial casings of corresponding size.

Hang in smoke house at low temperature for 8 hours, gradually raising temperature to 170°F. Keep at this temperature for 1 hour. Smoke 9 hours in all.

Take out of smokehouse and hang in sausage room temperature overnight, being sure not to let them hang in a draft. The best way to protect them from draft is to cover top and sides of smoke-tree with canvas. Then place in cooler.

Polish Style Sausage

Boneless Chucks (Cured)	40 lb.
Lean Pork Trimmings (Cured)	20 lb.
Pork Cheeks (Cured)	20 lb.
Reg. Pork Trimmings (Cured)	20 lb.
Dry Milk Solids	4 lb.
Shaved Ice	20 lb.
Salt	8 oz.
Black Pepper	8 oz.
Mace	2 oz.
Coriander	2 oz.
Garlic (Finely cut)	6 oz.

Grind beef through ⅛ inch plate, then place in silent cutter and chop. While chopping (about 5 minutes) add dry milk solids and ice. Grind all pork through ⅜ inch plate, then put in mixer and add chopped beef and spices. Mix thoroughly.

Stuff in beef middles or corresponding size of cellulose casing. Smoke, starting with low temperature and increase heat until desired color is obtained. Then cook 40–45 minutes at 160°F. Submerge in or spray with cold water, let hang in room temperature to dry, and cool. Then place in cooler.

Mortadella Style Sausage

Boneless Chucks	70 lb.
Beef Fat (Off Flank or Brisket, not Suet)	10 lb.
Lean Pork Trimmings	20 lb.
Dry Milk Solids	3 lb. 8 oz.
Red French Wine	4 qt.
Ice Water	6 lb.
Salt	3 lb.
Cure for Sausage	1 qt.
White Pepper	7 oz.
Gelatin	4 oz.
Coriander	2 oz.
Bay Leaves	4 oz.
Cinnamon	3 oz.
Whole Cloves	2 oz.
Mace	4 oz.

Place spices in muslin bag and put in wine. Heat wine for 20 minutes at 200°F., then strain wine and let cool.

Grind beef, beef fat, and pork through one-inch plate. Place in meat truck evenly. Dissolve gelatin in ice water, then add cure to water and sprinkle on meat in truck. Next add dry milk solids, salt, and white pepper. Then mix well.

When wine is cold, sprinkle over mixture in truck and mix again. Then run all through ⅛ inch plate and spread on pan 5–6 inches deep and let stand overnight for 12–15 hours. Then stuff in beef bung cap ends or corresponding size of cellulose casings. Bladders may also be used.

Let stuffed sausage hang overnight in cooler. Then hang in smokehouse with temperature about 120°F. Gradually increase temperature so that at the end of 8 hours it will be 170°F. Maintain this for 1 hour more. Rinse with hot water and put under cold shower. Hang in temperature of 45 to 50°F. for 3 days and then they are ready for use.

Salami Cotto (Cooked Salami)

Boneless Chucks	60 lb.
Trimmed Pork Cheek Meat	25 lb.
Regular Pork Trimmings	15 lb.
Dry Milk Solids	3½ lb.
Cold Water	6 lb.
Salt	2 lb. 14 oz.
Cure for Sausage	1 qt.
Cracked Black Pepper	7 oz.
Garlic Powder	2 oz.

Grind the beef through ⅛ inch plate and the pork through ⅜ inch plate. Place all meat in mixer. Sprinkle cure and water evenly over meat while mixer revolves. Then sprinkle dry milk solids and seasoning over mixture. Mix well. Spread in pans to a thickness of about 6 to 8 inches and cure for 48 hours.

Stuff in beef cap ends or corresponding size of cellulose casings. Hang sausage in smoke house and smoke at a temperature of about 120°F. Gradually increase temperature so that at the end of 7 hours, the temperature is 170°F. Maintain this temperature for 30 minutes and the sausage is finished.

After sausage is taken out of smokehouse, place under hot shower to rinse off grease on the outside of casings. Then give a cold shower and dry in sausage room temperature. When dry, place in cooler.

If smoke flavor is desired, use Hickory Sawdust, otherwise apply heat only.

Norwegian Style Sweet Bologna

Beef Trimmings	30 lb.
Boneless Chucks	30 lb.
Pork Trimmings	40 lb.
Dry Milk Solids	4 lb. 8 oz.
Dried Seedless Raisins	3 lb.
Salt	3 lb. 4 oz.
Cure for Sausage	1 qt.
White Pepper	5 oz.
Onion Powder	1 oz.

Grind beef and pork separately through a ¼ inch plate. Wash raisins clean in cold water. Put beef in silent cutter, add salt and cure, shaved ice, and dry milk solids. Chop 5 minutes. Then add pork and spices. When mixture is almost fine enough, add raisins and chop 2–3 revolutions. Add enough ice during chopping period to make mixture of right consistency. It should be chopped to about the same fineness as regular bologna.

Stuff in 1⅞ x 20 cellulose casings or wide beef middles. Hang in smokehouse and smoke for 2–2½ hours at 120–160°F. Cook at 160°F. for 45 to 60 minutes, depending upon diameter of casing. Then put under cold shower or immerse in cold water for about 10 minutes. Enough heat should be left in sausage so it will dry itself.

Strained honey can also be used in this sausage instead of raisins. When honey is used, heat 1½ pounds honey so it will run freely when poured and pour it evenly over the meat just before the chopping process is completed. About three revolutions of the chopper are sufficient to get the honey evenly distributed; otherwise the operation is the same as with raisins.

Skinless Frankfurters

Veal Trimmings	20 lb.
Beef Trimmings	15 lb.
Bull Meat	30 lb.
Pork Cheeks	35 lb.
Dry Milk Solids	5 lb.
Salt	3 lb.
Cure for Sausage Meat	1 qt.
Pepper	7 oz.
Mace	2 oz.
Mustard (Ground)	3 oz.
Onion Powder	1 oz.
Garlic Powder	¼ oz.

High-Grade Frankfurter

Veal Trimmings	45 lb.
Bull Meat	20 lb.
Regular Pork Trimmings	35 lb.
Dry Milk Solids	4½ lb.
Salt	3 lb.
Cure for Sausage Meat	1 qt.
White Pepper	7 oz.
Mace	2 oz.
Ground Mustard	4 oz.
Onion Powder	1 oz.
Garlic	¼ oz.
Paprika	2 oz.
Moisture	36 lb.

Sheep casings
144 lb. weight in smoke (1¾ hr.; 130–175°F.)
134½ lb. weight out smoke, in cook
6½% of smoke loss
137¼ lb. weight out cook
2% cook gain
131 lb. weight next morning
9% of processing and cooler loss.

Vienna-Style Sausage

Veal Trimmings	75 lb.
Regular Pork Trimmings	25 lb.
Dry Milk Solids	4½ lb.
White Pepper	6 oz.

Ground Anise	1 oz.
Ground Caraway Seed	1 oz.
Ground Mustard	3 oz.
Salt	3 lb.
Cure for Sausage Meat	1 qt.

Mayence-Style Sausage

Lean Pork Neck Trimmings	77 lb.
Pork Rinds	23 lb.
*Dry Milk Solids	8 lb.
Blood	20 lb.
Salt	3 lb.
Sodium Nitrate	2½ oz.
Sodium Nitrite	¼ oz.
Corn Sugar	6 oz.
White Pepper	5 oz.
Peppermint	3½ oz.
Ground Cloves	2½ oz.
Marjoram	2 oz.
Mace	1½ oz.

Run pork through a 1-in. plate. Cook the pork rinds until tender, and grind twice through $\frac{1}{16}$-in. plate. Put salt, nitrate, nitrite and corn sugar in 3 quarts of blood and stir well. Place the pork and pork rinds in the mixer and pour in the blood. Sprinkle in the dry milk solids and spices and mix well.

Fill hog stomachs with the mixture and cook at 190 to 200°F., for two to three hours, depending on the size of the stomachs. The sausage should be stirred slowly for 15 to 20 minutes. If it is left undisturbed the blood will settle on one side.

Liver Sausage

Hog Livers (Fresh Clean)	50 lb.
Pork Belly Trimmings	15 lb.
Pork Cheeks or Neck	

* Not over 1½ per cent fat.

Trimmings	35 lb.
Dry Milk Solids	3½ lb.
Salt	3 lb.
Cure for Sausage	1 qt.
Onions (Fresh Peeled)	4 lb.
White Pepper	6 oz.
Sweet Marjoram	1½ oz.
Mace or Nutmeg	2 oz.
Ginger	1½ oz.

Run livers and onions through $\frac{1}{16}$ inch plate, then place in silent cutter and chop fine. Run pork through ⅛ inch plate and add to liver in silent cutter. Then add salt, cure, dry milk solids and spices. Chop until fine.

Stuff into hog bungs. Cook at 165°F. until inside temperature reaches 150°F.; then reduce temperature to 160°F. and cook 30 minutes, and chill quickly. Hang up to dry.

If smoked liver sausage is desired, hang sausage in smoke house after it is dry, and smoke at low temperature until the desired color is attained. Care must be taken so that temperature will not get so high that sausage begins to drip. If this happens, the sausage will look streaked and the shrink will be considerable.

Braunschweiger-Style Liver Sausage

Hog Liver (Fresh or Frozen)	55 lb.
Skinned Pork Jowls	45 lb.
Dry Milk Solids	3½ lb.
Salt	3 lb.
Cure for Sausage	1 qt.
White Pepper	6 oz.
Mace	2½ oz.
Ground Cloves	¾ oz.
Marjoram	2 oz.
Ginger	1½ oz.

| Onion Powder | 2 oz. |

Grind livers through ⅛ inch plate, then put in silent cutter and chop very fine. While chopping livers, add cure and salt. If fresh onions are used, grind them with the livers.

When livers are fine enough (bubbles appearing on surface is an indication), add dry milk solids, then pork jowls that have been ground through a ¼ inch plate, and spices. Chop until fine enough.

Stuff into sewed hog bungs, then cook at 165°F. until inside temperature reaches 152–155°F. Then cool and hang in smokehouse. Smoke at low temperature for 6 hours.

Marble Liverwurst

Fresh Hog Livers	35 lb.
Veal Trimmings	10 lb.
Backfat	10 lb.
Fresh Hog Livers	15 lb.
Reg. Pork Trimmings	30 lb.
Dry Milk Solids	4 lb.
Salt	3 lb.
Cure for Sausage	1 qt.
White Pepper	6 oz.
Sage	1 oz.
Sweet Marjoram	1½ oz.
Ground Cloves	½ oz.
Onions (Fresh Peeled)	3 lb.

Cook 15 lb. of fresh, clean livers at 160°F. for about 25–30 minutes. When ready cut into ⅜–½ inch cubes. Cook backfat for about the same length of time and cube to same size.

Grind 35 lb. fresh hog livers, pork, veal trimmings and onions through 1/16 inch plate. Then chop this mixture in silent cutter until fine. During chopping add salt, cure, dry milk solids and spices.

Place this mixture in mixer and add cubed liver and backfat, mix until cubes are evenly distributed.

Stuff into sewed hog bungs. Cook at 160°F. for 1½ to 2 hours, depending upon diameter of bungs. Then chill quickly, hang to dry.

If a smoke flavor is desired, hang in smokehouse after drying and smoke at lowest temperature possible, until color is satisfactory.

Smoked Meatwurst

Bull Meat or Boneless	
Chucks	35 lb.
Beef Trimmings	30 lb.
Reg. Pork Trimmings	35 lb.
Dry Milk Solids	4 lb.
Salt	3 lb.
Cure for Sausage	1 qt.
White Pepper	6 oz.
Allspice	2 oz.
Mace	2 oz.
Coriander	1 oz.
Onion Powder	½ oz.
Garlic Powder	¼ oz.

Run beef and pork through a one inch plate then put in mixer. Sprinkle in 10–12 lb. ice cold water, cure, dry milk solids, salt and spices. Mix well and then run all through ⅛ inch plate.

Stuff in beef rounds and hang overnight or about 12 hours in 38–40°F. cooler.

Then put in smokehouse and smoke until the desired color has been obtained. Then cook 30 minutes at 160°F. Immerse or give cold water shower.

Bockwurst

Veal Trimmings	40 lb.
Beef Trimmings	25 lb.
Pork Trimmings	20 lb.

Lean Pork Trimmings 15 lb.
Dry Milk Solids   4 lb. 8 oz.
Salt   3 lb. 4 oz.
Cure for Sausage   1 qt.
White Pepper   7 oz.
Parsley (Chopped Fine)
                     4 bunches
Sage   1 oz.
Angostura Bitters   2 oz.
Fresh Eggs   2 doz.
Grind veal, beef, and pork through ⅜ inch plate. Chop beef, salt, cure, veal, and pork, in order named. After veal is added to chopper add dry milk solids. After pork is added, put in parsley and spices. Just before meat is chopped fine enough, add eggs that have been beaten previously and run several revolutions.

Stuff in extra wide hog casings or beef rounds. This is a fresh sausage and must be handled about the same as pork sausage.

Blood Pudding
Hogskins   15 lb.
Hog Snouts, Ears,
Weasands, Giblets,
  etc.   30 lb.
Backfat (Diced)   20 lb.
Hog Blood (or Beef
  Blood)   35 lb.
Dry Milk Solids   12 lb.
Salt   2 lb. 12 oz.
Black Pepper   6 oz.
Fresh Onions   1 lb.
Marjoram   2 oz.
Ground Cloves   1 oz.
Cinnamon   ½ oz.
Put hogskins and other materials, except backfat, in nets and cook until tender. Scald backfat and dice, rinse in hot water before mixing with other ingredients. Grind hogskins through fine plate,

other material through $\frac{3}{16}$ inch plate.

Place all ground material in mixer, then add blood that has been strained, dry milk solids, backfat and spices. Mix well and stuff into wide beef rounds, place them in kettle and cook 30–35 minutes at 160°F. Test with wire skewer; if no red color shows the pudding is cooked enough. Submerged in cold water or shower with cold water. Five minutes of either will be sufficient. Hang up to dry.

Liver Pudding
Hog Livers (Scalded)   35 lb.
Hogskins (Cooked
  Until Tender)   20 lb.
Hog Stomachs (Cooked
  Until Tender)   20 lb.
Beef Tripe (Cooked
  Until Tender)   10 lb.
Hog Gut Fat
  (Scalded)   15 lb.
Dry Milk Solids   12 lb.
Onions   4 lb.
Salt   3 lb.
White Pepper   6 oz.
Marjoram   1 oz.
Nutmeg   2 oz.
Place hogskins, livers, hog stomachs, beef tripe, gut fat, onions, and dry milk solids, in order named, in silent cutter. While chopping, add spices. Chop fine. Put in truck, add plenty of meat broth (it will take from 30–35 pounds of this) and mix well.

Stuff in beef rounds, then cook 30 minutes. Chill in cold water. When chilled, hang up to dry.

Liverwurst
Hog Livers (Scalded)   40 lb.

Hog Gut Fat
(Scalded)   15 lb.
Hog Tripe (Cooked
Until Tender)   15 lb.
Hogskins (Cooked
Until Tender)   20 lb.
Beef Tripe (Cooked
Until Tender)   10 lb.
Dry Milk Solids   4 lb.
Onions (Fresh Peeled)   4 lb.
Salt   3 lb.
White Pepper   6 oz.
Sweet Marjoram   1½ oz.
Ground Cloves   ½ oz.

Place hogskins in net, and add the beef tripe and hog tripe. Place all in a steam jacketed kettle. Pour just enough water over them to cover nicely.

Cook slowly until tender. When cooked enough, take out. Scald the livers and gut fat in the same broth for a few minutes.

After ingredients are so prepared, place hogskins in silent cutter and chop until they do not feel gritty to touch, then add liver, tripe, and onions. Chop until almost fine enough, then add hog fat, dry milk solids and spices, and chop until fine.

Place this mixture in meat truck and add a good portion of the broth, mix well. From 30 to 35 pounds of broth can be added. Stuff in beef rounds. Cook for 30 minutes, then chill in cold water. When chilled hang to dry.

Ring Bologna
Formula No. 1

Boneless Chucks   65 lb.
Reg. Pork Trimmings   35 lb.
Dry Milk Solids   5 lb.
Salt   3 lb.
Cure for Sausage   1 qt.

White Pepper   8 oz.
Allspice   2 oz.
Onion Powder   1 oz.
Garlic Powder (Op-
tional)   ½ oz.

Grind boneless beef and pork trimmings separately through a ⅜ inch plate. Chop beef in silent cutter, adding cure, salt, shaved ice and dry milk solids. Chop until almost fine, then add pork trimmings and spices and chop until fine enough. During chopping process add enough ice so that mixture is of right consistency.

Stuff into beef rounds, smoke, cook, chill, and hang in cooler.

This formula can be used for straight Bologna by stuffing into beef middles or cellulose casings of desired length.

No. 2

Boneless Chucks   35 lb.
Beef Trimmings   25 lb.
Pork Hearts   15 lb.
Reg. Pork Trimmings   25 lb.
Dry Milk Solids   5 lb.
Salt   4 lb.
Cure for Sausage   1 qt.
White Pepper   8 oz.
Allspice   2 oz.
Onion Powder   1 oz.
Garlic Powder (Op-
tional)   ½ oz.

Grind beef and pork hearts and pork trimmings separately through ⅜ inch plate.

Chop beef in silent cutter, adding cure, salt, shaved ice, and dry milk solids. Chop until almost fine, then add pork hearts, pork trimmings and spices. Chop until fine enough. During chopping process add enough ice so that mixture is of right consistency.

Stuff in beef rounds, smoke,

cook, chill, and hang in cooler. This formula may be used for straight Bologna, stuffing into beef middles or cellulose casings of desired length.

Bologna with Soya Flour

Beef Trimmings or Bull Meat	75 lb.
*Reg. Pork Trimmings	75 lb.
Pork Blade Fat, Cured	12 lb.
Sodium Nitrite	3/8 oz.
Salt	4 lb. 8 oz.
White Pepper	10 oz.
Coriander	3 oz.
Sugar	10 oz.
Ground Celery Seed	1 oz.
Nutmeg	3 oz.
Soya Flour	6 to 9 lb.
Ice	As required

Run the beef or bull meat through a No. 1 plate and then chop to a stiff dough, adding the pork trimmings just before the beef is finished. Mix in the nitrite, which should be dissolved in 1 pint of water, and the salt. Allow this to remain overnight in a box truck.

When the meat is placed in the chopper the following morning, add the spices and the soya flour. Finish by adding the 12 lb. of pork blade fat cut into cubes. Ham or backfat will do, but either must be cured. Soya flour must never be added before the meat is cured.

For bologna, the dough should be fairly stiff. Care must be taken not to burn the meat in the chopper.

* This may be one-half pork cheeks and one-half regular pork trimmings if desired.

Stuff and hang in workroom two hours before placing in smoke house.

Due to the great gain with soya flour, it is advisable to increase the spice mixture 10 to 15 per cent for best results.

Meat Preservative
(For bacon, ham or sausage)
British Patent 554,025

Sodium Chloride	86
Sodium Nitrite	3½
Sodium Sulfate	1½
Magnesium Carbonate	1
Magnesium Chloride	1
Calcium Sulfate	1
Calcium Nitrate	1
Calcium Carbonate	½

Cure for Sausage Meat

Sodium Nitrate	3 lb. 7 oz.
Sodium Nitrite	5 oz.
Dextrose (Corn Sugar)	10 lb.

Place above ingredients in a 5 gallon container (preferably glass), fill with water and dissolve. Use 1 quart of solution for each 100 pounds of meat.

This cure is referred to in all formulas given in this booklet and meets all federal and state regulations. Any other good cure may be used without impairing the quality of the finished product.

Meat Curing Salt
U. S. Patent 2,299,999

Magnesium Chloride	0.3–2.5
Water	0.5–3
Sugar	1–3
Sodium Chloride	
	To make 100

Pork-Sausage Seasoning

Ground Sage	5½

Savory	10½
Ground White Pepper	10½
Ground Black Pepper	14½
Ground African Ginger	5
Ground Nutmeg	3
Ground Cayenne Pepper	1

Preserving Lard

A sample of lard treated with 2 per cent Siam benzoin was in good condition after 18 months' exposure to air at normal temperatures, while untreated lard was rancid after three months.

Dried Salted Meat
British Patent 550,421

Minced lean meat is heated with 50 wt.-% of dilute hydrochloric acid (of such concentration that the pH falls to 1.5) so that it reaches 80°C. in 45 minutes, and is then neutralized to pH 6 with sodium hydroxide. The product is roller-dried to give a dried meat containing about 7% of sodium chloride.

Preservation of Crab Meat
U. S. Patent 2,155,308

Atlantic crab is precooked (e. g., for 3–6 minutes at 99–116°C.) and then cooled rapidly, so as to weaken the tissues by which the carapace, gills, and loosely-adhering body tissues of the body cavity are attached to the crab. These parts are removed, and the residue is cooked and canned.

Protein Food
U. S. Patent 2,155,417

A homogeneous mixture of animal (beef) blood and fresh skim milk is heated at 66–80°C. and spray-dried at 143–149°C. so as to produce a dry solid residue containing hemoglobin and casein in intimate association as the predominant protein materials; it contains:

Blood Solids	3¾
Water	5–6
Lactose	48
Ash	6–8
Protein	38–40

If used as a binder for sausage meats, loss of water on smoking is minimized.

Oxtail-Type Soup Cubes

Corn Flour	40 lb.
Potato Flour	40 lb.
Pea Flour	10 lb.
Bean Flour	10 lb.
Dried Onions	10 lb.
Dried Turnips	10 lb.
Dried Carrots	10 lb.
Salt	5 lb.
Ground Cinnamon	8 oz.
White Pepper	8 oz.
Beef Stock	10 lb.
Beef Extract	10 lb.
Boiling Water	5 lb.

Chop the dried onions, turnips and carrots fine. Place all the ingredients, except the beef extract, in an iron steam kettle fitted with a stirrer. Heat and stir continuously.

Dissolve the beef extract in the boiling water and let it drip from an overhead container into the mixture in the pan, continuing to heat and stir until dry.

Press into cubes and wrap.

Fish-Preserving Ice

Ice containing 0.05–0.1% sodium nitrate is very effective for keeping freshly caught fish in good condition.

Cooking (Fish) Deodorizer

The method for preventing disagreeable odor occurring in the cooking of sea fish consists in adding fruit pulp or a concentrate, distillate or dry product obtained from fruit to the fish during the cooking.

In order to obtain the pulp, cleaned fruits, such as pears, apples, apricots and the like are boiled. The pulp is stirred, before commencing the cooking of the fish, into the still cold oil mostly used for frying or into the water used for boiling. When the oil or water has been heated to the boiling point, the fish is introduced and cooked. Vegetable admixtures, like tomato pulp, may also be used.

CHAPTER VIII

INKS AND MARKING SUBSTANCES

Fluorescent Ink

Aesculin	10 g.
Sodium Salicylate	10 g.
Caustic Soda (1% Solution)	100 cc.

1 part of the above is added to 10 parts of ordinary ink to make it fluorescent.

Sympathetic Ink

Cobalt Chloride	10
Glycerin	2
Water	88

Writing with above solution disappears on drying; becomes visible on holding over heat.

Detecting Invisible-Ink Messages

Apply following with absorbent cotton pad to develop writing:

Potassium Iodide	4
Iodine	$\frac{1}{10}$
Salt	5
Aluminum Chloride	2
Glycerin	3½
Water	30

Black Writing Ink

Tannin	9
Iron Sulfate	4
Hydrochloric Acid	1
Aniline Blue	3
Acid Green Dye	0.8
Phenol	0.5
Water	780

Semi-Gallate Writing Ink

Trypan Blue	10 g.
Glycerol	10 g.
Gallic Acid	5 g.
Resorcinol	1 g.
Hydrochloric Acid, Normal	100 cc.

Hectograph Ink

Methyl Violet	100 g.
Glycerin	200 cc.
Alcohol	200 cc.
Water	1 l.

Duplicating Ink

U. S. Patent 2,155,861

Beeswax	4.4
Oleostearin	4.4
Mutton Tallow	26.7
Lard Oil	8.9
Crystal Violet	13.3
Brilliant Green	16.7
Magenta	6.7
Chrysoidine	18.9

Quick-Drying Printing Ink for Stamp Pads

Glycerin	2
Water	2
Butyl Carbitol	1

Black Stencil Ink

Carbon Black Pigment	28
Alkali Blue Toner	12
Lithographic Varnish	50
Paste Cobalt Drier	10

Grind pigments into varnish and drier mixed, using a roller or stone mill.

Black Marking Ink

Nigrosine Base	14 oz.
Cresylic Acid	3 qt.
Phenol	1 qt.

Mix at room temperature. Yields 1 gal.

Printing Ink
U. S. Patent 2,155,103

Boiled Linseed Oil	18.00
Litho Varnish	8.00
Cobalt Linoleate Drier	14.00
Petroleum Jelly	4.00
Triethanolamine Stearate	2.50
Lead-Manganese Acetate Resinate	1.27

Printing Offset Composition
U. S. Patent 2,142,667

A

Cellulose Acetate	1.0
Dichloroethane	5.0
Ethanol	2.5
Ethyl Lactate	1.1

A mixture of the above is sprayed onto freshly printed sheets, and on evaporation deposits particles on areas which keep the sheets slightly apart while the ink is drying.

B

Cellulose Ether	1–16 oz.
Organic Solvent (Highly Volatile)	1 gal.

A mixture of the above ensures more rapid evaporation.

Ink for Clock Numerals

Clean Powdered Lampblack	1 g.
Spike Lavender Oil	20 drops

Thin to proper consistency with spirit varnish, and apply with a soft smooth brush.

Ink for Ruling on Glass

India Ink	98
Lepages Mucilage	2

Glass Marking
Write with ordinary pen, using as ink a 10% solution of sodium silicate containing a small amount of Nekal BX as wetting agent. Then heat slowly up to dull red heat using a torch. There will be left a permanent mark on the glass resembling an etched mark.

China and Glass Ink
(Water Insoluble)

Bleached Shellac	30
Venice Turpentine	9
Turpentine	45
Red Mercuric Lamp-black, Sulfide or other pigment	16

Mix with gentle warming until smooth.

Glass Etch
The following solution is recommended for white and silk finish:

Hydrogen Fluoride (60–65%)	2
Ammonium Fluoride	4
Sodium Carbonate	1

Marking Porcelain
Write with Blaisdell China Marking Pencil No. 165-T. Heat slowly with torch until a permanent black metallic mark is left. Cool slowly.

Ceramic Stenciling Ink
U. S. Patent 2,318,124

Diethyl Phthalate	1
Venice Turpentine	33–37
Copaiba Resin	
To make	100

The above is air-drying.

Laboratory Ink

Silver Nitrate	30 g.

Aerosol (10% Soln.)   1   ml.
Ordinary Ink   1   ml.
Gum Arabic   0.5 g.
Water
          To make 100   ml.

India-Ink Thinner
Distilled Water   10 cc.
Conc. Ammonia
   Water   4 drops
Ethyl Alcohol
   (95%)   2 cc.

Mix all the above together and use to thin India ink which has become thick. It is also good for cleaning drawing instruments.

Erasing Fluid for Tracing Cloth
Acetic Acid (28%)   1
Alcohol   7

Rejuvenating Typewriter Ribbons
      German Patent 730,122
Ribbons are dampened with the following and dried:
Carbon Tetrachloride   6
Liquid Petrolatum   3

Transfer Carbon Paper
U. S. Patent 2,138,836

A carbon paper for use in a transfer process, having a color layer of initially porous and unglazed character, is made by forming a color paste of an oil-soluble salt of a basic dye as its principal color constituent, of a waxy substance, of a high-molecular fatty alcohol, and of a readily volatile solvent and applying the paste to a backing.

Formula No. 1
Nigrosine Base N   10–12
Stearic Acid   10

Higher Alcohols   10
Montan Wax, or Halowax 12

No. 2
Nigrosine Base N   10.0
Induline Base B   5.0
Nigrosine Jet, Spirit-
   soluble   10.0
Stearic Acid   5.0
Carnauba Wax   8.0
Mineral Oil   13.0
Triethanolamine
   Stearate   0.5

Red-Marking Crayon
Rhodamine B Stearate   1
Calcophen Red Y   1
Stearic Acid   20
Beeswax   5
Paraffin Wax   16
Carnauba Wax   7

Dissolve coloring materials in molten stearic acid. Then add the waxes, melt out, and pour into molds. Chill molds with cracked ice to remove.

Luminous Crayon
U. S. Patent 2,317,159

A luminous plastic writing stick is composed of 50 to 100 parts by weight of Japan wax, 25 to 75 parts of ozokerite, 50 to 300 parts of paraffin and 25 to 250 parts of petroleum jelly and a small amount of a luminous sulfide.

Hot-Metal Marking Crayon
U. S. Patent 2,294,403
Chlorinated
   Naphthalene   1.75–2.5
Sodium Nitrate   2–2.5
Titanium Dioxide   2.5

Heat and mix at 130°C.; mold and cut into rods.

CHAPTER IX

SKINS, LEATHER AND FUR

Egg-Yolk Substitute for Use in Tanning

Soybean Lecithin	15 lb.
Water	90 lb.
Sodium Bicarbonate	1 lb.

Melt the lecithin and gradually add the boiling water containing the sodium bicarbonate. Agitate in a mechanical mixer for 15 to 30 minutes until a uniform emulsion is formed.

The above emulsion is an excellent substitute for egg yolk in the fat liquoring of leather. It is of more uniform composition, and cheaper to use than egg yolk.

Tanning Extract
British Patent 548,594

A satisfactory substitute for chestnut extract is made by acidifying (to about pH 3) mimosa extract with 1–10% on weight of dry extract, of citric or tartaric acid.

Softening Dry Hides

Soaking dry hides for 2 days in 0.1% sodium polysulfide solution removes hardened cement substance and softens the hides until they equal green-salted stock.

Control of Hide and Skin Beetles

Sodium silicofluoride as an insecticide will afford excellent protection against attack by the Dermestid hide and skin beetle if the hides, etc., are dipped for 10–15 minutes in a liquid containing 0.5 per cent sodium silicofluoride and 0.01 per cent acetic acid. This solution will probably simultaneously inhibit the damage done by the skin Tineid moth.

Dehairing Hides

Ground and Sifted	
Sulfur	8
Lime	16–20
Water	10–12

Boil 4–5 hrs. with live steam; drum hides with this for 72 hrs.

Sole Leather Waterproofing

Tallow	25
Mineral Oil	10
Cod Oil	20
Paraffin Wax	20
Carbon Tetrachloride	25

Melt together the tallow and paraffin wax; add the mineral oil and cod oil; stir well while cooling, and when cooled to about 110°F., add the carbon tetrachloride and mix well.

The leather must be dried before this treatment.

Waterproofing Leather Emulsion
U. S. Patent 1,793,983

A cement for use in waterproofing leather may be made from an emulsion of 100 volumes of a 5% rubber solution in benzene, 40–50 volumes water or glycerin, and 0.25% sodium oleate soap. This cement may be used on wet leather without preliminary partial drying of both coatings.

Preventing Mold Growth on Leather

0.25–0.50%, on dry weight of leather, of following gives good protection:

p-Nitrophenol
p-Chloro-m-xylenol
p-Chlorometacresol
Pentachlorphenol
Tetrachlorphenol

Oil for Softening Leather Goods

Glue	3
Montan Wax	5
Synthetic Wax (Glyceryl Monostearate)	10
Sulfonated Neatsfoot Oil	40
Glycerin	0.5
Water	41.5

Soften glue in water, heat waxes just slightly above melting point, keep at same temperature while stirring in part of water; add sulfonated oil and keep stirring, add rest of water containing glue and softener.

For softening leather goods, preferably hard leather, like belts, etc., moisten soft cloth with mixture and rub in well. Go over oiled goods with dry soft cloth.

Leather Belt Stuffing
Formula No. 1

Train Oil	80
Stearin	20

No. 2

Castor Oil	60
Tallow	20
Stearin	20

Wide belts are stuffed cold and narrow belts hot.

Box Calf Oil

Mix into a light well refined mineral oil (Visc. 100 S.U.S. @ 100°F.) 0.1% of a brown oil soluble aniline dye, to make it bloomless. Take 50% of above oil and add 20% sulfonated neatsfoot oil and 30% sulfonated castor oil. This mixture gives luster to box calf leather if used in a thin film and rubbed in thoroughly.

Fine-Leather Dressing
(For books and desk leather goods)

Beeswax	20 g.
Diglycol Stearate	8 g.
Diglycol Laurate	1 g.
Cedarwood Oil	62 cc.
Lanolin	340 g.
Neatsfoot Oil	To make 1000 g.

Shoe-Bottom Filler
U. S. Patent 2,317,326

Binder:

FF Rosin	69.5
Mineral Oil	30.0
Oxalic Acid	0.5

Filler:

20/30 Cork Granules	11.75
14/20 Cork Granules	11.75

The rosin and mineral oil are placed in a heated kettle and stirred together at 275°F. until a homogeneous mixture is obtained. At that time ½ part of oxalic acid is added and thoroughly mixed with the rosin-oil mixture. The mixture is then cooled to approximately 250°F. for addition to the fillers. The required amount of cork granules are weighed and placed in a mixer. The binder is added and the whole quickly stirred. The binder thoroughly coats each granule of cork with a thin film without impregnation. After a satisfactory admixture is made, the material is removed from the mixer and formed into bricks or loaves of any suitable size and shape.

Wax for Leather Strips

Lanolin	15
Paraffin Wax	25
Montan Wax	30
Rosin W W	10
Gypsum	20
Caustic Soda (30° Bé)	2

Melt lanolin, paraffin, montan wax and rosin together, add soda. Mix well and add gypsum under constant stirring.

Shoe Box Toe Stiffening
U. S. Patent 2,331,095

Ethyl Cellulose	17½
Castor Oil	7½
Rosin	75

Shoe Stiffener
British Patent 548,638
Formula No. 1

Polyisobutylene and Rosin, Hydrogenated	90
Rubber	8
Paraffin Oil	2

The above are heated together for 1¾ hours at 260°C., and cooled to give a suitable impregnation compound.

No. 2
British Patent 548,638

Hydrogenated Rosin	90
Crepe Rubber	8
Plasticizer	2

Warm together and mix until uniform.

Coloring Used Shoes

For recoloring worn shoes, nitrocellulose lacquer applied with an atomizer is recommended. Composition of the lacquer is:

Nitrocellulose Colored Lacquer	100
Protective Varnish	30
Tritolyl Phosphate	4
Castor Oil	6
Butyl Acetate	90
Alcohol	80

The leather should first be cleaned, e.g., with green soap 50, ammonia spirits 50, water 900 and given a base coat with the above nitrocellulose lacquer.

Black Leather Dye Solution

Nigrosine Base	10
Oleic Acid	10
Aniline	5
Furfural	5

Dissolve with gradual heating (not above 110°C.). Cool and add:

Alcohol	20
Acetone	20
Benzol	90
Ammonia	1

Shoe-Sole Stain

Shellac	13
Casein	4
Trigamine	5
Water	130

Warm and stir until dissolved. Then mix in:

Water Soluble Nigrosine	5

Continue heating and add:

Crude Montan Wax	6
Carnauba Wax No. 3	2
Paraffin Wax	5
Stearic Acid	2

Bactericidal Bristles
U. S. Patent 2,304,478
Formula No. 1

100 grams of white bristles are immersed at room temperatures in a solution containing between 3% and 5% of silver nitrate and the bristles remain in the solution which may occasionally be stirred until a sample of the material shows a silver content of at least 8% of the weight of the bristles dried in air, after the sample has previously been washed with water. The so treated white bristles may either be directly worked after a short washing treatment or they may be watered for a longer time while repeatedly changing the washing water. The so treated bristles show a silver content of 7% of the weight of air dried bristles in spite of repeated watering.

No. 2

100 grams of bristles are treated as set forth in Formula No. 1 and after they show a silver content of about 12% without washing they are stored for 24 hours, after which they are washed. By storing the main amount of the silver is fixed in the bristles and thus even by a longer washing treatment only small amounts of soluble silver compounds may be removed and the bristles show a silver content of about 8%. Instead of using one continuous rest period it may be advantageous to repeat the washing treatment several times while interposing, every time, a short rest period.

Washing Sheepskins

The goods are brushed over with a strong solution of benzine soap, and then run through the washing machine for thirty to forty-five minutes. The subsequent rinsing with benzine should be very thorough, or the wool will retain a greasy handle.

If the wool is very dirty, and has been much felted by long wear, the skin must be wet washed, but care must be taken neither to wash nor to dry at too high a temperature. The quality of the skin must also be carefully examined. There are many, especially those taken from diseased animals and those which have been tendered in the tanning, which will not stand wet washing unless they are carefully sewn on to a strong linen cloth. Discretion should be used in incurring responsibility for such skins.

Over-heating, either in washing or in drying, makes the leather hard and brittle, and this is difficult, or impossible to remedy. Use soft water, e.g., condense water, for the washing. Begin by removing the coarser dirt with a weak liquor of soda and ammonia. Then wring and work by hand with a good neutral soap. It is unnecessary to use brushes, as the fingers can get down to the leather more easily and quickly than a brush. More and more soap is poured over the goods till it remains quite white. As long as there is dirt and grease in the wool the lather will feel sticky, and have a gray color.

Before each addition of fresh soap it is a good plan to rinse with weak soda, whereby considerable saving in soap is effected. The final rinsing, after completion of the washing, is done first with soda and then with clean water.

The whole series of operations is carried out on a bench on which the skins can be spread out flat.

After rinsing we come, with white skins, to the bleaching. Dyed skins must be soured to liven the color, using sulfuric acid for those which have been acid dyed, and acetic acid for those dyed with basic dyes.

Various bleaching processes are current. Some persons use peroxide of sodium, others permanganate or sulfurous acid. Sodium peroxide requires to be in experienced hands to be used with advantage, and it is easy to spoil everything by not using the proper quantity.

Permanganate bleaching is cheap and easy, but sulfur bleaching is, on the whole, the best.

In bleaching with a sulfur chest no rinsing is necessary as the more soap there is in the wool, the better the fumes of the burning sulfur act, but no dirty soap must be left behind in the wool. In using peroxide of sodium every trace of soap must be carefully rinsed out before bleaching.

Soap washing of skins should never be done at temperatures about 70°F., and all bleaching with peroxides must be done quite cold.

Drying is done at 70° to 75°F. in a drying-room, or in the open air. The skins are nailed on frames to dry under tension. When quite dry they are beaten with sticks and combed and brushed.

Aftertreatment with salt and alum is quite unnecessary if stearine is well rubbed into the leather before the wet washing, but in this case the dried skins must be solvent washed to remove stearine.

Curing Hairy Sheepskins

Salting prior to removing the fat gives a better cure, as the flesh side seems to set under the influence of the salt, so that the flesh tissues stand up better to the action of the scudding knife and the skins do not have a scraped appearance. The best cure is to salt the fresh, unwashed skins with 10 per cent salt, turn in the edges, fold in half down the backbone and roll for 24 hours; scrape off the fat with a curved scudding knife, re-salt with 10 per cent salt, roll again for 24 hours and dry in the shade. Salt sprout seems to be increased by washing or wetting the skins and decreased by rolling. Washing may add to the salt dissolved on the surface, and when the skin dries, this salt crystallizes out. On the other hand, rolling enables the salt to penetrate well into the interior.

Fur Carroting Solution
U. S. Patent 2,300,660
Formula No. 1

An aqueous solution is formed of:

Perchloric Acid
(68–70%)   4–9
Hydrogen Peroxide
("100 V")   14–20
Nitric Acid (40° Bé.) 1.5–3
No. 2

Nitric acid 12% or more, hydrogen peroxide, as needed, gelatin, 3% or less.

No. 3

U. S. Patent 2,330,813

Nitric Acid	3–40
Sulfuric Acid	3–60
Chloric Acid	20–260
Water	To make 1000

No. 4

British Patent 551,705

Sulfuric Acid	1.3–5.4
Chloric Acid	1.3–5.4
Nitric Acid	0.7–1.7
Hydrogen Peroxide	1.0–2.4
Water	To make 100

Dilute to density 1.0–1.07 before use.

No. 5

U. S. Patent 2,155,161·

20 pounds of mercury is treated with 80 pounds of nitric acid (density 1.383) for about 2 hours and 3.7 volume per cent of the resulting product is added to:

Ethyl Alcohol	6.2
Water	90.1

No. 6

U. S. Patent 2,309,254

Zinc Sulfate	3–8
Acetic Acid (28%)	8–15
Tannic Acid	6–8
Nitric Acid (1.53)	2–10
Sulfuric Acid (66° Bé.)	3–10
Hydrogen Peroxide (100 Vol.)	3–8

After-Chrome for Fur Felt Hats

Charge the bath with 1–3% sulfuric acid (according to the depth of the shade to be dyed and the acid still contained in the goods) and the requisite dyestuff; or for thicker shapes and hat bodies with 5–10% Glauber's salt, 1–3% sulfuric acid and the requisite dyestuff.

Enter the well wetted or boiled felts at 40°–50°C. (105°–120°F.), raise in ½ to ¾ hour to the boil, and dye at the boil for ½ to 1 hour. Then cool the bath down to 60°–70°C. (140°–160°F.), add the corresponding quantity of bichrome (about one-third of the quantity of dyestuff, and for half-milled felt, even in the case of deep shades, not more than 1.5%), raise again gradually to the boil, and finally boil for another ½ hour.

Dyeing Fur without Dyeing Skin

The flesh side of the dry skin should before dyeing be coated with tallow or a mixture of fat of a somewhat higher melting point, and when the fat has solidified, the skins are entered into the dyebath, the temperature of which should be at 5°C. (or 10°F.) lower than the melting point of the fat.

Another method of protecting the leather from becoming stained is resorted to particularly with manufactured articles of leather, and consists in covering the flesh side with a thick wheat meal paste and, if necessary, with paper also.

Another method is the following: Fasten the skins with the flesh side to a wooden board or on a frame, and dip the hair side into the dye liquor to the extent of their length but not with the grain.

Wax Finish for Furs

Dissolve 3 to 6 ounces of paraffin wax in 1 gallon petroleum cleaning solvent.

Approved cleaning solvent is preferable because of its safety during ordinary handling.

Precaution—Paraffin separates from the petroleum solvent at temperatures below 70°F. At 15°F. it is completely chilled out of the solvent.

This finish is used for the saturation of dry-cleaned furs to replace any oils removed and to make them water repellent. It is also sponged or sprayed on materials that are lifeless or lusterless after cleaning and drying to produce high gloss.

Fur Bleaching

The hair of the fur in its normal state has a film of lipid material which tends to prevent the uniform penetration and absorption of the catalyst and bleaching solutions. In order to eliminate this film, and also to bring the hair into a state suitable for efficient absorption of these solutions, the fur is given what is known as a "killing." In general the "killing" solution consists of a mild alkali, such as sodium carbonate, ammonia, borax, di-sodium phosphate and the like, frequently in conjunction with soaps, or with other surface-active agents, and sometimes with solvents. By immersion of the furs in the "killing" solutions the hair is degreased, and at the same time a degree of swelling takes place, proportional to the extent of absorption of water by the fiber. The more efficient the "killing" action, the more regular and uniform will be the absorption of the catalyst solution, and consequently the ensuing bleaching action. Because of the wide range of furs with their different hair characteristics,

it is important to adjust the "killing" treatment in accordance with the kind of fur being treated. Some types of "killing" solutions are the following:*

Typical Killing Baths Used Preliminary to Bleaching
Dip Killing

Wolf—Sal soda 4–7 grams per liter; 2 hours 75°–80°F.

Red fox—Sal soda 4 grams per liter; 2 hours 75°–80°F.

Squirrel—Sal soda 6 grams per liter; Savon (soap) 3 grams per liter; 2 hours 80°F.

Susliki — Ammonia (sp. gr. 0.90) 5 cc. per liter; 1 hour; 80°F.

Vicuna—Potassium carbonate 5 grams per liter; 2 hours; 80°F.

Flying squirrel—(1) Penetrant 0.75 grams per liter; 1 hour; 80°F.; then in: (2) Sal soda 6 grams per liter; Savon (soap) 3 grams per liter; 1 hour; 80°F.

Mole—5 cc. Ammonia (sp. gr. 0.90) 5 grams soda ash per liter; 2 hours; 80°F.

Rabbit—10 grams borax; 1 gram penetrant per liter; 2 hours; 80°F.

Rabbit — 10 grams disodium phosphate; 1 gram penetrant per liter; 2 hours; 80°F.

Brown moufflon—Soda ash 5 grams; formaldehyde 1 cc. per liter; 2 hours; 80°F.

Black dog—Bicarbonate of soda 15 grams per liter; 2 hours; 80°F.

Raccoon — Modified soda 5 grams; 1 gram penetrant per liter; 2 hours; 80°F.

Skunk—(1) Soda ash 10 grams

* "Killing" treatments may be applied by brush, or by immersion, or by a combination of both.

per liter; 2 hours; 80°F.; wash then in: (2) Ammonia (sp. gr. 0.90) 10 cc. per liter; 1 hour; 80°F.

Brush Killing

American opossum—30 grams sal soda per liter.

Muskrat—Soda ash 15 grams; Sal ammoniac 15 grams; Penetrant 2 grams per liter. (Dissolve ingredients separately and mix the solutions at room temperature.)

Raccoon—Caustic soda 5 grams per liter.

Application of the Catalyst
U. S. Patents 1,564,378, 1,573,200

A. Ferrous compounds as catalysts.

The most commonly employed catalyst is ferrous sulfate, although other ferrous compounds are also used, but to a much smaller degree. Solutions of ferrous sulfate tend to oxidize readily to the ferric state, and it is generally necessary to have a stabilizing compound present to prevent or retard such oxidation. Considerable variations are possible as to the composition of the catalyst solution with reference to both concentration and added ingredients. The following are some examples of catalyst solutions in current practice.

1. Ferrous sulfate 16 grams, per liter of water; ammonium chloride 16 grams, per liter of water.

2. Ferrous sulfate 40 grams, per liter; ammonium chloride 16 grams, per liter; cream of tartar 2.5 grams per liter.

3. Ferrous sulfate 25 grams, per liter; ammonium chloride 10 grams, per liter; tartar emetic 2.5 grams, per liter; glacial acetic acid 1.5 cc., per liter.

4. Ferrous acetate (iron liquor 20° Bé.) 50 cc. per liter.

As a general rule, the catalyst is applied to the furs by immersion. It may also be applied by the brush method although in practice such application is less common. The various types of fur tend to absorb the catalyst in differing degrees, and it is necessary to work out the particular conditions for each type of fur. In doing so, consideration must be given to a number of factors which have an important bearing on the fixation of the catalyst by the hair. The effects of these factors may be summarized as follows:

1. The amount of ferrous compound absorbed by the hair varies directly with the intensity of the "killing." This in turn may be directly co-related with the pH of the solutions used. However, if the hair after "killing" is adjusted to a constant pH somewhat below 7, the variation in amount of catalyst absorption due to the "killing" is greatly reduced.

2. Washing of the hair after treatment with the ferrous solution has only a very slight effect on the removal of iron fixed by the hair, indicating a rather stable iron-protein compound.

3. The use of small amounts of acid to stabilize the ferrous salt solution tends to reduce the amount of iron absorbed. The pH of the ferrous solution at which absorption is greatest is in the vicinity of 5. Adjustment of the pH of the hair after "killing," to lower values in the acid range, shows a

corresponding decrease in the amount of iron fixed.

4. Increasing the temperature of the ferrous salt solution, as well as increasing the duration of treatment, results in increased absorption of the ferrous compounds.

5. Increasing the concentration of the ferrous salt results in increased absorption of the catalyst by the hair, although the ratio is not a simple direct one.

6. Different sections of the individual hair fiber absorb the catalyst solution to varying extents, the greatest amount of absorption being at the basal section nearest the skin, with decreased absorption towards the hair tip. Where it is desirable to have a more uniform absorption throughout the length of the hair fiber, it is customary to give the tips of the hair a preliminary brush "killing" before the regular "killing."

Bleaching

The furs, after having been treated with the catalyst solutions, are ready to be bleached. For this, a simple solution of hydrogen peroxide may be used, or any one of a large number of peroxygen solutions containing various additions for the purpose of stabilizing, activating, or otherwise facilitating the bleaching action. The following are a few typical illustrations of bleach baths of practical value:

1. Hydrogen peroxide 3–6 volume; ammonia (sp. gr. 0.90) 7.5 cc. per liter.

2. Hydrogen peroxide 2–4 volume; sodium carbonate 2 grams per liter; ammonium chloride 2 grams per liter.

3. Hydrogen peroxide 2 volume; sodium pyrophosphate 5 grams per liter.

4. Hydrogen peroxide 5 volume; sodium silicate 42° Bé.—5 grams per liter.

5. Hydrogen peroxide 4 volume; ammonia (sp. gr. 0.90) 2 cc. per liter; isopropyl naphthalene sodium sulfonate 0.5 grams per liter.

6. Hydrogen peroxide 5 volume; ammonium persulfate 7.5 grams per liter; ammonia 7.5 cc. per liter. A like amount of ammonia is added each ½ hour for three hours.

7. Hydrogen peroxide 3 volume; potassium persulfate 7.5 grams; sodium perborate 7.5 grams; ammonia 7.5 cc.

These illustrations indicate a wide range of pH suitable for the bleach bath, in conjunction with the ferrous catalyst. In actual practice, this range is limited to about pH 6 to pH 10–11. In bleaching at lower pH, for example 4–5, the use of an acid bleach bath gives satisfactory results for certain types of bleaching, but it must be remembered that there is a tendency for the catalyst to be stripped from the fiber under such conditions. The general procedure is to immerse the skins in the bleach bath at a temperature of 70°–90°F., stirring frequently to assure uniform action. The duration of bleaching is usually 3–5 hours. While the dip process is the method chiefly employed, the other processes such as those previously indicated may be used as well.

The course of the bleach reaction shows some interesting phenomena, which may be described briefly as follows (based on observations and experiments made in connection with large scale fur bleaching operations).

1. During the first half hour or so of treatment very little bleaching effect can be noticed. However, there is a definite change in color of the catalyst on the fiber, indicating the formation of a ferric compound.

2. During this initial stage there is a sharp rise in the temperature of the bleach bath, after which the temperature remains approximately constant for several hours, and then tends to decline.

3. The initial period is also accompanied by a considerable consumption of the hydrogen peroxide present, after which the rate of decomposition, or utilization of the hydrogen peroxide, is much slower, and much more regular.

4. Varying the composition of the bleach bath, particularly with reference to the presence of stabilizing agents and additional activating agents, causes a corresponding variation in the initial action of the bleach bath, but all types of bleach bath tend to run parallel courses after this initial activity.

5. The average bleaching operation is complete within 4–5 hours, and it is a striking fact, that at the end of this period, a considerable amount of hydrogen peroxide still remains in the bleach solutions; in some cases more than half of that present at the outset of the bleaching. Attemps to start with bleach solutions containing only half or

slightly more, of the hydrogen peroxide normally used (other things being equal) do not give satisfactory bleaching results. It has been found practicable, however, to utilize the residual hydrogen peroxide by building up the used bleach solution to the original concentration with the addition of the requisite quantity of the other constituents. For certain purposes, it is also possible to conduct the bleaching over a greater period of time, using a lesser concentration of hydrogen peroxide, which is very well stabilized, and the pH of which does not exceed 8–8.5.

After the furs have been bleached to the desired degree of decolorization, they are hydroextracted, washed thoroughly, and may then be dried and finished as usual. The bleached furs are in general a pale beige color due to the presence of a basic ferric compound in the hair, which acts as a mineral dye or coloring matter on the bleached fur. This mineral dye may be of various shades of yellow tan, depending on the composition of the catalyst solution, as well as on the constituents of the bleach bath other than the hydrogen peroxide. Where it is desired to modify this color, the skins may be dyed directly, or after a preliminary mordanting treatment.

In order to achieve paler, or more delicate tints, or to obtain a bleached fur substantially free of the catalyst, the furs after removing from the bleach bath may be given one of several types of treatment. By immersion of the bleached skins in a solution of acid, the basic ferric compound may be dissolved out of the fiber almost completely. Acid compounds which may be used, are sulfuric, hydrochloric, the various organic acids such as acetic, oxalic, tartaric and citric, and some acid salts—such as sodium bi-sulfate. An acid which is of special interest is hydrofluoric, generally used in the form of an acid salt, such as ammonium bi-fluoride. These fluorine compounds have the interesting property of forming double salts with the basic ferric compound, such double salts being white.

Another type of treatment involves the use of reducing agents, which convert the catalyst from the ferric state into the ferrous condition, the fiber thus being substantially in the same condition as before the bleaching, that is, impregnated with the ferrous compound, which can act as a mordant for a dyeing operation. Any of the reducing compounds, such as sulfurous acid, sulfites, or hydrosulfites may be employed to accomplish this result. In some cases, the combination of the acid stripping method with the reducing treatment gives improved results.

CHAPTER X

LUBRICANTS AND OILS

Wire Drawing Lubricants
Formula No. 1
U. S. Patent 2,329,731

The metal to be drawn or formed by pressure is first immersed in an aqueous soap solution (0.025 to 0.5%) for at least one hour (suitably from 3 to 8 hours) prior to forming.

No. 2
U. S. Patent 2,349,708

Sulfur and wire-drawing soap are used in the proportions of from 2 to 4 parts by weight of sulfur and from 1 to 3 parts by weight of the soap.

Metal Can Lubricant
U. S. Patent 2,145,252

Lubricant for shaping lacquered sheet metal consists of a mixture of:

Glycerin	7.25
Ethanol	2.30
Wetting Agent	0.04
Water	90.40

Powder-Metallurgy Die
Lubrication
Formula No. 1

Dies are lined with flat-lying, overlapping metal flakes of so-called bronzing powders used in paint and ink manufacture. Suspended in a volatile carrier such as carbon tetrachloride and sprayed on the die walls, they form a thin, substantially impenetrable layer.

The lubrication quality may be improved by adding a fatty acid (as stearic acid) or a soap (as aluminum stearate) to the carrier. The lubricant may also be painted on the die wall or applied with an automatic wiper, preferably after each piece is ejected. Use of this lubricant reduces the pressure required for ejection as much as 95 per cent, and prevents lamination.

No. 2
U. S. Patent 2,276,453

Stearic Acid	25
Spermaceti	5
Lanolin	5
Borax	2
Water	25

Gasoline Line and Airplane Parts
Lubricant
(Insoluble in water, gasoline, oils and solvents)

Acrawax	3
Glyceryl Monoricinoleate (S125)	10

Gun Lubricant
U. S. Patent 2,271,044

Lubricating Oil (Light)	100.0

A mixture of:

2:6-Dimethylphenol 2:4:6-Trimethylphenol	0.5
Mono- and Di-Lorol Phosphate	2.5

Corrosionless Bearing Lubricant
U. S. Patent 2,145,970
Corrosion is inhibited in lubricating bearing surfaces, one of which consists of cadmium or copper, by incorporating in the lubricant a small proportion, 0.1–0.2% by weight, of isoeugenol, or 0.05–0.50% by weight of vanillin.

Non-Rusting Turbin Oil
U. S. Patent 2,342,636
Degras	0.05–0.2%
Aniline Disulfide	0.1%
Viscous Mineral Oil	
To make	100%

Air-Pump Lubricant
U. S. Patent 2,353,830
An air pump lubricant suitable for lubrication at elevated temperatures of the order of 500–1000° F. comprising about 80% water, about 15% mineral oil within the viscosity range of 60–80 Saybolt Universal seconds at 210° F., about 2.5% triethanolamine stearate, and about 2.5% free stearic acid.

Anti-Seize Thread Lubricant
U. S. Patent 2,311,772
Mineral Oil (Saybolt Viscosity > 140 sec. at 210° F.)	120
Lead Stearate, Fused	20–35

Anti-Seize Lubricating Paste
(For Aluminum Thread Joints)
U. S. Patent 2,311,772
Viscous Mineral Oil	120
Lead Stearate	20–35

Solid Lubricant
U. S. Patent 2,269,720
Petroleum Jelly	43.5

Candelilla Wax	54.0
± Lithopone	2.0

Lubricating-Oil Corrosion Inhibitor
U. S. Patent 2,296,433
Add
Benzyl Thiocyanate	0.005–0.5%

to a lubricating oil.

Anti-Corrosive Spindle Lubricant
Oleic Acid	3
Potash (26° Bé.)	7
Alcohol	2
Mineral Oil	88

Clock Lubricant
Hydraulic Oil, Low Pour Point	16
Sperm Oil	8

Non-Flowing Lubricant
(Metal to Rubber)
U. S. Patent 2,299,139
Potato Starch	2.0
Triethanolamine Oleate	6.0
o-Phenylphenol	0.1
Water	100.0

Oil-Base Well-Drilling Fluid
U. S. Patent 2,316,967
Refined Stove Oil	50
Ground Oyster Shells	33
Slaked Lime	4
Air Blown Asphalt	13

Plastic-Molding Lubricant
U. S. Patent 2,279,859
Adhesion between a mold and a heat-moldable plastic, which may be present, is prevented by interposing between their respective surfaces a carbonate which is decomposed by heat, evolving carbon dioxide; e.g., 1% aqueous

potassium bicarbonate is brushed on the mold and allowed to dry.

Rubber-Mold Lubricant

Cut 400 grams of butadiene rubber into 2 x 5 cm. pieces. Take 5 kg. of benzene; cover the cut rubber with a part of this benzene. As it swells, add more benzene. Keep the solution at 25° C. This stock solution has a rubber content of between 1:10 and 1:12. For use add 5 kg. more of benzene and 2.3 kg. of ground mica. The mica is added with constant stirring. The mixture is applied on the rubber and the metal surface with a brush. Before each use the mixture should be stirred thoroughly.

Ethyl Cellulose Molding Lubricant
U. S. Patent 2,349,134

Paraffin Wax	10–75
12-Hydroxy Stearin	90–25

Soluble Oil
U. S. Patent 2,303,136
Formula No. 1

Sodium Abietate	22
Oleic Acid	3
Water	6
Kerosene	To make 100

No. 2

Kerosene	140
Water	12
Alkali Metal Resinate	45
Free Oleic Acid	6

Metal-Cutting Oil
U. S. Patent 2,258,309
Formula No. 1

Partly Saponified Fat	20–35
Mineral Oil	35–50
Water	45–15
Calcium Phosphate	2–20

No. 2

Tall Oil	12.0
Low Viscosity Mineral Oil	82.6
Diethylene Glycol	2.0
Caustic Potash (45%)	3.4

By substituting 4% of the tall oil with oleic acid, a soluble oil, which can be used in hard water is obtained.

Forging Tool and Die Lubricant
U. S. Patent 2,345,198

Graphite	15 –20
Aluminum Stearate	1½–10
Light Lubricating Oil	To make 100

Emulsive Lubricant
U. S. Patent 2,345,199

Glyceryl Monooleate	73
Triethanolamine Oleate	22
Triethanolamine	5
Water	100

This gives a film forming lubricant suitable in a Timken test.

Metal-Quenching Oil
U. S. Patent 2,340,726

Mineral Oil	99
Paracumarone Resin	1

Leather-Packing and Gasket Lubricant
(Not affected by acids, water and hydrocarbons)

Glyceryl Monoricinoleate (S125)	65
Acraway	35

Flushing Oil
U. S. Patent 2,355,591

Water	10–35
Sodium Petroleum Sulfonate	5–15

Mineral Lubricating
Oil   To make 100

Belt Dressing

Heavy Cylinder Oil   65
Neatsfoot Oil   25
Rosin   6
Pine Oil   3
Prussian Blue   1
Warm together until dissolved.

Leather-Belt Lubricant

Train Oil or Degras   50
Bone Oil   20
Petrolatum   25
Wood Tar   5

GR-S Rubber-Belt Dressing

Raw Castor Oil   40
Blown Castor Oil   40
Corn Oil or Cottonseed
Oil   10

Penetrating Oil

No. 1 Fuel Oil   20 lb.
Spindle Oil   10 lb.
Chlorinated Solvent sufficient to
make 5 gal.

Stopcock Lubricant

Vistanex Polybutene
(No. 6 grade or
equivalent)   15
Yellow Petrolatum   85

Vistanex and petrolatum are
blended at 100–110°C. with occasional stirring.

High-Temperature Stopcock Grease

Lubricating Oil   85–75
Lithium Stearate or
Aluminum Dis-
tearate   15–25
Heat to 200°C.

Organic-Vapor-Proof Stopcock Lubricant

Before making up, all materials
are dried in vacuo at 70° to 80°C.
This treatment concentrates glycerol from 94% to better than 99%
in about 4 hours and a McLeod
gage on the system shows a pressure of 10^{-4} mm.

The most successful lubricant is
1 to 3% of medium viscosity polyvinyl alcohol and 15 to 20% of
mannitol, in glycerol. After the
ingredients have been pasted in the
cold, the mixture is carefully
heated to about 130°C. and held
there with continuous stirring until the dispersion is uniform and
complete. Stirring, when crystals
first appear after the mix cools, is
beneficial in keeping the mannitol
finely divided. Although the product is rather dry in appearance, it
behaves well after repeated turning of the stopcock.

The mannitol may be replaced
with about 40% of sucrose. This
preparation behaves well without
the polyvinyl alcohol. Sucrose
crystals will usually appear in the
supersaturated solution after about
two days' standing, and stirring
for a short time will keep them in
a fine state of division.

Hydrocarbon-Resistant Stopcock Lubricant

A solution of cellulose acetate is
made by heating 7.5 g. of Cela-
nese, cut into small pieces, in 45 g.
of tetraethylene glycol. After 4
hours at 140°C., with frequent stirring, the solution appears homogeneous. Citric acid (30 g.) is
heated on an oil bath to 190° and
the cellulose acetate solution

added. Heating is continued at 180–190°C. for 90 minutes.

In order to remove dissolved water, the solution is immediately poured into a previously heated glass jar in a desiccator and the desiccator evacuated as rapidly as foaming permits. The dehydration has little effect on the final consistency.

Extracting Fish Liver Oil
U. S. Patent 2,325,367

200 parts of finely ground bluefin tuna liver is intimately mixed with 50 parts of wheat germ flour and 4 parts of 45% aqueous caustic potash is added. The mixture is then stirred for about one hour while heating at about 80°C. Care is taken to exclude air from the reaction mixture during the heating step. The mass is cooled to room temperature and then extracted three times with ethylene dichloride. The combined extracts are filtered, dried and the solvent removed under reduced pressure. The resulting oil is superior in color, odor and taste and vitamin A potency to oils produced from the same type of livers by other processes.

Bleaching Oils and Fats
U. S. Patent 2,158,163

Oil or molten fat (tallow) is treated with the following in concentrated aqueous solution at about 60°C.

Hydrogen Peroxide	0.3–1.5
Sodium Nitrite	0.4–2.0

Preventing Discoloration of Higher Fatty Acids
U. S. Patent 2,162,542

Add 0.01–0.02% oxalic acid to higher fatty acids.

Chapter XI

MATERIALS OF CONSTRUCTION

Plasticizing Concrete
Add to water used and mix until dissolved:

Glue	0.5
Hydrochloric Acid	2.0

Concrete Made with Sea Water
U. S. Patent 2,336,723
Using Portland Cement the following is added:

Sodium Silicate	1.25
Calcium Chloride	0.50
Potassium Alum	0.50
Kaolin	1.25

Concrete Road Protective

Linseed Oil	1 pt.
Kerosene	1 pt.
Water	8 pt.
Soap	½ oz.

Apply 1 gal. per sq. yd. in warm dry weather to prevent scaling due to later use of calcium chloride for ice removal.

Hydraulic Cement
Canadian Patent 420,086

Calcined Gypsum	95.4
Litharge	2.5
Gum Arabic	1.0
Soda Ash	0.1
Soluble Starch	1.0

Fine Plastic Building Cement
U. S. Patent 2,176,862

Portland Cement	50
Silica	35
Ground Pumicite	10

Natural Mohave Silicate	5

Building Cement
Formula No. 1
German Patent 731,173

Magnesium Chloride Solution (Sp. gr. 1.2)	100 lb.
Calcium Hydroxide Solution	10 lb.

Mix and add:

	Parts by Volume
Filler	3:9
Calcined Magnesite, Ground	1

Mold and dry.

No. 2
U. S. Patent 2,307,270

Powdered Cement Clinker	95–97%
Hydrated Ferrous Sulfate	3–5%

Oxychloride Cement

Calcined Magnesium Oxide	11
Fine Sand	22
Sand	67

To each 100 grams of the dry mix is added 15 cc. of 24 Bé. magnesium chloride solution.

Increasing Fluidity of Cement Mix
U. S. Patent 2,169,980
1–1½ pints sulfite liquor is added per sack of cement.

234

Masons' Mortar
U. S. Patent 2,164,871

Portland Cement Clinker	47
Precipitated Calcium Carbonate	30–47
Clay	20–3
Gypsum	3

Artificial Building Stone
U. S. Patent 2,155,531

A composition for molding under pressure comprises:

Shale	85
Bituminous Binder	5
Hot Water, to a thick consistency	
Melted Rubber	5
Pigment	5

Preparing Cement Floors for Painting

Very smooth floors should first be etched with 10% hydrochloric acid, scrubbed with water and thoroughly dried before painting.

Waterproofing Cement Flooring
German Patent 732,109

Gypsum	2.5–3.5
Cement	0.5–1
Sand	5–7

Cement Floor Hardener

First clean floor mechanically and then scrub with soap suds. Then apply, as a first coat:

Zinc Fluosilicate	2 lb.
Magnesium Fluosilicate	8 lb.
Water	32 gal.

Spread with mop and after 3–4 hrs. apply:

Zinc Fluosilicate	2 lb.
Magnesium Fluosilicate	8 lb.
Water	5 gal.

After this coating has dried mop with water.

White "Black"-Board
German Patent 726,117

Wood, metal or slate is baked at 150°C. and coated with following:

Oil Modified Alkyd

Resin	9–10
Hydroterpinol	12–15
Titanium Dioxide	10–15
Kaolin	5–30

Gasoline- and Kerosene-Resistant Plaster

The following mixtures which are resistant and impermeable to gasoline and kerosene, are recommended as plaster for tanks constructed of brick.

Formula No. 1

Magnesium Oxide	1
Sand	2

Mix with magnesium chloride of density 1.16–1.18.

No. 2

Dolomite Powder	1
Sand	3

Casein Drying Oil, Emulsion with addition of 2% of Alum.

No. 3

Portland Cement	1.0
Sand	2.0
Casein	0.2
Calcium Oxide Powder	0.1
Water	

No. 4

Sand	10
Calcium Oxide Powder	3
Waterglass	1

Mix with casein drying oil emulsion.

Waterproofing Cast Gypsum
British Patent 545,805

Precast articles of gypsum plaster, composed of:

Calcium Sulfate	80
Lime	10
Calcium Carbonate	10

can be rendered more resistant to running water by immersing the articles in aqueous solutions of ammonia, sodium or potassium salts of phosphoric or oxalic acid.

Catalyst for Quick-Setting Anhydride Plaster
British Patent 554,952

Potassium Sulfate	1–4
Zinc Sulfate	1
Aluminum Sulfate	1

½–4% of this catalyst is used on the weight of anhydride.

Retarding the Hardening of Plaster of Paris

To the water to be used with the plaster, add 10% of casein glue.

Separating Fluid for Dental Plaster of Paris Casts

Water Glass	50
Water	50
Suitable Dye	As desired

Suitable dyes for use in this solution are rhodamine, fluorescein and eosine red.

High-Strength Brick Tile
U. S. Patent 2,302,988

Blast-Furnace Slag, Ground	50
Lime Sludge	25
Cement, Hydraulic	15
Calcium Chloride	4
Barium Carbonate	3
Calcium Stearate	2

Fireproof Thermal Insulation
U. S. Patent 2,284,400

| Bubble Slag | 55 |
| Asbestos | 5 |

| Zinc Oxide | 4 |
| Sodium Silicate | 35 |

Sound Insulation
U. S. Patent 2,301,986
Formula No. 1

A dry fibrous mass to be combined with water for producing a plaster for absorbing and deadening sounds is formed of:

Mangled Cellulose	6 cu. ft.
Fibered Asbestos	1 cu. ft.
Powdered Titanium Dioxide	1 lb.
Powdered Dry Soap	4 oz.
Powdered Copper Sulfate	1 lb.
Dry Water-Soluble Binder (Such as Dextrin or Glue)	5 lb.
Cream of Tartar	3 oz.
Acid Sodium Carbonate	3 oz.

No. 2
Canadian Patent 419,229

Granulated Cork (10–30 Mesh)	204
Pulverized Cork	50
Asbestos Fiber	330
Cut Back Asphalt Binder	To suit

Refractory
U. S. Patent 2,155,858

Olivine is mixed and pulverized with 12% of dolomite containing 40% of calcium carbonate, and is then mixed with sodium silicate as bond.

Clay Refractory
U. S. Patent 2,261,400

Flint	30
Talc	30
Georgia Clay	35
Domestic Ball Clay	5

The above mixture is bonded by wetting the mixture, mixing it with 85% phosphoric acid, 4–6 parts, and chromic anhydride, 1–2 parts, digesting, drying, and grinding the mixture, molding it into desired shapes, and firing at about 315°C.

Zircon Refractory
U. S. Patent 2,303,304
An aqueous slip is made of:

Zircon	94–96
China Clay	6–4

Cast, dry, bake and fire. After trimming, impregnate with phosphoric acid and fire at a higher temperature.

Fusion Cast Refractory
U. S. Patent 2,154,318

Aluminum Oxide	99–97%
Calcium Fluoride	1–3%

Refractory Crucible
Canadian Patent 416,460

Plastic Fire Clay	85–95
Fused Silica	5–15

Temper with water, mold, dry and fire.

Hard Refractory Non-Acid
Brick
U. S. Patent 2,301,402

Pyrophillite	5
and/or Sericite	5
Calcined Magnesium Oxide	100
Kaolin	3

Refractory
U. S. Patent 2,158,034

Zircon	40
Iron-Silicon	60

is mixed with water to a fluid condition and digested at 93°C. for 10 hours with 6 parts of phosphoric acid; the product is dried and then hardened by heat.

Dental Investment (Refractory)
U. S. Patent 2,152,152

Silica	95–90
Magnesium Oxide	5–10

Phosphoric Acid (10% solution) sufficient to make mass moldable.

Refractory Coating for Metal
or Ceramics
German Patent 730,883

Sodium Silicate (40–42° Bé.)	85
Manganese Dioxide	15

Electric Furnace Luting Lining

Dunite, Powdered	74
Magnesite, Powdered	20
Clay, Fireproof	4
Sodium Silicate	2

Mix and grind solids to pass through 1 mm. sieve and moisten with silicate before use.

Firebrick
U. S. Patent 2,300,683

Clay	90
Alumina	2½
Iron	2½
Graphite	2½
Carborundum	2½

Steatite Body

Talc (Sierramic #1)	85.3
Barium Carbonate	9.0
Kentucky Ball Clay (Airfloated #4)	5.2
Bentonite (Wyoming)	0.5

Ceramic Switch Composition
U. S. Patent 2,153,000

Powdered, Fused Magnesia	65–70

Clay   35–30
This is resistant to shock and
has a low firing-shrinkage.

Ceramic Electrical Insulator
Formula No. 1
U. S. Patent 2,168,230
Pulverized Talc   70–95
Antimony Oxide   1–10
Clay   0–20
No. 2
British Patent 557,811
Talc   43.7
Kaolin   10.9
Magnesium Carbonate   13.6
Barium Carbonate   31.8
Shape and fire at 1100–1300°F.

High-Frequency Ceramic
Insulation
U. S. Patent 2,328,410
Magnesium Beryllium
Titanate   30–35
Titanium Dioxide   65–60
Beryl   5
Mix together and grind to fine
powder. Mold under pressure with
a little gum arabic solution in
water and then fire.

Ceramic Binder Lubricants
Formula No 1
Dextrin and Starch   3
Water   94
Cook together until dispersed.
Then add:
Stearic Acid, Powdered   3
No. 2
Polymerized Glycol
Stearate   1
Water   2

Multiple Ceramic Flux
(For Whiteware)
U. S. Patent 2,261,884
Nepheline Syenite   50

Calcium Fluoride   15
Barium Carbonate   5
Sodium Silicofluoride   5
Talc   10
Whiting   5
Ulexite   10

Form for Slip Casting Ceramics
U. S. Patent 2,303,303
Calcium Sul-
fate · ½ H$_2$O   60 (Vol.)
Wood Flour   40 (Vol.)

Ceramic Denture Base
U. S. Patent 2,341,998
Feldspar   65–80
Amorphous Silica   10–20
Borax Glass   5–15
This fuses at 1500–1600°F.

Ceramic Pigment
(Yellow-Brown)
U. S. Patent 2,251,829
Chromium Sesquioxide   2.5
Tungsten Trioxide   7.5
Titanium Dioxide   90.0
1–5% of an alkali or alkaline-
earth oxide (calcium oxide) may
also be incorporated.

High-Temperature-Resistant
Ceramic Glaze
U. S. Patent 2,170,387
Powdered Zircon   90–50%
Phosphoric Acid   10–50%
Powdered Ferro
Silicon   1–5%

Pink Ceramic Underglaze Stain
U. S. Patent 2,243,033
Manganese Carbonate   5–25
Aluminum Hydroxide   94–74
Calcium Fluoride   1–5
An intimate mixture of the
above is calcined at 850–1300° for
1–5 hours.

Clear Vitreous Enamel for Metal
U. S. Patent 2,337,103

Borax	8.0
Feldspar	22.4
Quartz	26.0
Cryolite	12.4
Soda Ash	1.6
Sodium Nitrate	3.0
Calcium Carbonate	5.1
Sodium Titanium Silicate	10.0

Vitreous Undercoat for Enamelled Iron
U. S. Patent 2,293,146

A frit is made from:

Borax	35.5
Felspar	26.0
Quartz	20.5
Soda Ash	6.6
Sodium Nitrate	5.3
Fluorite	6.1

Grind and mix 1000 parts of above with:

Clay	70.0
Borax	5.0
Antimony Oxide	12.5
Calcium Molybdate	12.5
Water	55.0

Permanent Marking of Mica
British Patent 558,550

A. Flux

Lead Oxide	70
Boron Oxide	20
Silica	10

B. Stain

Ferrous Oxide	25
Chromium Oxide	25
Manganese Dioxide	12
Cobalt Oxide	38

4 parts of A and 1 part of B are mixed and applied to mica which is heated to 560–580°C.

Clear, Colorless Glass Batch
U. S. Patent 2,252,131

Sand	1000
Soda Ash	400
Calcium Carbonate	250
Potassium Nitrate	10
Arsenic Trioxide	2
Tellurium	2

Low-Expansion Glass
U. S. Patent 2,148,621

Silica	55
Alumina	15
Lead Oxide	15
Magnesium Oxide	15

A glass made from this is suitable for sealing to tungsten and molybdenum.

Optical Glass Batch
U. S. Patent 2,294,077

Silica	360
Calcium Carbonate	80–110
Sodium Nitrate	50–70
Soda Ash	63–68
Sodium Oxide	7–10

Optical Glass Batch
U. S. Patent 2,294,844

Aluminum Phosphate	65–50.6
Potassium Dihydrogen Phosphate	14.5–0
Magnesium Dihydrogen Phosphate	27–21.6
Calcium Dihydrogen Phosphate	0–27

Ruby Glass
U. S. Patent 2,174,554

Glass is colored ruby by adding to the batch:

Copper	0.25
Sodium Cyanide	1.25

High-Silica Glass Batch
U. S. Patent 2,247,331

Sand	700–800
Felspar	50–130
Calcined Borax	100–160

A glass (80–87% silicon dioxide) free from gas-forming constituents and of low viscosity is formed. The ingredients are preferably melted in an electric furnace.

Boro-Silicate Enamel
U. S. Patent 2,337,103

Borax	8.0
Feldspar	22.4
Quartz	26.0
Cryolite	12.4
Soda Ash	6.1
Sodium Nitrate	3.0
Calcium Carbonate	5.1
Sodium Titanium Silicate	10.0

Cork Substitute

Glue	100
Glycerin	75
Peanut Hulls, Ground	100
Saponin	1–2
Formaldehyde	2
Water	350

Asbestos Roofing Paper

Asbestos Fiber	100
Bitumen	100

Warm together and mix until uniform and plastic. Pass through rolls and calenders.

Light-Weight Asphalt Aggregate
British Patent 554,950

Wood Flour	25
Asphalt	55
Grit	20

Heat and mix in an oil jacketed tank to decompose the wood flour. Gas generated makes mass very porous and light.

Improving Adhesion of Asphalt to Wet Surfaces
U. S. Patent 2,286,244

Mix into the asphalt:

Tall Oil	0.5–2
Aluminum Sulfate	0.1–0.25

Salt-Water Pit Lining
U. S. Patent 2,348,320

Bitumen	10–30
Soil (Earth)	50–70
Sawdust	30–70

Wood Preservative
Formula No. 1

Potassium Dichromate	5
Copper Sulfate	3
Pyroarsenic Acid	1

No. 2

Sodium Dichromate	5
Copper Sulfate	3
Pyroarsenic Acid	1

No. 3

Chromic Acid Anhydride	29
Copper Carbonate	13½
Arsenic Acid	1

The toxic reaction products deposited in the wood are approximately the same for all three salts. The second is slightly cheaper than the first. The third is much more expensive but leaves no residual soluble salts in the wood after fixation. The component salts in the mixture are readily soluble in water. A 7% solution when injected into pine poles will yield a final retention of 1 pound of dry salts per cubic foot when an empty-cell process is employed in treatment. After injection in the wood a chemical reaction occurs between these soluble salts resulting in the formation of water-insoluble compounds. This is due to the reducing action of the wood substance. The rate and degree of fixation increase with time and temperature. After 28 weeks of seasoning 98% of the injected salts are insoluble as determined by standard leaching tests. Absorption is selective, as the fixation of the individual components is not the same either in rate or ratio. This must be taken into account in maintaining the strength of the treating solution. Laboratory and field tests indicate that wood treated with greensalt is highly resistant to decay. The solution is non-corrosive to treating equipment.

No. 4
Canadian Patent 416,657

Creosote, Coal-Tar	45
Bentonite	10
Sodium Fluoride	27¼
Potassium Dichromate	11¼
Dinitrophenol	4½

Before use, add water to form a paste.

No. 5
U. S. Patent 2,344,019

Solid Bitumen	2½
Creosote Oil	5
Sodium Fluoride	5
Potassium Bichromate	1
Dinitrophenol	¼

No. 6
U. S. Patent 2,296,401

Arsenic Trioxide	5–30
Sodium Fluoride	30–70
Dinitrobenzol	2–10
Ammonium Linoleate	2–8
Turpentin	10–20

Water, to make a thin paste.

No. 7
British Patent 546,436

Arsenic Pentoxide and Water	5
Chromium Trioxide	15
Cupric Hydroxide	6
Water	300–3000

No. 8
U. S. Patent 2,152,160

Sodium Silicofluoride	>0.6
Zinc Chloride	>0.6
Water	To make 100

Fireproofing Wood
British Patent 546,256

Wood is impregnated with a water solution containing:

Sodium Dichromate	2.24
Copper Sulfate	2.24
Zinc Chloride	5.00
Boric Acid	2.00

Chapter XII

METALS AND ALLOYS

Iron and Steel Pickling

Low carbon, hot rolled steel sheet and strip	Mill scale is removed by the action of a warm, dilute solution of sulfuric acid. Bath characteristics: cleaning agent, about 10% sulfuric acid* temperature, 150–190°F.
Heat-treated steel parts	Furnace scale may be removed by the following bath: cleaning agent, 8–10% sulfuric acid temperature, 145–155°F.
Steel forgings	Forging scale is more difficult to remove, and the following descaling bath may be used: cleaning agent, 8–15% sulfuric acid temperature, 150–160°F.
Stainless steel parts	If the parts have a coating of heavy oxide, the following bath may be used: cleaning agent: 7% sulfuric acid (by vol.) 2% hydrochloric acid (by vol.) temperature, about 150°F. When the parts have a light scale, this mixture may be used: cleaning agent, 11% hydrochloric acid (by vol.) for every gallon of this aqueous solution add ¼ lb. nitric acid temperature, about 150°F.

---

* All acid pickle bath concentrations are expressed in per cent by weight in an aqueous solution.

Inhibitors, usually synthetic organic chemicals, are added to acid pickle solutions to retard the attack of the acid on the clean metal areas without seriously reducing the rate of rust or scale removal. Counterbalancing the added cost of the inhibitor and the somewhat increased pickling time are the savings caused by the reduction in metal loss and in amount of acid used; hydrogen embrittlement is also decreased.

Passivating agents are added to pickle solutions to retard the attack on fixtures and tank and yet allow efficient pickling action. If stainless steel equipment is used, 1 to 5% anhydrous ferric sulfate added to the acid bath will extend its useful life.

Light gage sheet steel | Light scale can be removed and sheet steel etched prior to drawing with a pickle of the following composition:
2–4% sulfuric acid
2–4% iron sulfate

Iron and steel parts (especially heat-treated parts) | Electropickling is used to remove scales that are difficult or impossible to remove with still pickling. The work is made cathodic in a bath of the following characteristics:
concentration, 10–20% sulfuric acid
current density, 10–150 amp. per ft.^{2}
temperature, 50–150°F.
time, 1–3 min.

In one commercial process, this is followed by an anodic treatment under the following conditions:
concentration, 40–50% sulfuric acid
current density, 100–150 amp. per ft.^{2}

Steel Pickling
U. S. Patent 2,155,854

Iron or steel sheets are pickled at 82–93°C. in a solution containing:

Sulfuric Acid   18–20
Iron Sulfate   13–16

Pickle liquor is continuously withdrawn and cooled to <28°C. to separate iron sulfate crystals and reduce the concentration to about 13%; the sulfuric acid content is made up and the liquor returned for re-use.

Stainless Steel Pickling Solution
U. S. Patent 2,337,062

	By Volume
Sulfuric Acid	3–30
Nitric Acid	4–30
Hydrofluoric Acid	1–2
Water	To make 100

Stainless Steel Pickling

Ferric Sulfate	6.0–12
Hydrofluoric Acid	1.5– 3

A solution of the above at 71–

82°C. effects smooth pickling of 18–8 stainless steel.

Pickling Solution for Chrome and Chrome Nickel Steel
U. S. Patent 2,172,041

Hydrofluoric Acid	0.5–10
Chromic Acid	5–20
Sulfuric Acid	1–10
Water	To make 100

Pickling Agent for Copper-Beryllium
U. S. Patent 2,284,743

Sodium Hydroxide	40–50
Sodium Cyanide	5–10
Water	To make 100

Passivation of Stainless Steel Wire

The usual passivating solution is made up to contain about 20% by volume of concentrated nitric acid in water and works best when heated to 120°F. The duration of the treatment should be about 20 minutes.

For further treatment of the

original lot, try a pickle in a solution made up to contain 25 parts by volume of hydrochloric acid in 100 parts of water, to which is added concentrated nitric acid to the extent of about 3 parts by volume. This solution is used at about 140°–160°F. and is pretty active, so that a short time of immersion may suffice to brighten the steel. (This solution is suggested only to *salvage* those parts that you want to brighten—it is not meant or recommended for ordinary passivation as a substitute for the straight nitric acid solution described previously.)

Pickling Corrosion Inhibitor
U. S. Patent 2,355,599
Add 0.1–2% alphatrioxymethylene to the pickling solution.

After-Treatment of Pickled Metals
U. S. Patent 2,257,133
After acid-pickling, metal objects are rinsed first in a hot bath containing:

Soap	0.125–0.275
Sodium Hydroxide	0.275–0.550
Wetting Agent	0.020–0.500
Sodium Hypochlorite	0.010–0.040

and then rinsed in a bath, at 60–100°C., containing the above constituents and also glycerin, glycol, or an ammonium soap 0.25–0.625%. They are dried without swilling.

Rust Inhibitor or Reinhibitor for Internal Combustion Engines

Borax (Granular)	95.5–96.5
Mercapto-Benzothiazole	3.5– 4.5

These mixed ingredients may be used as a rust inhibitor for automobile, tractor, and truck cooling systems. 1½ ounces per each gallon of cooling water is needed. It may also be used as a reinhibitor with ethylene glycol type antifreeze which has been salvaged and re-used a second or third time.

Rendering Stainless Steel Resistant to Sea Water
U. S. Patent 2,172,388
Stainless steel is treated in a bath of:

Iron Chloride	100 g.
Hydrochloric Acid	5 cc.
Water	1 l.

A protective surface film is formed.

Cleaning and Protecting Metal
U. S. Patent 2,302,510
Iron, steel or zinc is degreased and protected by spraying with a water solution of:

Fuller's Earth	867
Zinc Dihydrogen Phosphate	95
Copper Sulfate	21
Sodium Nitrate	17

Prevention of Corrosive Effect of Brine on Refrigerating Machinery
Use 125 lb. of sodium dichromate per 1000 cu. ft. calcium chloride brine or 200 lb. per 1000 cu. ft. of sodium chloride brine.

The dichromate is first dissolved in a little warm water and added to the brine with agitation before the latter enters the cooling system. If the brine is neutral or acid its pH should be increased to 8 by caustic soda solution.

Rustproofing
Formula No. 1

Borax	1½ oz.
Water	32 oz.

This solution will inhibit the formation of rust on iron or steel with which it is contacted. May be effectively used on closed cooling water systems or grinding steel circuits. Brillo or steel wool or razor blades will not rust if kept in a small container of this solution, when not being used.

Note: This solution is slightly corrosive to aluminum, brass, magnesium and copper.

No. 2

Iron or steel parts may be stored indefinitely without corrosion if stored in closed containers immersed in a solution of 0.25% of potassium chromate in methanol.

No. 3

Sodium Bichromate	0.1
Abopon (Sodium Borophosphate)	10.0
Water	89.9

Spray on cleaned ferrous metals.

No. 4
U. S. Patent 2,284,241

Mineral Oil	10
Octadecyl Acid Maleate	2
Naphtha, Low Flash	86

No. 5
U. S. Patent 2,291,460

Palm Oil, Crude	3–20
Turpentin	1–7
Petroleum	To make 100

(Flash Point 250°F.)

No. 6
U. S. Patent 2,301,983

The surface is cleaned and treated with:

Lead Nitrate	10 g.
Nitric Acid (70%)	20 cc.

then treated with a solution of:

Arsenic Pentoxide	6
Chromic Oxide	½
Chromium Trioxide	3

Finally it is dried.

No. 7
U. S. Patent 2,276,353

Iron, steel, zinc, aluminum or their alloys are immersed for 5 min. in a boiling bath of either of the following:

a

Chromium Trioxide	2½
Sodium Silico-fluoride	25
Water	1000

b

Chromium Trioxide	10
Sodium Nitrate	12½
Manganese Silico-fluoride	25
Water	1000

No. 8
U. S. Patent 2,302,643

Phosphoric Acid (75%)	10	cc.
Sodium Sulfite	3	g.
Chromium Sulfate	0.6	g.
Water	To make 1	l.

No. 9
(Phosphatizing Bath)
U. S. Patent 2,298,280

Immerse in N/5 solution of zinc or manganese dihydrogen phosphate (determined by phenolphthalein) containing 0.4% of hydroxylamine at a temperature of 15° to 80°C.

(Purification of Phosphating Baths)
German Patent 719,550

To the phosphate baths for zinc-aluminum alloys is added an acid fluoride, for example, a fluosilicate. The amount added is approximately 2 g. per liter.

Prevention of Corrosion of Metal in Contact with Chloroform

Chloroform at $-70°F$. is very corrosive on galvanized iron, brass, copper, etc. This solvent in conjunction with dry ice is often used as a non-inflammable refrigeration medium in place of acetone. If metal equipment is used to convey chloroform in such applications, the corrosion problem is bad. It can be eliminated almost entirely by the addition of 5% by volume of any low cloud point mineral oil.

Preventing Moisture Absorption and Corrosion

1. Silica gel is a solid, chemically inert dehydrating agent that prevents corrosion by absorbing the moisture from the air inside the package.

2. When silica gel is used, an enclosing moisture barrier of approved design must be provided to keep out atmospheric moisture. This moisture barrier may be either a flat sheet wrapper, prefabricated bag or other container constructed of moisture vapor impervious material.

3. Any packing, liners, wood blocking or other hygroscopic dunnage used inside the barrier contains moisture which must be taken care of by increasing the amount of dehydrating agent shown in the table by an amount equal to at least one-half of the weight of the packing, liners, wood blocking, or other hygroscopic dunnage. It is obvious from the above that in the design of the package, the hygroscopic dunnage used inside the moisture barrier should be kept to a reasonable minimum. Once dunnage has been taken care of in accordance with the above requirement it is not necessary to further increase the quantity of silica gel needed if the pack is to be protected for longer than six months, since once the silica gel has dehydrated the dunnage it is then thoroughly safe.

Weight of Silica Gel Required to Prevent Corrosion

These quantities comply with the requirements of various government specifications for dehydration-packing protection over a six month protection period.

(Requirements of the individual agencies are listed below)

Area in Sq. In.	Silica Gel Required Weight	Area in Sq. In.	Silica Gel Required Weight		Area in Sq. In.	Silica Gel Required Weight	
10	5 g.		Lb.	Oz.		Lb.	Oz.
50	15 g.	1500	1	1	11000	7	9
	Lb. Oz.	1600	1	2	12000	8	4
100	1	1700	1	3	13000	8	15
125	2	1800	1	4	14000	9	10
150	2	1900	1	5	15000	10	5
175	2	2000	1	6	16000	11	0
200	3	2250	1	9	17000	11	11

Area in Sq. In.	Weight Silica Gel Required Lb. Oz.		Area in Sq. In.	Weight Silica Gel Required Lb. Oz.		Area in Sq. In.	Weight Silica Gel Required Lb. Oz.	
250	3		2500	1	12	18000	12	6
300	4		2750	1	14	19000	13	1
350	4		3000	2	1	20000	13	12
400	5		3500	2	7	21000	14	7
450	5		4000	2	12	22000	15	2
500	6		4500	3	2	23000	15	13
600	7		5000	3	7	24000	16	8
700	8		5500	3	13	25000	17	3
800	9		6000	4	2	26000	17	14
900	10		6500	4	7	27000	18	9
1000	11		7000	4	13	28000	19	4
1100	12		7500	5	3	29000	19	15
1200	13		8000	5	8	30000	20	11
1300	14		9000	6	3	31000	21	5
1400	15		10000	6	14	32000	22	0

Silica gel is available in two types of cloth bags, a regular jean cloth and a paperlined, laminated dustproof jean cloth bag. Bag sizes commonly available are 5 g., 1 oz., 2 oz., 3 oz., 4 oz., 6 oz., 8 oz., 1 lb., 2 lb., and 5 lb.

The amount of silica gel required to dehydrate dunnage enclosed within the package is equal to half the weight of enclosed dunnage in all cases except U. S. Army Signal Corps packs where some individual specifications call for a weight equal to weight of enclosed dunnage. Packs containing wood blocking should be protected by a weight of silica gel equal to the weight of the wood.

Corrosionproofing Magnesium Alloys

Formula No. 1

Immerse the part (or apply the solution locally) for 1 min. at room temperature in a bath of composition:

Sodium Dichromate 1.5 lb.
Concentrated Nitric
Acid (sp. gr.
1.42) 1.5 pt.
Water To make 1.0 gal.

After immersion the part is rinsed in cold water, then dipped in hot water and allowed to dry. The resultant coating on clean surfaces is of matte to brassy, iridescent color.

It may also be applied as a spray on parts and assemblies which are too large to put into available tanks. On large parts, the coating may also be applied by brushing on a generous amount of solution for the recommended time.

No. 2

This treatment provides maximum salt water and marine atmospheric protection. When combined with paint coatings, it offers the simplest and best treatment for maximum protection with negligible dimensional change. It consists essentially of two steps applied as follows, after proper degreasing:

Step 1. Immerse the parts for 5 min. in an aqueous solution containing 15 to 20% by weight of hydrofluoric acid at room temperature. Rinse in cold running water.

An alternative step, for use on wrought products only, is to immerse the parts for 15 min. in an aqueous solution containing sodium, potassium, or ammonium acid fluoride at room temperature; then rinse in cold running water. This is advantageous because aluminum inserts, rivets, and such like, are not materially attacked; it is also more economical. It should not be used on castings of any type, as inferior protection will result.

Step 2. Boil parts for 45 min. in an aqueous solution containing 10 to 15% sodium dichromate. After boiling, the parts are rinsed in cold running water, followed by a dip in hot water to facilitate drying.

The addition of calcium fluoride or magnesium fluoride to the dichromate bath will improve corrosion resistance, promote film formation, and insure more uniform coating. These fluorides are only very slightly soluble and may be added conveniently by suspending the salt in the bath in canvas bags. Either of these salts will provide the correct amount for proper film formation. When fluoride is added to the chromate bath the treatment time can be reduced to 30 min. If it is not convenient to suspend a bag of calcium fluoride or magnesium fluoride in the solu-

tion, 0.5% by weight of the fluoride may be added initially to the bath; additional amounts can be added from time to time as required, although this is seldom necessary in actual operation.

Bright Magnesium Finish
U. S. Patent 2,287,049
Dip in following solution:
Phenol   3–5%
Oxalic Acid   3–5%
Sulfuric Acid (2–6%
    solution) To make 100%

Gray Finish for Magnesium
U. S. Patent 2,138,794
A uniform dark gray finish is imparted by scratch-brushing, pickling in dilute nitric acid and immersing in a boiling bath containing:
Sodium Chromate   9
Chromium Sulfate   1
Immerse for 0.5–60 minutes, rinse and dry.

Pre-Painting Treatment for Magnesium

Treatments of magnesium surfaces are largely for the purpose of improving paint adhesion and have in themselves little protective value. Two common treatments are the chrome pickle and the dichromate boil.

In the chrome pickle process the part is immersed for 30 sec. to 2 min. in a solution of 1.5 lb. per gal. of sodium dichromate and 1.5 pints per gal. of nitric acid, at a temperature of 50°C. (122°F.). The work is then rinsed in cold water. This process removes about 0.0006 in. per surface.

Where tolerances are small, the

dichromate boil is suitable. Following cleaning, the work is given a pre-treatment dip in a 15 to 20 per cent (by weight) hydrofluoric acid solution. The work is then immersed for at least 45 min. in a boiling 10 per cent sodium dichromate solution. The pH should be maintained between 4.2 and 5.5 by adding chromic acid. This treatment is suitable for all magnesium alloys except Dowmetal M and AM3S.

Non-Aqueous Metal Cleaning Bath
U. S. Patent 2,353,026

Sodamide	1–25
Sodium Hydride	1–20
Sodium Hydroxide	
	To make 100

This melted and applied in a fused state, to the metal, and the excess is later washed off with water.

Stripping Oxide Film from Aluminum
U. S. Patent 2,353,786
Immerse in:

Sulfuric Acid	8.2
Phosphoric Acid	5.2
Chromic Acid	2.0
Water	84.6

Then rinse in clear cold water and then in hot water.

Pre-Galvanizing Coating
U. S. Patent 2,276,101
Steel strip or wire is coated with molten flux of:

Borax	90
Calcium Fluoride	10

Heat at 790°C. for 3 min.; cool and strip off flux before galvanizing.

Aluminum Pre-Painting Treatment
Formula No. 1
A treatment which serves the same purpose as chromadizing consists in applying a solution containing, on a volume basis, butyl alcohol, 40 per cent; isopropyl alcohol, 30 per cent; 85 per cent phosphoric acid, 10 per cent; water, 20 per cent. This is applied with a rag or by dipping or spraying. After 1 or 2 min. the surface is scrubbed lightly with a brush, rinsed thoroughly, dried, and primed immediately.

No. 2
U. S. Patent 2,137,988
To render the surface of aluminum alloy articles adherent to paint, etc., they are cleaned in 3–5% sodium hydroxide solution for 2–10 minutes at 60–70°C., rinsed, dipped in 10% by volume nitric acid, rinsed, immersed in 15–30% by volume hydrochloric acid for 2–10 minutes at 50–80°C., rinsed, cleaned in sodium hydroxide solution to remove the gray film, rinsed, dipped in 5–100% by volume nitric acid to whiten the surface, washed well, and dried.

Silver Tarnish Inhibitor
British Patent 395,491
After thorough degreasing, the articles are immersed for one minute in a solution containing 0.5 g. chromic acid in 1000 cc. of water, and subsequently rinsed and dried.

Repairing Damaged Galvanized Coatings
Remove all rust and dirt then

apply, along with a zinc chloride and ammonium chloride flux:

Zinc 8
Lead 1
Tin 91

Melt with an acetylene torch and allow to cool.

Tinplate Protective Coating
Canadian Patent 410,224

To provide resistance to surface staining and to preserve a new-bright appearance in contact with food products, tin plate is treated with a hot aqueous solution containing substantially the following:

Sodium Phosphate 0.86
Sodium Chromate 0.84

Treat for not over 2 minutes and rinse.

Coloring Cadmium Brown

Dip for 2–8 min. in:

Potassium Dichromate 7.5 g.
Nitric Acid (36° Bé.) 3.8 g.
Water 1 l.

at 60–75°F.

Coloring Stainless Steel
U. S. Patent 2,172,353

Black Coloring:

Nitric Acid 10–14
Sulfuric Acid 36–50
Water 40–50

Brown Coloring:

Ammonium Vanadate 4–6
Sulfuric Acid 7–11
Water 19–23

In each case treat for 1 hr. at 85–93°C.

Black Coating of Stainless Steel

Clean and degrease then immerse molten sodium dichromate (730–750°F.) for 15–20 min. Re-

---

move, cool and wash well with water.

The coating formed are very tenacious to bending, stretching, etc.

Bluing Steel

Immerse in molten mixture of:

Sodium Nitrate 55
Sodium Nitrite 45

Hold at 250–300°C for ½ hr. Preliminary polishing improves gloss of finish.

Cold Tinning Compounds
U. S. Patent 2,144,798

Tin 4 oz.
Mercury 4 lb.

Rub together to make an amalgam and triturate with:

Copper Sulfate 2
Mercuric Chloride 1
Carborundum (80 mesh) 1
Glass Powder 5
Glycerin 1½
Water To make a thick paste
Zinc Chloride 1¼

Hot Tinning
British Patent 546,179

Harder tin deposits are obtained by dipping the metal in molten tin containing:

Cobalt 0.1
Nickel 0.2
Bismuth 5.0
Zinc 8.2

Tinning Copper
U. S. Patent 2,159,510

Stannous Chloride 50
Caustic Soda 56
Sodium Cyanide 50
Water 1000

Immerse metal in above solution, drain and heat to 230°C. Repeat as desired.

Hot Lead Coating Alloy
Formula No. 1
U. S. Patent 2,262,304

A lead alloy suitable for coating iron, copper or steel articles by hot-dipping contains:

Tin	20.00
Silver	0.10
Copper	0.10
Bismuth	0.25

No. 2
Lead Alloy for Hot Coating Steel
U. S. Patent 2,298,237

Antimony	1
Zinc	2
Tin	5
Lead	2

Hot Metal Coating on One Side Only
U. S. Patent 2,137,464

To coat steel strip with tin (or other metal) on one face only by hot-dipping, a resist is applied to the other face by bringing it in contact with a rubber roller which dips into a solution containing:

Chromium Tri-oxide	12.5 lb.
Sulfuric Acid, Concentrated	6.0 lb.
Hydrochloric Acid, Concentrated	4.0 lb.
Sodium Sulfate	4 oz. per 5 gal.

Metallic Spray Coating
U. S. Patent 2,161,104

Zinc Dust	100
Zinc Chloride (20% Ale-Solution)	50–75
Arsenic Trioxide	2½

Immersion Tin-Plating of Cast Iron

Nickel Sulfate	200 g.
Stannous Sulfate	13 g.
Tartaric Acid	8 g.
Nickel Chloride	30 g.
"Nacconol" NR (Wetting Agent)	1 g.
Water	To make 1 l.

The solution is maintained at 150°F. The cast iron is immersed in the solution for 1 to 5 hours and rinsed thoroughly.

Chromium-Plating Cast Iron

Degrease with gasoline and burnish with chalk.

Plate with following at 2.5 amp./sq. dm. at 68°C. for 2.5 hrs.:

Chrome Oxide	150 g.
Sulfuric Acid	1.5 g.
Water	To make 1 l.

Chromium-Plating Bath
U. S. Patent 2,136,197

The bath contains (in 400 gal.):

Chromic Anhydride	500
Manganese Dioxide	40
Potassium Permanganate	15
Manganese Dichloride	5
Sodium Potassium Tartrate	5

Palladium-Plating Bath
U. S. Patent 2,335,821

A process for electrodepositing ductile thick coatings of palladium comprises electrolyzing an aqueous bath having a palladium-ion content greater than 10^{-8} and consisting of about 25 to 50 grams of palladium per liter as the chloride, about 50 cc. to 700 cc. of concentrated hydrochloric acid per liter, and about 2.5 to 50 grams of ammonium chloride per liter at a temperature of about 20°C. to 90°C.

Aluminum-Plating
U. S. Patent 2,170,375

Aluminum Chloride	20
Aluminum Bromide	20
Benzol	88
Ethyl Bromide	57

Use aluminum anodes at 20°C with cathode C.D. of 1.5 amp./dm². Ethyl bromide is added from time to time as needed.

Coating Aluminum
U. S. Patent 2,146,838

A colorless, hard, adherent, adsorptive coating is produced on aluminum by immersing the metal for 10–30 minutes at 70–100°C. in 1–2% aqueous sodium aluminate containing, as stabilizer (a) sodium silicate 1–2%, (b) glycerin or glucose 10–25%, (c) tannic acid 1–2% of the sodium aluminate content.

Nickel-Plating Bath

Nickel Sulfate		
Crystals	100	g.
Sodium Sulfate	150	g.
Boric Acid	20	g.
Sodium Chloride	15	g.
Water	To make 1 l.	

Nickel-Plating on Magnesium

Recent work has shown that magnesium alloys can be plated with nickel or silver. For either process the first step is cathodic cleaning in either a suitable proprietary cleaning solution or one made up of the following approximate composition operated at 160 to 190°F.

Sodium Carbonate ($Na_2CO_3 \cdot 10H_2O$)	7.4
Trisodium Phosphate ($Na_2PO_4 \cdot 12H_2O$)	14.5
Sodium Hydroxide (NaOH)	3.7
Sodium Metasilicate (Na_2SiO_3)	7.4
Wetting Agent (Sodium or Ammonium Lauryl Sulfate)	0.9

A current density of 50 to 200 amp./ft.² is applied for from 1 to 2 min. Cleaning is followed by thoroughly rinsing the parts in first hot and then cold water.

For nickel plating the work is then given about a 5-sec. dip in the following bath:

Chromic Anhydride, C.P. (CrO_3)	15.8
Nitric Acid, C.P. (HNO_3) (Conc.)	10.4
Sulfuric Acid, C.P. (H_2SO_4) (Conc.)	0.007

Operated at approximately 70°F. or room temperature.

To eliminate chromic acid carry-over into the next solution, the work must be thoroughly rinsed in cold water (use a spray or running water where possible). It is then given a second etching treatment of from 1 to 5 min. in a solution composed as follows:

Hydrofluoric Acid, 52% HF	22.0
Nitric Acid, C.P. (HNO_3) (Conc.)	1.1

Operated at approximately 70°F. or room temperature.

Rhenium-Plating
U. S. Patent 2,138,573

Deposits having Brinell hardness 250, and resistant to hydrochloric acid are obtained from acid baths of:

Formula No. 1

Potassium Rhenate 11.0 g.
Sulfuric Acid,
Concentrated 3.5 g./l.
(pH 0.9) at current density 10–14 amp./dm.² and 25–45°C., or from alkaline baths:

No. 2

Hydrogen Rhenate 40 cc.
Sodium Carbonate,
Acid 40 cc.
Ammonium Sulfate 20 g./l.
(pH 8.6), current density 13–17 amp./dm.², 80–90°C

Bright Zinc-Plating
U. S. Patent 2,157,129

To a bath containing:
Zinc Cyanide 10 oz.
Sodium Cyanide 9 oz.
Sodium
Hydroxide 10 oz./gal.
is added ¼ oz. of sodium thiosulfate per gallon to render it suitable for bright barrel-plating of hollow articles, using a current of 200 amperes at 6 volts.

Manganese-Plating

Manganese Sulfate 100–200 g.
Ammonium Sulfate 75 g.
Glycerin 50 cc.
Water To make 1 l.
pH 7–7.7; c.d. 10–15 amp./dm.² at 15–20°C.
Solution must be agitated to get a good deposit.

Simple Copper-Plating
Steel is dipped in a hot solution of:
Copper Chloride 7
Sulfuric acid (d. 1.12) 20

Phosphoric Acid 30
Water 43
then rinse and dry quickly.

Brush Electroplating
Economical electroplating can be done by using only enough of any of the plating solutions, except for chromium, to wet a few thicknesses of cloth wrapped around a rod or piece of the metal to be plated. This metal should be connected to the carbon of a battery or positive pole and the object to be plated to the zinc or negative pole. The wet cloth is then constantly rubbed over the cleaned surface to be plated. If one dry cell does not give a coating, use 2 or 3 connected in series, i.e., the zinc of one to the carbon of the next one. This method makes possible the plating of large objects that could not be put into a plating tank and smaller objects can be plated with the preparation of only a small amount of the plating solution. Only thin deposits can be made, but their appearance is excellent.

Silver Mirrors on Glass
Clean the glass thoroughly with soap and water until the glass does not show "water break" (irregularities in the running off of rinsing water). Then lay the glass on a flat surface and cover with dilute stannous (tin) chloride while preparing the following solutions.

A

Distilled Water 25 cc.
Granulated Sugar 2 g.
Nitric Acid (Conc.) 10 drops
Boil for five minutes and cool before using.

B

Distilled Water	40 cc.
Silver Nitrate	2 g.
Sodium Hydroxide	1 g.

C

| Distilled Water | 15 cc. |
| Silver Nitrate | 1 g. |

Add just enough ammonium hydroxide to solution B to redissolve the brown precipitate. Then add solution C to B until a slight darkening is produced.

Pour off the stannous chloride and rinse the glass several times with distilled water. Do not dry. Lay on a flat surface and pour on a freshly made mixture of one part of A to four parts of B. Just enough of the solution should be used to cover the surface without any running off over the edges. The silvering should begin at once and should be complete in 15–20 minutes.

Caution: Solution B or any mixture containing it may explode after standing a few hours. Discard such solutions at the end of a work period.

Stainless Steel Anodic Polishing
U. S. Patent 2,348,517

Glycine	7/15
Water	8/15
Phosphoric Acid (85%)	5
Sulfuric Acid (96%)	4

The steel is made the anode and the current is regulated to effect polishing.

Electrolytic Metal Polishing
(Zinc, Copper or Their Alloys)
U. S. Patent 2,330,404

Water	100
Chromic Acid	12½
Sodium Dichromate	37½
Acetic Acid	12½
Sulfuric Acid	10

Metal is made anode in above bath using a current density not less than 200 amp./sq. ft.

Electropolishing Steel
U. S. Patent 2,338,321

Sulfuric Acid	5–80%
Phosphoric Acid	5–80%
Chromic Acid	0.5–20%

The combined acid strength should be between 50–90%. The steel is used as anode in above solution with c.d. of 50–1000 amp./sq. ft.

Metal Electrocleaning Bath
British Patent 547,592

Sodium Metasilicate and Sodium Carbonate	77
Sodium Hydroxide	19
Sodium Oxalate	3
Glue	1
Water	To suit

Copper-Deplating of Steel
The metal forming the anode is thoroughly cleaned and stripped in a solution of:

Copper Cyanide	3½	oz.
Sodium Cyanide	5	oz.
Soda Ash	1	oz.
Water	1	gal.

$pH = 12$

Work at 35–40 amp./ft.^{2} at 175°F. The current density at cathode should be 5–15 amp./ft.^{2}

Stereotype Surfacing
Iron surfacing of stereotypes is accomplished in the following manner:

a. *Cleaning.* The stereotype surface is first scrubbed with a solution of caustic soda, 2 to 4 oz./gal. (15 to 30 g./l.) using a soft-haired brush, to remove dirt and grease. (A solution of sodium cyanide or trisodium phosphate can be used also.) This is followed by rinsing and then cleaning as anode for about ½ minute at 6 volts in a warm (120°F.) (49°C.) salt bath, consisting of 8 oz./gal. (60 g./l.) sodium chloride and 1 oz./gal. (7.5 g./l.) ammonium chloride. After the anodic treatment the stereotype is scrubbed and rinsed at the same time until the dark surface smudge is removed.

b. *Electroplating.* After cleaning, the stereotype is placed in an iron bath and plated with 0.00075 in. to 0.001 in. (0.019 to 0.025 mm.) of iron. To avoid pitting, a vertical reciprocating motion of the stereotype while plating is recommended. Usually pitting can be sufficiently retarded by lifting the stereotype occasionally during the plating period and "swishing" up and down to remove clinging gas bubbles.

More or less bright deposits can be obtained with the sulfate-chloride bath when the pH is below 3.0, but these deposits are quite often rough for thicknesses greater than 0.001 in. (0.025 mm.).

After experimenting with a number of addition agents, it was found that smooth, semi-bright deposits could be obtained at the higher pH values by addition of 2 to 4 cc./l. of o-cresol sulfonic acid and 0.25 to 0.5 g./l. of Duponol "ME" to the sulfate-chloride composition.

Recovering Silver from Stripping Solutions

To the stripping solution add a strong salt solution until no more white silver chloride precipitates.

Dissolve the silver chloride precipitate in sodium cyanide. This must be done under a hood or out in the open if the acid has not been washed out completely, since the reaction between any residual acid and the cyanide will result in the evolution of hydrocyanic acid fumes which are highly poisonous.

The silver can be plated out on thin sheets of silver or on steel using steel or carbon anodes and can then be remelted as you are doing at present.

Use eight ounces of sodium cyanide for each gallon of final solution. When all, or most of the silver has been plated out, the cyanide solution can be used to dissolve more silver chloride precipitate and it will be necessary to make only occasional additions of sodium cyanide to balance decomposition losses.

Silver-Plating Magnesium

The work is first cleaned cathodically as described for nickel plating, and is rinsed in cold running water. It is then anodized for 3 to 6 min. in a solution containing trisodium prosphate (Na_3PO_4. $12H_2O$) 25 per cent, at a current density of 275 to 375 amp./ft.² The electrolyte is heated to 140 to 170°F. This produces a stable, adherent, smooth gray anodic film. The anodized parts are then given a cold water rinse and the copper strike in a bath of the following composition:

Copper Cyanide   0.67 oz./gal.
Potassium
   Cyanide   13.3 oz./gal.

Temperature   68 to 77°F.
Current
    Density   270 amp./ft.²
After the magnesium is completely covered with copper, it is given a thorough cold water rinse and silver plated in the following electrolyte:
Boric Acid   1.3 to 4.0 oz./gal.
Silver Cyanide   1.1 oz./gal.
Potassium
    Cyanide   1.5 oz./gal.
Anodes, High Purity
    Silver Current
    Density 1.8 to 4.3 amp./ft.²
Temperature   68 to 77°F.
Anode to cathode ratio 1 to 6 max., larger anodes produce poor deposits.
The deposits are hard, smooth and white. Thicknesses under 0.001 in. are quite adherent.

Silver-Plating without Electricity
Solution A:
    Silver Nitrate   10 g.
    Water   300 cc.
    Mix until dissolved.
Add just enough ammonia to clear up milkiness and then add:
    Water   To make 1000 cc.
    Bring to a boil.
Solution B:
    Silver Nitrate   2 g.
    Water   996.3 cc.
    Rochelle Salt   1.7 g.
    Water   To make 1000.0 cc.
    Bring to a boil.
Mix equal parts of solutions A and B and immerse cleaned articles in it until silver deposit is uniform.

Macroscopic Brass Etching
Ferrous Chloride
    (10% Sol.)   1

Chromic Acid Solution
    Saturated with Salt   1
Acetic Acid (20%)   2

Metallographic Etchant for Silver Solders
When a metallographic specimen of silver solder is etched with the commonly used ammonium hydroxide and hydrogen peroxide reagent, the dark and light constituents of the microstructure are brought into good contrast but the dark constituent is very likely to be overetched with the result that the details of the structure are lost.

A much more satisfactory etchant both from the standpoint of convenience and results, is a 2% ferric chloride solution. This acts slowly enough to permit the structural details of the dark constituent to be revealed clearly. The desired degree of contrast is obtained by controlling the etching time, which is generally in the range of 5 to 30 sec.

Etching Solution for Beryllium
    Sulfuric Acid   15
    Water   85
The solution is used at 125°F. The metal is immersed in the solution for 5 seconds, removed and rinsed quickly.

Non-Ferrous Metallographic Etch
    Ferric Chloride   5 g.
    Ethyl Alcohol   96 ml.
    Hydrochloric Acid
        (Conc.)   2 ml.
The ferric chloride is first dissolved in alcohol, and then the solution is acidified by the concentrated hydrochloric.

Etching Magnesium Metallographic Specimens

The choice of etchants used for micro examination is based more on the physical condition of the alloys than on their composition. For sand-cast, permanent mold-cast or die-cast metal in the as-cast condition, and for all of the alloys in the aged condition, the "glycol" etchant is perhaps the best. It has the following composition:

Ethylene Glycol	75
Distilled Water	24
Concentrated Nitric Acid	1

The freshly polished specimen is immersed face up in the etchant for 5 to 15 sec., washed well in running water, then in alcohol and dried in a blast of air.

To show grain boundaries in the solution-heat-treated castings and most of the wrought alloys, the "acetic-glycol" etchant is used. It is composed of:

Ethylene Glycol	60
Distilled Water	19
Glacial Acetic Acid	20
Concentrated Nitric Acid	1

For estimating the amount of massive compound in heat-treated castings or in wrought metal the "phosphopicral" etchant is used. This etchant stains the solid solution and leaves the compounds white. It consists of the following:

Orthophosphoric Acid	0.7 cc.
Picric Acid	4.0 g.
Ethyl Alcohol (95%)	100 cc.

The specimen is immersed in the etchant face up for about 10 to 20 sec. or until the polished surface is darkened. The specimen is then washed in alcohol and dried

or it can be washed in alcohol, then in water, then alcohol again and dried. Washing directly in water will lighten the stain, and the contrast between the white compound and the darkened solid solution will be lessened. This etchant improves with use.

For revealing the grain boundaries in the Dowmetal FS–1 (Mg, 3.0 Al, 1.0 Zn, 0.3 Mn) alloy sheet the "acetic-picral" can be used. It also is excellent for macro grain size determinations. This etchant must be made up fresh each time it is used, but it can be prepared readily by mixing the two following solutions:

Saturated Picric Acid in 95% Ethanol	100
Glacial Acetic Acid	10

Acid Etching Resist
U. S. Patent 2,168,756

Thermoplastic "Bakelite" Resin	21
Carnauba Wax	5
Talc	6
Aluminum Stearate	3
Ground Mica	60
Oil-Soluble Red Dye	0.9

Testing for Traces of Grease or Finger Prints on Metals

Iron or aluminum is dipped in 3% copper sulfate solution; zinc in 0.1% copper sulfate solution. A coating of copper appears on metal not covered by grease or print.

Identifying Cadmium, Tin and Zinc Coatings

Immerse the sample in a solution of one part commercial hydrochloric acid and one part water.

If a rapid reaction takes place the metal is zinc. If no obvious reaction occurs, hold a piece of Armco or electrolytic iron in contact with the sample beneath the surface of the acid solution. If rapid gas evolution takes place at the interface of the iron and acid, the metal is cadmium. If no obvious gas evolution of this nature occurs, the metal is tin.

Powdered Lead Coated Iron
British Patent 544,840

Lead Oxide	12
Ferrous Oxide	88

A mixture of the above is reduced with charcoal 25 lb. at 816–1150°C., to produce a coherent, loosely bonded mass, which may be pulverized, yielding lead-coated iron powder.

Powdered Iron
British Patent 545,057

Aqueous potassium hydroxide or sodium hydroxide containing more than 5% of sodium chloride and having density 1.35–1.6 is electrolysed above 100°C. with iron anodes and silver-plated copper or nickel-plated steel cathodes, using an anode, current density of less than 5 amp./dm.^{2} and a cathode, current density of more than 2 amp./dm.^{2}. The cathode deposit can be readily ground to a fine powder containing more than 99.5% of iron.

Powdered Metals (Copper)
U. S. Patent 2,259,457

Copper chips 113.6 are ground in ball mills with a solution containing:

Saponin	0.02
Salicin	0.05
Ammonia, Aqueous (d. 0.88)	14.20

The mill is emptied at intervals, and the suspension of ground metal separated from the chips and replaced by fresh solution.

Sponge Iron Reducing Agent
U. S. Patent 2,248,735

Coke Breeze	100
Sodium Nitrate, or Potassium Nitrate	8

Interaction is improved by the initial oxidation step in which the nitrate and coke yield carbon monoxide which starts the reduction of the ferric oxide.

Metal Carbides
British Patent 550,133

Tungsten is mixed with 7.25% of carbon and ball-milled for 7 hours, then sieved (through 80 mesh), placed in a molybdenum boat, and heated for 3 hours at 1425–1450°; the product is cooled and sieved through 250-mesh silk bolting cloth, and, if it is to be stored, packed in an atmosphere of carbon dioxide. Tungsten carbide containing 5.8% of combined carbon and 0.03% of free carbon is produced.

Pulverizing Cemented Tungsten Carbide
U. S. Patent 2,138,672

Alloys, e.g., of tungsten carbide with cobalt 3–25%, are heated at 1600–1700°C. for 30 seconds in a closed carbon tube; the spongy product may be ground.

Diamond Embedding Composition
German Patent 720,005

Iron, Powdered	80

Tin, Powdered 20

This is solidified under heat and pressure.

Calorizing Metal
U. S. Patent 2,279,268

A zinc soap, e.g., zinc stearate, oleate or palmitate, is dissolved in a paint vehicle, e.g., solvent naphtha, and granular aluminum is added. A suitable mixture comprises zinc soap 21.4 + solvent 78.6% mixed with 48.6% by weight of aluminum. Ferrous articles are coated, the coating is dried, and the whole fired at 600–700°C.

Case-Hardening Iron or Steel
British Patent 548,275

Steel is heated at 800–1000° in mixture of:

Salt	280
Coal Dust	160
Soap	60
Charcoal	844
Dried Blood	676
With or without	
Bone Meal	60
Soot	60

Case-Hardening Compound
British Patent 548,275

Salt	280
Coal Dust	160
Soap	60
Charcoal	60
Dried Blood	676

High-Speed Steel Hardening Bath
U. S. Patent 2,299,186

Barium Chloride	12
Sodium Cyanide	33
Soda Ash	43
Potassium Carbonate	8

Calcium Fluoride 4

Use at 480–560°C.

Metal-Tempering Compound

Potassium Nitrate	¼	oz.
Zinc Sulfate	¼	oz.
Powdered Alum	½	oz.
Olive Oil	2	oz.
Soap	½	lb.

Best results are obtained from this mixture when it is made up fresh, and must be sealed when not in use.

Case-Hardening (Small Articles)

Many times a hard surface is required on objects of wrought iron, such as small chisels, punches, etc. This hardened surface can be obtained in a number of ways:

First, the punch, or object is heated red hot in contact with carbon or charcoal. This heating process permits some of the carbon to enter the pores of the iron and transforms it into steel.

A second method consists of heating the object to a bright redness and sprinkling with potassium cyanide (which is a deadly poison), then returning the object to the fire and after a few minutes cooling in water by immersion. *Be exceedingly careful in handling cyanide as even the fumes are very poisonous.*

Small articles may be hardened on the outside by coating them with a paste of arsenious acid (poison), powdered leather, horn, or bones; or other nitrogenous bodies with hydrochloric acid, and then heating them to bright redness in a muffle furnace.

Resurfacing Cold-Rolled Low-
Carbon Steel
U. S. Patent 2,295,204
Pickle in:

Sulfuric Acid (66%)	8
Water	95
Ferric Sulfate	4

Hardened Aluminum Alloy
Casting
U. S. Patent 2,263,823

Molten aluminum alloy containing hardeners, e.g., copper, silicon,
iron, and zinc, is cast and removed
from the mold as soon as it is solid
(>455°C.) and quenched so that
the hardeners are, at least in part,
in solid solution. The alloy is re-
heated for ¼ to 3 hours at a temperature 455 to 565°C. at which
the hardeners go into solid solution, quenched to the ageing temperature (>232°C.), and aged.

Permanent Magnet Alloy
Formula No. 1
U. S. Patent 2,347,817

Nickel	10–30
Aluminum	5–12
Vanadium	2–10
Cobalt	5–15
Iron	To make 100

No. 2
German Patent 707,516

Carbon	0.0– 0.4
Nickel	9.0–38.0
Aluminum	7.0–18.0
Columbium and/or Tantalum	0.5–11.0
Iron	To make 100

No. 3
U. S. Patent 2,349,857

Nickel	10–30
Aluminum	5–10
Vanadium	2–10
Cobalt	5–15

Copper	0.1–10
Iron	To make 100

No. 4
(High initial permeability)

Nickel	76–78
Copper	5–6
Molybdenum	3½–4½
Manganese	½–1
Iron	To make 100

No. 5
U. S. Patent 2,264,038
Steel containing:

Cobalt	35
Nickel	18
Titanium	8
Aluminum	6

is made by melting together iron,
nickel, and cobalt, and then adding the aluminum and titanium.
The alloy is cast very hot, into a
sand mold, stripped from the mold
as soon as it is solid, air-cooled,
heated for about 5 hours at 650°C.,
and then slowly cooled.

Non-Corrosive Bearing
Formula No. 1
U. S. Patent 2,257,313

Cadmium-silver or copper-lead
alloy bearing liners are protected
from corrosion by coatings produced by immersion in aqueous
0.75% by weight phosphoric acid
containing manganese dioxide 1
ounce per gallon at 93°C., or in
1–5% aqueous boric acid or borax
containing manganese dioxide, or
in phosphoric acid containing
manganese dioxide and iron fil-
ings.

No. 2
U. S. Patent 2,341,550

Gold	65–99
Lead	1–35

Hardened Lead Bearing Alloy
U. S. Patent 2,264,251
Formula No. 1
Calcium (Primary Hardener)	0.08
Tin (Oil Corrosion Inhibitor)	4.50

No. 2
Calcium (Primary Hardener)	0.6
Cadmium (Secondary Hardener)	1.0
Tin (Oil Corrosion Inhibitor)	3.0

Cadmium Bearing Alloy
German Patent 719,978
Silver	0.1–2
Copper	0.1–0.5
Tellurium	0.1–2
Cadmium	To make 100

Iron Bearing Alloy
German Patent 714,396
Lead	15
Antimony	⅕–5
Iron	To make 100

Lead Alloy
High permanent hardness and Bondability
U. S. Patent 2,299,711
Tin	½–10
Antimony	6–18
Silver	1–10
Copper	< ½
Lead	To make 100

Non-Corrosive Lead Alloy
U. S. Patent 2,252,104
Lead alloys, suitable for storage-battery terminals, ships fittings, etc., contain:
Tin	5.5
Antimony	5.3
Copper	1.6
Arsenic	0.1

Non-Discoloring Tin Foil
U. S. Patent 2,151,302
Tin foil which is not discolored by contact with foodstuffs (cheese) comprises a foil of tin or tin-antimony alloy coated with tin containing:
Cadmium	0.5
Aluminum, Magnesium, or Calcium	0.3

Casting and Rolling Magnesium Alloy
U. S. Patent 2,340,795
Aluminum	4–6
Zinc	1–2
Manganese	0.1–1
Titanium	0.5–1
Magnesium	To make 100

Electric Thermostat
U. S. Patent 2,317,018
An electrical resistor of laminated thermostatic metal comprises high and low expanding laminae. The high expanding lamina is an alloy containing about 15% chromium, about 4.25% aluminum, about .5% manganese, and the balance iron. The low expanding lamina is a nickel-iron alloy comprising essentially nickel in an amount falling within a range of 35 to 42% by weight and the balance iron.

Electric Fuse Alloy
U. S. Patent 2,293,762
Tin	40–60
Cadmium	20–35
Silver	15–25
Copper	2–5

This alloy is suitable for fuses to blow at small currents.

Electrical Contact Alloy
Formula No. 1
U. S. Patent 2,138,599
Substitutes for high-platinum or iridium contact alloys contain:

Silver	60–67.0
Copper	12–13.4
Nickel	5– 5.6
Palladium	25–14.0

No. 2
U. S. Patent 2,154,700
A mixture of chromium powder 90–99 and tin powder 10–1 is ground, moulded to shape, hot-pressed at 250°C. and heated in vacuum at about 800°C.

Spark Plug Electrode Alloy
German Patent 734,494

Chromium	5–30
Nickel	<60
Manganese	<20
Silicon	<4
Aluminum	<5

Resistor Alloy
German Patent 734,854

Chromium	<40
Cerium	0.02–1.2
Iron	To make 100

Electric Resistance Alloy
U. S. Patent 2,293,878

Chromium	10–30
Aluminum	<5 ⎫ Total <6
Copper	<3 ⎭
Total	<6
Nickel	To make 100

Electric Wire Resistance Alloy
U. S. Patent 2,242,865

Nickel	40
Manganese	40

Copper	10
Iron	10

Heat-Hardenable Copper Alloy
U. S. Patent 2,357,190

Nickel	0.5–5
Silicon	0.1–1.25
Cerium	0.05–0.5
Copper	93–99

Hardening Copper
U. S. Patent 2,252,604

Copper Phosphide	1.0
Salicylic Acid	0.5
Pumice Powder	0.5

Add this to 4 parts of molten copper. After 5–8 minutes, a graphite rod is held beneath the metal for 5 minutes and the melt cast.

Corrosion-Resisting Coating for Copper
U. S. Patent 2,272,216

Zinc Tetra-Hydrogen Orthophosphate	4.2
Ammonium Peroxy-disulfate	0.8

The coating is washed in aqueous chromic anhydride, phosphoric acid, oxalic acid, or a salt of iron, aluminum, or chromium.

Cleaning and Covering for Molten Copper Alloys
German Patent 734,391

Powdered Slag	30
Ground Chamotte	60
Soda Ash	5
Salt	5

Tin-Free Gear Bronze

Antimony	7.5
Nickel	2.0
Copper	90.5

Cast at 1200°C.

Refining White Metal Alloys
U. S. Patent 2,150,353

Aluminum is eliminated from lead and/or tin alloys by stirring the molten metal at 590°C. with a mixture containing:

Sodium Carbonate	42.5
Sodium Chloride	17.0
Potassium Chloride	25.5
Sodium Hydroxide	15.0

Hard Platinum Alloy
British Patent 546,897

The hardness of alloys of platinum ($>82\%$) with other metals is considerably increased by addition of 0.5–1% of tin; e.g., platinum with rubidium 4%, tin 0.5% has a Vickers hardness of 146 compared with 115 for the 4.5% rubidium-platinum alloy, and addition of 0.5% of tin to 5% iridium-platinum alloy raises the hardness from 75 to 102.

Hard Non-Corrosive Gold Jewelry Alloy
U. S. Patent 2,169,592

Gold	41.67
Copper	40.45
Cobalt	0.50
Nickel	1.00
Silver	7.67
Zinc	8.71

Dental Copper Amalgam

This process is generally known as "immersion plating." An electrolyte and iron and mercury electrodes are the only materials required.

Composition of Electrolyte:

Copper Sulfate	200
Sulfuric Acid	20
Water	1000

The electrolyte is placed in a glass or glazed container. 50 g. of pure mercury is poured into the bottom of the container and a steel or iron strip, weighing at least 50 g., is placed in an upright position in the container in contact with the mercury.

The reaction is permitted to proceed for a week or more. The resulting copper amalgam will contain from 15 to 35% copper.

Dental Filling Alloy

Silver	68.5
Tin	26.5
Copper	4.0
Zinc	1.0

The copper and silver are first melted in a suitable ceramic crucible, then the tin is added and, finally, just before the melt is cooled the zinc is added with vigorous stirring. The melt is kept just above the melting point to prevent excessive oxidation. A small addition of powdered graphite also helps to prevent oxidation. After casting into suitable form, such as solid rods, or cylinders, the casting may be comminuted to the desired fineness by cutting or shaving on an ordinary lathe.

Dental Casting Alloy
Formula No. 1
U. S. Patent 2,156,757

A nickel alloy which can readily be remelted and cast contains:

Chromium	5–30
Cobalt	10–50
Boron	1–10
Tungsten	1–4
Molybdenum	2–8

No. 2
U. S. Patent 2,135,600

Cobalt with Chromium	30
Tungsten	15

Dental Model Alloy

Silver	64.0
Tin	31.0
Copper	4.0
Zinc	1.0

Use same method of preparation as given for the Dental Filling Alloy.

Dental Inlay Alloy
(Gold Substitute)

Copper	52.5
Zinc	46.0
Aluminum	1.5

Corrosion-Resistant Denture Alloy
German Patent 737,032

Silicon	0.1–0.5
Manganese	0.2–0.5
Magnesium	4–10
Aluminum	To make 100

Dental Clasp Alloy
U. S. Patent 2,304,416

Gold	50–70
Platinum	2–9
Palladium	1–6
Copper	9–15
Zinc	0.1–2
Silver	13.5–26

Solder for Zinc-Aluminum Alloys
Formula No. 1
German Patent 727,651

Zinc	0.5–20
Bismuth	<6
Tin	15–50
Lead	To make 100

No. 2
German Patent 732,319

Cadmium	20–55
Zinc	½–10
Lead	To make 100

Solder for Aluminum Bronze

Zinc	52
Aluminum	18
Tin	30

This solder is inexpensive. Owing to the content of aluminum and zinc it is possible to build up a solid alloy with the bronze. The solder is produced in graphite crucibles, in which aluminum is first melted under a cover of coal, at 750°C. The required amount of zinc is then added in small quantities, and mixed thoroughly. Finally tin is added. After complete mixing, the alloy is poured into molds.

Solder for Aluminum Foil
U. S. Patent 2,335,615

Bismuth	500–550
Lead	60–120
Tin	180–200
Cadmium	80
Silver	50

Powdered Solder for Aluminum and Stainless Steels

Zinc Chloride	1
Ammonium Chloride	4
Solder (Tin 30, Lead 70) Powdered	1

Aluminum Solders
U. S. Patent 2,252,409
Formula No. 1

Zinc	8.0
Nickel	0.2
Tin	To make 100.0

No. 2

Zinc	8.0
Cadmium	0.5
Nickel	0.2
Tin	To make 100.0

No. 3

Zinc	8.0

Cadmium	0.5
Manganese	0.2
Tin	To make 100.0

No. 4

Zinc	5.0
Nickel	0.1
Cadmium	1.0
Manganese	0.1
Tin	To make 100.0

No. 5

Zinc	8.0
Manganese	0.2
Tin	To make 100.0

No. 6

Zinc	5.0
Nickel	0.1
Manganese	0.1
Tin	To make 100.0

Hard Solder
U. S. Patent 2,310,231

Sodium	½–5
Silver	40–60
Zinc	17–21
Cadmium	11–15
Copper	16–20

Brazing Solder
U. S. Patent 2,355,067

Silver	10–15
Cadmium	10–15
Copper	55–70
Zinc	7–12
Sodium	0.05–1.0
Phosphorus	0.04–1.5

Gas Meter Solder

Tin	38
Antimony	2
Lead	60

Soft Solder for Tinplate

Lead	87.5
Arsenic	0.5
Antimony	12.0

Tinless Soft Solders

Lead	90
Cadmium	8
Zinc	2

Soft Solder

Lead	91.00
Cadmium	7.00
Zinc	1.20
Tin	0.50
Copper	0.30
Phosphorus	0.08

Low-Tin Solder
Formula No. 1

Tin	32½
Silver	½
Antimony	1
Lead	66

This solder is better than a 40–60 solder.

No. 2

Lead	68
Tin	30
Antimony	2

Soft Solder
British Patent 552,330

Soft solder having a high creep-resistance at elevated temperature consists of lead alloyed with 1.5% of silver and 5% of tin, the proportion of tin being dependent on that of the silver. The solder can be used with rosin flux, satisfactorily wets copper and copper alloys, and has but little solvent power for copper.

Soldering Flux
Formula No. 1
U. S. Patent 2,279,828

| Petrolatum (Crude or Refined) | 50 |
| Petroleum Mahogany Sulfonic Acid | 10 |

Ammonium Chloride-
Zinc Chloride
Eutectic   40
             No. 2
   British Patent 557,816
Rosin   24
Cetylpyridinium
   Bromide   0.2–8
Alcohol   To make 100

Hard Soldering Flux
U. S. Patent 2,174,551
Boric Acid   42.10
Potassium Carbonate   1.33
Potassium Bifluoride   7.70
Potassium Fluoride   27.20
Potassium Silico-
   fluoride   0.77
Potassium Boro-
   fluoride   4.63
Water   20.00

Soldering Flux for Galvanized
            Iron
    U. S. Patent 2,160,195
Stannous Chloride   9½
Zinc Chloride   3
Ammonium Chloride   ¾
Water to give a density of 1.35.
Hydrochloric Acid, 2–3% may
be added.

Foundry Mold Composition
     Formula No. 1
   U. S. Patent 2,322,638
Silica Flour   65–95
Sodium Silicate   4–18
Pitch   2–18
Water, per 100 of above   8–17
             No. 2
   U. S. Patent 2,322,667
Silica Flour   85–95
Sodium Silicate   15–2
Water, per 100 of above   5–25
             No. 3
Quartz Sand   96.0

Milled Bentonite   3.2
Sulfite Cellulose Lye   0.8
             No. 4
Sand   77
Reworked Mold Sand   20
Fireclay   13
This is gas free after first min-
ute of use.

Foundry Mold and Core
   U. S. Patent 2,304,751
Zircon, Granular   100
Coke, Granular   50
Core Oil   0.1–0.75

Magnesium Alloy Foundry Sand
Sand, Washed &
   Dried Quartz   800
Bentonite   10
Ammonium Fluosilicate   26
Sulfur   7
Boric Acid   10
Diethylene Glycol   2
Water   10–16

Mold-Facing Sand
   U. S. Patent 2,348,155
*Sand   80
†Bentonite   12
†Pitch   4
†Sea Coal   4

Casting Mold Coating
Ground Chalk   65–70
Coke   25–30
Clay   5–8
Sulfite Liquor   1–2

Ingot Mold Coating
   U. S. Patent 2,289,709
Linseed Oil   1.5
Cumarone Resin   2.0

* Sharp type capable of withstanding
temperature up to 3000°F.
† Dry powdered.

Naphtha 3.5
Graphite 2.0

Insulating Compounds for
Molten Cast Steel
Formula No. 1
U. S. Patent 2,250,009

Coke Breeze, or Gas-
producer Flue Dust 60
Blast-Furnace Slag 40

No. 2

Coke Breeze, or Gas-
producer Flue Dust 85
Vermiculite, Exfoliated 15

To prevent piping in cast in-
gots, spread over the surface.

Dental Investment
U. S. Patent 2,333,430

Calcium Sulfate 10–25%
Feldspar 30–80%
Silica Grog 10–45%

Foundry Mold Wash
U. S. Patent 2,270,770

Graphite or Silica
Flour 25–50
Suspend this in:
Water 25–60

Alcohol 90
Methanol 5
Ethyl Acetate 5 } 50–80
Aviation Gasoline 1

Bentonite ½–5
Powdered Rosin 1–10

Mold Composition for Making
Complicated Precision Castings
of Aluminum or Brass
Formula No. 1

Refined Plaster of Paris,
or Stucco 60
Silica (150 mesh or finer) 40

No. 2

Refined Plaster of Paris,
or Stucco 50
Terra Alba 10

Silica (150 mesh or finer) 40

No. 3

Refined Plaster of Paris,
or Stucco 60
Silica (150 mesh or finer) 37
Fibrous Talc or Asbestos 3

This formula produces a plaster
that has greater mechanical
strength, after drying, than the
two previous ones.

Water is added to the above
formulae, mixed to a consistency
of thick cream, and poured over
the pattern. To eliminate air bubbles, the mold should be vibrated
while the mass is still plastic. The
setting time of the formulae can
be accelerated by adding sodium
chloride, potassium chloride, or
alum, or can be retarded by adding
soluble citrates, phosphates, or acetates in small amounts (about
1%). The molds should be dried
at 500–900°F., for aluminum, and
900–1200°F. for brass in a still air
oven. Drying time is between 8–12
hours, depending upon the
amount of water used in preparing the composition. They are allowed to cool to 250–400°F. before
the metal is poured into them. The
heating and cooling of the molds
should be done slowly, otherwise
they will crack. If strict control
is exercised in the preparation of
the mold composition and the drying, castings can be consistently
cast within two thousandths of an
inch.

Parting Compound to be Sprayed
on Patterns before Pouring
Plaster

Soluble Oil, Cutting Oil,
Castor Oil, or Coconut
Oil 5

Benzene or Carbon
Tetrachloride   95

Non-Slip Stair Tread Coating
U. S. Patent 2,301,721
Abrasive grains are used as a
homogeneously distributed aggregate with a sintered bond formed
of:

Copper	76
Nickel	20
Tin	4

Welding Rod
Formula No. 1
U. S. Patent 2,349,945

Tin	1–8
Nickel	0.1–1
Lead	3–20
Copper	To make 100

Welding-Rod Alloy
No. 2
U. S. Patent 2,160,423

Cobalt	26.28
Chromium	15.42
Molybdenum	0.32
Tungsten	9.51
Vanadium	1.83
Carbon	1.41
Iron	To make 100

Brazing Alloy
U. S. Patent 2,245,327
Copper-base brazing alloys
which can be extracted and hot-
worked contain:

Phosphorus	5.6
Arsenic	4.2

Melts at 660°C.

Brazing Alloy for Brass,
Copper and Steel

Copper	68–84
Cadmium	3–24
Silver	<3.9
Phosphorus	2–4

Brazing Alloy for Welding
to Iron

Copper	80.0
Tin	17.3
Nickel	1.5
Silicon	0.5
Iron	0.5

Copper Brazing Wire
German Patent 707,261

Antimony	0.30–0.4
Misch Metal	0.05–0.1
Phosphorus	0.05–0.1
Nickel	0.10–0.2
Copper	To make 100

Welding Rod for Red Brass

Copper	97.5
Silicon	1.5
Zinc	1.0

White Metal Welding
Composition
U. S. Patent 2,333,989

Zinc	34.8
Tin	60.6
Aluminum	1.5
Stearic Acid	1.5
Paraffin Wax	1.5

Welding Electrode Coating
Formula No. 1
U. S. Patent 2,150,925
Mild steel or alloy rod is coated
with a mixture containing:

Aluminum Oxide	15
Manganese Carbonate	10
Titanium Dioxide	50
Silicon Dioxide	15
Sodium Silicate	10

No. 2
U. S. Patent 2,271,351

Calcium Carbonate	50
Calcium Fluoride	90
Iron-Manganese	36
Potassium Metasilicate	97

Water, to make a suitable dip.

The coated rods are dried in a warm, humid atmosphere (dry bulb at 104.4–87.8°C.; wet bulb at 54.4–68.3°C.).

Aluminum Welding Rod Coating Formula No. 2
British Patent 554,239

Rods are dipped in:

Sodium Chloride	30
Potassium Chloride	30
Lithium Fluoride	15
Sodium Aluminum Fluoride	25

Apply at 450°.

No. 2
U. S. Patent 2,337,714

Sodium Chloride	30
Potassium Chloride	30
Lithium Fluoride	15
Cryolite	25

Melt together at a temperature just below melting point of the aluminum rod, which is dipped into it and withdrawn.

Welding Rod Flux Coating
U. S. Patent 2,164,775

Titanium Dioxide	32
Zirconium Oxide	14
Wood Flour	12
Slip Clay	32
Ferromanganese	10
Sodium Silicate (44%)	9

The above coating should be about 8–20% of weight of rod.

Brazing Flux
(For brass, copper, monel and stainless steel)

Potassium Fluoride	35
Borax	10
Boric Acid	45
Water	To make 100

Mix well and heat to 185°F.

---

then cool while stirring to form a smooth paste.

Nickel-Chrome Alloy Welding Flux
British Patent 551,915

Calcium Fluoride	15
Calcium Hydroxide	17
Boric Acid	41
Sodium Silicate	45

Gold Melting Flux
U. S. Patent 2,281,528

Potassium Bitartrate	45
Sodium Chloride	50
Charcoal	5

Copper Welding Flux Formula No. 1
U. S. Patent 2,284,619

Borax	36
Sodium Bifluoride	23
Sodium Silicofluoride	19
Antimony Trifluoride	1
Phosphoric Acid	13
Trisodium Phosphate	8

Mix with water to form a paste, before use.

No. 2
U. S. Patent 2,284,619

Borax	36
Sodium Bifluoride	23
Sodium Fluosilicate	19
Antimony Fluoride	1
Phosphoric Acid	13
Trisodium Phosphate	8

No. 3
Flux for Welding Copper-Base Materials

Boric Acid	3
Sodium Nitrate	2

Flux for Welding Aluminum

Sodium Chloride	30
Potassium Chloride	25

Sodium Fluoride	20
Lithium Chloride	25

Aluminum or Magnesium Welding Flux
British Patent 557,522

Lithium Chloride	25–40
Magnesium Chloride	10–15
Calcium Chloride	2–7
Potassium Fluoride	15–35
Potassium Chloride	30–45
Sodium Chloride	0–5
Potassium Bromide	<10

Flux for Welding Steel Contaminated with Sulfur

Sulfur affects steel in several ways. It may combine with the iron chemically to form sulfides; it may diffuse into the metal; or it may adhere to the surface of the metal. Steel so contaminated is difficult to weld. Steel contaminated with sulfur may be welded with an oxygen acetylene flame with aid of a welding rod provided with a flux consisting by weight of 25 parts manganese dioxide; 15 parts borax; 15 parts ground silica; 5 parts kaolin.

Flux for Welding Iron-Base Alloys

Sodium Nitrate	2
Sodium Carbonate	1
Borax	1

Non-Adhering Weld Spatter Coating
Formula No. 1
U. S. Patent 2,250,940

Linseed Oil Acids	20
Tung Oil	20
Rosin	20
Phenolic Resin	5

Glycerin	13
Phthalic Anhydride	22

No. 2

Make a solution of about three bars of yellow laundry soap in five gallons of water. Cut the soap into chips in about three gallons of water and heat to near boiling point, until soap is thoroughly dissolved. Add the other two gallons of water and allow to cool until lukewarm, then add gradually with constant agitation of solution, about one pint of castor oil.

Flux for Melting Non-Ferrous Metals
U. S. Patent 2,279,565

Anhydrous Borax	1
Air Slaked Lime	1–2

Non-Corrosive Magnesium Flux
U. S. Patent 2,296,396

Lithium Fluoride	6.00
Lithium Chloride	15.75
Sodium Chloride	39.10
Potassium Chloride	39.10
Potassium Dichromate	0.05

Magnesium Refining Flux
British Patent 546,981

Ammonium Chloride	107
Sodium Chloride	24
Magnesium Oxide	40

are ground together and heated to produce a flux containing:

Magnesium Chloride	74.0
Sodium Chloride	17.4
Magnesium Oxychloride	6.8

42% of this flux is mixed with:

Sodium Chloride	26
Magnesium Fluoride	32

Aluminum Scrap Melting Flux

Sodium Chloride	50
Potassium Chloride	50

Chapter XIII

PAINT, VARNISH, LACQUER AND OTHER COATINGS

Exterior House Paints

The generally accepted best exterior house paint formulation is a compromise between a formulation for longest life and a formulation which is self-cleaning and gives best appearance through its life. It is also very important that an exterior house paint should fail by slow erosion and give a good repaint surface free from cracking, flaking and peeling. The following is a type of formulae for white paint, which has good durability, a good appearance throughout its life and gives a good repaint surface.

Outside White House Paint Formula No. 1

Chalking-Type Rutile Titanium Dioxide	116	lb.
Asbestine	270	lb.
35% Co-Fumed Leaded Zinc	463	lb.
Basic Carbonate White Lead	116	lb.
Pale Refined Linseed Oil	59	gal.
Q-Bodied Linseed Oil	6.5	gal.
Petroleum Spirits	6.75	gal.
Paint Drier	2.5	gal.

No. 2

Titanium Dioxide	150	lb.
Leaded Zinc Oxide 35%	350	lb.
Metro-Nite	750	lb.
Linseed Oil (Raw)	67	gal.
Two-Hour Kettled Linseed Oil	4	gal.
Lead-Cobalt Drier	5	gal.
Mineral Spirits	7.5	gal.

The following formula is a similar type of white house paint with a lower oil content, designed to conserve oil during the war period.

No. 3

Chalking-Type Rutile Titanium Dioxide	116	lb.
Asbestine	270	lb.
35% Co-Fumed Leaded Zinc	463	lb.
Basic Carbonate White Lead	116	lb.
Pale Refined Linseed Oil	32	gal.
*Cut Z-4 Linseed Oil	19.75	gal.
Petroleum Spirits	31.5	gal.
Cooked Paint-Type Drier	1.12	gal.

An exterior primer undercoat designed to control penetration has been very satisfactory for improving the durability of finishing coats of exterior paints when used on new wood or on badly weathered old paint jobs. In many cases, one coat of primer undercoat and one coat of regular ex-

* 80% solids by weight in petroleum spirits.

271

terior house paint has showed equal durability to three coats of the regular exterior house paint. The following is a typical formula for this type of primer undercoat. It can be tinted with colors-in-oil for use under tinted house paints.

White Primer-Undercoat

Titanium Barium		
Pigment	354	lb.
Basic Carbonate		
White Lead	354	lb.
5X Asbestine	100	lb.
3X Asbestine	135	lb.
Litharge	9	lb.
Raw Linseed Oil	28	gal.
Litharge-Treated		
Linseed Oil.		
Viscosity U-W.		
70% N.V.	14	gal.
5-Gal. Ester Gum-		
Dehydrated Castor		
Oil Varnish	7	gal.
Petroleum Spirits	32	gal.
6% Cobalt Naph-		
thenate Drier	4	fl. oz.

White house paint should not be tinted. The chalking of the white will cause too much color change. A formulation specially designed to give satisfactory color retention is given below. This formula can be tinted with colors-in-oil to give the desired shade.

House Paint Base for Tints

Chalk-Resistant		
Titanium		
Dioxide	179	lb.
3X Asbestine	272	lb.
35% Co-Fumed		
Leaded Zinc	452	lb.
Raw Linseed Oil	59.25	gal.

Q-Bodied Linseed

Oil	5.75	gal.
Petroleum Spirits	7	gal.
Paint Drier	3	gal.

Brick and Stucco Paints

This type of paint gives a desirable low gloss, waterproof finish for use over brick, stucco and masonry surfaces.

Brick and Stucco Primer

Titanium Barium		
Pigment	354	lb.
Basic Carbonate		
White Lead	354	lb.
5X Asbestine	100	lb.
3X Asbestine	135	lb.
Litharge	9	lb.
Raw Linseed Oil	28	gal.
Litharge-Treated		
Linseed Oil,		
Viscosity U-W.		
70% N. V.	14	gal.
5-Gal. Ester Gum-		
Dehydrated Castor		
Oil Varnish	7	gal.
Petroleum Spirits	32	gal.
6% Cobalt Naph-		
thenate Drier	4	fl. oz.

Brick and Stucco White Paint

Chalking-Type		
Rutile Titanium		
Dioxide	116	lb.
Asbestine	270	lb.
35% Co-Fumed		
Leaded Zinc	463	lb.
Basic Carbonate		
White Lead	116	lb.
Pale Refined		
Linseed Oil	32	gal.
*Cut Z-4 Linseed		
Oil	19.75	gal.
Petroleum		
Spirits	31.5	gal.

Cooked Paint-
Type Drier   1.125 gal.

Brick and Stucco Paint Base
for Tints
Chalk-Resisting
Titanium
Dioxide   179   lb.
3X Asbestine   272   lb.
35% Co-Fumed
Leaded Zinc   452   lb.
Raw Linseed
Oil   29½   gal.
*Cut Z-4 Linseed
Oil   25   gal.
Petroleum Spirits 18¼   gal.
Paint Drier   1⅝   gal.

Exterior Trim and Trellis Paints

Trim and trellis paints are designed for use as trim paints and
for general exterior use on lawn
furniture, porch furniture, toys,
etc. These paints will retain a high
gloss over a long period of time.

Black Trim and Trellis Paint
(Oil Type)
Formula No. 1
Carbon Black   21   lb.
Litharge   4   lb.
Z-2 to Z-3 Dehydrated
Castor Oil   21   gal.
50 gal. Ester Gum
Varnish   55   gal.
Petroleum Spirits 21   gal.
8% Lead Naphthenate
Drier   1   gal.
2% Manganese Naph-
thenate Drier   4   pt.
2% Cobalt Naph-
thenate Drier   6   pt.

---

* 80% solids by weight in petroleum
spirits.

No. 2
(Alkyd Type)
Carbon Black   22   lb.
Litharge   4   lb.
Raw Linseed Oil   10   gal.
Long Oil Alkyd   77   gal.
Petroleum Spirits 9   gal.
Lead Naphthenate   1   gal.
Manganese
Naphthenate   1⅜   gal.
Cobalt Naphthenate ½   gal.

No. 3
(Red)
Toluidine Red   75   lb.
Litharge   3   lb.
Raw Linseed Oil   9   gal.
Long Oil Alkyd   75⅜   gal.
Mineral Spirits   10   gal.
Manganese
Naphthenate   ⅜   gal.
Cobalt Naphthenate ⅛   gal.

No. 4
(Red)
Red Iron Oxide   36.8   lb.
Asbestine   9.2   lb.
Raw Linseed Oil   27.6   gal.
Kettle Bodied
Linseed Oil
(Viscosity Q)   6.9   gal.
Duraplex E-73   11.5   gal.
Mineral Spirits   6.7   gal.
Cobalt Naphthenate
Drier 6%   .2   gal.
Manganese Naph-
thenate Drier 6%   .2   gal.
Lead Naphthenate
Drier 16%   .9   gal.

No. 5
(Blue)
Resinated Phthalo-
cyanine Blue   38   lb.
Ti-Pure R-610
(Chalk-Resistant

Titanium Di-
oxide) 68 lb.
Litharge 2⅖ lb.
Raw Linseed Oil 8½ gal.
Long Oil Alkyd 76 gal.
Petroleum Spirits 8 gal.
Manganese
Naphthenate 1¼ gal.
Cobalt Naphthenate ⅜ gal.

No. 6
(Green)

Dark Chrome
Green 125 lb.
Litharge 3 lb.
Raw Linseed Oil 8 gal.
Long Oil Alkyd 81¾ gal.
Petroleum Spirits 7¾ gal.
Manganese
Naphthenate ½ gal.
Cobalt Naphthenate ⅛ gal.

No. 7
(Orange)

Chrome Orange 346 lb.
Litharge 3 lb.
Raw Linseed Oil 8 gal.
Blown Oil Alkyd 76 gal.
Petroleum Spirits 8 gal.
Manganese
Naphthenate ⅜ gal.
Cobalt Naphthenate ⅛ gal.

No. 8
(White)

Titanium Pigment
(Non-Chalking) 25 lb.
Rezyl 823–1
(50% Solids) 31.4 lb.
Rezyl 419–1
(60% Solids) 26.9 lb.
Mineral Spirits 11.8 gal.
Cobalt Naphthenate
(6%) 0.2 gal.
Anti-Skinning
Agent 0.1 gal.
Nuact (Drier) 0.1 gal.

Wagon and Tractor Enamels

Wagon and Tractor Enamels are designed to give fast drying, durable finishes for all farm implements, lawn furniture, etc.

Wagon and Tractor Enamel
(Black)
Formula No. 1

Carbon Black 24 lb.
Phenolic Resin-
Dehydrated Castor
Oil Varnish
(30 gal.) 97 gal.
Lead Naphthenate
(24%) 3 pt.
Cobalt Naphthenate
(4%) 1½ gal.

No. 2
(Red)

Dark Para Red 39 lb.
Calcium Carbonate 71 lb.
30-gal. Varnish—
½ Pentarythritol
Esterified Rosin;
½ Ester Gum—
½ Linseed—
½ Dehydrated Castor
Oil 89 gal.
Lead Naphthenate
Drier (8%) 1 gal.
Manganese Naph-
thenate Drier
(2%) 2 gal.
Cobalt Napthenate
Drier (2%) 2½ gal.

Traffic Paints
Formula No. 1
(White)

XX Zinc Oxide 217 lb.
Lithopone 681 lb.
Asbestine 113 lb.
Aluminum
Stearate 5 lb.

19-gal. Ester Gum-
Chinawood Oil
Varnish (Thinned
with 75% Petro-
leum Spirits –25%
Xylol)   64   gal.
Cold Cut East India
Varnish (43%
Solids) Thinned
with 80% Petro-
leum Spirits—
20% Hi-Flash
Naphtha   26   gal.
Hi-Flash Naphtha   1¼   gal.
No. 2
(Yellow)
Chrome Yellow   368   lb.
Zinc Oxide   164   lb.
Asbestine   204   lb.
Diatomaceous
Silica   82   lb.
20-Gal. F-7 Amberol
—50% Linseed
—50% Dehydrated
Castor Varnish   60¼   gal.
VM&P Naphtha   12¼   gal.
Lead Naphthenate
Drier (8%)   2   gal.
Cobalt Naphthenate
Drier (2%)   2   gal.

Congo Resin Traffic Paints
Formula No. 1
Vehicle
(1) "Slack Melt"
Congo Resin   100
(2) Kettle Bodied Lin-
seed Oil ("Z" Visc.)   100
(3) VM&P Naphtha
To 50% Non-volatile
content
(4) Driers Equivalent
To 0.3% Pb., 0.1% Co.
on weight of oil
Heat resin and oil to 580°F. If
"pill" is clear on glass when cold,

allow to cool sufficiently and add
(3) and (4).
Paint
(1) Titanium Dioxide—
Chalking Type   0.90
(2) Barytes   2.25
(3) Asbestine   0.45
(4) Mica   0.45
(5) Pumice   0.45
(6) Above Vehicle   3.12
Grind (1), (2), (3) in a portion of vehicle. Add remainder of
vehicle, and (4) and (5) and stir
in thoroughly.
Lithopone (such as Albalith
#351 or equivalent) may be sub-
stituted for titanium dioxide and
barytes on a hiding power basis.
The zinc sulfide and barium sulfate
content of lithopone replaces the
combination of titanium dioxide
and barytes called for in the
formula. This holds true for
all traffic paint formulae given
here.

No. 2

Vehicle
(1) Batu Scraped or
Black East India Bold   69
(2) "Fine Melt" Congo
Resin   31
(3) Kettle Bodied Lin-
seed Oil ("Z" Vis-
cosity)   82
(4) VM&P Naphtha
To 40% non-volatile content
(5) Driers Equivalent
To 0.5% Pb., 0.02% Co. on
the weight of oil
Heat (1) and (2) to 610°F.
Allow to cool to 560°F. Add (3).
Re-heat to 540°F. Allow to cool
to 300°F. Add (4), and then (5).
Paint
Same formula as for No. 1, except using above vehicle.

Exterior Metal Paints

It is necessary, for satisfactory results, to thoroughly clean metal surfaces to remove grease, oil, rust, etc. It is advisable to treat ferrous metal surfaces with a rust preventative, such as the following, before priming.

Metal Protection (Phosphatizing)
U. S. Patent 2,293,716

Nitric Acid (d. 1.26)	400
Sulfuric Acid (75%)	1075
Zinc Oxide	420
Sodium Chlorate	500
Copper Carbonate	2½
Water	To make 5000

Use 500 lb. above per 1000 gal. at 46–77°C.

Pickling and Rustproofing for
Iron
U. S. Patent 2,294,571

Phosphoric Acid	1400
Water	1500
Alcohol	750
Propyl Alcohol	190
Hydrogen Peroxide (20 vol.)	75
Titanium Tetrachloride	40

Red Lead Primer
Formula No. 1

Dry Red Lead (95%)	1435	lb.
Raw Linseed Oil	44½	gal.
Turpentine or Petroleum Spirits	33	gal.
Cobalt Naphthenate Drier (2%)	3	gal.

No. 2

Dry Red Lead (95%)	400	lb.
Red Oxide	235	lb.
Asbestine	265	lb.
Barytes	227	lb.

Boiled Linseed Oil	26	gal.
Cut Z-4 Linseed Oil	22	gal.
Petroleum Spirits	19½	gal.
Lead Manganese Cooked Paint Drier	2½	gal.

Galvanized Iron Primer

Flaked Metallic Lead Paste	5⅓	lb.
Long Oil Pure Alkyd Varnish (45% Solids)	7¹⁄₁₀	pt.
Cobalt Naphthenate Drier (6% Cobalt)	1¼	oz.

The above formula produces one gallon of paint and has given excellent service as a primer for untreated galvanized iron.

A black primer may be made by adding to the above formula, 1.9 ounce of carbon black. The carbon black should be ground into the alkyd varnish before the addition of the metallic lead paste.

Zinc Dust—Zinc Oxide Primer

Zinc Dust	632
XX Zinc Oxide	158
Raw Linseed Oil	170
Z-1 Linseed Oil (Heat-bodied)	19
Mineral Spirits	13
Cooked Drier	8

Zinc Chromate Primer (Navy)

Zinc Chromate	280	lb.
Titanium Dioxide	76	lb.
Raw Sienna	26	lb.
Asbestine	127	lb.
Mineral Spirits	32½	gal.
Phenolic Resin— ½ Linseed, ½ Wood Oil Varnish (25 gal. oil length)	25¾	gal.

Alkyd Resin Solution
(Spec. 52-R-13) 21 gal.
Lead Naphthenate
(24%) 2½ gal.
Cobalt Naph-
thenate (6%) 1¼ fl. oz.
Manganese Naph-
thenate (6%) 1¼ fl. oz.

Zinc Tetroxy Chromate— Iron Oxide Primer

Zinc Tetroxy
Chromate 560 lb.
Iron Oxide 300 lb.
Raw Linseed Oil 55½ gal.
Kettle Bodied "Q"
Linseed Oil 7¼ gal.
*Mineral Spirits and
Drier 12 gal.

Zinc Tetroxy Chromate Primer

Zinc Tetroxy
Chromate 855 lb.
Raw Linseed Oil 55½ gal.
Kettle Bodied "Q"
Linseed Oil 7¼ gal.
*Mineral Spirits
and Drier 12 gal.
Use all of the kettle bodied oil
and as much of the raw linseed oil
as required to make paste, grind
and reduce with the remainder of
vehicle.

Zinc Chromate Primer

Zinc Yellow 29.15
Magnesium Silicate 5.15
Aluminum Stearate
(10% gel in Xylol) 3.92

* 5% of liquid drier should be sufficient. High percentages of drier may
tend to make a puffy paint. If this
should occur, the consistency may be
reduced by using ¼% soluble litharge
(based on the weight of pigment) in
making up the paste and reducing the
percentage of liquid drier.

Maleic Anhydride 0.31
Rezyl 728–5 (50%
in Xylol) 46.15
Bakelite BK3962X
(50% Xylol) 9.35
Solvent:
90% Aromatic Petro-
leum Naphtha
10% Butanol 9.83
Lead Naphthenate
(24%) 0.16
Cobalt Naphthenate
(6%) 0.29

Zinc Chromate—Iron Oxide Alkyd Primer

Zinc Chromate Pig-
ment 173.0
Red Iron Oxide 975.0
Zinc Oxide 216.0
Magnesium Silicate 216.0
Calcium Carbonate 389.0
China Clay 194.0
*Alkyd Resin Solution
(40% Solids) 2572.0
Mineral Spirits 177.0
Cobalt Naphthenate 3.4
Lead Naphthenate 21.4
Anti-Skinning Agent 5.1
Grind in ball mill.

Alkyd Resin Primer

Ammonium Ferrous
Phosphate 72½
Zinc Tetroxy
Chromate 217½
Asbestine 108¾
Aluminum Stearate 1½
Xylene Alkyd Resin
Solution 398½
Xylene 243½

* Linseed oil-modified glycerol phthalate resin, containing 32% phthalic anhydride by weight and free of rosin
and phenol.

Lead Naphthenate
(24%) 5¾
Cobalt Naphthenate
(6%) 3

Transparent Flexible Metal Primer
U. S. Patent 2,248,961

Medium Oil Varnish	17.6
Beeswax	1.6
Paraffin Wax	0.3
Boiled Linseed Oil	9.6
Mineral Spirits (B.P. 300–400°F.)	70.9

Red Lead Paint for Structural Steel

Red Lead (92%)	50.0
Asbestine	23.0
Raw Linseed Oil	20.5
Japan Drier	3.0
Mineral Spirits	3.5

Carbon Black Paint for Steel

Carbon Black	6.0
Red Lead	1.5
Black Iron Oxide	22.5
Raw Linseed Oil	63.0
Japan Drier	3.0
Mineral Spirits	4.0

Heat-Resisting Black Enamel

Carbon Black	2.1
Milori Blue	0.7
Rezyl 823–1 (50% Solids)	57.1
High Flash Naphtha	16.8
Zinc Naphthenate (8%)	0.39
Lead Naphthenate (24%)	0.35
Cobalt Naphthenate (6%)	0.48
Manganese Naphthenate (6%)	0.24
Nuact (Drier)	0.08

Black Bridge Paint

Amorphous Graphite (35% Pure Graphite)	35
Red Lead	9
Raw Linseed Oil	48
Driers and Thinners	8

Metal-Protective Synthetic Coating

Brown Iron Oxide	22.6
Zinc Yellow	3.9
Zinc Oxide	2.6
Magnesium Silicate	5.8
Whiting	17.6
Rezyl 823–1 (50% solids)	42.0
Mineral Spirits	18.5

Anti-Corrosive Paint for Structural Steel

Red Lead Paste (7% Linseed Oil)	100	lb.
Leafed Metallic Lead Paste	9	lb.
Micalith or Mineral Spirits Paste	6	lb.
Raw Linseed Oil	2	gal.
Drier	1	pt.
Turpentine	1.5	pt.

This formula when properly applied makes an excellent anti-corrosion formula for structural steel, which dries rapidly, prevents separation of pigment over rivet heads and away from edges and most important eliminates "crawl," including top coat "crawl."

Black Coating Steel
U. S. Patent 2,289,443

Lithium Hydroxide	1.72
Sodium Nitrate	0.15
Sodium Acetate	0.22
Sodium Nitrite	0.12
Water	60.00

Isopropyl Alcohol   40.00
Triton 720   0.20
Apply to surface; dry and heat to about 430°.

Black Baking Enamel

A baking enamel with high gloss, good adhesion to metal, and unusual mar-resistance and resistance to water and soap can be made as indicated below:

Super Spectra
(Carbon) Black   14
Enamel Liquid
(See below)   225
Solvesso #2   165

Grind the above in pebble mill and add:

Enamel Liquid
(See below)   210
Cobalt Naphthenate
Drier (6%)   1¾
Manganese Naphthenate
Drier (6%)   1¾

Reduce this enamel with about 1 part Xylol to 10 parts enamel to spray. Bake about 2 hr. at 250°F. in a convection oven. In infra-red baking, 15 to 20 min. at an energy density of 2 to 3 watts per square inch on sheet metal will usually suffice.

Enamel Liquid
Super-Beckasite #3005   90
Tung Oil   105

Heat to 450°F. in an open kettle and hold 15 min. Add:

Beckosol #18 (100%
solids)   239
High Flash Naphtha   45
Mineral Spirits   390

Marine Paints

Anti-Corrosive Shipbottom Paint
Zinc Oxide   186

Venetian Red   93
Silica   93
WG or N Rosin   145
Coal-Tar Naphtha   380
Coal Tar   47½
Manganese Linoleate   129

Anti-Fouling Shipbottom Paint
Zinc Oxide   238
Silica   78
Asbestine   72
Cuprous Oxide   145
Mercuric Oxide   45
WG or N Rosin   248
Coal-Tar Naphtha   338
Coal Tar   63
Pine Oil   39

Anti-Fouling Copperbottom Paint
for Wooden Vessels
WG or N Rosin   330
Hydrogenated Methyl
Abietate   165
Coal Tar Naphtha   98
Mineral Spirits   110
Cuprous Oxide   660
Diatomaceous Silica   110

Light Gray Deck Paint
Non-Chalking Titanium
Dioxide   105
Asbestine   315
Dipentine   21
Alkyd Resin Solution
(Spec. 52-R-13)   187
Phenolic Resin—
½ Linseed; ½ Wood
Oil Varnish (25-
gal. Oil Length)   219
Mineral Spirits   184
Lead Naphthenate
(24%)   6.6
Cobalt Naphthenate
(6%)   4.4

Tint with lampblack ground in alkyd resin solution.

Red Deck Paint

Indian Red	238	lb.
Red Lead (97%)	92	lb.
40-gal. Pentalyn Ester Linseed Oil Varnish	76	gal.
Mineral Spirits	10½	gal.
Lead Naphthenate (8%)	2	gal.
Cobalt Naphthenate (2%)	½	gal.

Black Hull Paint

Carbon Black	36¾	lb.
Litharge	7	lb.
Asbestine	138	lb.
40-gal. Pentalyn Ester Linseed Oil Varnish	19	gal.
Linseed Oil	51¼	gal.
Mineral Spirits	22	gal.
Lead Naphathenate (8%)	2	gal.
Manganese Naphthenate (2%)	7	pt.
Cobalt Naphthenate (2%)	7	pt.

Light-Gray Hull Paint

XX Zinc Oxide	389	lb.
Barium Base Titanox Pigment	198	lb.
Asbestine	72	lb.
Linseed Oil	18	gal.
40-gal. Pentalyn Ester Linseed Oil Varnish	44	gal.
Mineral Spirits	16	gal.
Lead Naphthenate (8%)	2	gal.
Manganese Naphthenate (2%)	5	pt.
Cobalt Naphthenate (2%)	3	pt.

Marine Outside White

Titanium Pigment	628
Zinc Oxide	280
Raw Linseed Oil	515
Mineral Spirits	29
Ultramarine Blue in Oil	½
Paint Drier 86D	76

Navy Chromate Primer

Zinc Chromate	280	lb.
Titanium Dioxide	76	lb.
Raw Sienna	26	lb.
Asbestine	127	lb.
Mineral Spirits	32½	gal.
Phenolic Resin— ½ Linseed, ½ Wood Oil Varnish (25-gal. Oil Length)	25¾	gal.
Alkyd Resin Solution (Spec. 52-R-13)	21	gal.
Lead Naphthenate (24%)	2½	gal.
Cobalt Naphthenate (6%)	1¼	fl. oz.
Manganese Naphthenate (6%)	1¼	fl. oz.

Fire-Resisting Canvas Preservative Paint

Antimony Oxide	187 lb.
Titanium Dioxide	47 lb.
Zinc Borate	47 lb.
Aluminum Stearate	1 lb. 12 oz.
Chlorinated Paraffin (42%)	16 gal. 2 pt.
Alkyd Resin (50% Solids)	15 gal. 7½ pt.
Chlorinated Paraffin (60%)	5 gal. 6 pt.
Varnolene	50 gal.

Lead Naphthenate	
(8%)	6½ pt.
Manganese Naphthenate (2%)	3¼ pt.
Cobalt Naphthenate (2%)	3¼ pt.

Marine Interior Flat Paint

Rutile Titanium Dioxide	382	lb.
Titanium Calcium Pigment	109	lb.
Zinc Oxide	169	lb.
Magnesium Silicate	85	lb.
Antimony Oxide	100	lb.
Alkyd Resin Solution (52-R-13)	28.37	gal.
Petroleum Spirits	23.25	gal.
Heavy Petroleum Spirits	22.85	gal.
Lead Naphthenate Drier (24%)	2.5	pt.
Cobalt Naphthenate Drier (6%)	1.0	pt.

Can be tinted with color, ground in oil, or varnish.

Marine Interior Enamel

Titanium Dioxide	220	lb.
Zinc Oxide	38	lb.
Titanium Calcium Pigment	106	lb.
Soya Lecithin	8	lb.
Alkyd Resin Solution (52-R-13)	68.75	gal.
Dipentine	2.94	gal.
Petroleum Spirits	17.23	gal.
Lead Naphthenate Drier (8%)	0.83	gal.
Cobalt Naphthenate Drier		

(2%)	0.29	gal.

Can be tinted with colors ground in oil or varnish.

Camouflage Coatings

These coatings are for use on tanks, trucks, guns and other standard field equipment.

Quick-Drying Enamel (Camouflage)

Chromium Oxide	16.6
Red Oxide	4.7
Antimony Sulfide	4.4
Magnesium Silicate	14.9
Diatomaceous Silica	9.9
Rezyl 823–1 (50% Solids)	42.5
VM&P Naphtha	17.0
Lead Naphthenate (24%)	3.0

Lusterless Sanding Enamel

Titanium Dioxide	26.0
Yellow Iron Oxide	11.3
Basic Lead Chromate	4.0
Magnesium Silicate	15.2
Lampblack	0.2
Rezyl 823–1 (50% Solids)	37.7
Mineral Spirits	21.9
Lead Naphthenate (24%)	0.2

Lusterless Olive Drab Enamel

Hydrated Yellow Iron Oxide	379.0
Lampblack	42.0
CP Medium Chrome Yellow	149.0
Magnesium Silicate	697.0
Diatomaceous Silica	548.0
Barium Sulfate	295.0

*Alkyd Resin Solution
(40% Solids)   2104.0
Mineral Spirits   549.0
Cobalt Naphthenate   4.3
Manganese Naph-
thenate   1.4
Lead Naphthenate   6.8
Anti-Skinning Agent   1.7

Congo Resin Ammunition Paint,
Lusterless
Vehicle:
"Slack Melt" Congo
Resin   100
Kettle Bodied Linseed
Oil ("Z" Visc.)   234
Mineral Spirits   545
Driers, equivalent to 0.3%
Pb., 0.1% Co., and 0.02%
Mn., on weight of oil.

Heat the resin and 90 lb. of oil
to 580°F. Add the remainder of
the oil. Heat to 580°F. Hold for
desired viscosity. Allow to cool sufficiently to add the mineral spirits
and then the driers.
Paint:
Composite Olive Drab
Pigment #M-8002
(or equivalent)   152
Above Vehicle   100

Congo Resin Vehicle for Gasoline-
Soluble and Gasoline-Removable
Paint
"Fine Melt" Congo Resin   70
Mineral Oil   30
Gasoline or Petroleum
Naphtha   100

Dissolve the resin and oil in the
solvent.

In order to reduce the effect of
oxidation on removability of the
paint film, the ratio of pigment to
non-volatile of vehicle should be
kept as high as possible, and antioxidants should be used in small
amounts.

Waterproof Coatings
Thermosetting Waterproof Coating for Raincoats, Hospital
Sheeting and Tents

	Primer Coat	Alternate	Top Coats
	XE–5131	XE–5132	XE–5133
Vinylite Resin XYNC (Dry Basis)	15.0	5.0	8.8
Bakelite Resin Solution XJ–16320 (Wet Basis)	....	42.0	21.0
Spirit Soluble Black	1.0	....	....
Carbon Black	1.0	....	....
Iron Oxide Black	....	3.2	3.2
Iron Oxide Yellow	....	2.3	2.3
Lead Titanate	....	1.9	1.9
Whiting	....	7.6	7.6
"Acrawax" C	....	*	*
Raw Castor Oil, Cold-Pressed	15.0	10.0	10.0
"Flexol" Plasticizer 3GO	15.0	2.5	2.5
Oleic Acid	....	0.6	0.6

---

* Linseed oil-modified glycerol phthalate resin, containing 32% phthalic anhydride by weight and free from rosin and phenol.

| | Primer Coat | Alternate | Top Coats |
	XE–5131	XE–5132	XE–5133
Butanol	....	12.5	12.5
"Synasol" Solvent	38.0	6.8	20.6
Hydrogenated Petroleum Naphtha			
("Solvesso" Solvent No. 1)	15.0	5.6	9.0

* Approximately 2/100 of 1 per cent (2 to 3 oz. per 1000 lb. solution) of Acrawax C mixed pigments before grinding.

The primer coat is a highly plasticized coating designed to control the penetration of subsequent coatings into the cloth. Approximately one-half ounce per square yard on a dry basis should be used. The high percentage of plasticizer included in the formula prevents loss of "hand" due to fiber lockage while the use of a dye in addition to the carbon black improves the hiding power of the coating. The primer requires only sufficient baking to remove solvents before the top coats are applied. When using light duty mixing equipment, it is suggested that eight parts of "Solvesso" solvent No. 1 be added to the XE–5131 formula. Firm anchorage is obtained even when the coating is applied from a very viscous solution.

The formulations given contain sufficient pigment and filler to yield adequate hiding and a deep olive drab color with a coating of two to three ounces per square yard. Using other types of pigments will, of course, necessitate a change in pigment-resin ratio and this in turn will be reflected in the performance of the coating.

The pigments may be dispersed in the plasticizer on a stone mill or in a pebble mill with equivalent results. With a pebble mill, a small amount of petroleum naphtha may be added to adjust the viscosity. Since "Acrawax" C is not readily soluble in the solvents used, it is advisable to disperse it simultaneously with the pigment in the plasticizer.

Raw cold-pressed castor oil is recommended as the largest portion of the plasticizing ingredients for these cloth coatings not only because of its availability but also because of the good exposure tests obtained with it. It is important that a cold-pressed grade of castor oil be used since other types may separate from the coating. The low temperature flexibility of the coatings can be improved by the substitution of "Flexol" Plasticizer 3GO for all or part of the castor oil and oleic acid in both primer and finish coats. This is accompanied, however, by some increased tackiness at elevated temperatures and may cause a slight decrease in water resistance. The inclusion of "Acrawax" C, as indicated above, decreases tackiness at elevated temperatures. The proportion of plasticizer should be varied slightly to fit the variation in oil absorption of the actual pigments used.

The lower aliphatic alcohols are the best solvents for the Vinylite resin XYNC—Bakelite resin XJ–16320 mixture, but the viscosity may be lowered by the judicious use of non-solvents such as acetone or

"Solvesso" No. 1. A "Synasol" solvent—"Solvesso" No. 1 mixture yields a solution of low viscosity. For many coaters, however, the evaporation rate of this mixture may be too rapid. In this case, the "Synasol" solvent may be replaced with butanol or high boiling alcohols as in formulas XE–5132 and XE–5133. When applied in heavy coats under conditions of high humidity, it is advisable to thin with further quantities of hydrocarbons (hydrogenated naphthas or aromatics, but not aliphatic hydrocarbons) since otherwise the condensation of moisture in the coating may cause the vinyl butyral resin and castor oil to separate.

A bake after each coat of two minutes at 180°F. is sufficient to remove the solvents and partially cure the coating. If the equipment permits, however, the use of a higher drying temperature is preferred, since it permits a shorter bake, prevents any tendency of the plasticizer to "sweat" before the final cure, and also prevents adhesives from loosening the bond of the coating to the cloth.

A final "vulcanizing" bake of ½ hour at 275°F. for XE–5132 or 1 hour at 275°F. for XE–5133 is required to complete the insolubilization of the coating. Higher temperatures and longer baking time will render the coating still more resistant to solvents and extreme exposure conditions without decreasing the flexibility or other properties. In most instances, this final bake may be advantageously delayed until the fabric is cemented into the final product, using the same bake to insolubilize the cement.

After application of the primer and one or two top coats, the coating should be faced or calendered to obtain a smooth surface. Press or roll temperatures should remain below 180°F. in order to avoid "strike-through" during the facing operation.

A cement suitable for bonding the uncured Vinylite and Bakelite resin-coated cloth may be prepared from a formulation similar to the cloth coating. A typical formulation which has yielded satisfactory results is:

Vinylite Cement
XL–5246

Vinylite Resin XYHL (Dry Basis)	9
Bakelite Resin XJ–16530 (Wet Basis)	50
Dibutyl Phthalate	18
"Staybelite" (Hydrogenated Rosin)	18
Acetone	5

This formulation yields a solution of a pasty consistency suitable for "finger" spreading. Thinning with suitable solvents such as alcohols or esters is required where brushing is to be used. Often, better cementing is obtained if the coated cloth is brushed with butanol or ethanol before applying the adhesive. Butanol-ethanol mixtures may also be used where solvent cementing of the uncured coating is preferred. The cemented seams can be cured along with the cloth coating during the final baking operations.

Flexible Coating for Cloth

Methyl Methacrylate	2.50 lb.
Potassium Persulfate	0.05 oz.
Soap	5.60 oz.
Water	10.00 lb.

The materials are placed in a closed container and agitated at 50°C. for 1 day or may be stirred in an open vessel fitted with a reflux condenser. The product is an emulsion of a flexible polymer.

Cloth is coated by passing it through this emulsion. The dried coating is resistant to water and many organic solvents.

Waterproof Awning Paint

Crude Beeswax	1	lb.
Rosin	1	lb.
Non-Drying Vegetable Oil	0.5	gal.
Paint, Outdoor House	1.25	gal.
Volatile Mineral Spirits (Painters' Naphtha)	3.75	gal.

Beeswax and rosin are melted in heated vegetable oil. Mixture is added to paint with stirring; when cooled sufficiently, it is diluted with naphtha. Allow to dry a week between coats. Do not fold for storage until thoroughly dry.

Moistureproof Coating

Araclor 5460	96
Paraffin Wax	4

Flame-Resisting Paints
Formula No. 1
British Patent 556,395

Phenol Formaldehyde Varnish	20
Triethanolamine	4
Sodium Silicate Solution	40
Potassium Silicate Solution	40
Asbestos	20
Kaolin	10

No. 2

Lithopone	932.04
Diatomaceous Silica	99.57
Water	326.61
2,3,4,6—Tetrachloro-phenol, Sodium Salt	3.98
Ammonium Hydroxide (28%)	15.93
Casein	7.96
"Slack Melt" Congo	47.81
Alkali Refined Linseed Oil	107.54
Oleic Acid	31.86
Pine Oil	19.91

Water Emulsion Paints for Exterior Use

Properly designed water-thinned paints for exterior use on brick and masonry give remarkable service. They can be applied over a damp surface. For exterior use they give service roughly equivalent to that of a good exterior oil paint. For interior use they are especially suitable for sealing plaster. They are equally suitable for the final coats on plaster. They dry rapidly, have very little odor, and are not inflammable.

These paints are made up as heavy pastes, to be reduced to the proper consistency with water. One pass through a loosely set roller mill is sufficient after a thorough mixing in a pony or paste mixer.

Formula No. 1
(White)

Titanox B–30 WD	170
Asbestine	110

Beckasol Emulsion

#1500	100
Water	50
Cobalt Naphthenate Drier (6%)	½

No. 2
(Red)

Red Iron Oxide	125
Asbestine	75
Beckasol Emulsion #1500	150
Water	15
Cobalt Naphthenate Drier (6%)	¾

No. 3
(Green)

Chromium Oxide	125
Asbestine	117
Beckasol Emulsion #1500	200
Water	20
Cobalt Naphthenate Drier (6%)	1

No. 4

a	20% Casein Solution	14.14
	Ammonium Linoleate S	5.00
	Water	116.00
b	Varnish Base	50.67
	Ammonium Hydroxide	1.50
c	Hydrated Aluminum Silicate	218.00

Drier added to give 0.4% lead and 0.05% cobalt on non-volatile oil.

Mixing Procedure:

1. Place casein solution in mixer.

2. Add ammonium linoleate S to casein solution (first dissolve the ammonium linoleate in small amount of hot water).

3. Add drier solution and ammonium hydroxide to varnish; incorporate varnish slowly into casein solution.

4. Add water gradually to emulsion.

5. Add pigment slowly after vehicle is well mixed.

This emulsion has been aged for 2 weeks at 150°F. in covered cans and remained stable without break.

For general use, reduce volume 1 to 1 with water.

a Casein solution formulated with Protovac 8979.

b Varnish base contains 2% linseed fatty acids.

c Further pigmentation required for coverage and tinting, such as diatomaceous earths, titanium dioxide, lithopone, etc.

Emulsion Stone Paint

A. Emulsion:

Gelatin	9.0
Gum Arabic	4.5
Cresylic Acid	1.5
Potassium Oleate	0.5
Water	49.2
Copal Ester	8.0
Raw Linseed Oil	10.0
Standoil (Sp. Gr. 0.96)	17.0
Cobalt Naphthenate (6% Co.)	0.3

B. Paint:

Emulsion (Above)	13.0
Raw Linseed Oil	3.9
Standoil	3.9
Dehydrated Castor Oil	3.9
Cobalt Naphthenate	0.3
Water	32.0
Whiting	25.0
China Clay	4.0
Titanium Dioxide	4.0
Lithopone	4.0
Barytes	6.0

After soaking in water a few

hours, or over night, the gum arabic and gelatin are dissolved by gentle heating in a water-jacketed kettle. The preservative is added. The oil phase is made in the usual way by dissolving the copalester in the linseed oil. The varnish, heated to about 200°F. is run into the water solution under constant stirring. After all of the varnish has been added the batch is allowed a short rest at intervals. This intermittent stirring is preferable to continuous stirring and produces a better emulsion. The finished emulsion should be aged at least 24 hours. To produce the paint, the emulsion is thoroughly mixed with the oils and driers, the pigments are added gradually, beginning with the lighter ones, and the addition of the water is delayed until last. With modern pigments no grinding should be necessary.

Waterproofing for Cement and Stucco
U. S. Patent 2,167,300

Soap	5
Varnish	2½
Alcohol	15
Naphtha	7½
Water	70

Sealing and Waterproofing Porous Masonry
U. S. Patent 2,290,707

Prior to application of paint, apply the following:

Tung Oil	250
Paraffin Wax	32
Carbon Tetrachloride	214
Petroleum Naphtha	416

Water Paint
Many times, one is confronted with the problem of compounding a water paint to be used for certain purposes. An easily mixed one can be made by the use of the following mixture of materials:

Gelatin or Liquid Glue	10
Water Glass	10
Glycerin	2
Iron Oxide (Pulverized)	5
Lamp Black (Pulverized)	20

Water, to make a pint or a pint and a half, depending on the consistency desired.

The above forms a black paint, but one can use a water soluble red pigment if desired, in the same proportions, for a red paint.

Cement-Water Paint
Formula No. 1

Portland Cement	65
Hydrated Lime	25
Titanium Dioxide or Zinc Sulfide	3–5
Calcium or Aluminum Stearate	0.5–1.0
Calcium Chloride	3–5

No. 2

Portland Cement	80
Hydrated Lime	10
Titanium Dioxide or Zinc Sulfide	3–5
Calcium or Aluminum Stearate	0.5–1.0
Calcium Chloride	3–5

Congo Resin Camouflage Paints
(Emulsion Type)
A. Varnish Base:

"Slack Melt" Congo	100
Alkali Refined Linseed Oil (Unbodied)	156
Linseed Oil Fatty Acids	5.12

Heat resin and 90 parts of oil to 580°F. Add the remainder of the oil. Re-heat to the point where "pill" sample is clear on glass when cold. Allow to cool to 300°F. Add the fatty acids; filter.

B. Casein Solution:

Casein	28
Water	417
Ammonium Hydroxide (28%)	16.7
Ammonium Alginate	2.56
Dowicide F	1.28

Add the casein to 160 parts of water at room temperature, using continuous stirring. Allow to stand for about 30 minutes. Add 1.3 parts of ammonium hydroxide and all of the Dowicide F. Heat in a steam jacketed kettle to 55°C. and hold at that temperature for about 30 minutes, using continuous agitation. Allow to cool to room temperature. In the meantime, disperse the ammonium alginate in the remainder of the water, to which then add the remainder of the ammonium hydroxide. When the casein solution has cooled to room temperature, add to it the solution of alginate and ammonium hydroxide.

C. Pigmentation (Olive Drab):

Ramapo Blue #173 (or Equivalent)	60
Irox Yellow	190
Red Oxide #1087 (or Equivalent)	253
Asbestine XXX (or Equivalent)	792

D. Paint Paste:

(1) Varnish Base (A)	261.12
(2) Casein Solution (B)	465.54
(3) Pigment Mixture (C)	1295
(4) Water	301

Add the casein solution to the varnish base in small portions, making certain that each portion has been thoroughly incorporated before adding the next. Use agitation continuously. Add the water and mix thoroughly. Add the pigment mixture slowly and in small portions, making certain each portion has been thoroughly incorporated before adding the next. Grind loosely on a roller or Buhr stone mill.

To prepare the finished paint, add ½ gallon of water to one gallon of paste.

Oleo-Resinous Emulsifiable Paint

(May be reduced with water, gasoline or naphtha.)

A. Varnish:

"Slack Melt" Congo Resin	100
Alkali Refined Linseed Oil	195
Mineral Spirits	598

Heat the resin and 90 parts of oil to 580°F. Add the remainder of the oil. Heat to 580°F. Hold for desired viscosity. Allow to cool sufficiently and add the mineral spirits.

B. Aerosol (Wetting Agent) Solution:

Aerosol O T	70
Denatured Alcohol	30

Dissolve the Aerosol in the alcohol.

C. Pigmentation:

Titanium Dioxide R–510 (or Equivalent)	56
Asbestine	128
China Clay	128

D. Paint Paste

(1) Varnish (A)	183
(2) Soya Lecithin	3

(3) Aerosol Solution
(B)   14
(4) Pigment Mixture
(C)   312

Disperse the soya lecithin in the Aerosol solution. Add this mixture to the varnish, using agitation. To about 100 parts of this mixture, add the pigment. Mix thoroughly and run loosely through a roller or Buhr stone mill. Add this paste to the remainder of the vehicle and mix thoroughly.

The paint is prepared by reducing the finished paste with equal parts of water or gasoline or other petroleum solvents by volume.

Congo Resin Fire-Retardant
Paint
(Emulsion Type)

A. Varnish Base
"Slack Melt" Congo
Resin   100
Alkali Refined Linseed
Oil   234
Oleic Acid   73
Pine Oil   50

Heat the resin and 90 parts of oil to 580°F. Add the remainder of the oil. Heat to 580°F. (until "pill" sample is again clear on glass when cold). Allow to cool to 300°F. Add the oleic acid and then the pine oil.

B. Casein Solution:
Casein   5
Ammonium Hydroxide
(28%)   11
Dowicide F.   1
Water   200

Add the water slowly to the casein, using agitation. Add the Dowicide and allow it to dissolve. Then add the ammonium hydroxide.

C. Pigmentation:
Albalith #332 (or
Equivalent)   564
Dawson Clay   60

D. Paint Paste:
(1) Varnish Base (A)   128
(2) Casein Solution
(B)   217
(3) Pigment Mixture
(C)   624

Add the varnish base to the casein solution slowly and in small portions, making sure that each portion is thoroughly incorporated before adding the next. Good agitation is required. Add the pigment mixture and mix well. Grind loosely on a roller mill or Buhr stone mill.

To prepare the finished paint, add ½ gallon of water to one gallon of paste.

Exterior Varnishes
Spar Varnish
Formula No. 1

For the utmost resistance to water, chemicals, sunlight and weathering, this spar varnish has no equal. More expensive than most, it still is more economical in the long run than lower quality less costly spars.

Heat-Reactive Pure
Phenolic Resin
(Bakelite 3360)   100 lb.
Raw Tung Oil   32 gal.
Dehydrated Castor
Oil (Synthenol)
Viscosity Z–2   8 gal.
Turpentine   5 gal.
Mineral Spirits   57 gal
Cobalt Naphthenate
Drier (6%)   1¼ lb.
Lead Naphthenate
Drier (24%)   6 lb.

Heat tung oil to 330°F. and add half the resin broken into walnut-size lumps. Reheat to 330°F. and add the remainder of the resin. Now heat slowly to 480°F., stirring until the resin is in complete solution and the reaction foam subsides. Hold 20 minutes at 480°F. and cut off fire. Check with the castor oil. Stir well. Reduce immediately and add cobalt and lead driers.

Viscosity — G (Gardner-Holdt System).

Non-Volatile Content—50%.

Yield—110 gal.

No. 2

This varnish is excellent for exterior exposure. It is designed for use on boats, lawn furniture, doors and other surfaces that must resist weathering and repel water. Properly pigmented, it makes good weather-and-wear-resisting enamels.

Modified Phenolic Resin (Amberol M–88)	100	lb.
Raw Tung Oil	15	gal.
Alkali Refined Linseed Oil	15	gal.
Dehydrated Castor Oil (Synthenol) Viscosity Z–2	20	gal.
Litharge	4	lb.
Dipentine	5	gal.
Mineral Spirits	56	gal.
Cobalt Naphthenate Drier (6%)	4	lb.
Manganese Naphthenate Drier (6%)	3	lb.

Heat resin and linseed oil in open kettle to 580°F. and hold about one hour for heavy body. Add tung oil, heat to 500°F., and hold about 5 minutes. Add castor oil and litharge and reheat to 530°F. Hold for heavy string (about 30 min.). Cut off fire, cool to 475°F. and thin. Then add cobalt and manganese driers.

Viscosity — E (Gardner-Holdt System).

Non-Volatile Content—55%.

Yield—117 gal.

No. 3

Dyphenite V 13133	100
Dehydrol	385
Mineral Spirits	405
Lead Naphthenate (24%)	10
Cobalt Naphthenate (6%)	4

Heat Dyphenite V 13133 and 300 parts Dehydrol to 570°F. Hold about 1½ hr. for first indication of string. Check with 85 parts Dehydrol. Cool to 400°F. and reduce. Add lead and cobalt driers last.

No. 4

Phenac X–687	10.00
Raw Tung Oil	17.90
Linseed Oil	6.80
Castor Oil	1.07
Mineral Spirits	17.85
Turpentine	2.38
Dipentine	2.38
Butanol	1.19
Lead Naphthenate Drier (24%)	0.20
Manganese Naphthenate Drier (6%)	0.08
Cobalt Naphthenate Drier (6%)	0.04
Zinc Naphthenate Drier (8%)	0.31
Anti-Skinning Agent	0.05

Chemical-Resistant Spar Varnish

Dyphene V 13080	100

Varnish Makers Lin-	
seed Oil	75
China Wood Oil	200
Mineral Spirits	250
Dipentine	50
Lead Naphthenate	
(24%)	5
Cobalt Naphthenate	
(6%)	2

Heat Dyphene V 13080 and linseed oil to 560°F. Hold 10 minutes. Add china wood oil. Heat to 460°F. Hold about 30 minutes for required viscosity. Cool to 400°F. and add mineral spirits and dipentine. Add lead and cobalt driers last.

Clear Automobile Varnish

Falkyd Solution A3	
(35% Phthalic Anhydride, Linseed Oil Modified Alkyd)	100 gal.
Union Solvent #30 (Petroleum Spirits)	33 gal.
Lead Naphthenate Drier (24%)	50 fl. oz.
Cobalt Naphthenate	35 fl. oz.

Traffic Paint Varnish

This liquid for traffic paint has been found by extensive testing to give a very durable traffic paint. It dries quickly to a hard, firm, abrasion-resistant film.

Modified Phenolic Resin (Amberol F-7)	100 lb.
Raw Tung Oil	8 gal.
Four-Hour Heat Bodied Linseed Oil (Visc. X)	9 gal.
Sugar of Lead	3 lb.
Turpentine	7 gal.

Mineral Spirits	38 gal.
Benzol	12 gal.
Cobalt Naphthenate Drier (6%)	1 lb.

Heat resin, tung oil, and half the linseed oil in an open stainless steel kettle to 570°F. Check with the remainder of the linseed oil and hold at 500°F. for about ½ hour. Add sugar of lead. Cut off fire and reduce. Finally, add cobalt drier.

Viscosity—A–C (Gardner-Holdt System).

Non-Volatile Content—40%.

Yield—82 gal.

Synthetic Resin Varnishes
Formula No. 1
Oil Length: 25 gal.

BR–17700 Resin	100 lb.
Linseed Oil, Alkali Refined	194 lb.
Mineral Spirits	270 lb.

Heat the oil and resin to 585°F. in one hour and hold about 110 minutes for body. Cool uniformly to 465°F. in one hour and thin.

No. 2
(Oil Length: 40 gal.)

BR–17700 Resin	100 lb.
Linseed Oil, Alkali Refined	311 lb.
Mineral Spirits	351 lb.

Heat the oil and resin to 585°F. in one hour and hold about 150 minutes for body. Cool uniformly to 465°F. in one hour and thin.

No. 3
(Oil Length: 25 gal.)

BR–17700 Resin	100
Dehydrated Castor Oil	194
Mineral Spirits	270

Heat the oil and resin to 585°F. in one hour and hold about 45 minutes for body. Cool uniformly to 465°F. in one hour and thin.

Aluminum Vehicles

One and one half to two pounds of aluminum paste or powder can be added to one gallon of the following vehicles. Varnish-grade paste òr powder will give a very bright aluminum finish for general use. Where a very smooth, chrome-like finish is desired, use one pound of super-fine, lining-grade aluminum paste to one gallon of the following vehicles.

Ready-Mixed Aluminum Alkyd Vehicle for Brush, Spray or Dip

Falkyd Solution C5D (29% Phthalic Anhydride, Fish Oil Modified Alkyd)	15 gal.
Solvesso #3	15 gal.
Xylol	7½ gal.
Cobalt Naphthenate	8–10 fl. oz.

Clear Varnish for Heat-Resistant Aluminum Paint

Falkyd Solution C3 (35% Phthalic Anhydride, Fish Oil Modified Alkyd)	73
Solvesso #3 (Petroleum Spirits)	27½
Xylol	22½
Cobalt Naphthenate	¼

Wall Coating Vehicle

Rosin	100 lb.
Z_6 Bodied Linseed	22 gal.
a. { Lime	11 lb.
{ Raw Linseed Oil	3 gal.
Mineral Spirits	130 gal.
Kerosene	5 gal.

Measure the Z_6 linseed into the kettle by weight or by means of a yard stick. Then add the rosin and heat to 500°F. Then add (a) slowly because of foaming. The temperature is then advanced to 575°F. or until foaming becomes persistent and held for an almost dry pill. The batch should be watched closely from 525 to 575°F. and after, until the batch has cooled to 500°F., or lower, and should preferably be away from the fire box in the stall. The persistent foam must be beaten down with a fire whip or broken by a tiny intermittent stream of water. Contrary to what many may think, this is not dangerous, and has no bad effect on the varnish, the water being immediately turned to gas or steam before it goes beneath the surface. Care must be exercised, however, not to stand close enough to the kettle to get burned by occasional spatterings of oleo-resin. The batch may be rapidly cooled to 425°F. and thinned with a saving of considerable time.

Congo Resin Treatment and Congo Vehicles

"Slack Melt" Congo Resin
(Using No. 2 Unassorted Congo Resin)

Crack the resin to pieces of about walnut size. Heat it at such a rate that 580°F. is reached in about 45 minutes, 600°F. in about 47 minutes, 610°F. in about 53 minutes, 615°F. in about 56 minutes, and then hold at 610 to 615°F. until the last lumps ("floaters") are just melted or nearly so. (The time required for holding at 610 to 615°F. is about 15 minutes). Allow to cool and strain through wire netting of about ⅛" mesh or continue with the manufacture of the varnish directly by adding oil to the processed resin at about 580°F. An uncovered kettle is used throughout.

Even "slacker melts" can be obtained if the holding period at 610 to 615°F. is reduced. However, the advantage of doing this in a varnish plant is questionable because comparatively large amounts of unmelted resin will require removal by filtration. This will result in reduced yield of usable "slack melt" resin. It is felt, therefore, that although the method described here does not give as "slack" a resin as may be obtained, it is nevertheless the optimum method for varnish plant use when taking all factors into consideration; "slackness of melt," greatest yield, speed of manufacturing, reduction of unmelted resin ("floaters") to a minimum, and ease of filtration.

A laboratory batch of a 12½ gallon oil length using the above processed resin and a kettle bodied linseed oil of "Z" viscosity (Gardner-Holdt), prepared by heating resin and oil to 580°F. in 40 minutes, allowing to cool and reducing to 50% non-volatile content with Sunoco Spirits, has a viscosity of "H" to "I" (Gardner-Holdt) and a color of 14 to 15 (Gardner). Higher viscosity may be obtained by holding at 580°F.

"Fine Melt" Congo Resin

Heat the raw Congo resin at about the same rate as described for "slack melt" resin but heat to 650–670°F., and hold at that interval of temperature until all foaming subsides.

Congo Resin Varnishes and Vehicles for Paints and Enamels

For the manufacture of varnishes which are to be used as such or are to be used as vehicles for coatings which do not contain zinc oxide or basic pigments of equal reactivity, "slack melt" Congo is strongly recommended over "fine melt" Congo. In the case of "slack melt" varnishes, as much slow agitation as possible should be employed.

Some typical varnish and vehicle formulae are as follows:

Congo Linseed Oil
(8 Gal. Length)
Formula No. 1

Rubbing Varnish; very fast drying and baking materials where flexibility is not of prime importance; for furniture and other interior applications.

(1) "Slack Melt"
   Congo Resin   100
(2) Kettle Bodied
   Linseed Oil ("Q"
   Viscosity)   80
(3) Petroleum Naphtha
   To 50% non-volatile content
(4) Driers   As desired
Average is 0.3% Pb., 0.1% Co., and 0.02% Mn. on weight of oil.

Heat (1) and (2) to 540–580°F. Take "pill" samples frequently from 540°F. upwards. As soon as a "pill" sample is clear on glass when cold, stop heating. (Do not heat higher than 580°F.) Allow to cool sufficiently to add the naphtha and then the driers.

No. 2

General utility varnish. Vehicle for enamels for interior use; fast drying and baking materials with fair to good flexibility.

(Congo Linseed Oil,
15–25 Gal. Oil Length)

(1) "Slack Melt"
Congo Resin   100 lb.
(2) Kettle Bodied
Linseed Oil ("Z"
Viscosity)   120–200 lb.
(3) Petroleum Naphtha
To 50% non-volatile content
(4) Driers   As desired

Average is about 0.3% Pb., 0.1% Co., and 0.02% Mn., on weight of oil.

Heat resin and 90 lb. of oil to 580°F. Add the remainder of the oil. Reheat to 580°F. (until cold "pill" is again clear on glass). If necessary or desired, hold for body. (Caution: Beware of gelatin at lower oil length.) Allow to cool sufficiently to add the naphtha and the driers.

No. 3

General Utility Varnish. Vehicle for enamels containing zinc oxide or other basic pigments; vehicle for primers for metals (zinc chromate, zinc tetroxy-chromate, red lead, etc.).

(Congo Linseed Oil,
25 Gal. Oil Length)

(1) "Fine Melt"
Congo Resin   100
(2) Kettle Bodied
Linseed Oil ("Z"
Viscosity)   200ᐧ
(3) Petroleum Naphtha
To 50% non-volatile content
(4) Driers   As desired

Average is 0.3% Pb., 0.1% Co., and 0.02% Mn. on weight of oil.

Heat (1) and (2) to 500°F. Allow to cool sufficiently to add (3) and then (4).

No. 4

Varnish for exterior use and vehicle for paints and enamels of all types for exterior use.

(Congo Linseed Oil,
40–45 Gal. Length)

(1) "Slack Melt"
Congo Resin   100 lb.
(2) Kettle Bodied
Linseed Oil ("Z"
Viscosity)   320–360 lb.
(3) Petroleum Naphtha
To 50% non-volatile content.
(4) Driers   As desired

Average is 0.3% Pb., 0.1% Co., and 0.02% Mn., on weight of oil.

Heat the resin and 90 lb. of oil to 580°F. Add the remainder of the oil. Heat to 580°F. Hold for desired viscosity. Allow to cool sufficiently to add the naphtha, and then the driers.

No. 5

Extremely fast drying varnish for interior purposes.

(Congo Linseed Oil,
4 Gal. Length)

(1) "Slack Melt"
Congo Resin   100 lb.
(2) Alkali Refined
Linseed Oil
(Unbodied)   32 lb.
(3) VM&P Naphtha and high solvency solvents to desired non-volatile content. (About 75% VM&P Naphtha and 25% Turpentine, Solvesso, or Hi-Flash Naphtha.)

Heat (1) and (2) to 520°–580°F. Take frequent "pill" samples from 520°F. upwards. As soon as "pill" is clear on glass

when cold, allow to cool sufficiently
to add (3).

Note: As non-volatile content
is decreased, the material tends to
"cloud." A non-volatile content of
about 30% non-volatile can be
satisfactorily obtained. To reduce
viscosity (if desired) and to increase naphtha reducibility, some
"fine melt" Congo may be incorporated with the "slack melt."

Congo Resin Varnish Sealer
(Congo Linseed Oil, 30 Gal. Oil Length)

(1) "Slack Melt"		
Congo Resin	100	lb.
(2) Kettle Bodied		
Linseed Oil ("Z"		
Viscosity)	240	lb.
(3) Petroleum Naphtha		
To desired non-volatile		
content		

(4) Driers equivalent to 0.3%
Pb., 0.1% Co., and 0.02% Mn. on
the weight of oil.

Heat the resin and 90 lb. oil to
580°F. Add the remainder of the
oil. Heat to 580°F. Hold for desired viscosity. Allow to cool sufficiently to add the naphtha, and
then the driers.

Interior Flat Wall Finishes

On new wood, plaster, or composition board, it is the usual practice to apply one coat of sealer, one
coat of undercoat and one coat of
flat wall finish.

Sealer

Titanox RCHT	200	lb.
Whiting	200	lb.
Asbestine	100	lb.
Aluminum Stearate	5	lb.
25-Gal. Modified		
Phenolic—De-		

hydrated Castor		
Oil Varnish		
(Viscosity E,		
Solids 50%)	70	gal.
Mineral Spirits	15	gal.
Lead Naphthenate		
Drier (16%)	½	gal.
Cobalt Naphthenate		
Drier (6%)	⅜	gal.

Undercoater
Formula No. 1

Titanox C	300	lb.
Whiting	125	lb.
Asbestine	75	lb.
Zinc Stearate	10	lb.
45-Gal. Ester Gum		
—Linseed Oil		
Varnish (Vis-		
cosity P–R,		
Solids 40%)	23	gal.
Gloss Oil (65%		
Solids)	4	gal.
Mineral Spirits	6½	gal.
Japan Drier	1½	gal.
Yield—55 Gal.		

No. 2

Rutile Titanium Cal-		
cium Pigment	720	lb.
Calcium Carbonate	174	lb.
25-Gal. Pentaery-		
thritol Esterified		
Rosin Varnish (⅔		
Dehydrated Castor—		
⅓ Linseed)	49	gal.
Petroleum Spirits	16	gal.
Lead Naphthenate		
Drier (8%)	1	gal.
Cobalt Naphthenate		
Drier (2%)	1	gal.

Flat Wall Paint
Formula No. 1

Titanox C	800	lb.
Asbestine	200	lb.
Aluminum Stearate	3	lb.

Flat Wall Liquid   60 gal.
Mineral Spirits   38 gal.
Ivory Soap Solution
(2%)   3 gal.
Yield—137 gal.
The flat wall liquid is a 50% solids varnish. The varnish is a 30-gallon length ester gum varnish. The paint is ground on a stone mill. The soap solution is added to the mix after grinding.

No. 2

Titanox C   800   lb.
Natural Whiting   200   lb.
Litharge   10   lb.
Aluminum Stearate 5   lb.
Blown Soya Bean
Oil (Z–6)   8   gal.
Kettle Bodied Lin-
seed Oil (Z–2)   8   gal.
Refined Linseed Oil 7½ gal.
Ester Gum Solution
(60% Solids)   9½   gal.
Kerosene   10   gal.
Mineral Spirits   43¼   gal.
Cobalt Naphthenate
(6%)   ¼   gal.
Yield—125 gal.

No. 3

Rutile Titanium Cal-
cium Pigment   500   lb.
Whiting   140   lb.
Asbestine   100   lb.
Calcium Lineolate
Pulp   40   lb.
30-Gal Ester Gum
Linseed Oil
Varnish   22½ gal.
Z–1 Heat Bodied

Linseed Oil   9½ gal.
Refined Linseed Oil 3   gal.
Heavy Mineral
Spirits   12   gal.
Mineral Spirits   17   gal.
Cooked Drier   3   gal.
The flat wall paint formulae, given above, can be tinted with desired colored pigments ground in oil or varnish to produce all shades.

Interior Semi-Gloss Finish

Surfaces to be finished with semi-gloss should be prepared with the same type of primer and undercoat as given under flat wall paints. The following semi-gloss formula can be tinted with colored pigments ground in oil or varnish to produce all desired shades:

Semi-Gloss White Enamel and Base for Tints

Rutile Titanium Cal-
cium Pigment   620 lb.
Calcium Carbonate 95 lb.
50-Gal. Ester Gum-
Linseed Varnish   44 gal.
50% Ester Gum Cut   6 gal.
Petroleum Spirits   25 gal.
Lead Naphthenate
Drier (2%)   2 pt.
Manganese Naph-
thenate Drier (2%) 1 pt.
Cobalt Naphthenate
Drier (2%)   2 pt.
For tint shade use C. P. tinting colors in varnish.

Interior Gloss Paints

Surfaces to be finished with interior gloss should be prepared by applying one coat of primer and one coat of undercoat as shown under flat wall paints. The following interior gloss paint formulae can be tinted with colored pigments ground in oil or varnish to produce any desired shade.

Interior Gloss Paint
Formula No. 1

Ti-Cal R–20	392.000 lb.	14.48 gal.
Suspenso Whiting	110.000 lb.	4.93 gal.
Aluminum Stearate	1.200 lb.	0.12 gal.
Ultramarine Blue	0.078 lb.	------
KV–91 (given below)	540.800 lb.	72.75 gal.
Mineral Thinner	54.200 lb.	8.35 gal.
Cobalt (6%)	1.900 lb.	0.24 gal.
Lead (24%)	3.200 lb.	0.33 gal.

Soya Lecithin (2.5#/100 gal.) gives better gloss than aluminum stearate, when used as a suspension agent.

No. 2

Ti-Cal R–20	420.000 lb.	15.50 gal.
Suspenso Whiting	88.000 lb.	3.91 gal.
Aluminum Stearate	1.200 lb.	.12 gal.
Ultramarine Blue	.078 lb.	------
KV–91 (given below)	540.800 lb.	72.75 gal.
Mineral Thinner	54.200 lb.	8.35 gal.
Cobalt (6%)	1.900 lb.	.24 gal.
Lead (24%)	3.200 lb.	.33 gal.

No. 3

Titanium Calcium Pigment	250 lb.
Lithopone	400 lb.
Treated Calcium Carbonate	50 lb.
30-Gal. Ester Gum Dehydrated Castor-Linseed Varnish (75% Dehydrated Castor-25% Linseed)	52 gal.
Pale Refined Oil	12 gal.
Petroleum Spirits	6 gal.
Z–2 Bodied Linseed Oil	5 gal.
Lead Naphthenate Drier (8%)	1 gal.
Manganese Naphthenate Drier (2%)	6 pt.
Cobalt Naphthenate Drier (2%)	6 pt.

No. 4

Rutile Titanium Dioxide	185
Atomite (Calcium Carbonate)	398
Zinc Oxide	65
Litharge	5
Spar Varnish	96
Kettle Bodied Linseed Oil (Z–1)	210
Thinner and Drier	242

This enamel has pale initial color and good color retention, especially if the spar varnish is chosen for freedom from after-yellowing. The enamel dries fast and has a very high gloss. When thoroughly dry it will withstand repeated washings with soap and water.

Varnish KV–91

Ester Gum 3664-A	100
Dehydrated Castor Oil (Z–3 Visc.)	320

Mineral Thinner   345

Run dehydrated castor oil at 585°F. in 35–45 minutes; at 385°F. hold 30 minutes for 8–10 inch string from cold glass. Take off fire.

Add ester gum (watch foaming). Cool to 450°F.

Add mineral thinner.

Interior Quick-Dry Enamels

Interior quick dry enamels are commonly used for interior trim, woodwork, furniture, toys, etc. Surfaces should be prepared with one coat of primer and one coat of undercoat before applying the finish coat of enamel. Formulae for primer and undercoat are given under flat wall finishes.

Non-Yellowing White Quick
Dry Enamel
Formula No. 1

Titanium Dioxide (LO)	29.3
Zinc Oxide	3.2
Rezyl 880–1 (60% Solids)	54.4
Mineral Spirits	9.7
Solvesso #3	3.3
Cobalt Naphthenate Drier (6%)	0.1

Grind pigments in Rezyl on roller mill; thin, and add drier. Reduce slightly with mineral spirits to brush or spray.

No. 2

Titanium Dioxide, Chalk Resistant	575 lb.
Zinc Oxide #15, Black Label	32 lb.
Falkyd Solution B41 (32% Phthalic Anhydride, Soya Bean Fatty Acid Modified Alkyd)	58 gal.

Grind together on 3 roller mill and thin with:

Falkyd Solution B41	45	gal.
Falkyd Solution A3	59	gal.
Mineral Spirits	45	gal.
Hi-Flash Naphtha	5	gal.
Lead Naphthenate (24%)	1¼	gal.
Cobalt Naphthenate (6%)	½	gal.

No. 3

Titanium Dioxide	270	lb.
Falkyd Solution B34 (35% Phthalic Anhydride, Soya Bean Fatty Acid Modified)	12½	gal.
Hi-Flash Naphtha	6¼	gal.

Grind on 3 roller mill and thin with:

Falkyd Solution B34	57½	gal.
Cobalt Naphthenate (6%)	½	gal.

Quick-Dry Enamel Base for
Tints

Rutile Titanium Calcium Pigment	205	lb.
Rutile Titanium Dioxide	120	lb.
20-Gal. Pentaerythritol Esterified Rosin-Dehydrated Castor Varnish	84	gal.
Lead Naphthenate Drier (8%)	2¼	gal.
Manganese Naphthenate Drier (2%)	1½	gal.
Cobalt Naphthenate Drier (2%)	½	gal.

For Tint Shades use C. P. colors in varnish.

Quick-Dry Red Enamel
(Air drying; for brushing or
spraying)
Formula No. 1

Toluidine Red	100 lb.
Falkyd Solution A3	35 gal.

Grind on 3 roller mill and thin
with:

Falkyd Solution A3	65	gal.
Mineral Spirits	2½	gal.
Cobalt Naph-thenate (6%)	36	fl. oz.
Lead Naphthenate (24%)	85	fl. oz.

No. 2

Toluidine Red	82	lb.
20-Gal. Pentaery-thritol Esterified Rosin-Dehydrated Castor Varnish	82	gal.
Petroleum Spirits	6	gal.
Lead Naphthenate Drier (8%)	2	gal.
Manganese Naph-thenate Drier (2%)	1½	gal.
Cobalt Naphthenate Drier (2%)	1½	gal.

Quick-Dry Green Enamel
(Air drying; for spraying or
brushing)
Formula No. 1

Light Chrome Green	50 lb.
Medium Chrome Green	75 lb.
Falkyd Solution A3 (35% Phthalic Anhydride, Lin-seed Oil Modi-fied Alkyd)	25 gal.

Grind together on 3 roller mill
and thin with:

Falkyd Solution A3	80 gal.

Solvesso #3	10½ gal.
Cobalt Naph-thenate (6%)	25 fl. oz.
Lead Naphthenate (24%)	85 fl. oz.

No. 2

Dark Chrome Green	105 lb.
Calcium Carbonate	25 lb.
20-Gal. Pentaery-thritol Esterified Rosin-Dehydrated Castor Varnish	95 gal.
Lead Naphthenate Drier (24%)	4 pt.
Cobalt Naphthenate Drier (2%)	4 pt.
Manganese Naph-thenate Drier (6%)	3 pt.

Four Hour, Brushing, Black
Enamel

(1) Coresin Black Paste	90 lb.
(2) Falkyd Solution A5D	51 gal.
Phthalic Anhy-dride, Linseed seed Oil, Modi-fied Alkyd (29%)	
(3) Mineral Spirits	27 gal.
(4) Cobalt Naph-thenate (6%)	70 fl. oz.
(5) Manganese Naphthenate (6%)	35 fl. oz.
(6) Lead Naph-thenate (24%)	1 gal.

Mix (2)–(6) thoroughly and
add slowly, while mixing well, to
(1).

Blue Quick-Dry Enamel

Rutile Titanium Cal-cium Pigment	150 lb.
Lithopone	100 lb.

Resinated Copper
Phthalocyanine
Blue   10 lb.
Lampblack   1 lb.
20-Gal. Pentaery-
thritol Esterified
Rosin-Dehydrated
Castor Varnish   88 gal.
Manganese Naph-
thenate Drier
(2%)   1½ gal.
Lead Naphthenate
Drier (24%)   4 pt.
Cobalt Naphthenate
Drier (2%)   4 pt.

Floor Enamels

It is advisable to treat new wood floors with a sealer before painting. The following is a type of sealer for this purpose.

Liquid Wood Sealer
Rezyl 823–1 (50%
Solids)   42.65
VM&P Naphtha   24.90
Terpene Type Solvent
(Terpene B)   3.24
Lead Naphthenate
(24%)   0.21
Cobalt Naphthenate
(6%)   0.27

Green Floor Enamel
Medium Chrome
Green   98 lb.
Calcium Carbonate   30 lb.
25-Gal. Modified
Phenolic Linseed-
Dehydrated Castor
Varnish (80%
Castor, 20% Lin-
seed)   90 gal.
Lead Naphthenate
Drier (8%)   1½ gal.
Manganese Naph-

thenate Drier
(2%)   1½ gal.
Cobalt Naphthenate
Drier (2%)   2½ gal.

Brown Floor Enamel
95% Mineral Hydrated
Iron Oxide   100 lb.
Red Iron Oxide   50 lb.
Calcium Carbonate   25 lb.
Aluminum Stearate   8 oz.
25-Gal. Modified
Phenolic Linseed-
Dehydrated Castor
Varnish (80%
Castor, 20% Lin-
seed)   . 89 gal.
Lead Naphthenate
Drier (8%)   1¼ gal.
Manganese Naph-
thenate Drier
(2%)   1½ gal.
Cobalt Naphthenate
Drier (2%)   1½ gal.

Gray Floor Enamel
Lithopone   170 lb.
Rutile Titanium Cal-
cium Pigment   82 lb.
Lead Titanate   62 lb.
25-Gal. Modified
Phenolic Linseed-
Dehydrated Castor
Varnish (80%
Castor, 20% Lin-
seed)   89 gal.
Lead Naphthenate
Drier (8%)   1½ gal.
Manganese Naph-
thenate Drier
(2%)   1½ gal.
Cobalt Naphthenate
Drier (2%)   1½ gal.

Interior Emulsion-Type Finishes

In recent years the resin emulsion-type finishes have become an important factor in interior wall finishes. These paints can be reduced with water to application consistency. They can be tinted with water dispersed colored pigments.

Water-Reducible Oil-Type Paste Paint
Formula No. 1

This is a combination drying oil-casein paste paint of the type which has recently become popular for interior use. One gallon of paste is thinned with about one-half gallon of water for use. This paint covers solidly in one coat over plaster or wallpaper. It produces a flat finish which is as washable after aging 30 days as most straight oil-type paints. This paint dries in one to two hours and is almost odorless.

Rutile Titanium	
Dioxide	50
Lithopone	400
China Clay	75
Muriatic Casein	20
Trimol 80	67
Dowicide G	4½
Pine Oil	2
Water	245

Weight per gal.—15.2 lb.

Use only water-dispersible pigments. Put the casein in a pony mixer and add 125 lb. warm water. Mix well and allow to soak 2 hours. Dissolve Dowicide in remainder of water and add slowly, mixing constantly. Then add Trimol 80 and pine oil and mix one hour. Work in the pigments and mix well. Grind once through a loosely-set three-roll mill.

Colors may be produced by adding small amounts of dry alkali-resisting pigments to the mixture in the pony mixer. Do not use chrome greens, yellows, and oranges.

No. 2

Rutile Titanium	
Dioxide	200 lb.
Clay (Emulsion	
Paint Type)	360 lb.
Mica (Aratone	
#270 or Equal)	100 lb.
Emulsion	73½ gal.

Add pigments to the emulsion. Mix thoroughly and pass through loosely set roller mill.

Emulsion
Water Phase

Sodium Alginate	1.3
Water at 150°F.	296
Dowicide G	5.9
Dowicide A	5.9
Pine Oil	4.9
Anti-Foam H or	
Foamex	4.9
Muriatic Casein	47.4

Oil Phase

Ester Gum (Pale)	27.2
Alk. Ref. Linseed Oil	50.9
Z–3 Heat Bodied Lin-	
seed Oil	55.5
Linseed Fatty Acids	2.8
Wetting Agent (Du-	
ponol ME Dry or	
Equal)	0.56
Cobalt Linoleate (6%)	0.74
Lead Linoleate (13%)	6.1
Ammonium Hydroxide	
(28%)	1.2

Dilution

Cold Water	330 parts

A Oil Phase:

1 Dissolve the ester gum in refined linseed oil at about 10-gal. oil length by heating to 375°F. Cool.

2 Blend the ester gum varnish

with the rest of the oil phase—adding the ammonia last, and make sure that the mixing is thorough.

B Water Phase:

1 Add the alginate to hot water (about 94 parts at 150°F. or higher) in a small container and let it swell.

2 Run the remainder of the hot water (200 parts at 150°F. or higher) into the main emulsion mixing tank. (An agitated, insulated, steam-jacketed kettle is desirable.)

   a. Add the Antifoam H and pine oil.

   b. Add the Dowicides.

   c. Add the casein gradually, with constant stirring.

   d. Add the alginate suspension —after the casein has become wetted—and continue stirring until all the casein granules have been dissolved.

Resin Emulsion Paint

C Emulsion:

1 Add the oil phase, A, gradually to the water phase, B, allowing each addition to become well mixed — and continue stirring after the final addition until the emulsion is thoroughly uniform.

2 Dilute the emulsion gradually with the remaining cold water (330 parts) and continue stirring until the emulsion is uniform.

(Important Note: Operations B and C1 require heat, which is supplied by the use of hot water. If allowed to cool, additional application of heat will be needed.)

No. 3

Ti-Pure R–110	47.5
Ponolith LR–W	406.5
*Clay	260.5
Special Soya Protein Solution	656.0
Cobalt Naphthenate (6%)	0.17
Zinc Naphthenate (8%)	0.41

Special Protein Solution

Special Soya Protein	127
Borax	3
Ammonium Chloride	13
Sodium Fluoride	10
Poly Z Linseed Oil	64
Dowicide G	5
Pine Oil	13
Water	636

Wt./gal. 8.71 lb.

Protein (by wt.) 14.6%.

Non-Volatile (by wt.) 26.98%.

Procedure: Add ammonium chloride, sodium fluoride, pine oil and special soya protein to about 250 parts of water and soak for 30 minutes. Then add borax, Dowicide G, linseed oil and remainder of the water and heat to 165°F. while stirring constantly. Cool to room temperature. Better results are obtained if the emulsion is allowed to stand overnight before pigmenting.

A second paint formulated with the same vehicles, as above, but which has higher brightness and higher wet and dry hiding power, is given below:

No. 4

Ti-Pure R–110	67.5
Ponolith LR–W	386.0
Clay	260.5
Special Protein	

---

* Low binder absorption clay such as Dawson from United Clay Mines Corp., or No. 7 Air-Floated Clay from Georgia Kaolin Co., or Peerless #15 from R. T. Vanderbilt Co.

Solution   656.0

Cobalt Naphthenate
(6%)   0.17

Zinc Naphthenate
(8%)   0.41

Congo Resin Emulsion Paint
Formula No. 1
For Interior Use
(Specification TT–P–88 Type)

A Varnish Base

1 "Fine Melt" Congo
  Resin   100

2 Kettle Bodied Linseed
  Oil ("Z" Visc.)   200

3 Linseed Oil Fatty
  Acids   9

Heat 1 and 2 to 550°F. Allow
to cool to 300°F. Add 3. Filter.

B Casein Solution

Casein   17.4
Water   125.2

Ammonium Hydroxide
(28%)   1.0

Dowicide F   0.4

Superloid or Equivalent
(Ammonium Al-
ginate)   0.5

Add casein to 97.2 parts of
water at room temperature, using
continuous agitation. Allow to
stand for 30 minutes to permit the
casein to soften and swell. Add
ammonium hydroxide and Dowicide F. Heat in a steam-jacketed
kettle to 55°C., and hold for about
30 minutes, using agitation. In
the meantime, dissolve the Superloid in 28 parts of water. When
the casein solution has cooled to
room temperature, add the Superloid solution.

C Ammonium Hydroxide
Solution

Ammonium Hydroxide
(28%)   4

Water   41

Add the ammonium hydroxide
to the water and stir.

D Pigmentation

Albalith #332 (or
Equivalent)   195

Cryptone ZS–830 (or
Equivalent)   75

Celite #110 (or
Equivalent)   30

Emulsion Paint Paste

1 Varnish Base A   71
2 Casein Solution B   144.5
3 Ammonium Hydrox-
  ide Solution C   45
4 Pigmentation Mix-
  ture D   300
5 Water   40

Add 2 to 1 slowly in small portions, always making certain that
each portion has become thoroughly incorporated before adding the next. Add 3 and mix well.
Then add 4 and 5. After thorough agitation, grind loosely
through a roller or Buhr stone
mill.

To prepare the finished paint,
add ½ gallon of water to each gallon of paste.

No. 2

A Varnish Base

"Slack Melt" Congo   100

Kettle Bodied Linseed
Oil ("Q" Viscosity)   320

Heat resin and 90 parts of oil
to 580°F., using as much slow agitation as possible. Examine for
cold clear bead on glass. Add remainder of oil and reheat to
560°F.

B Water Phase of Emulsion

Ammonium Alginate
(Superloid or
Equivalent)   1.33

Water	296.0
Dowicide G or F	5.93
Dowicide A	5.93
Pine Oil	4.90
Antifoam (Antifoam H or Foamex)	4.90
Muriatic Casein	47.40

1 Heat about 96 parts of water to 150°F. or higher. Add the alginate and allow to swell.

2 Heat about 200 parts of water to 150°F. or higher in an emulsion mixing tank. Add the antifoam, pine oil, and preservatives. Add the casein gradually with constant stirring.

3 Add 1 to 2 and continue stirring until a smooth uniform material is obtained.

C Oil Phase

Varnish Base A	133.6
Linseed Oil Fatty Acids	2.78
Wetting Agent (Nacconol NR or Equivalent)	0.56
Lead Drier (13% Pb)	6.10
Cobalt Drier (6% Co)	0.74
Ammonium Hydroxide (28%)	1.17

Add the fatty acids, wetting agent, driers and ammonium hydroxide to the varnish base.

D Emulsion Vehicle

Water Phase B	366.39
Oil Phase C	144.95
Water	330.00

Add the oil phase, C, gradually to the water phase, B, and mix until emulsion is uniform. Add the water gradually and stir until the emulsion is uniform.

E Emulsion Paint Paste

Ponolith HC (or Equivalent)	450
Ti-Pure R–110 (or	
Equivalent)	100
China Clay (Low absorption, such as Dawson or #7 Air-Floated)	128
Emulsion Vehicle D	626

If water dispersing pigments are preferred, Ponolith HC-W may be used instead of Ponolith HC, and Ti-Pure R–300 instead of Ti-Pure R–110.

Add the pigments to the emulsion, using agitation and grind loosely.

To prepare the finished paint, add 1 volume of water to 2 volumes of paste.

No. 3

(Lower Film Brightness than No. 2; High Wet Hiding, Particularly suggested for Tinted Emulsion Paints.)

A Varnish Base (Same as No. 2)
B Water Phase (Same as No. 2)
C Oil Phase (Same as No. 2)
D Emulsion Vehicle (Same as No. 2).
E Emulsion Paint Paste

Ti-Pure R–110 (or Equivalent)	189.0
China Clay (Same as No. 2)	346.5
Mica (Same as No. 2)	100.0
Emulsion Vehicle "D"	626.0

Mix as Formula No. 2.

To prepare the finished paint, add 1 volume of water to 2 volumes of paint paste.

Emulsion Paint Tint
(Cream)

Ethyl Alcohol	15
Butyl Alcohol	10
"Carbitol"	15
Iron Oxide	20

Diglycol Stearate   10
Water   To make 100

Interior Varnishes

High-Grade General-Purpose
     Varnish
Modified Phenolic
  Resin (Beckacite
  1123)   100 lb.
Four-Hour Heat
  Bodied Linseed Oil
  —Viscosity X   10 gal.
Dehydrated Castor
  Oil (Synthenol)
  Viscosity Z–2   10 gal.
Segregated Fish Oil
  (Celesterol) Vis-
  cosity Z   5 gal.
Steam Distilled Wood
  Turpentine   6 gal.
Mineral Spirits   40 gal.
Lead Naphthenate
  Drier (24%)   5½ lb.
Cobalt Naphthenate
  Drier (6%)   3½ lb.
Heat resin, castor oil, and lin-
seed oil in open kettle to 565°F.
Hold about one hour, or until the
material gives a ropy string from
the stirring spatula. Add the segre-
gated fish oil and reheat to 550°F.
Remove from fire. Cool to 475°F.
and thin with turpentine and min-
eral spirits. Finally, add naph-
thenate driers.
Viscosity E-G (Gardner-Holdt
System).
  Non-Volatile Content 50%
  Yield   77 gal.

Floor and Deck Varnish

This is a high grade, spar-type
varnish suitable for all exterior
uses. It can be used on porches,
floors, boats, and on furniture exposed to the weather. This varnish
also makes an excellent vehicle for
floor and deck enamels. It is suit-
able for all colors and tints except
white. When the varnish alone is
to be applied, it is brushed or
sprayed without thinning.
Phenol Modified Pen-
  taerythritol Resin
  (Pentalyn M)   100 lb.
Dehydrated Castor
  Oil (Castung)
  Viscosity G   20 gal.
Four-Hour Heat
  Bodied Linseed Oil
  Viscosity X   15 gal.
Mineral Spirits   48 gal.
Lead Naphthenate
  Drier (24%)   1 lb.
Zinc Naphthenate
  Drier (8%)   7 lb.
Cobalt Naphthenate
  Drier (6%)   3 lb.
Heat resin and castor oil in open
kettle to 585°F. Hold about one
hour. Then add linseed oil and
reheat to 585°F. Hold about one
hour. Remove from fire. Cool to
470°F. and thin with mineral
spirits. Finally, add lead, zinc and
cobalt driers.
Viscosity   E   (Gardner-Holdt
System)
  Non-Volatile Content 55%
  Yield   90 gal.

Low-Cost Floor and Trim Varnish

This varnish is fast-drying,
glossy, very pale in color, low in
cost and suitable for all indoor
uses. It will not discolor and is
fairly tough and resistant to
scratching. However, it is not as
waterproof or as resistant to alcohol, soaps, and cleaners as a better
grade of varnish. It is brushed or
sprayed without reduction.

Gum Rosin (I) 100 lb.
Unslaked Builder's
(Dolomite) Lime 5 lb.
Dehydrated Castor Oil
(Synthenol), Visc.
Z–2 10 gal.
Activated Linseed Oil
(Esskol), Visc. Z 5 gal.
Segregated Fish Oil
(Celesterol), Visc.
Z 10 gal.
Lead Acetate 5 lb.
Mineral Spirits 47 gal.
Cobalt Naphthenate
Drier (4%) 5 lb.

Heat rosin to 500°F. in open kettle and stir in lime. Stir until the mixture thickens. Add the castor and linseed oils and heat to 585°F. Hold 30 minutes at 585°F. Add segregated fish oil and heat to 575°F. Hold 20 minutes. Remove kettle from fire and cool to 555°F. Stir in sugar of lead. Allow batch to cool to 475°F. and add mineral spirits; slowly at first with good stirring, then rapidly. Finally, add cobalt naphthenate.

Viscosity G (Gardner-Holdt System).

Non-Volatile Content 50%
Yield 78 gal.

Penetrating Sealer for Floors
(Mopping Varnish)

This is applied to new floors with mops. It impregnates and strengthens the wood and repels water. Two coats, applied 24 hours apart, make a good finish and will not discolor the wood. Applied to gymnasium floors, the finish will not show rubber-burn.

Modified Phenolic
Resin (Beckacite
1123) 100 lb.

Four-Hour Heat-
Bodied Linseed Oil,
Visc. X 5 gal.
Dehydrated Castor Oil
(Synthenol), Vis-
cosity Z–2 5 gal.
Selectively Refined
Fish Oil, Vis-
cosity Z 5 gal.
Litharge 1 lb.
Mineral Spirits 63 gal.
Cobalt Naphthenate
Drier (6%) 1 lb.
Manganese Naph-
thenate Drier (6%) 1 lb.

Heat resin, litharge, linseed and castor oils in open kettle to 550°F. and hold one hour. Add fish oil and hold at 530°F. for 15 minutes. Cut off fire, cool to 475°F. and thin. Then add cobalt and manganese driers.

Viscosity A (Gardner-Holdt System).

Non-Volatile Content 35%
Yield 85 gal.

Pale-Gloss Varnish for Art Work

This varnish dries rapidly, is extremely pale in color, does not after-yellow, and has good adhesion and high gloss. It is valuable for the protection of paintings, photographs, and other art work from discoloration and staining due to handling, atmospheric fumes, etc. It may be applied by brushing or by spraying without thinning further.

Teglac Z–152 100 lb.
Tung Oil 10 gal.
Dehydrated Castor Oil
(Synthenol), Vis-
cosity Z–2 20 gal.
Dipentine 5 gal.
Mineral Spirits 47 gal.

Lead Naphthenate
Drier (16%)   7 lb.
Cobalt Naphthenate
Drier (4%)   1¾ lb.
Heat resin, tung oil and 15 gal. castor oil in closed kettle under carbon dioxide atmosphere to 550°F. Hold until the material becomes stringy. Cut off fire. Then add remainder of the castor oil and mix well. Cool to 480°F. and add dipentine and mineral spirits. Finally, stir in driers
Viscosity E-G (Gardner-Holdt System).
Non-Volatile Content   50%
Yield   90 gal.

Furniture-Rubbing Varnish
Thermally Processed
Congo Resin   100 lb.
Four-Hour Heat
Bodied Linseed Oil,
Viscosity X   4 gal.
Dehydrated Castor
Oil (Synthenol),
Viscosity Z–2   4 gal.
Mineral Spirits   31 gal.
Lead Naphthenate
Drier (24%)   ½ lb.
Cobalt Naphthenate
Drier (6%)   1 lb.
Calcium Naphthenate
Drier (4%)   1½ lb.
Prepare the thermally processed Congo from raw Congo as follows: Heat the resin in an open kettle to 650°F. for 1½ hr. Hold until foam subsides and hot resin runs off paddle like hot oil. Pour out in pans to cool. Break up when cold.

Cook the varnish as follows: Heat resin and oils in open kettle to 580°F. for 1¼ hours. Hold until a drop on glass is very hard and has little tack (about 20 minutes). Cool to 450°F. and thin with mineral spirits. Then add naphthenate driers.
Viscosity G (Gardner-Holdt System).
Non-Volatile Content   45%
Yield   47 gal.
This varnish is suitable for furniture finishing. It can be dull rubbed, or rubbed and polished to a very high gloss.

Synthetic Rubbing Varnish
This rubbing varnish is used on furniture. It sets up quickly and dries hard enough overnight to be rubbed. It has good toughness and adhesion, and is extremely mar-resistant.
Hard Modified
Phenolic Resin
(Beckacite 1123)   50 lb.
Hard Maleic Resin
(Amberol 800)   50 lb.
Dehydrated Castor
Oil (Synthenol),
Viscosity Z–2   8 gal.
Segregated Fish
Oil (Celesterol),
Viscosity Z   4 gal.
Litharge   1 lb.
Turpentine   5 gal.
Mineral Spirits   41 gal.
Cobalt Naphthenate
Drier (6%)   ½ lb.
Manganese Naph-
thenate Drier (6%) 1 lb.
Heat phenolic resin and castor oil in open kettle to 570°F. and hold for heavy string from stirring rod. Add fish oil, litharge, and maleic resin and melt out. Reheat to 525°F. and hold about 20 min. Cut off fire, cool to 460°F. and reduce with turpentine and mineral

spirits. Then add cobalt and manganese driers.

Viscosity G (Gardner-Holdt System).

Non-Volatile Content   40%
Yield   67 gal.

Wall-Sizing Varnish

This varnish is used for sealing plaster and masonry surfaces prior to painting. It may be mixed with the first coat of wall paint, or used alone.

I Wood Rosin	200 lb.
Unslaked Dolomite Lime	10 lb.
4-Hour Kettle Bodied Linseed Oil, Viscosity X–Z	10 gal.
Selectively Refined Fish Oil, Viscosity Z	10 gal.
Litharge	1¾ lb.
Mineral Spirits	47 gal.
Cobalt Naphthenate Drier (6%)	2¾ lb.

Heat rosin to 500°F., add litharge, and sift in lime slowly. When the mixture begins to thicken add the linseed oil and heat to 565°F. Hold until the material becomes quite thick (about ½ hour) and add the fish oil. Reheat to 550°F. and hold 15 minutes. Cool to 475°F. and reduce. Finally, add cobalt drier.

Viscosity G (Gardner-Holdt System).

Non-Volatile   55%
Yield   85 gal.

Insulating Impregnating Varnish
U. S. Patent 2,295,958

Cellulose Acetate	50
"Vinsol"	50
Acetone	75

White Architectural Enamel Liquid

This varnish makes an exceptionally good white enamel liquid for interior use. It dries quickly, has a high gloss, and, most important, really stays white.

Hard Maleic Resin (Amberol 801–P)	100 lb.
Dehydrated Castor Oil (Synthenol), Viscosity Z–2	10 gal.
Raw Tung Oil	10 gal.
Kettle Bodied Soya Bean Oil (Viscosity Z)	10 gal.
Lead Acetate	2 lb.
Mineral Spirits	53 gal.

Heat resin and tung oil in closed kettle under inert atmosphere to 550°F. Hold 15 min. Add castor oil, reheat to 550°F., and hold 15 min. Then add soya bean oil and reheat to 550°F. Hold for heavy string (about 30 min.). Cut off fire, add lead acetate, and cool to 475°F. Then stir in mineral spirits.

Viscosity J (Gardner-Holdt System).

Non-Volatile Content   50%
Yield   90 gal.

Architectural White Enamel

Titanium Dioxide	300
Zinc Oxide	25
Dyal V 7893	350
Mineral Spirits	345
Lead Naphthenate (24%)	2½
Cobalt Naphthenate (6%)	2½

Grind titanium dioxide and zinc oxide in 200 parts of Dyal V 7893 and 30 parts of mineral spirits.

Thin down the paste with remainder of ingredients.

Alkyd Blending Varnish

This varnish is designed to blend with architectural alkyd resins to make a vehicle for white enamels. Use about one part of varnish to two to four parts of alkyd. The varnish vastly improves the gloss of the finished enamel and improves the hardness and thorough drying. It is non-yellowing and so will not spoil the good initial color or color retention of the alkyd enamel.

Hard Maleic Resin (Amberol 801–P)	150 lb.
Dehydrated Castor Oil (Synthenol), Viscosity Z–2	15 gal.
Lead Acetate	2 lb.
Mineral Spirits	42 gal.

Heat resin and 12 gal. castor oil in closed kettle under inert atmosphere to 570°F. Hold until the mass gives a heavy string from the stirring rod (about 15 min.). Check the remaining 3 gal. castor oil, heat to 550°F., and hold 20 min. Cut off fire, stir in lead acetate, and allow to cool to 500°F. Then add the mineral spirits.

Viscosity J-K (Gardner-Holdt System).

Non-Volatile Content	50%
Yield	70 gal.

General-Purpose Grinding Oil, Mixing Varnish and Paint Oil

A Refined Lignin Liquor	176 lb.

*Overpolymerized

* This is heat-polymerized linseed oil of such viscosity that in a Gardner-

Linseed Oil	88	lb.
B Calcined Magnesia	3⅛	lb.
V.M. Linseed Oil	10½	lb.
Mineral Spirits	12	gal.
C Hydrated Lime	14¼	lb.
Mineral Spirits	3	gal.
Lead Naphthenate (24%)	2¼	lb.
Manganese Naphthenate (6%)	2¼	lb.
Cobalt Drier (4%)	2¼	lb.

Yield 101 gal.

Raise A to 400°F.

Cream B to a slurry and work in to the batch, dispersing well.

Attain 525°F. Hold it until clear (1 to 5 minutes).

Let cool naturally to 425°F.

Add 12 gal. of mineral spirits. Stir well.

Hold batch at 360°F. and lime with C (stirred to a slurry).

Stir 15 minutes at 360°F. Fire off.

Special Paints and Coatings
Cold Wall Glaze
German Patent 733,804

Casein	1
Water	5

Soak well and add:

Barium Sulfate	1

Holdt tube the bubble moves only the diameter of one bubble while the bubble in the "Z–6" Gardner-Holdt tube traverses the entire length of the tube.

Binder Volume	33%
Oleoresin	38.8%
Specific Gravity	0.865 (7.2 lb./gal.)
Viscosity	E
Color	15 to 16
Drying Time	9¼ hours

This vehicle is stated to show high durability for its cost ($0.35) per gal. and shows a drying time of 9 hours.

Cement   5
Alum   ½
Mix and apply as coating for inside and outside walls.

Foundry-Mold Coating
U. S. Patent 2,282,349
Powdered Zircon   100
Bituminous Coke   1–50
Bentonite   1–5
Suspend the above in water.

Coating for Asphalt Drums
U. S. Patent 2,293,249
The interior of drum is sprayed with the following and heated to 205°F. before filling.
Clay   4
Water   4
Calcium Chloride   ½–2

Coating or Lining for Iron Pipes
British Patent 558,492
Cracked and Blown
  Asphalt   70
Talc (200 mesh)   30

Protective Coating for Aluminum
U. S. Patent 2,294,334
Aluminum is immersed for 30 min. in a boiling solution of:
Soda Ash   24
Sodium Dichromate   2
Water   11
Wash and heat for 1 hr. in steam at 95 lb./in.²

Acoustic Sound-Deadening Coating
U. S. Patent 2,166,236
Rubber   10–40
Wood Flour   40
Asphalt   30–70
Rosin   5–10
Colloidal Clay   5–20

Protective Coating for Lithographic Plates
U. S. Patent 2,331,245
Dextrin   10–15
Methylcellulose   1–6
Water   To make 100

Corrosion and Abrasionproof Coating for Petroleum Refinery Equipment
U. S. Patent 2,298,079
Lead Monoxide   100
Glycerin   27½
Naphthenic Acid   1
After applying, cover with ground stoneware, flint or sand.

Non-Leafing Aluminum Paste Paint
British Patent 555,789
Aluminum Powder,
  Leafing   65.0
Thinner (Naphtha)   33.5
Lead Naphthenate   1.5
Mix without grinding.

Coating Composition for Ice Trays
Canadian Patent 412,729
In ice trays the metallic portions that are to come into contact with ice cubes are sprayed with a heated mixture of about 1 gal. solvent, 3.5 oz. carnauba wax and 3.5 oz. paraffin.

Hot-Melt Coating, Moistureproof
U. S. Patent 2,297,709
Ethyl Cellulose
  (<40 centipoises)   10
Hydrogenated Castor Oil   44
Phenol-Formaldehyde
  Resin (Oil-Soluble)   18
Paraffin Wax   28

Paint or Lacquer "Stop-Off"
Coating

Whiting	1 lb.
Water	1 qt.
Glycerin	1 oz.

Paint or spray on surface to be shielded, just prior to painting. When paint coating is dry, remove above coating with wet cloth or razor blade.

Oilproof Barrel Coating

Glue	30
Water	65
Bentonite (200 mesh)	5

pH is adjusted to 10 with sodium hydroxide. The water is first heated to 150°F. and the glue added slowly with very gentle stirring. When all the glue is dissolved add the bentonite by sifting. A small percentage of sodium hydroxide will raise the pH to 10 which may be checked with any common indicator. A viscous jelly-like coating is obtained which, when rolled in a barrel while still hot, with the above mixing procedure, will give a thick even coating. Thinner, or less jelly-like, coatings may be made by lower pH adjustments.

Oilproof Coating for Paper
Containers
U. S. Patent 2,301,048

Invert Sugar Syrup	100
Powdered Mica	15–30
Starch	2–10

Gilder's Size
German Patent 706,755

The size contains pigments and adhesive made of:

Glue	60
Sugar Solution (70%)	25

Calcium Nitrate	15
Formaldehyde	5

Semi-Gloss Polishing Paint
Canadian Patent 406,927

Aluminum or Magnesium Silicate	55–80
Flatting Varnish	15–25
Boiled Linseed Oil	5–12
Vegetable Wax	0.5

Non-Stick Coating for Asphalt
Containers

Citrus Pectate Pulp	3
Sodium Pyrophosphate	0.14
Soft Water	100

Seventy-five parts of the water are heated to boiling, and the sodium pyrophosphate and pectate pulp added. Dispersion is then carried out by agitating as previously described. This 4% dispersion may be cooled and diluted by addition of the remaining 25 parts of water.

Again, spraying produces the best results, brushing or use of a doctor blade are suitable methods, and roll coating is the least successful. Fairly good results were achieved using a 5% dispersion and coating with a roller.

Paper, coated with the 3% pectate sol and dried, will have on its surface an extremely thin film (¾ lb. to 1½ lb. per 1000 sq. ft.) which adheres enough to permit folding for package fabrication. Packages made from paper so treated may be filled with melted asphalt which after cooling will part readily from the container. The package contents, moreover, will have a clean appearance since the only contaminant will be the extremely thin and transparent film of pectate that served to separate the asphalt from the paper.

Asphalt Drum Lining
Canadian Patent 413,867

The interior of the drum is coated by spraying a slurry of:

Clay	4
Water	4
Calcium Chloride	½–2

Then heat to 400°F. This coating prevents adhesion of asphalt to drum.

Fusible Coating Composition
U. S. Patent 2,299,144

Asphalt	13
Gilsonite	67
Montan Wax	14
Paraffin Wax	6

Wine Tank Coating

Gilsonite	25
Paraffin Wax	75

Melt together until uniform and apply very hot.

Protective Coatings for Stone Top Tables
Formula No. 1

Paraffin Wax (High-Melting)	1 lb.
Kerosene	1 qt.
Raw Linseed Oil	2 qt.
Lampblack	¼ oz.

No. 2

Paraffin Wax (High-Melting)	½ lb.
Carbon Tetrachloride	1 qt.
Naphtha	1 qt.
Lampblack	¼ oz.

Use either formula. Apply and allow solvent to evaporate. Rub down with soft rag. Several coats should be applied for best results.

Casein Powder Paints

These paints are supplied in powder form. They are mixed with water to a creamy consistency for application. Two formulae are given below, both for white paints. Tints are made by simply incorporating a small amount of alkaliproof pigment in the powder.

Formula No. 1

100 Mesh Casein	10.0
Hydrated Lime	7.2
Lithopone	25.5
Asbestine	23.2
China Clay	32.4
Pine Oil	1.5
Dowicide G	0.2

No. 2

100 Mesh Casein	12.2
Hydrated Lime	7.3
Rutile Titanium Dioxide	26.5
China Clay	51.8
Pine Oil	2.0
Dowicide G	0.2

Fruit Protective Coatings
U. S. Patent 2,333,887
Formula No. 1

Cellulose Acetobutyrate	10
Ethylene Dichloride	90

No. 2

Sodium Silicate (1:325)	10
Water	80
Hydrochloric Acid (10%)	10
Aerosol O.T. (10%)	5

No. 3

Ethyl Cellulose	10
Ethylene Dichloride	30
Ethyl Alcohol	60

No. 4

Casein Solution	90
Magnesium Carbonate	2
Water	8

No. 5

Mazein (Zein)	10
Ethyl Alcohol	50
Water	3

No. 6

Chlorinated Rubber (5 c.p.s.)	10
Ethylene Dichloride	90
Ethanol	20

No. 7

Manila Gum DBB	8
Ethanol	92

No. 8

Casein	10
Borax	5
Morpholine	2
Water	100

No. 9

Corn Starch	1.000
Sodium Hydroxide	.125
Water	64.000

No. 10

Methyl Cellulose	5
Water	300

No. 11

Glossy Tack-free Coating over Latex.

Mazein (Zein)	10.0
Ethyl Alcohol	20.6
Oleic Acid	6.5
Triethanolamine	3.5
Water	120.0

No. 12

Cellulose Acetobutyrate	4.32
Ethylene Dichloride	44.10
Morpholine	1.00
Oleic Acid	3.25
Water	47.33

No. 13

Cellulose Acetobutyrate	6.35
Ethylene Dichloride	43.00
Triethanolamine	1.59
Oleic Acid	3.17
Dibutyl Phthalate	1.27
Water	44.60

No. 14

Ethyl Cellulose	5.00
Ethanol	37.40
Oleic Acid	5.00
Morpholine	1.20
Water	50.28

No. 15

Ethyl Cellulose	10
Diethylene Chloride	100
Triethanolamine	2
Oleic Acid	15
Water	150

No. 16

China Wood Oil	100.000
*Lead Naphthenate	0.160
*Manganese Naphthenate	0.084
*Cobalt Naphthenate	0.084
Triethanolamine	17.300
Oleic Acid	32.700
Water	1200.000

No. 17

Melamine Resin	50.0
Oleic Acid	26.2
Triethanolamine	13.8
Water	500.0

No. 18

Coumarone-Indene Resin	50.0
Oleic Acid	20.0
Triethanolamine	4.2
Mineral Spirits	15.0
Water	500.0

No. 19

Modified Phenolic Resin	25.0
Oleic Acid	20.0
Triethanolamine	4.2
Mineral Spirits	15.0
Water	500.0

No. 20

Alkyd Resin	50.0
Oleic Acid	20.0
Morpholine	18.3
Water	515.0

* Calculated on metal content.

No. 21

Alkyd Resin	50
Dibutyl Phthalate	4
Oleic Acid	4
Triethanolamine	2
Carbon Tetrachloride	50
Zinc Naphthenate (8%)	1
Water	200

No. 22

Alkyd Resin	26.4
Morpholine	3.0
Water	250.0

No. 23

Polymerized Methyl Methacrylate	13.12
Polymerized Butyl Methacrylate	13.12
Dammar Resin	8.00
Dibutyl Phthalate	5.77
Benzoyl Peroxide	0.22
Dupanol ME (Wetting agent)	2.39
Water	57.38

No. 24

Chlorinated Rubber	5.0
Benzol	50.0
Oleic Acid	9.7
Triethanolamine	5.1
Water	250.00
Casein	10.0
Morpholine	10.0

No. 25

Chlorinated Rubber	5.0
Benzol	50.0
Dibutyl Phthalate	2.0
Oleic Acid	9.7
Triethanolamine	5.1
Water	170.0

No. 26

Chlorinated Rubber	4.125
Benzol	45.870
Morpholine	1.500
Oleic Acid	2.000
Water	50.000

No. 27

Chlorinated Rubber	5.0
Morpholine	9.2
Oleic Acid	5.0
Water	250.0

No. 28

Chlorinated Rubber (25 c.p.s.)	10
Dibutyl Phthalate	5
Benzol	49
Trigamine Stearate	3
Water	210

No. 29

Rosin W.G.	70
Beeswax	2
Sodium Carbonate	7
Water	221

No. 30

East India Chips	50.0
Oleic Acid	20.0
Morpholine	18.3
Water	500.0

No. 31

Kauri Chips	21.6
Oleic Acid	8.7
Morpholine	7.9
Water	217.0

No. 32

Carnauba Wax	13.6
Oleic Acid	2.9
Morpholine	2.5
Water	81.0

No. 33

Paraffin Wax	25.0
Oleic Acid	13.1
Triethanolamine	6.9
Water	315.0

Cleaning Paint Brushes

Several small portions of cleaning solvent, such as gasoline, coal oil, or turpentine, used in succession are much more economical and efficient than the same total quantity of cleaner used at one time.

Work the brush (immediately after finishing a job) in just

enough solvent, in a can, to wet well the bristles. Then rub off as much as possible of the liquid in the bristles on paper or a wall. Discard the liquid in the can, put in the same quantity of fresh solvent and repeat the process. Continue the repetition until the solvent remains essentially clean. Then, after getting as much of the solvent as possible out of the bristles, wash with soap and water. Wet the brush with water and rub it vigorously on a bar of laundry soap. Continue until an excellent lather forms with the soap and until the wash water is no longer milky. Brushes cleaned in this way will have, after drying, soft bristles almost as good as when new.

Filling Scratches on Varnished Furniture

If scratches on varnished furniture are quite deep, they may be filled by running varnish into them with a fine pointed brush, after cleaning with turpentine, and being careful to use a matching color of pigmented varnish or varnish stain, if the surface is tinted. If the scratch is not deep, it may be more readily touched up by careful brushing with the varnish or stain, and similar treatment may be applied to spots. If the surface is really gouged, the gouged places should be filled with plastic wood, which then should be coated with varnish or varnish stain.

After the repairs indicated, the surface should be polished with paste wax to help restore uniform appearance.

For badly damaged furniture, complete refinishing is the best remedy.

Preventing Skinning of Paint in Cans
U. S. Patent 2,325,380

The surface of the paint, in can, is sprayed with a very thin coating of:

Castor Oil	90
Eugenol	10

Fire-Resistant Coating
British Patent 545,840

A fire-resisting coating composition, for application to readily combustible fibrous materials such as wood, consists of the following mixture:

Bentonite as an aqueous suspension (1 part in 5½ parts of water)	50
Asbestos Powder	25
Sodium Silicate (sp. gr. 1.7)	25

Wood Filler

Beeswax	10
Rosin	10
Sawdust	To suit

Melt the beeswax and rosin and stir in enough sawdust so that a hard mass is obtained on cooling.

Wax Wood Filler

Paraffin Wax	50
Montan Wax	10
Rosin	40

The wax and rosin mixture is colored either with mineral colors or coal tar dyes.

Wax Furniture Filler

Japan Wax	30
Paraffin Wax	10
Wool Fat (Neutral)	10

Rosin   40

The waxes are simply melted together.

Painting Glazed Tile

Mix thoroughly together one part of floor varnish and four parts of flat wall paint. Both paint and varnish should be of high quality materials. Apply one good coat to the tile surface. This will dry with a fair gloss which may be lessened —after the paint has become thoroughly dry—by a light rubbing with fine steel wool. Follow this with a coat of heavy paste paint reduced to the desired consistency with flatting oil, if you wish a flat or nearly flat finish. If a higher gloss is desired, add a larger proportion of varnish.

Olive Drab Wrinkle Enamel

Base pigment paste
(Ground in 2 passes, 3-roller mill)

	By Weight
Mirasol 404	50.00
Medium Chrome Yellow	5.00
Lampblack	1.75
Red Oxide	0.75
Asbestine 3X	42.50

Enamel	By Weight
Above paste	36.00
Mirasol 404	46.00
Xylol	9.84
Lactol Spirits	6.56
Mn Nuolate (4%)	1.60

Black Wrinkle Enamel

Base pigment paste
(Ground in 2 passes, 3-roller mill)

	By Weight
Mirasol 404	45.00
Vulcan Carbon Black	2.75

Blanc Fixe "W"   52.25

Enamel	By Weight
Above paste	34.55
Mirasol 404	46.45
Xylol	10.47
Lactol Spirits	6.98
Mn Nuolate (4%)	1.55

Phosphorescent Paints

This type of paint is activated by visible light. It continues to glow in the dark for a considerable period of time after the light source is removed. It is useful for switches and switch plates, moldings and corners, house numerals, and other articles which it is desirable to see in the darkness. Such paints are applied by brushing over a neutral white enamel. For exterior use, a protective clear finish coat must be applied.

Formula No. 1

Phosphorescent Pigment	
(ZnS–2301)	400 lb.
Zinc Palmitate	13 lb.
Alkyd Resin (Glyptal 2475)	76 gal.
Manganese Naphthenate Drier (6%)	2 gal.

Yield—85 gal.

Do not grind this paint. Simply mix in the pigments thoroughly. Grinding will reduce the luminescent properties of the pigment. The paint is an extremely heavy paste which must be reduced with mineral spirits for brushing or spraying.

No. 2

Acryloid B–73	44 gal.
Solvesso #2	56 gal.
Phosphorescent Pigment	400 lb.

Clear Protective Coat for Phosphorescent Paint

| Acryloid B–73 | 85 | Solvesso #2 | 21 |

Wood Stains
Formula No. 1

Selected acid dyes, dissolved in water, form the most light-fast stains that can be used on wood furniture. These dyes should be dissolved in hot water, using from one to three ounces avoirdupois of the mixed dyes per gallon of water. The exact amount to use depends upon the depth of color desired. Always mix and store these stains in glass, enameled or earthenware containers, never use metal containers.

These stain powders are made by blending the dyes as follows:

Dye	Walnut	Cherry	Oak	Maple	Mahogany Red	Brown
Metanil Yellow Conc.	..	79	75	87	..	15
Orange 11 15	15	15	15	10	55	45
Azo Rubine	..	..	..	..	40	..
Crocein Scarlet MOO	..	4	..	1	..	15
Nigrosine WSB	5	2	10	2	5	25
Sap Brown	80	..	..	..	..	..

Color variations may be obtained by varying the proportions of dyes in the above formulations.

No. 2
(For Staining Packing Boxes and Cases)

Dupont Orange RO	23
Pontamine Yellow SX	35
Pontamine Black E Double	42

This is used from a water solution containing one lb. per 10 gal. to which has been added 4 lb. 3 oz. sodium pentachlorphenate. After dipping the wood panels in the stain and preservative, drain and, if desired, dip in a solution of copper sulfate crystals containing 3 lb. 5 oz. of the crystals in 10 gal. of water.

Lacquers
White Automobile or Refrigerator Lacquer

¼ Sec. RS Nitrocellulose (30% Alcohol)	11.3
Rezyl 99–4 (50% in Toluol)	23.7
Dibutyl Phthalate	2.4
Titanium Dioxide	8.9
Ethyl Acetate	9.0
Butyl Acetate	18.5
Butyl Alcohol	10.1
Toluol	11.1
Petroleum Base Lacquer Diluent	5.0
Non-Volatile 31.1%	

Load all ingredients into pebble mill and grind about 30 hours, or until smooth.

Clear Furniture Lacquer

¼ Sec. RS Nitrocellulose (30% Alcohol)	6.6
Amberol 801	7.0
Rezyl 387–4 (60% in Toluol)	11.7
Dibutyl Phthalate	1.0
Ethyl Acetate	11.2

Butyl Acetate   20.0
Butyl Alcohol   9.4
Toluol   20.1
Petroleum Base
  Lacquer Diluent   13.0
    Non-Volatile 19.6%
Viscosity   A–C
Spray with little or no reduction.

Flat Lacquer
U. S. Patent 2,312,309
Less than 1% Carnauba Wax is dissolved in a hot spraying lacquer which is applied at 55–120°C.

Clear Lacquer for Metal or Glass
Acryloid B–72 (40%
  non-volatile)   60
Dibutyl Phthalate   2
Xylol   38
Prepare by mixing.
Brush at this consistency: Reduce 4:1 with xylol to spray.
This lacquer has remarkable adhesion to polished metals and glass, will not discolor, and has excellent outdoor durability.

Clear Transparent Silver Lacquer
Film Scrap Solution
  (18% Solids)   39.0
Rezyl 99–5   3.4
Dibutyl Phthalate   1.0
Butyl Acetate   37.0
Xylol   19.6
Make up by slow mixing, until dissolved.
This lacquer should be reduced about equal parts with lacquer thinner to spray. Reduce about two parts lacquer with one part lacquer thinner to brush.
This lacquer is suitable for use on any polished metal, gold, silver, brass, copper, aluminum, etc., to

prevent it from tarnishing. It is not suitable for outside exposure. This is also an excellent lacquer for photographic prints, to prevent staining and soiling by handling.

Clear Metal Lacquer
(Exterior)
¼ Sec. RS Nitrocellu-
  lose (30% Alcohol)   6.2
Beckasol 1323 (50%
  in Toluol)   26.1
Dibutyl Phthalate   2.6
Ethyl Acetate   10.2
Butyl Acetate   11.1
Butyl Alcohol   8.7
Methyl Ethyl Ketone   9.1
Toluol   18.5
Petroleum Base
  Lacquer Diluent   7.5
    Non-Volatile 20%

Artificial Leather Coating Bases
Formula No. 1
Nitrocellulose, RS
  150-Sec.   40
Castor Oil, Raw   60
      No. 2
Nitrocellulose, RS
  150-Sec.   40
Castor Oil, 2-AC   60
      No. 3
Nitrocellulose, RS
  150-Sec.   37
Castor Oil, Raw   31.5
Castor Oil, No. 15   31.5
      No. 4
Nitrocellulose, RS
  150-Sec.   33.3
Blended Oil   66.7
      No. 5
Nitrocellulose, RS
  150-Sec.   40
Castor Oil, Raw   30
Paraflex RG2   30

No. 6

Nitrocellulose, RS	
150-Sec.	44
Castor Oil, Raw	42
Tricresyl Phosphate	14

No. 7

Nitrocellulose, RS	
150-Sec.	44
Castor Oil, Raw	42
Tributyl Phosphate	14

No. 8

Nitrocellulose, RS	
150-Sec.	40
Castor Oil, 2-AC	42
Butyl Acetyl Ricinoleate	18

No. 9

Nitrocellulose, RS	
150-Sec.	39
Blended Oil	30.5
Butyl Acetyl Ricinoleate	30.5

No. 10

Nitrocellulose, RS	
150-Sec.	30
Castor Oil, Raw	45
Titanium Dioxide	25

Data

Spew Point, °F	160
Cold-Crack Point °F.	0
Fade-Ometer Life, in hrs.	116
Flex Life (in terms of 1000 folds):	
Original	292
After 4 mos.	4.5
% loss	98.5

No. 11

Nitrocellulose, RS	
150-Sec.	30
Castor Oil, 2-AC	45
Titanium Dioxide	25

Data

Spew Point, °F.	175
Cold-Crack Point °F.	5
Fade-Ometer Life, in hrs.	116
Flex Life (in terms of 1000 folds):	
Original	75.8
After 4 mos.	5
% loss	93.4

No. 12

Nitrocellulose, RS	
150-Sec.	16.9
Ethyl Cellulose N-100	16.9
Castor Oil, Raw	41.2
Titanium Dioxide	25

Data

Spew Point, °F.	190+
Cold-Crack Point ° F.	−12
Fade-Ometer Life, in hrs.	162
Flex Life (in terms of 1000 folds):	
Original	40
After 4 mos.	4.5
% loss	88.8

No. 13

Nitrocellulose, RS	
150-Sec.	16.9
Ethyl Cellulose N-100	16.9
Castor Oil, 2-AC	41.2
Titanium Dioxide	25

Data

Spew Point, °F.	190+
Cold-Crack Point °F.	−10
Fade-Ometer Life, in hrs.	162
Flex Life (in terms of 1000 folds):	
Original	24
After 4 mos.	5
% loss	79.2

No. 14

Nitrocellulose, RS	
150-Sec.	15.7
Ethyl Cellulose N-100	15.7
Castor Oil, Pale No. 4	43.6
Titanium Dioxide	25

Data

Spew Point, °F.	190+
Cold-Crack Point °F.	10
Fade-Ometer Life, in hrs.	230
Flex Life (in terms of 1000 folds):	
Original	60
After 4 mos.	6
% loss	90

No. 15

Nitrocellulose, RS 150-Sec.	14.6
Ethyl Cellulose N-100	14.6
Castor Oil, Pale No. 16	45.8
Titanium Dioxide	25

Data

Spew Point, °F.	190+
Cold-Crack Point °F.	23
Fade-Ometer Life, in hrs.	116
Flex Life (in terms of 1000 folds):	
Original	140
After 4 mos.	18
% loss	87.2

Book Lacquer

The following composition has been very successfully used by libraries for the coating of fabric covered books to prevent mildew and the attack of roaches and other crawling insects:

Medium Viscosity

Ethyl Cellulose	600 g.
Ester Gum	100 g.
Castor Oil	300 g.
Benzol	3200 cc.
Methanol	800 cc.
Steam Distilled Pine Oil	600 cc.

Brushing Linoleum Lacquer

5–6 Sec. RS Nitrocellu- lose (30% Alcohol)	6.6

1110 Beckacite	9.4
P–238 Beckasol	20.9
Methyl Ethyl Ketone	4.5
Butyl Acetate	7.7
Butyl Alcohol	3.0
Butyl Cellosolve	9.4
Toluol	38.5

Wood Lacquer
U. S. Patent 2,168,040

Nitrocellulose (3 sec.)	9.1
Dammar, Dewaxed	13.9
Alcohol	20.4
Toluol	21.6
Butyl Acetate	20.0
Ethyl Acetate	9.0
Butyl Alcohol	3.0
Cocoanut Oil	3.0

Cellulose Acetate Lacquer
Swiss Patent 222,545

Cellulose Acetate	20
Dibutyl Phthalate	4
Tricresyl Phosphate	6
Castor Oil	6
p-Toluene Sulfonamid	6
Methyl Acetate	150
Methyl Alcohol	10
"Cellosolve"	10

Cheese-Wrapper Foil Lacquer
Canadian Patent 367,128

Dammar	8
Copal	5
Ethyl Cellulose	4
Plasticizer	6
Benzene	15
Ethyl Acetate	25
Butanol	10
Alcohol	20
Toluol	10

Avoiding Stickiness of Lacquer
Coated Paper

Add 1% Duponol M.E. on weight of solids in lacquer.

Lacquer for Moistureproofing Cellophane
U. S. Patent 2,280,829

Nitrocellulose (10 sec.)	6.70
Paraffin Wax (M.P. 60)	0.15
Dibutyl Phthalate	2.90
Dammar	1.50
Alcohol	2.90
Acetone	1.45
Toluol	33.10
Ethyl Acetate	51.00
Water	0.30

Gold Lacquer

Fast Spirit Yellow 2R Conc. Dye	0.7 oz.
Fast Spirit Orange R Conc. Dye	0.1 oz.
Fast Spirit Brown G Dye	0.2 oz.
Methyl Alcohol	3.5 qt.
Butyl Acetate	1.0 pt.
Clear Metal Lacquer	1.0 gal.

Dissolve dyes in methyl alcohol, add butyl acetate, and then stir into the metal lacquer. The metal lacquer should be a nitrocellulose base material. The above formula yields 2 gallons. This lacquer is used on sheet steel to give the appearance of gold or brass.

Flat Lacquers

To get a smooth finish on furniture where a high gloss is not desired, flat lacquers are used. Below are given formulae for a dull, flat lacquer and a high sheen or semigloss lacquer. Intermediate sheens can be obtained by blending the two materials.

Formula No. 1
Dull Sheen

¼ Sec. RS Nitrocellulose (30% Alcohol)	9.0
Rezyl 99–4 (50% in Toluol)	12.5
Amberol 801	6.5
Dibutyl Phthalate	2.5
Castor Oil (Raw)	1.3
Zinc Stearate	3.1
Ethyl Acetate	5.0
Butyl Acetate	22.1
Butyl Alcohol	9.2
Acetone	2.0
Denatured Alcohol	3.1
Toluol	16.9
Petroleum Base Lacquer Diluent No. 2	6.8

Semi-Gloss

¼ Sec. RS Nitrocellulose (30% Alcohol)	8.5
Rezyl 99–4 (50% in Toluol)	12.3
Amberol 801	8.3
Dibutyl Phthalate	2.5
Castor Oil (Raw)	1.3
Zinc Stearate	.8
Ethyl Acetate	5.6
Butyl Acetate	22.1
Butyl Alcohol	9.2
Acetone	2.0
Denatured Alcohol	3.1
Toluol	16.8
Petroleum Base Lacquer Diluent	7.5

Load all ingredients into pebble mill and run until pigment is ground fine (about 24 hours). Reduce each lacquer with about one-third its volume of the following lacquer thinner to spray:

Lacquer Thinner
Formula No. 1

Butyl Acetate	26
Ethyl Acetate	8
Methyl Ethyl Ketone	15
Butanol	12½
Petroleum Naphtha	38½

No. 2

Ethyl Acetate	5
Butyl Acetate	25
Butyl Alcohol	15
Toluol	25
Petroleum Base	
Lacquer Diluent	30

Crystallizing Lacquer
U. S. Patent 2,344,191

A

Acetanilide	14–18 lb.
Alcohol	6–10 gal.

B

Ethyl Cellulose	150–170 lb.
Mineral Spirits	115–120 gal.
Butanol	18–22 gal.

For use take:

A	4–6 gal.
B	6–8 gal.
Amyl Alcohol	1–4 gal.

Frosting Glass

Sandarac Resin	6½
Denatured Alcohol	30
Xylol	16
Toluol	47½

Clean glass thoroughly and air dry; then spray using 30 lb. pressure, holding gun about 20 inches from the work.

Filling and Binding Lacquer
U. S. Patent 2,357,573

Nitrocellulose Lacquer	48
Calcined Gypsum	48
Lacquer Solvent	2–10

Flame Resistant Lacquer Coating

Cellulose Nitrate	
(RS 5–6 sec.)	60.0
Raw Castor Oil	80.0
Tricresyl Phosphate	40.0
Magnesium Ammonium Phosphate	80.0

Titanium Dioxide	10.0
Luxol Fast Brown R	0.3
Carbon Black	0.5

Solvent

Denatured Ethanol	24.0
Tollac	180.0
Butyl Acetate	216.0

Perspirationproof Lacquer Base
U. S. Patent 2,170,187

Nitrocellulose	52
Alkyd Resin	18
Dibutyl Phthalate	15

Dissolve in any usual lacquer solvent and then stir in:

Dehydrated Powdered Silica Gel	15

Protective Lacquer for Lenses

Vinsol Resin	2 lb.
Vinylite AYAF	2 lb.
Methanol	1 gal.

Dipping Lacquers for Glass Identification

Green:

Air-drying Nitrocellulose Lacquer	1 qt.
Acetone	3 qt.
Auramine O dye	⅔ oz.

Green: Substitute Brilliant Green for Auramine O in above.

Blue: Substitute same amount of Alizarine Fast Blue RB

Red: Substitute same amount of Oil Scarlet 6G

Violet: Substitute same amount of Gentian Violet

Purple: Substitute same amount of Rhodamine B.

The mixture should be stirred until the dye has dissolved and any insoluble matter then filtered off.

Airplane Dope
(Lacquer)

Airplane dopes based on cellulose acetate have the important advantages of stability and non-flammability. A typical formulation is given below:

Cellulose Acetate	10.1
Triphenyl Phosphate	1.1
Ethyl Acetate	21.3
Toluol	12.0
Acetone	20.3
Diacetone Alcohol	10.0
Methyl Ethyl Ketone	25.2

Clear Transparent Metal Lacquer for Exterior Exposure

Low Viscosity Hercose AP	9
*50% Rezyl 14 Solution	8
Dibutyl Phthalate	2
Butyl Acetate	30
Isopropyl Acetate	20
Butanol	10
Xylol	5
Toluol	16

Make up the Rezyl 14 Solution by melting the resin and solvents together in a steam bath. Make up the lacquer by first stirring the Hercose in the butyl and isopropyl acetates mixed, until a clear solution is obtained. Then add the remaining ingredients.

The finished lacquer is of brushing consistency. It should be thinned slightly with ethyl acetate for spraying.

This lacquer has remarkable weather resistance. It can be used on any polished metal surface to prevent it from tarnishing, indoors as well as out.

* Rezyl 14 Solution:	
Rezyl 14 (Solid)	50
Xylol	25
Ethyl Acetate	25

Stable Zein-Rosin Solution
U. S. Patent 2,277,891

Dried Zein	100
Anhydrous Ethyl Alcohol	237
Rosin	100

The rosin is first dissolved in the ethyl alcohol and the zein dissolved in the ethyl alcohol-rosin solvent.

Prolamine (Zein) Plastic Lacquers
U. S. Patent 2,340,913
Formula No. 1

Zein	45
Rosin	20
Plasticizer (Santicizer 8)	25
Denatured Alcohol	270

No. 2

Rosin	175
Zein	390
Diethylene Glycol	150
Santicizer 8	110
Denatured Alcohol	175
Titanium Dioxide	70

No. 3

Zein	20
Gum Mastic	20
Denatured Alcohol	140
Benzol	20

No. 4

Zein	30
Denatured Alcohol	95
Dibutyl Tartrate	8
Copal	40

No. 5
U. S. Patent 2,250,041

Anhydrous Zein	100
Ester Gum	50
Dibutyl Tartrate	50
Triacetin	25
Toluol	100
Anhydrous Ethyl Alcohol	200

Imitation Shellac Lacquer
Formula No. 1

1. Dissolve 33 grams of zein in 67 grams of 92% ethyl alcohol.
2. Dissolve 33 grams of rosin in 67 grams of 92% alcohol.
3. Mix 2 parts of the rosin solution with 1 part of the zein. Add 1.5% p-toluene sulfonic acid based on the total solution.

No. 2

Alcohol Soluble

Maleic Resin	24 lb.
Zein	6 lb.
Denatured Alcohol	70 lb.

Dissolve at room temperature with the aid of a mixer. The resin should be broken into small lumps before attempting to dissolve. This formula yields 14 gallons.

This product is excellent for use on floors, bowling alleys, and many other places where only shellac has been used in the past. It dries quickly, and is very hard and tough. It is not brittle and will not chip.

Adhesive Scratch-Resisting
Primer
U. S. Patent 2,293,558

42 parts of shellac and at least 58 parts of polyvinyl butyraldehyde resin and cellulose nitrate in the proportion by weight of between about 65 parts of polyvinyl butyral resin to 35 parts of cellulose nitrate and 25 parts of polyvinyl butyral resin to 75 parts of cellulose nitrate is a good base.

Paint Remover
U. S. Patent 2,168,024

Mineral Spirits	80–75
Turpentine	15
Butyl Lactate	6
Dibutyl Phthalate	1

Finish for Laboratory Table-Tops

The new unpainted wooden table or bench top is treated with furfuryl alcohol applied with paint brush. This coating is allowed to soak in for a half hour or more, and one coat of technical hydrochloric (muriatic) acid quickly applied. The acid causes the alcohol, together with chemically reactive solid substances in the wood, to copolymerize and form an impregnated mass of chemical-resistant, insoluble material. On account of the somewhat lacrimatory vapor produced at this stage, it is desirable for the operator not to try to use the room until the tables are dry.

The resulting greenish-gray surface is now rubbed briskly with cloth soaked in boiled linseed oil. A fine smooth black finish is thus obtained.

Chapter XIV

PAPER

Tub Sizing for Paper
U. S. Patent 2,348,685

Borax	1–3
Sodium Perborate	¼–2
Corn Starch	To make 100
(Rich in alpha amylose)	
Water	To suit

Heat at 200°F. until no further change in viscosity results.

Rosin Size (30% Free Rosin)
(For Paper Sizing)

1 liter of tap water is heated to boiling, 25.2 grams of sodium carbonate added, followed by 200 grams of powdered Nelio rosin.

Alumina Sol
(For Paper Sizing)

To 4 grams of sodium aluminate in 3 liters of water alum is added until a pH of 9.00 is reached. The system is diluted to a total volume of 18 liters and, after the floc has settled, the supernatant liquor is removed and replaced with fresh water. This washing is continued until the floc is free of sulfate ions. After removal of the last supernatant liquor, the floc is heated to 60°C. with rapid stirring. A small quantity of aluminum chloride is added which causes the precipitate to peptize.

Lime in Place of Caustic Soda
in Making Paper Size

Mix 100 kg. of rye flour with 700 l. of water at 20–5°, heat to 40°, add gradually 8 kg. calcium oxide in 100 l. of water and, after 10 to 15 minutes, heat nearly to the boiling point.

Defoamer for Pulp and Paper
Stock
U. S. Patent 2,304,304

Paraffin Wax	0.12–1.5
Diglycol Laurate	0.25–2.5
Water	To make 100

Pest-Resistant Paper
Australian Patent 113,158

Formaldehyde	5
Ammonium Hydroxide	2
Glycerin	5
Casein	5
Water	83

Several sheets of paper are put together and in between is placed one sheet coated with the above solution, to which is added some blue coloring material. The resulting laminated sheet is then coated on both sides with another glycerin-containing solution consisting of:

Formaldehyde	10
Glycerin	2
Boric Acid	1
Sulfuric Acid	1
Water	86

The resulting product is then suitable for manufacture into packing boxes, bags, wrapping materials and related packaging products.

Anti-Mist Paper for Glass
U. S. Patent 2,333,794

An article for simultaneously dry cleaning and polishing glass surfaces or the like and applying to it an anti-mist film is made by impregnating tissue paper with one-half per cent by weight of pure sodium stearate. The soap is distributed in a relatively dry state over the paper carrier in such a way as to permit it to be transferred to the glass surface, there to act as an anti-mist, during the process of dry cleaning and polishing.

Paper Transparentizing Solution
(Non-inflammable)

Coconut Oil   1
Heat then stir into a mixture of:
Petrolatum Liquid, Light   1
Naphtha   1
Carbon Tetrachloride   1
Stir very well for about 10 minutes.

Non-Tarnish Paper
(For protecting silver-ware)
A

Silver Nitrate Crystals   5 lb.
Water   25 lb.
B

Oxalic Acid   2.65 lb.
Water   25 lb.
Add A to B, then add just enough 26° Bé. ammonium hydroxide to dissolve the precipitate. Finally make up to 80 lb. solution.

Put on paper, wring out, and dry. If this solution is to be put on cloth, add to the final solution 25 lb. Swift's Carton Glue. This cloth or paper is used to wrap silver to prevent it from tarnishing.

Paper Softening
U. S. Patent 2,268,674

A dry or partly-dried paper web is passed through a bath containing:
Urea   20
Glycerin   10
Talc   10
Water   60
N-Hydroxymethyl-
    pyridinium Chloride   1

Paper Glaze Coating

Kaolin   100
Water   To suspend above
Add to a solution made by boiling:
Glue   16–20
Gelatin   1.8–4
Glycerol   0.16–0.26
Water   To suit
Then add:
Alum   0.4–0.536
Bluing   0.125–0.54
Before applying, mix warming with:
Milk   2.5

Non-Slip, Non-Hygroscopic Adhesive Coated Paper
U. S. Patent 2,167,711

Paper is coated with a water soluble adhesive and then dipped in:
Triethanolamine   2.8
Hot Water   250.0
which has been stirred into:
Mineral Oil   7.0
Chlorinated Diphenyl   3.0
Oleic Acid   7.3

Coating or Laminating Paper
U. S. Patent 2,275,957

A dense, substantially non-porous surface film (0.001–0.01 in. thick) is formed on a paper carrier-web moving at 80 ft. per minute by means of calender rolls at 120–140°C. with a stearate lubricant.

Vinyl Resin	80.6
Dibutoxy Ethyl Phthalate	17.0
Carnauba Wax	1.3
Calcium Stearate	0.8
Calcium Hydroxide	0.3

Paper Laminating
U. S. Patent 2,229,028

Zein	100
Ethyl Alcohol (95%)	275
Benzol	10
Blown Castor Oil	10
Formaldehyde	5

The paper sheets are dipped in the above solution for 2 seconds, the impregnated sheets dried at 60–75°C. (149–167°F.), and the dried sheets superposed and subjected to pressures between 1200 and 2000 pounds per square inch at temperatures between 110–160°C. (230–320°F.) for 2 to 15 minutes.

Paper Pulp Oilproofing
U. S. Patent 2,301,048

Corn Syrup	100
Mica	15–30
Glue or Starch	2–10

Grease-Resistant Paper
U. S. Patent 2,256,853

Paper is coated with a mixture of:

Clay	87.7
Polyvinyl Alcohol	1.3
Maize Starch, Thin-Boiling, Cooked	11.0
Formaldehyde	

Grease-Resistant Glue Coatings

The insolubility of high-grade glues and gelatins, and of Arlex, in oils and fats opens up a new field of use in container coatings. The flexible glue may be coated by a spreader on materials like paperboard for subsequent production of boxes for greasy but non-fluid products. Alternatively, finished paper containers may be coated with a sprayed-in lining of flexible glue composition.

Formula No. 1

Glue	11.1
Glycerin	13.3
Cane Sugar	3.1
Anti-Foam (Foamex)	0.9
Water	71.6

No. 2

Glue	9.1
Arlex	13.1
Cane Sugar	3.0
Anti-Foam (Foamex)	0.8
Water	74.0

Both compositions have a higher melting point and better blocking resistance than similar formulas containing no sugar. A high grade glue or gelatin should be employed.

Oilproofing Paper Containers
U. S. Patent 2,301,048

A composition is used, for forming an oil-resistant film upon the walls of a paper-pulp container, comprising:

Non-crystallizing Sugar Syrup	100
Comminuted Graphite, Mica or Aluminum	15–30
Glue, Starch, Gelatin, Flour or Gluten	2–10

Moistureproofing for Cellulose Foil
U. S. Patent 2,311,831

Nitrocellulose	40
Paraffin Wax	2
Ethyl Acetate	400
Toluol	160
Alkyd Resin	10
Dibutyl Phthalate	30

Heat and stir at 40°C. until uniform.

Paraffin Wax Sizing and
Waterproofing
(Emulsion for Paper)
U. S. Patent 2,172,392

Paraffin Wax	4

Melt and add with agitation to hot dispersion of:

Casein or Soybean Protein	1–2
Ammonium Oleate	½
Water	To suit

Wetproof Cigarette Paper
U. S. Patent 2,348,324
Cigarette paper is coated with 0.05–0.2 lb. dry aluminum stearate per 1000 sq. ft. of cigarette paper.

Heat-Sealable Waxed Paper
U. S. Patent 2,337,939
Thin super calendered paper is impregnated with:

Hydrogenated Castor Oil	30–50
Spermaceti	10–30
Ethyl Cellulose	14–22

Waxed Paper
U. S. Patent 2,280,216
Paper rolls are successively soaked in a boiling solution of wax in carbon tetrachloride:

Carnauba Wax	1.5
Beeswax	2.0
Carbon Tetrachloride	150.0

The solution is prepared at 45–50°C. with removal of carbon tetrachloride between the soakings by heat and suction.

Improving Transparency of
Waxed Paper
A 1% solution of ammonium oleate is used to pre-treat paper, prior to overcoating with paraffin wax to give additional transparency.

Milk Bottle Cap Coating
Canadian Patent 414,415
Paper is impregnated with:

Isomerized Rubber	100
Paraffin Wax	12½
Hydrogenated Rosin	25

Deinking Newspapers

Old Newspapers	3000
Soda Ash	60
Bentonite	225
Stanolind No. 250	6
Slaked Lime	30
Water	67000

Operate at 35°C.

Chapter XV

PHOTOGRAPHY

Home-Sensitized Photographic Paper

Paper is coated (in dark) with:

Tartaric Acid	8 gr.
Silver Nitrate	9 gr.
Ferric Ammonium Citrate	40 gr.
Gelatin	6 gr.
Water	1 oz.

Sensitized Photographic Paper
British Patent 546,637

A photographic quality paper is coated with a mixture of:

Di-p-dimethylanilino-phenyl Methane	2.5
Sodium Acetate	0.4
Sodium Nitrite	0.8

in 95% ethyl alcohol (250 parts); after drying, the paper is sensitized by an additional coating with ferric chloride in water. Exposure of the dried sensitized paper under a negative to a mercury-vapor lamp produces a green positive image which may be substantially fixed by exposure to moist ammonia vapor.

Gelatin Composition for True-to-Scale Prints

Gelatin	20	g.
Oxgall, Powdered	10	g.
Ferrous Sulfate	1.5	g.
Water	200	cc.
Glycerin	10	cc.

Universal Developer

The following (M–Q) formula is for both films and papers:

Distilled Water (125°F.)	2.00	qt.
Metol (Elon)	125.00	gr.
Sodium Sulfite	10.00	oz.
Sodium Pyrosulfite	0.25	oz.
Hydroquinone	1.25	oz.
Sodium Carbonate	6.50	oz.
Potassium Bromide	0.50	oz.
Distilled Water		
	To make 1 gal.	

This developer gives full development in 4–6 minutes at 70°F. on film, with full emulsion speed. Twenty minute development will not fog the film nor materially increase contrast, but merely increases the density of both highlights and shadows to give increased emulsion speed.

Low-Contrast Developer
U. S. Patent 2,321,348

Hydroquinone	15
Sodium Sulfite	30
Sodium Hydroxide	240
Potassium Bromide	9
Alum	320
Water	1000

Stable Developer

Amidol	50
Potassium Metabisulfate	200
Sodium Sulfite	100

Water   To make 1000
For use dilute with 5–20 times of water.

Economical Photographic Developing

Since the keeping qualities of concentrated developer solutions are limited, it is recommended that the preserving solution (sodium sulfite, sodium acid sulfite, potassium pyrosulfite) and the accelerating solution (sodium carbonate, caustic alkali) be stored separately. Instead of using a mixture of the 2 solutions they can be used one after the other. The compositions of the 2 baths for a hard and a soft developer for use in this manner are as follows:

Formula No. 1
Hard Developer

Metol	3.0
Hydroquinone	7.5
Potassium Pyrosulfite	30.0
Potassium Carbonate	2.0
Potassium Bromide	2.0
Water	1000.0

Soft Developer

Metol	6.0
Hydroquinone	3.0
Potassium Pyrosulfite	30.0
Potassium Bromide	1.0
Water	1000.0

No. 2
Hard Developer

Potassium Carbonate	100
Sodium Sulfite	10
Potassium Bromide	2
Water	1000

Soft Developer

Potassium Carbonate	70
Sodium Sulfite	10
Potassium Bromide	1
Water	1000

No washing between the 2 baths is necessary. Treatment in each bath is for 3 minutes for plates, films or paper.

No. 3
(Two solution type)
Solution A

Metol	5 g.
Sodium Sulfite	100 g.
Hydroquinone	5 g.
Water	1 l.

Solution B

Potassium Carbonate	100 g.
Water	1 l.

The film is bathed in solution A for 2 minutes at 20°C., then for 1 minute in solution B. If the film has been correctly exposed, the image will not appear until the film has been placed in solution B. If a film is over-exposed, the image will appear in solution A, and solution B is omitted.

Non-Dichroic Fog Developer

Metol	$2\frac{1}{2}$
Hydroquinone	$1\frac{1}{4}$
Sodium Sulfite	25
Sodium Bicarbonate	15
Sodium Hyposulfite	$\frac{1}{2}$
Water	1000

Fine-Grain Developer
Formula No. 1
(Modification of DK 20)
Solution A

Metol	5.0 g.
Sodium Sulfite	100 g.
Potassium Thiocyanate	1.0 g.
Potassium Bromide	0.5 g.
Water (Distilled)	1.0 l.

Solution B

Borax	5.0 g.
Distilled Water	1.0 l.

Bathe 4 minutes in A and then

4–6 minutes in B at 20°C. without intermediate washing.

No. 2
Fine-Grain Developer without Speed Loss

Solution A

Pyrocatechol	13.0 g.
Sodium Sulfite Septahydrate	4.0 g.
Distilled Water	100.0 cc.

Solution B

| Sodium Hydroxide | 15.0 g. |
| Distilled Water | 150.0 cc. |

Solution C

| Potassium Thio-cyanate | 1.0 g. |
| Distilled Water | 30.0 cc. |

Prepare the developing solution by adding 14 cc. of A, 8 cc. of B, and 10 to 20 drops of C to 500 cc. distilled water. Develop 10–12 minutes at 20°C. with frequent agitation.

No. 3

Sodium Sulfite	65 g.
o-Phenylenediamine	8 g.
Metol	8 g.
Potassium Pyrosulfite	7 g.
Water	700 cc.

Developing time 12–13 minutes at 18°C. This developer is non-poisonous, does not stain the fingers and does not yield spotted negatives.

No. 4

Metol	5.00
Sodium Sulfite	100.00
Borax	0.67
Potassium Sulfocyanate	1.00
Potassium Bromide	0.50
Water	1000.00

Copper Sulfide Intensification of Film or Prints

Bleach negative in a bath, at 20°C., made by dissolving 200 g.

potassium ferricyanide in 1 l. distilled water. The negative is then bathed for 5 minutes in a bath made by dissolving 100 g. copper (ous) chloride dihydrate in 1 l. distilled water. Next, bathe 15 minutes in water, then in a dilute solution, 0.5 to 1.0%, sodium sulfide until no further darkening takes place, and finally wash thoroughly in water. The resulting negatives are stable for an indefinite time.

Intensifying Baths for Negatives
Formula No. 1

Solution A

Potassium Ferri-cyanide	100 g.
Potassium Bromide	100 g.
Water	1 l.

Solution B

| Sodium Sulfide Nonahydrate | 40 g. |
| Water | 100 cc. |

(Dilute 1 to 30 for use.)

Bleach the image in solution A, then bathe in solution B without intermediate washing.

No. 2

Solution A

Potassium Citrate	70 g.
Copper Sulfate Pentahydrate	6 g.
Water	1 l.

Solution B

| Potassium Ferricyanide | 6 g. |
| Water | 1 l. |

Mix equal parts of A and B, and bathe the film in the mixture for 5 minutes, then in 10% hypo for 10 minutes, wash for 10 minutes, then bathe for 10 minutes in a bath consisting of:

| Basic Aniline Red | 3 g. |
| Acetic Acid | 12 cc. |

Water   1 l.

Finish the processing by a 10 minute wash in water.

No. 3

Potassium Dichromate	5 g.
Potassium Bromide	4 g.
Chrome Alum	3 g.
Hydrochloric Acid	2 cc.
Water	1 l.

Bathe film in this bath until desired intensification has been attained, then wash in water until the yellow color has been eliminated.

No. 4

Bleach film in a bath consisting of:

Conc. Hydrochloric Acid	2 cc.
Potassium Dichromate	3 g.
Chrome Alum	5 g.
Potassium Bromide	4 g.
Water	1 l.

Next, wash the film thoroughly, then bathe in a bath consisting of:

Sodium Sulfide Nonahydrate	1 g.
Sodium Metabisulfite Pentahydrate	1 g.
Water	100 cc.

and complete the intensification by a final wash in water.

Intensifiers
Formula No. 1

A

Potassium Ferricyanide	100
Potassium Bromide	100
Water	1000

B

Sodium Sulfide, Crystalline	40
Water	100

B is diluted 1:30 for use. A and B are used one after the other.

No. 2

A

Potassium Citrate	70
Copper Sulfate, Crystalline	6
Water	1000

B

Potassium Ferricyanide	6
Water	1000

A and B are mixed in the ratio 1:1. After treatment in this bath and fixing in 10% sodium thiosulfate, the image is washed 10 minutes and placed for 10 minutes in a bath containing 3 g. Basic Aniline Red (or Thioflavine Red) and 12 cc. acetic acid per liter of water.

No. 3

Potassium Dichromate	5 g.
Potassium Bromide	4 g.
Chromium Alum	3 g.
Hydrochloric Acid	2 cc.
Water	100 cc.

A solution of the above is diluted 1:9 for use. After treatment in this, the plate of film is washed in running water or 2% sodium carbonate solution until the yellow color has disappeared.

Gelatin-Hardening Solutions
Formula No. 1

Dissolve 8 g. tannic acid in 250 cc. warm water. When it is completely dissolved, cool at 20°C. and add 30 cc. formalin. Bathe film in the bath until desired degree of hardness is attained.

No. 2

Chrome Alum	2.5 g.
Water	300 cc.

Dissolve the chrome alum in warm water, cool to 20°C., then add 8 cc. formalin. Bathe film in bath until desired degree of hardness is attained.

No. 3

Formalin	5 cc.
Water	100 cc.
1% Sodium Carbonate (1 g. sodium carbonate in 100 cc. of water)	5 cc.

This hardening bath renders the film particularly resistant to hot water.

Photographic Hardener
(Proof against boiling water)

Formaldehyde	1
Sodium Carbonate (1:20 Solution)	20

Photographic Fixing and Hardening Bath
British Patent 548,000

Mix the following materials in order at 60°C., cooling to 25°C., and filtering if necessary.

Solution A:

Anhydrous Sodium Sulfite	48
Sodium Thiosulfate	1750
Guanidine Nitrate	330
Water	700

Solution B:

Acetic Acid	300
Boric Acid	30
Potassium Alum	70
Sodium Sulfite, Anhydrous and Water	800

Then 30 parts of urea are added to A, and B is poured into A.

Acid Hardening Fixer

Of the two formulas given, No. 1 is the most expensive, but keeps well; No. 2 is ultra-rapid but as it does not keep well it should be discarded after a one-time use. The cost of No. 1 can be lowered somewhat by using 100 cc. of ammonium thiosulfate instead of 185 cc. and by substituting potassium chrome alum instead of aluminum hexahydrate. The two formulas are as follows:

Formula No. 1

Water	700 cc.
Ammonium Thiosulfate (60% Sol,)	185 g.
Anhydrous Sodium Sulfate	12 g.
Acetic Acid (99%)	9 cc.
Boric Acid	7.5 g.
Aluminum Chloride Hexahydrate	12.5 g.
Water To make 1000	cc.

No. 2

Water	700 cc.
Ammonium Thiosulfate (60% Sol.)	185 cc.
Sodium Sulfite	15 g.
Sulfuric Acid (5%)	80 cc.
Potassium Chrome Alum	15 g.
Water To make 1000	cc.

Removal of Yellow Spots on Prints

Bathe print 1 hour in cold water.

Bathe 10 minutes in 10% alum.

Bathe 10–20 minutes in a hot solution consisting of:

Sodium Thiosulfate Pentahydrate	20 g.
Alum	5 g.
Water	100 cc.

Wash thoroughly in cold water.

Restoring Yellow Prints

Yellow prints are in most cases caused by incomplete washing. They may be restored as follows:

Bleach the print, using a yellow

safe-light, in the following bleaching solution:

Potassium Perman-
ganate .... 1.5 g.
Sodium Chloride .... 4.0 g.
Acetic Acid (28%) .... 12 cc.
Water .... 250 cc.

Wash the print, then clear in a 2% sodium metabisulfite (pyrosulfite solution), wash thoroughly, then redevelop in bright light with an ordinary paper developer, and finally wash.

Photographic Bleaching Solution (Substitute for Potassium Ferricyanide)

Ferric Alum Solu-
tion (10%) .... 50 cc.
Tartaric Acid Solu-
tion (15%) .... 15 cc.
Conc. Hydrochloric
Acid .... 5.0 cc.

Immerse film at 20°C. until the silver is thoroughly bleached.

Normalizing Overdeveloped Negatives

Two solutions for reducing the excessive contrast in overdeveloped negatives:

Formula No. 1

Paraquinone .... 1 g.
Water .... 100 cc.
Sulfuric Acid .... 3 cc.

No. 2

Potassium Perman-
ganate .... 1 g.
Water .... 1000 cc.
Sulfuric Acid .... 5 cc.

Followed by a bath of:

Sodium Pyrosulfite ...... 2 g.
Water .... 400 cc.

Reduction is slow, and it is necessary to guard against over-reduction, since the final effect is seen only after clearing in the fixing bath.

Reduction of Overdeveloped Films

Formula No. 1

Quinone .... 1 g.
Water .... 100 cc.
Sulfuric Acid .... 3 cc.

The film is bathed in this solution until sufficient reduction has been attained, then washed thoroughly in water.

No. 2

Solution A

Potassium Perman-
ganate .... 1 g.
Water .... 1 l.
Sulfuric Acid .... 5 cc.

Solution B

Sodium Metabisulfite .... 2 g.
Water .... 100 cc.

The film is bathed in A until the desired reduction has been attained, and then it is immersed in B to stop the reduction.

Photographic Resist
British Patent 547,382

Pale Crepe Rubber 48.0
Titanium Dioxide or
Lithopone 4.0
Fast Helio Red 8.0
Paraffin Wax 1.0
Magnesium Peroxide 0.5
Carbon Tetrachloride 64.0

Eliminating Abrasion Fog on Photographs

If abrasion is noticed on photographic papers after development, it can be removed by 1 minute immersion in a bath of 1 l. water and 5 cc. of a solution of 10 g. potassium iodide, 5 g. iodine and 100 cc. water with subsequent fixing. If abrasion fog is noticed only after fixing, the affected areas are treated with a cotton wad saturated with a solution of 60 cc. water, 20 cc. alcohol and 30 drops ammonia (aqueous).

Rapid Drying of Negatives
Rapid drying of negatives on glass support.

After washing the negatives are dipped in a 10% formalin solution and then dried at 80 to 90°C.

Rapid Drying by Salt Wash
Immerse the negative after washing in a 25% epsom salt solution for 1 minute, wipe off excess solution, then immerse in alcohol or benzene for 30 seconds and dry in a gentle current of air.

Rapid Drying by Organic Solvents
After washing immerse the negative in 80% alcohol for 1 minute, then dip in benzene or carbon tetrachloride for 10 seconds, and dry in cold air. The negative will dry in about 1 minute.

Non-Curling Photographic Negatives
U. S. Patent 2,341,485
Developed and fixed negatives are treated with a 2–7½% aqueous urea solution.

Preventing Drying Spots on Negatives
After washing the negative, bathe for 15 seconds in a 1% saponin solution, rinse briefly, then hang up to dry.

Preventing Curl in Photographic Prints
Use a final wash of 1 quart of Glucarine B to 50 gal. water.

Restoring Shrunken Photographic Film
U. S. Patent 2,319,660
Shrunken film is exposed to vapor of following solution:

Camphor	3
Alcohol	10
Propanol	8½
Butanol	8½
Dimethyl Phthalate	11
Water	19

Anti-Foam in Photo Solutions
1 oz. Foamex to 500 gal. of developer is used.

Fog Prevention
Fog may be caused by a number of factors. To reduce the fog caused by scattered light, developer action, or aerial oxidation, add 25 to 100 cc. of a 0.2% benzotriazole solution to 1.0 liter of developer.

Fog Removal
Slight fog may be removed from negatives by bathing them in a bath consisting of:

Thiourea	10.0 g.
Citric Acid	5.0 g.
Water	500 cc.

Improving X-Ray Photographs
If X-ray film is immersed, after washing, in 0.1% Invadine N solution for 5 minutes, it dries uniformly without the formation of water spots. By addition of 1 part 1% Invadine N solution to 9 parts developer, uniform streak-free development is obtained. Scratched

film is improved by 30 minute bathing in 1% Invadine N solution.

Photographic Etching Bath

This bath will etch a silver image in gelatin leaving imagewise gelatin that may subsequently be dyed.

Solution A

Copper Sulfate	
Pentahydrate	100 g.
Potassium Bromide	10 g.
Nitric Acid	15 cc.
Water	1 l.

Solution B

Hydrogen Peroxide	
(20%)	200 cc.
Water	1 l.

Mix A and B and add 2 l. water just before using. Bathe film in this bath until the gelatin has been etched to the film base.

Photographic Print-Out Emulsion

This emulsion may be coated on any paper that is not too porous.

Solution A

Tartaric Acid	4 g.
Silver Nitrate	4 g.
Ferric Ammonium	
Citrate	20 g.
Water	150 cc.

Solution B

Gelatin	10 g.
Water	100 cc.

Add solution A to solution B and coat on the paper.

Photographic Printing-Out Papers

Printing-out papers can be obtained by bathing brown-developing gaslight papers in one of the following solutions:

Stannous Chloride	20 g.
Water	100 cc.

Potassium Nitrite	10 g.
Water	200 cc.
Hydroquinone	4 g.
Sodium Sulfite	20 g.
Potassium Bromide	2 g.
Water	200 cc.

Gaslight papers can also be developed in daylight for 5 minutes in a developer consisting of:

Sodium Sulfite	20 g.
Amidol	2 g.
Water	300 cc.

After careful washing bleach to silver chloride, by red light, in a solution of:

Copper Sulfate	4.5 g.
Sodium Chloride	9.0 g.
Citric Acid	3.0 g.
Water	300.0 cc.

Kallitype Silver Printing

Ordinary drawing paper is bathed in a 5% silver nitrate solution and then dried. The paper is then sensitized by bathing for 5 minutes in a bath consisting of:

Ferric Oxalate	20.0 g.
Silver Nitrate	8.0 g.
Water	120.0 cc.

After drying the sensitized paper is printed in sunlight and developed in a solution consisting of:

Borax	14.0 g.
Potassium Sodium	
Tartrate	56.0 g.
Water	600 cc.

After developing the print is fixed in hypo and washed. A brownish toned print is obtained. It can be converted to black by a gold toner if desired.

Toning Agents

The tone of photographic prints

can be controlled to a certain extent by the type of developing solution used.

Reddish-Tone Developer
Formula No. 1

Sodium Sulfite	40.0 g.
Glycin	6.0 g.
Hydroquinone	6.0 g.
Sodium Carbonate	30.0 g.
Potassium Bromide	2.0 g.
Water	1.0 l.

The print is given a long exposure and short development for reddish tones. Shorter exposures and longer developments result in blacker tones.

No. 2

Sodium Sulfite	70.0 g.
Potassium Carbonate	90.0 g.
Potassium Bromide	2.0 g.
Hydroquinone	25.0 g.
Water	1.0 l.

The reddish tone is increased by increasing the dilution of this developer.

Blue-Tone and Blue-Black-Tone Developer

Metol	3.0 g.
Sodium Sulfite	40.0 g.
Hydroquinone	12.0 g.
Sodium Carbonate	75.0 g.
Potassium Bromide	0.8 g.
Water	1.0 l.

This developer produces blue black images. By adding 2 to 5 cc. of a 1% nitrobenzimidazole solution for each 100 cc. of the developer, a blue tone will be obtained.

Lithographic Plate Desensitizing Etch
U. S. Patent 2,333,221

Basic Chromium Sulfate	40–80 g.
Phosphoric Acid	6–18 g.
Water	500 cc.
Gum Arabic Solution (14° Bé.)	4–12 cc.

Lithographic Image Coating
U. S. Patent 2,233,573

Sodium Thiosulfate, Aqueous	30 cc.
Water	70 cc.
Potassium Acid Tartrate	30 gr.

An albumin image on a metallic lithographic base-plate is sharpened and protected by coating with a solution of the above ingredients.

Photoengraver's Cold Enamel

A

Water	80 cc.
Ammonium Dichromate	6 g.
Ammonia (d. o. 91)	10 cc.
Alcohol (95%)	20 cc.

B

Water	360 cc.
Ammonia	40 cc.
Shellac	35 g.

Photoengraver's Cold Enamel Lacquer

Aniline Dye	10–15 g.
Rosin	60 g.
Ammonia	12 cc.
Denatured Alcohol	750 cc.

Albumin (Lithographic) Substitute
Formula No. 1

100 grams casein; 1,350 cc. water; 15 cc. liquid conc. ammonia; and 150 cc. ammonium bichromate, 20% solution. The casein dissolves fairly readily in water upon the addition of ammonia, but the solution may be

heated to 160°F. if more speed is necessary. A whirler speed of 120 r.p.m. may be used. A trace of ammonia or a weak solution of carbonate of soda in the water will aid in development.

No. 2

2 ounces by weight of pure leaf gelatin soaked in 30 ounces by measure of water. Dissolve by placing vessel containing mixture into hot water; then add cold water to make 65 ounces of mixture; add 1.5 ounces by weight of ammonium dichromate and 0.5 ounce of concentrated ammonium hydroxide. Strain, coat and expose in the usual way. Extra care is needed in developing, since the image is a little softer in that stage than the usual albumin image. If albumin is available, a 50-50 solution of albumin and gelatine can be used.

Lithographic Paper Coating

Zein	382
Water	1,600
Sulfonated Tall Oil	75
Caustic Soda	15

To this mixture add a solution of:

Lead Acetate	20
Water	200

Lithographic Printing Plate Treatments
U. S. Patent 2,250,516

A For desensitizing plates before printing, a mixture of equal volumes of the following is used:

1. Ammonium Dichromate 1
   Water 2
2. Phosphoric Acid 0.5

Aqueous Arabogalactan (20%)	8.0

B For damping during printing, a mixture of 1 ounce of the above mixture and 2 gallons of water is used.

C For protection after printing:

Aqueous (20%) Arabogalactan	20
Water	100

Photographic Prints on Oxidized Aluminum

Photographic cyanotype pictures are obtained by immersing oxidized aluminum plates in a solution of:

Ferric Ammonium Citrate	25 g.
Potassium Ferricyanide	20 g.
Water	200 cc.

Then drying, exposing under the negative, developing with water and fixing. Silver pictures are preferably toned in a solution of 10 cc. of 1% gold chloride and 1 g. calcium carbonate in 100 cc. water (German patent 607,012, British patent 407,830). A suitable physical developer is the following mixture:

A	Metol	10 g.
	Citric Acid	50 g.
	Water	500 cc.
B	Silver Nitrate	2 g.
	Water	20 cc.

(A:B = 100:1)

An iron developer consists of:

A	Potassium Oxalate	380 g.
	Water	1000 cc.
B	Iron Sulfate	100 g.
	Citric Acid	1 g.
	Water	3800 cc.

(A:B = 3:1)

Photographic Substitutes

Critical material	Substitute
Potassium Hydroxide	Sodium Hydroxide
Potassium Carbonate	Sodium Carbonate
Hydroquinone	Pyrocatechol
Mercury Intensifier	Copper Intensifier
Farmer's Reducer	Substitute reducer:

Copper Sulfate-penta-
hydrate   30 g.
Sodium Chloride   30 g.
Water   1 l.

Add concentrated ammonium hydroxide until the precipitate first formed just redissolves. Then add 1 l. of 20% hypo.

Farmer's Reducer   Substitute reducer:

Potassium Dichromate   5 g.
Sulfuric Acid   10 g.
Water   1 l.

Acetic Acid Short Stop	Sodium Metabisulfite
Citric or Tartaric Acid	Oxalic Acid
Boric Acid (In fixing bath)	Sodium Metabisulfite
Hydroquinone Developer	Substitute developer:

Solution A

Sodium Sulfite-septa-
hydrate   150 g.
Sodium Carbonate   100 g.
Water   1 l.

Solution B

Pyrogallol   20 g.
Oxalic Acid   6 g.
Water   500 l.

For use mix A and B.

Restoring Faded Documents or Detecting Falsified Documents

Moisten the document, blot, and then cover with a few drops of 25% 8-hydroxy quinoline in 6% acetic acid. After standing for a few minutes, wash with water and dry. The residual iron from the ink will then be visible.

Preventing Silver Nitrate Stains

If the hands or other exposed parts are rinsed soon after the suspected contact with a concentrated solution of sodium thiosulfate ("hypo"), the effect is completely eliminated.

Blueprint Paper
Formula No. 1

A sized paper is coated with:

Sodium Ferrocyanide	1
Ammonium Oxalate	32
Potassium Ferricyanide	6

Glue   26
Water   To make 1000
It is then dried and overcoated
with:
Ammonium Ferri-
    Oxalate   175
Ammonium Oxalate   32
Potassium Ferricyanide   18
Water   To make 1000
              No. 2
Solution A
  Red Potassium
    Prussiate   5 g.
  Potassium Dichromate ¼ g.
  Distilled Water   25 cc.
Solution B
  Iron and Ammonium
    Oxalate   50 g.
  Neutral Potassium
    Oxalate   5 g.
  Sodium Chloride   ⁴⁄₁₀ g.
  Oxalic Acid   3 g.
  Distilled Water   210 cc.

Lacquer for Blueprints, Direct
    Prints, Maps, etc.
These prints can be written on
with writing ink and pencil and
the writing can be removed easily
with a damp rag.
  Nitrocellulose Lacquer   5 g.
  Ethyl Acetate   2 g.
  Benzene   3 g.

Correction Fluid for Blueprints
  Oxalic Acid   2 g.
  Neutral Potassium
    Oxalate   10 g.
  Distilled Water   100 cc.
  Gum Arabic   1 g.

Red Writing Fluid for Blueprints
Add a red writing ink or a red
aniline dye to correction fluid.

Correction Fluid for Vandyke
            Prints
  Mercuric Chloride
    (Poisonous!)   2 g.
  Water   100 cc.

Developer for B W Prints
Solution A
  Phloroglucinol   1 g.
  Water   100 cc.
Solution B
  Sodium Carbonate   5 g.
  Hypo   5 g.
  Sodium Acetate   5 g.
  Sodium Chloride   5 g.
  Water   100 cc.

Recovery of Silver from
        Fixing Baths
This method makes use of exhausted developer and only requires the addition of inexpensive
sodium hydroxide.
Add 25 cc. exhausted developer
to 15 cc. of exhausted fixing solution then add 10 cc. 2% sodium
hydroxide. Mix thoroughly and
allow to stand for at least a day
when the silver may be filtered
from the solution and washed.

Photographic Flashlight Powder
    British Patent 548,963
  A mixture of:
  Sodium Nitrite   1
  Magnesium Powder
    (100-mesh)   0.5–1.5
is thoroughly ball-milled and then
mixed with an additional amount
of 100-mesh magnesium powder
equal to 1–2 times the magnesium
content of the ball-milled mixture.

Chapter XVI

POLISHES

Paste Wax Polish
(Auto Wax)

Carnauba Wax	10.0
Candelilla Wax	5.0
Yellow Ozokerite	9.5
Yellow Beeswax	8.5
Stearic Acid	.5
Pine Oil	.5
Wood Turpentine	2.0
Kerosene	44.0
Mineral Spirits	20.0

(Floor and Furniture Wax)

Carnauba Wax	4.0
Candelilla Wax	2.0
Yellow Ozokerite	3.0
Yellow Beeswax	3.0
Paraffin Wax	10.0
Pine Oil	1.0
Wood Turpentine	10.7
Mineral Spirits	55.1
VM&P Naphtha	11.2

Melt waxes in steam bath or steam-jacketed kettle. Add solvents and heat until clear, if necessary. Cut off heat and cool to about 130°F. Pour into containers and allow to solidify uncovered.

Liquid Wax Polish

Bleached Montan Wax	5
Ozokerite	2
Paraffin Wax	6
Carnauba Wax	2
Turpentine	80
Diglycol Oleate	3

Melt waxes at 100°C. and then cool to 85°C. Add turpentine and diglycol oleate. Stir till cold.

Auto Polish
Formula No. 1

A

Paraffin Wax	55
Candelilla Wax	20
Durocer	20
Rezo Wax B	55

B

Turpentine	45
Naphtha	205

Melt A at 90°C. until uniform and then add B.

No. 2
U. S. Patent 2,274,509

Carnauba Wax	
(No. 3 Refined)	6.67 oz.
Oleic Acid	0.52 oz.
Potassium Hydroxide	.13 oz.
Japan Wax	.27 oz.
Triethanolamine	.31 oz.
Powdered Borax	.13 oz.
Ammonia (26°) in 3.5	
oz. water	100 drops
Water	26 oz.
Shellac Solution in	
5¼ oz. water	5⅜ oz.
Booster Solution in	
5¼ oz. water	2¾ oz.
Water	21 oz.

The booster is made by adding 5 oz. casein to 1 qt. water, mixing with a solution of ½ oz. potassium hydroxide in 2 oz. hot water, then adding 1 oz. of strong ammonia, 0.8 oz. of zinc sulfate in 2 oz. of boiling water, and finally ¼ oz. of yellow pine oil. The mixture is stirred until thick and 10 oz.

cold water added. By omitting the final water (21 oz.), a no-rub floor polish is obtained. The emulsion is maintained by the potassium hydroxide and oleic acid, which form soap.

No. 3

Water	16.35
Soap	4.05
Glycerin	6.75
Kerosene	19.30
Dibutyl Phthalate	3.55
Abrasive	50.00

Windshield Glass Polish and Cleaner
U. S. Patent 2,296,097

Feldspar	12
Calcium Carbonate	8
Sodium Bicarbonate	¾
Bentonite	3

Add sufficient water to make a thin cream before applying.

Glass Polish
U. S. Patent 2,322,066

Boil about four ounces of comminuted castile soap in about one cup of water, pour the solution into 6.25 pounds of whiting, together with 1.5 ounces of aqueous ammonia, one ounce of olive oil and 0.5 ounce of oil of sassafras. The mass is mixed and kneaded until it has a relatively stiff moldable consistency.

Oven Polish

Ozokerite	38
Paraffin Wax	513
Lampblack	175
Graphite	600
Carnauba Wax	10
Benzine	800
Turpentine	700

Silver Polish

Soap Flakes	10
Hot Water	200
Santomerse S	20
Tetrasodium Pyrophosphate	5
Swift's Carton Glue	50
Sodium Thiosulfate	50
Snow Floss (Diatomaceous Earth)	40

Polish for Gold and Soft Metals

Soap	20–25
Coconut Oil	1
Precipitated Chalk	25
Kieselguhr	8
Glycerin	40–45
Lemenone (Artificial Lemon Oil)	1

The kieselguhr and chalk serve as abrasives, the Lemenone as a perfume, the oil to impart a certain amount of sheen to the metal, and the glycerin as a lubricating dispersing medium.

Metal Polishing Cloth

Stir a suspension of 100 g. of calcium carbonate, 40 g. of kieselguhr, and 8 g. of rouge in 1 l. of water and impregnate the cloths. Press out the excess liquid and dry the cloths at about 120°F. Then immerse the cloths in a hot 10 per cent solution of hard soap. Squeeze out excess fluid and dry again.

Chromium Polish

Powdered Soap	3
Hot Water	53
Distilled Olein	5
Ammonia (10%)	3
Denatured Alcohol	16
Tripoli	20

The ingredients should be mixed in the order given.

Leather Polish

Carnauba Wax	40
Montan Wax	60
Paraffin Wax	80

Heat until molten and add a solution of:

Oil Soluble Dye	4
Molten Stearic Acid	15

Add molten mass slowly with good stirring to:

Turpentine	300

Allow to cool before using.

Liquid Leather Polish

Crude Montan Wax	10
Carnauba Wax	3
Candelilla Wax	3
Ozokerite	1
Paraffin Wax	5
Diglycol Oleate	25
Water	70
Turpentine	90
Dye Color	5

Paste Leather Polish

Bleached Montan Wax	10
Crude Montan Wax	5
Candelilla Wax	4
Carnauba Wax	2
Paraffin Wax	5
Diglycol Stearate	3
Water	50
Turpentine	70
Oil Soluble Color	5

Shoe Polish Paste
(For Tubing)

Candelilla Wax	50
Stroba Wax	25
Glyceryl Monostearate	30
or Diglycol Stearate	
Varsol	250
Water	225

Melt waxes; cool to 90°C.—add Varsol and water at 90°C.

Wax Shoe Polish

Durocer	20
Rezo Wax B	25
Paraffin Wax	55
Candelilla Wax	20
Turpentine	45
Sovasol #4	205

Melt and mix.

This makes a paste which gives a good shine. The solvent does not squeeze out when the mass is pressed and the mixture makes a firm jell. It should be poured at 90°C.

Colorless Shoe Polish

Durocer (Synthetic Wax)	11 oz.
Candelilla Wax	11 oz.
Paraffin Wax	80 oz.
Sovasol #4	175 fl. oz.

Warm together and mix.

This produces a firm jell, on cooling, which polishes readily, is homogeneous and is not tacky.

Shoe Creams

Shoe creams are essentially emulsions of wax and a wax solvent in water, the stability of which is maintained by a small amount of soap. Two typical commercial neutral shoe creams suitable for use on very fine leather contain the following ingredients:

Formula No. 1

Carnauba Wax	6
Paraffin Wax	4
Turpentine	15
Water	70
Hard (Ordinary) Soap	5

No. 2

Carnauba Wax	10
Paraffin Wax	10
Turpentine	15

Water 47
Soft (Potash) Soap 3

The waxes are melted together and mixed with the turpentine. The solution is then poured slowly into a boiling solution of the soap in water, meantime stirring vigorously to produce a stable emulsion. Stir until nearly ready to set and run into a container.

No. 3

Paraffin Wax 700
Carnauba Wax 500
Montan Wax, Crude 400
Nigrosine Base 100
Candelilla Wax 100
Shellac Wax 100
Ozokerite 100
Turpentine 8000

No. 4

1. Candelilla Wax 25
2. Stroba Wax 12½
3. Diglycol Stearate S 15
4. Toluol 125
5. Water (Boiling) 112½

Warm 1, 2, and 3 until melted. Keep temperature at 90–100°C. and add 4. Add 5 slowly with good mixing and continue mixing until temperature falls to 40°C. then pour into tubes.

Furniture Polish
Formula No. 1

Spindle (Mineral) Oil 20
Chinawood Oil 10
Varnolene 60
Trigamine Stearate 13
Denatured Alcohol 4
Ammonia 1
Water 120

Mix well with a high-speed stirrer.

This makes a stable milky emulsion which polishes easily.

No. 2

Water 400
Albasol AR (Emulsifier) 5
Powdered Bentonite 2.5
Gum Karaya 2.5
Formalin (40%) 3
Methyl Salicylate 0.5
Pine Oil 1.5

"Two-Tone" Furniture Polish

A

Varsol (Mineral Spirits) 100
Stove Oil (Kerosene) 55
Turpentine 15
Boiled Linseed Oil ⅒
Amyl Acetate ½
Soudan Orange To color

B

Water 100
Methanol 45
Glycerol 10
Acetic Acid 5
Tomato Red To color

Equal volumes of solutions A and B are placed in the bottles. This gives two liquid layers of attractively contrasting colors. If too little linseed oil is used the two layers separate too quickly after shaking. If too much linseed oil is taken the polish dries dull.

Shake well and apply with a clean cloth.

Liquid Furniture and Floor Polish

Carnauba Wax 3.0
Glyco Wax A 3.0
Candelilla Wax 2.8
Paraffin Wax 3.3
Linseed Oil Soap .1
Wood Turpentine 12.5
Mineral Spirits 65.1
VM&P Naphtha 10.2

Put turpentine, mineral spirits,

and naphtha in container and mix with high speed mixer. Add linseed oil soap. Melt waxes and add slowly. Mix until cold.

Floor Wax Remover
U. S. Patent 2,327,495

Morpholine Oleate	0.1
Water	99.9

Apply hot and leave on for three minutes. Scrub lightly to loosen wax and wash away.

Bright Drying Floor Polish
(Emulsion) (Rubless)
Formula No. 1

A

Candelilla Wax	12.5

Heat to 200°F.

B

Triethanolamine	1.8
Linolenic Acid	1.7
Borax	1.0
Water	83.0

Heat to boiling.
Pour A into B stirring well.

No. 2

"Fine Melt" Congo	120
Caustic Soda	3
Morpholine	12
Water	500
Water	100
Carnauba Wax	20
Stearic Acid	6
Triethanolamine	3
Water	100
Water	400

To 500 parts of water, add 3 parts of sodium hydroxide and 12 parts of morpholine. Heat this solution to 85–90°C. Using continued stirring, add the "fine melt" Congo slowly, taking about 15–20 minutes to add the entire amount. It is very important that the resin be ground to powder size.

Hold at 85–90°C. for ½ hour, during which time 100 parts of water are added. The resin should now be completely dissolved.

In a separate container, melt together the carnauba wax and the stearic acid, and add slowly to the Congo mixture, always continuously stirring. Then add the triethanolamine and about 100 parts of hot water slowly, keeping the temperature around 80°C. The remaining 400 parts of water are added more quickly and the polish is allowed to cool to room temperature; water is added to compensate for evaporation loss, and the polish is then filtered.

This polish exhibits excellent gloss, good leveling properties, and very good water resistance.

No. 3

A polish with slightly less gloss and water resistance than No. 2.

"Fine Melt" Congo	120
Caustic Soda	5
Morpholine	6
Water	500
Water	100
Carnauba Wax	20
Stearic Acid	6
Triethanolamine	3
Water	100
Water	400

Same method as in No. 2.

The use of a protective colloid like casein is found to increase compatibility as well as the drying time of Congo polishes.

The stock casein solution is prepared according to the following formula:

Casein	500
Water	2360
Borax	75

Pine Oil	10
Phenol	5

The casein is soaked in 1900–2000 parts of water at room temperature for ½ hour, with stirring. The mixture is then gradually heated. When the temperature reaches 52–55°C., the borax is added either as a solid or dissolved in about 150 cc. of water. Heating is continued and the solution completed by holding the batch at about 75°C. for ½ hour, with stirring. Heat is removed, the phenol and pine oil are stirred into the solution, and the remainder of the water is added. When cooled sufficiently, additional water is added to make up for evaporation losses.

The casein solution is incorporated in the polish just after the addition of the resin and after the temperature has been held at 85–90°C. for ½ hour.

No. 4

"Fine Melt" Congo	120
Caustic Soda	3
Morpholine	12
Casein Solution	60
Water	500
Water	100
Carnauba Wax	20
Stearic Acid	6
Morpholine	3
Water	100
Water	400

Same method as in No. 2.

By increasing the amount of morpholine to dissolve the resin, the gloss of a polish is increased.

No. 5

"Fine Melt" Congo	120
Caustic Soda	10
Triton W-30	5
Casein Solution	60
Water	500

Water	100
Carnauba Wax	20
Stearic Acid	6
Morpholine	3
Water	100
Water	400

Same method as in No. 1.

This polish has very good gloss and leveling properties but shows a decrease in water resistance as compared to the other polishes prepared with morpholine. Triton W-30 is a leveling and wetting agent. Any similar agents may be used in its place.

No. 6

"Fine Melt" Congo	120
Caustic Soda	3
Morpholine	24
Casein	60
Water	500
Water	100
Carnauba Wax	30
Stearic Acid	7
Morpholine	3
Water	100
Water	400

Same method as in No. 2.

The slight increase in carnauba wax helps greatly in improving the drying time of this polish.

Floor Wax
Formula No. 1

Paraffin Wax	80
Ozokerite	45
Acrawax C	35
Carnauba Wax	25
Benzine	500
Turpentine	300

No. 2

Yellow Beeswax	180
Paraffin Wax	45
Turpentine	135
Benzine	180

No. 3

Paraffin Wax	600
Ozokerite	60
Bleached Montan Wax	20
Benzine	900
Turpentine	300
Pine Needle Oil	20

Diamond Abrasive
British Patent 557,714

A shaped compact of diamond dust and copper powder is sintered by immersion at 800°C. in a bath of:

Sodium Chloride	1
Calcium Chloride	2

The absorbed salts are washed out of the pores by water after cooling.

Synthetic Abrasive
U. S. Patent 2,138,799

Abrasive granules (Moh hardness 7.5–8.5) are made by fusing a mixture of the following for 1 hour at 1205°, cooling, and crushing the melt.

Garnet Dust	200
Borax	100
Kaolin	10

Auto Rubbing Compound

This paste rubbing compound is used for rubbing down fresh coats of lacquer or synthetic enamel. Apply with a damp cloth or waste. Rub until perfectly smooth and free from dirt specks, orange peel, etc.

Air Floated Rose Tripoli	33½
Thin Mineral Oil	6¼
Kerosene	1¾
Pine Oil	1
Oleic Acid	3½

Triethanolamine	1¾
Water	22

Put water and triethanolamine into pan of a dough mixer. Premix mineral oil, kerosene, pine oil, and oleic acid and add slowly, with good stirring. Then work in the tripoli until a smooth and uniform paste is obtained.

Buffing Compounds

A

Double Pressed Stearic Acid	366
Avirol WS (Wetting Agent)	22
Pyrophyllite, 200 mesh	336
Air Floated Tripoli	336

Apply "A" to all except the edge of the "coloring" wheel, using that dry edge as a wiper section.

Should the work require an even higher luster, then add a touch of red rouge, "B," to the dry edge of the wheel.

B

Double Pressed Stearic Acid	180
Avirol WS	20
No. 00 Red Iron Oxide	800

For spinning work the brass colored red with "A" and "B" cannot be surpassed in luster and freedom from scratches by the use of a "lime" composition.

If, on some non-spinning jobs, or nickel plate, a "lime" rouge is desired, formula "C" is suggested:

C

Double Pressed Stearic Acid	200
Tallow Stearin	30
Avirol WS	20
Calcined Dolomite	750

For "coloring" chromium:

D

Double Pressed Stearic	
Acid	120
Tallow Stearin	30
Avirol WS	20
Levigated Aluminum	830

When an intermediate quality finish is required at a minimum cost and without a "cutting down" operation, a "cut and color" composition similar to "E" has proven very satisfactory.

E

Double Pressed Stearic	
Acid	310
Tallow Stearin	15
Avirol WS	20
Double Ground Tripoli	680

F

This compound is applied to hardwood novelties on a buffing wheel or in a tumbling barrel. It is applied directly over the stained wood, and produces a high-luster, very smooth finish. It gives the luster of a resin finish with the slip and feel of a wax finish.

Candelilla Wax	30 lb.
Paraffin Wax	25 lb.
Yellow Beeswax	5 lb.
Diatomaceous Silica	2 lb.
Mineral Spirits	50 lb.
Oil Soluble Orange Dye	$5/16$ oz.
Triethanolamine	5 oz.
Trihydroxyethanyolamine Stearate	12 lb.
Water	65 lb.
*Resin Solution	31 lb.
Ammonium Hydroxide (28%)	3 lb.

Put water, triethanolamine, and trihydroxyethylamine stearate in a steam-jacketed kettle and bring to 200°F. Melt waxes in another steam-jacketed kettle, stir in dye, silica and mineral spirits and bring to 200°F. Add this mixture to the water solution slowly, with vigorous stirring. Pre-mix resin solution and ammonia and add to the above solution. Cut off steam and stir occasionally until cold to prevent stratification. It will solidify to a smooth, soft paste. Yield 215 lbs.

Cream Buffing Wax

This compound is used with flexible-shaft mechanical buffers to produce a high polish on automobile finishes. It is non-scratching and leaves a protective wax film. The material is a heavy liquid cream emulsion which is very stable.

Carnauba Wax	4
Yellow Beeswax	1¼
Yellow Ozokerite Wax	¾
Paraffin Wax	2
Mineral Spirits	18¾
Diatomaceous Silica	2¾
Water	164½
Trihydroxyethylamine Stearate	6

Put water in steam-jacketed kettle, add stearate, and heat to 165°F. Melt waxes and mineral spirits to 165°F. in water bath and stir in silica. Then add hot wax solution to water solution slowly with vigorous agitation. Mix occasionally until cold.

---

*Make the resin solution as follows:

Amberol 750	20 lb.
Denatured Alcohol	20 lb.

Break the resin into lumps and dissolve in the alcohol by mixing.

Chapter XVII

PYROTECHNICS AND EXPLOSIVES

Ammunition Primer
Formula No. 1
U. S. Patent 2,327,867

Lead Hypophosphite	8–10
Lead Nitrate	10–12
Lead Styphnate	28–33
Barium Nitrate	14–23.5
Powdered Glass	30
Trinitroresorcinol	0.5–2

No. 2

Normal Lead Triazoacetate	10
Lead Styphnate	32
Lead Sulfocyanate	8
Lead Nitrate	30
Glass	20

No. 3
U. S. Patent 2,175,826

Lead Nitrate	3.3
Calcium Dihydrogen Phosphate	1.7

Non-Combustible Blasting Primer
British Patent 560,227

Potassium Chlorate	53.0
Sodium Salicylate	14.5
Ammonium Oxalate	32.5

Stable Explosive Black Powder
U. S. Patent 2,167,849

Sugar	24
Potassium Nitrate	33
Boiling Water	6

Dissolve and add:

Coal Dust	10

Mix until uniform and add:

Powdered Potassium Chlorate	33

Mix and grain through a screen.
Dry at a low temperature.

Blasting Powder
U. S. Patent 2,168,030

Black Powder	74.8
Ammonium Nitrate	19.8
Urea	0.5
Powdered Charcoal	0.7
Ground Wheel Cake	4.2

Oil Well Explosive Charge
U. S. Patent 2,299,907

Nitroglycerin	50
Ammonium Nitrate	25
Ammonium Chloride	25

Demolition Explosive
U. S. Patent 2,333,275

Nitrostarch	50
Barium Nitrate	40
Coal Dust	3
Aluminum Powder	3
Dry Dicyandiamid	1
Graphite	1½
Paraffin Wax	1½

Pyrotechnic Flare
U. S. Patent 2,149,314

Magnesium Powder	50–54
Castor or Linseed Oil	1–4
Barium Nitrate	36–40
Strontium Nitrate	6–8

Green Pyrotechnic Torch
This torch in a size of ¾″ diam. and 12 inches long, exclusive of the handle socket, will burn 10 minutes with a fairly good color and does not choke up.

Barium Nitrate	40
Potassium Chlorate	12
Shellac	4
Petrolatum	1

Sift the barium nitrate and rub in the petrolatum thoroughly; sift and mix the potassium chlorate and shellac; add barium nitrate and petrolatum; mix well and sift and mix twice more.

Colored Rocket Smoke Flares
Formula No. 1
Yellow

Potassium Chlorate	33
Lactose	24
Auramine	34
Chrysoidine	9

No. 2
Red

Paranitraniline Red	60
Potassium Chlorate	20
Lactose	20

No. 3
Green

Auramin	15
Indigo	26
Potassium Chlorate	33
Lactose	26

No. 4
Red

Red Antimony Sulfide	55
Powdered Sulfur	15
Powdered Potassium Nitrate	30

No. 5
Orange

Orange Oil Soluble Aniline Dye	10 lb.
Light Mineral Lubricating Oil	1 gal.
Tetrachlorethane	1 pt.

White smokes can obviously be prepared with formula No. 5 omitting the coloring agent, i.e., it may be made from any low viscosity motor oil to which may be added ammonium chloride in finely divided form, which when admitted to the exhaust manifold of any internal combustion engine will be readily vaporized and discharged through the tailpipe as a very dense opaque white smoke. Black smokes are made by burning fuel oils with restricted air, and produce dense black clouds due to the presence of finely dispersed carbon particles in the atmosphere.

Colored Smokes
The following compositions will generate colored smokes when they are blown onto an electric or gas heated hot plate. The density of the smoke will be determined by the heat energy capacity of the plate and the volume of these mixed powder compositions that are blown on the plate at any one interval.

Formula No. 1
Yellow

Auramine (Oil Soluble Yellow)	48
Powdered Sugar	21
Powdered Potassium Chlorate	31

No. 2
Red

Oil Soluble Red Dye	52
Powdered Sugar	27
Powdered Potassium Chlorate	21

No. 3
Purple

Purple Dye No. 80260 (General Dyestuffs)	50
Powdered Sugar	20
Potassium Chlorate	30

No. 4
Blue

Blue Dye No. 80073 (General Dyestuffs)	46
Powdered Sugar	20
Potassium Chlorate	34

Chapter XVIII

PLASTICS, RESINS, RUBBER, WAX

Cartridge Wax
Formula No. 1

Beeswax	1
Copaiba Balsam	1

No. 2

Carnauba Wax	70
Beeswax	20
Ceresin	10

Cheese Coating Wax
U. S. Patent 2,299,951

Paraffin Wax	50
Rubber Hydrochloride	5–15
Microcrystalline Paraffin Wax	To make 100

Ski Wax
For Ironing In:

Paraffin Wax (m.p. 125°F.)	60
Ceresin	16
Palm Oil	14
Talc	10

For Rubbing In:

Paraffin Wax (m.p. 125°F.)	60
Refined Wool Grease	6
Carnauba Wax	4
Montan Wax	18
Rosin	12

Batik Wax
Formula No. 1

Rosin	70
Beeswax	20
Japan Wax	10

No. 2

Rosin	60

Beeswax	15
Ceresin	15

No. 3

Rosin	60
Japan Wax	20
Tallow	20

No. 4

Paraffin Wax	60
Ceresin	10
Japan Wax	30

Coopers' Wax

Paraffin Wax	50
Tallow	30
Beeswax	5
Venice Turpentine	5
Talc	10

Modelling Wax
Formula No. 1

Paraffin Wax	45
Beeswax	30
Tallow	14
Petrolatum	3
Rosin	8
Oil-Soluble Dye	3

No. 2

Beeswax	80
Venice Turpentine	16
Sesame Oil	4

No. 3

Bleached Beeswax	70
Venice Turpentine	22
Sesame Oil	8

No. 4

Bleached Beeswax	80
Stearin	20

No. 5
For Artificial Fruits

Bleached Beeswax	71
Mutton Tallow	11
Rosin	11
Pigment (Cinnabar, etc.)	7

Ironing (Laundry) Wax
Formula No. 1

Paraffin Wax	66
Japan Wax	34

No. 2

Paraffin Wax	60
Ceresin	10
Stearin	20
Japan Wax	5

Gilders' Wax for Fire Gilding

Beeswax	32
Copper Sulfate	8
Borax	1
Verdigris	8

Copper Engraving Wax
Formula No. 1

Beeswax	70
Mastic	4
Syrian Asphalt	26

No. 2

Beeswax	50
Mastic	13
Syrian Asphalt	25
Amber	12

Cable Wax
Formula No. 1

Bleached Ozokerite	25
Paraffin Wax	70
Rosin	5

No. 2

Paraffin Wax	70
Japan Wax	2
Carnauba Wax	2
Rosin	18

No. 3

Paraffin Wax	75

Refined Montan Wax	2
Rosin	23

Grafting Wax
Formula No. 1
(Hand)

Rosin	8
Beeswax	4
Beef Tallow	1½

Mix all the above materials by melting, stir thoroughly and pour the mixture into cold water. Knead and pull the wax under water, until it reaches the proper consistency then remove from the water and wrap in oiled paper.

No. 2
(Brush)

Rosin	4 lb.
Beeswax	1 lb.
Raw Linseed Oil	¼ pt.
Powdered Charcoal	½ lb.
Tar	½ lb.

Melt the above all together, and use as needed, applying with a rather stiff bristle brush, while in the melted state.

Gravure Wax
Formula No. 1

Beeswax	80
Venice Turpentine	20

No. 2

Beeswax	66
Tallow	34

Shoemakers' Thread Wax
Formula No. 1

Crude Montan Wax	20
Beeswax	30
Paraffin Wax	32
Rosin	8
Petroleum Pitch	10
Oil-Soluble Nigrosine	
	To suit

No. 2

Beeswax	20
Paraffin Wax	70
Rosin	10

Saddler's Wax
Formula No. 1

Beeswax	10
Venice Turpentine	1
Rosin	2

No. 2

Bleached Beeswax	8
Rosin	6
Olive Oil	1

No. 3

Bleached Beeswax	6
Rosin	3
Tallow	1

Inhibiting Discoloration of Paraffin Wax
U. S. Patent 2,301,806

Addition is made of about 0.01–5% of diamyl phthalate, tributyl phosphate or other ester of an alkyl monohydric alcohol and an aliphatic hydroxy acid, phthalic acid or an oxygen acid of phosphorus.

Carnauba Wax Substitute
U. S. Patent 2,255,242
Formula No. 1

The process is relatively simple and consists in melting together a fat, such as stearin or palmitin, having a melting point above 20°C., with a coumarone-indene resin having a melting point above 100°C. and which is completely soluble in mineral spirits. The optimum mixtures range from 2:3 parts of the one ingredient to 3:2 of the other.

No. 2

A wax prepared by melting together shellac wax (85 parts) with beeswax (10), heating to 210–230°, gradually adding 5 parts of sal (*Shorea robusta*) dammar, and straining and cooling has d_{15} 0.993, m.p. 83°, acid value 6.3, sapon. value 85.6 and complies with the specifications for carnauba wax, e.g., for making carbon papers.

Lanette Wax SX Substitute

Flaked Cetyl Alcohol	9
Duponol C	1

Candle Coatings

Dissolve 0.05 grams Sudan color in a melt of 25 grams stearic acid and add a melt of 75 grams paraffin wax. Where extreme fastness of light, sublimation, and aging is required, 2 grams of fast candle color is used instead of 0.05 grams Sudan color.

Modifiers for Paraffin Waxes

The inclusion of Staybelite esters in paraffin waxes increases their gloss, transparency, and adhesion to various supporting substrata. The effect on physical properties, as measured by degree of hardness, of the inclusion of 5% of each of the Staybelite esters on 125°F. M.P. paraffin, is given below.

Penetration of Modified 125°F. M.P. Paraffin Wax

Type Modification	At 32°F.	At 77°F.
Control (100% Paraffin Wax)	14	22
Staybelite Ester:		
No. 10 (5%)	14	21
No. 1 (5%)	15	25

No. 2 (5%)   18   27
No. 3 (5%)   18   28

It will be noted that Staybelite ester No. 10 causes relatively little change in the hardness or embrittlement of paraffin, its chief function being to add gloss, transparency, and adhesion.

Depending on the ester chosen, Staybelite esters No. 1 and No. 2 show a definite softening action on paraffin, while Staybelite ester No. 3 exerts exceptionally powerful plasticizing action on this wax. It is noteworthy that on exposure at 0 to 3°C. the Staybelite ester No. 3 modified paraffin film after storage for 24 hrs. showed no exudation of the plasticizer. Indeed, it appears that this ester may have some effect in increasing the low temperature flexibility of paraffin. The indicated use of these resins in preparing wax adhesives, sizes, coatings, and laminating formulations is evident.

It should be pointed out that Hercules Staybelite esters No. 1, 2, 3, and 10 show less tendency to sludge in molten paraffin blends than similar unhydrogenated rosin esters.

Dental Base Plate Wax
Formula No. 1

Paraffin Wax	70
Beeswax	20
Carnauba Wax	4
Gum Dammar	6

No. 2

Paraffin Wax	80
Acrawax C	5
Rosin	5

Dental Inlay Wax

Paraffin Wax	18

Carnauba Wax	3.5
Candelilla Wax	1
*Resin	2.5

Dental Adhesive Wax

Flexowax C	11
Beeswax	11
Ozokerite	13
Venice Turpentine	2

Dental Carving Wax

Paraffin Wax	30
Ozokerite	30
Montan Wax	20
Carnauba Wax	20

Dental Investment (Casting) Wax

Ozokerite 61%, paraffin 29% and rosin 10%. Another is composed of 70% ceresin and 30% beeswax. A third is made of 2 parts beeswax and 1 part rosin with a small amount of Venice turpentine. A fourth is comprised of 1 pound of beeswax, 1 ounce of Venice turpentine and a few drops of glycerin.

Dental Duplicating and Impression Compound
Formula No. 1

For ten units of impression compound, mix:

Alginic Acid (Insoluble)	40
Calcium Sulfate ($CaSO_4 \cdot 2H_2O$)	50
Magnesium Oxide	20
Borax	5
Trisodium Phosphate	30
Powdered Wax (Acrawax C)	50

---

* The resin can be a phenol modified rosin, or a petroleum resin. Carnauba wax can be replaced by synthetic waxes like Acrawax C.

Pass through a number 100 sieve.

Package for use as needed.

Sift one-tenth of the above quantity of powder into 40 to 60 cc. water at 65 to 75°F. and stir rapidly about 1½ minutes. Place in the dental tray and take the impression. The set will be satisfactory in 3 to 4 minutes. Hotter water hastens and colder water retards the set.

Wash the impression in cold water and immerse for 10 to 15 minutes in a solution of manganese sulfate (75 grams dissolved in ½ liter of water). Rinse and blow or blot off the adhering water. Pour the model immediately.

When used as a duplicating material use water at 50°F. Treat surface of model with a thin coat of petrolatum. Vibrate the impression material to place in the same manner as for agar compounds. The set will be slower by reason of the use of colder water.

No. 2

U. S. Patent 2,325,051

Powdered Carob Seed Gum	10
Powdered Calcium Borate	¼
Precipitated Calcium Carbonate	89¾

Electrolytic Wax Plating

U. S. Patent 2,215,143–4

40 g. of beeswax is dispersed in 2 l. of a 1% silicate solution. The silicate has the effect of increasing the throwing power of the solution. A current density of 30 to 40 amp. per sq. ft. is employed for a period of about 6 seconds, the particles traveling toward the anode. The deposit is then dried and heated to a continuous coating.

Sulfur Plastic

U. S. Patent 2,174,000

Sodium Sulfide	165 g.
Sulfur	135 g.
Water	800 g.

Dissolve and heat to 90°C. Add slowly, with stirring:

Formaldehyde	500 cc.

Filter; wash; dry and mold.

Self-Hardening Plastic

U. S. Patent 2,163,243

Paracoumarone Resin	2–5
Chlorinated Rubber	1–4
Toluol	3–15
Wood Flour	0.5–15

Molded Cork and Cane Stalk Product

U. S. Patent 2,155,429

A wet mixture of a granular cork or vegetable base with a fluid binder is molded at 143°/40 lb. per square inch for 10–30 minutes. To ground sugar-cane stalks (1.5 lb.) are added 225 g. of an aqueous dispersion of 100 g. of black blood-albumin in 500 g. of water and 100 g. of latex (25–35% of rubber solids).

Acrylic Plastic Molding Composition

U. S. Patent 2,326,543

Methyl Methacrylate Copolymer	100
Styrene or Vinyl Acetate	Up to 33
Polyvinyl Acetal	33–300

Polymerization Inhibitor for Methacrylates

U. S. Patent 2,299,128

0.1% of acetamide, ammonium

carbonate, ammonia, pyrrol, piperidine, ammonium carbamate, hexamine or aldehyde ammonia.

Melamine Resin
British Patent 556,142

Formaldehyde   2–2.5 mol.
Melamine   0.25 mol.
Dicyanodiamide   1.00 mol.

Heat at 50°C. for 6 hr. after adjusting pH to 8–9 with guanadine carbonate.

Ethyl Cellulose
Canadian Patent 410,522

1 part by weight of chipped wood pulp is covered with 14 parts of 50% sodium hydroxide solution, soaked 7 hours, treated in a high-pressure autoclave with 4.5 parts of ethyl chloride for 12 hours at 80–140°C., residual ethyl chloride, by-product ethyl ether and ethyl alcohol are distilled off, and excess alkali and sodium chloride are removed by washing with water.

Plastic Injection Molding Powder
U. S. Patent 2,326,812

Ethyl Cellulose   65–90
Dammar   5–35
Dibutyl
  Phthalate   Less than 17½

Polystyrene Sponge

Styrene   4 oz.
Ethylene Dichloride 10 oz.
Benzol Peroxide   1 g.

The solution is heated with stirring at about 60°C. for 6 hr. The syrup is then poured into appropriate shaped containers which are placed in boiling water for about 1 hour to evaporate the solvent leaving a sponge of plastic.

It is preferable to use a container having a small vent at the top. The sponge may be used for insulation or in life preservers, etc.

Stabilizing Polystyrene
U. S. Patent 2,304,466

Stabilization to light and air is effected by the incorporation of about 0.3–3% of a polyhydric phenol such as p-tert-butyl catechol.

Inhibiting Polymerization of Styrene
U. S. Patent 2,320,859

Add about ½% or less of 1,5-dinitroanthraquinone.

Single Stage Phenol-Formaldehyde Resin

Phenol   100 parts
Formalin   105 parts
Calcium Hydroxide   2 parts
Aromatic Amine
  (Aniline)   0.1 mole

The calcium hydroxide is used in water suspension. The phenol should be warmed to melt it or used in water solution. All the reactants are placed in a suitable flask connected for vacuum distillation. Heat slowly to boiling at atmospheric pressure. The reaction evolves heat. As soon as the boiling point is reached, distill under vacuum with the temperature below 80°C. Samples of the resin may be withdrawn from time to time to determine the extent of dehydration and the brittleness of the product when cold. Continued heating will form an insoluble infusible product.

Cast Phenolic (Plastic) Dye
U. S. Patent 2,156,442

Water	1000
Diethylene Glycol, or	75
Diethylene Glycol	
Ethyl Ether	70
Sodium Chloride	150
Sodium Orthophosphate	30
Sodium Mono Hydrogen	
Orthophosphate	10
Basic Dye	2

Plastic Molding Powder

Soybean Meal	400
Phenol	762
Barium Hydroxide	
Hydrated	100
Formaldehyde (37%)	986
Wood Flour	800
Stearic Acid	10
Calcium Stearate	10
Zinc Sulfide	50
Dye (Toluidine Red)	50
Hexamethylene-	
tetramine	94

The soybean meal is mixed with the phenol and the alkaline catalyst, so that the particles of meal may become thoroughly permeated. Aqueous formaldehyde (1.5 moles per mole of phenol) is then added and the mixture is heated in a closed steam-jacketed mixer, 15 minutes at 15 pounds per square inch gage steam pressure in the jacket (250°F.) to start the reaction and one hour at a jacket temperature of 190°F. and inside temperature of 175–185°F. to complete the formation of a two-stage resin. The wood-flour filler, stearic acid lubricant, dye, and pigment are added; the mixture is cooled and the accelerator, either hexamethylenetetramine or paraformaldehyde (equivalent to 0.5 mole formaldehyde per mole phenol), is added in concentrated aqueous solution with thorough mixing. The batch is dried to a moisture content of 3–6 per cent and is then suitable for treatment on the calender rolls.

The resulting powdered mixture is worked on the calender rolls, with the cold roll at 150° and the hot roll at 200°F. Small portions of each batch are first rolled for various times and tested on a Rossi-Peaks flowmeter to determine the variation of flow and of curing time with length of rollings. The main portion of each batch is then rolled sufficiently to give the optimum flow and curing time; for standard molding powders and for most of the tests the flow chosen is 1.0 inch at 500 pounds pressure and 150°C., or 1.5 inches at 700 pounds pressure and 150°C.

Emulsion Polymerization of Vinyl Acetate

Water	50.00
Aerosol OT	2.50
Potassium Metabisulfite	0.05

Dissolve and add:

Vinyl Acetate	10.00

Stir.
Heat for 1 hour at 60–65°C.

Moistureproof Coating for Acetate Sheet
U. S. Patent 2,166,711

Candelilla Wax	40
Ester Gum	60
Toluol	662

Apply at 55°C. by roller coating to give a finished coating of 0.01 in.

Improving Stretch of Cellulose Acetate
British Patent 555,805

Acetate yarn or foil is treated in:

Cyclopentanone	38
Gasoline	100

at 25°C. for ½ hour; wash with ether and scour in 0.02% soap solution at 60°C. for 5 minutes.

Polymerizing Butadiene
U. S. Patent 2,151,382

Butadiene alone or mixed with butene is polymerized by contact with a solution of boron trifluoride in water (1.25 mols.) at 0–50°C.

Tough Bituminous Composition
U. S. Patent 2,289,229

Ethyl Cellulose	25
Gilsonite	65
Castor Oil	10

Injection Molding Shellac Compound

Shellac	300
Jute	200
Kaolin (or Barytes)	100 (300)
Pigment	15
Calcium Stearate	9

The mixed components are kneaded in dilute aqueous ammonia, dried, run through steam-heated rollers, crushed, cured at 85–90°C. for 1.5 hours and molded.

Shellac Substitute for Records
British Patent 555,520

Rosin Residue (Petroleum Insoluble)	60–80
Ethyl Cellulose	10–30
Castor Oil	5–30

Flexible Plastic Piping
U. S. Patent 2,340,866

The following is extruded under heat and pressure:

Polyvinyl Alcohol	100
Water	50
Glycerin	35
Formamide	5

Thermoplastic, Tailor's Dummy
U. S. Patent 2,329,207

A knitted fabric is impregnated with a melted mixture of:

Candelilla Wax	30
Rubber	10
Beeswax	10
Rosin or Polystyrene	50

This softens and becomes plastic when heated above 130°F. and sets below 100°F.

Moistureproof Coating for "Cellophane"
U. S. Patent 2,342,209

Cyclized Rubber	24
Paraffin Wax (m.p. 60°C.)	3
Polybutene (m.w. 1,000)	3

Milk Bottle Cap Coating
U. S. Patent 2,325,168

Isomerized Rubber	100
Paraffin Wax	12½
Hydrogenated Rosin	25

Electrical Insulating Filling Compound
U. S. Patent 2,154,276

Asphalt	58
Arochlor (55–70% Cl)	17
Petroleum Oil	25

Electrical Insulation Formula No. 1
U. S. Patent 2,288,322

Zinc Oxide	6–14

Asbestos   60–50
Fused Acidic Copal
          To make 100
Heat together, with mixing.
              No. 2
     British Patent 555,904
Polymerized Rosin   44.44
Gum Accroides   44.44
Castor Oil   8.88
Stearic Acid   2.22

Chemically-Resistant Packing
Gasket
Vinyl Chloride Resin   62
Dibutyl Phthalate   22
Graphite   16
Sleeve
Vinyl Chloride Resin   100
Dibutyl Phthalate   60
Microasbestos   40
Calcium Stearate   1

Plastic Stuffing Box Packing
     U. S. Patent 2,330,502
Granulated Cork   3–4
Mica   3–4.2
Polymerized Isobutyl-
  ene (m.w. 10,000–
  –16,400)   7–10
Petrolatum   1

     Resin-Impregnated Wood
          Formula No. 1
Dimethylolurea   18.78
Urea   2.85
Water   78.37
The solution is prepared by dissolving the compounds in water
at a temperature of 120°F., with
stirring. It can be stored for several months at room temperature
without resinification.

If a higher solids content is desired, this may be obtained by using higher temperatures. For example, a 30% solids solution can

be obtained by heating a mixture
of :
              No. 2
Dimethylolurea   28.17
Urea   4.28
Water   67.55
at 140°F. without stirring. The
resulting solution should be stored
at 100°F. The higher the temperature, the more rapidly resinification of dimethylolurea occurs. At
100°F. and with 30% solids content, a small amount of precipitate
may form in one to two weeks.

If the temperature of the wood
to be impregnated with the 30%
urea—dimethylolurea solution is
lower than 100°F., the impregnating solution will cool and precipitate and prevent maximum possible impregnation. If it is not possible to heat the wood to 100°F.,
the impregnating solution should
be heated to a correspondingly
higher temperature.

          Bow Rosins
        Formula No. 1
Violin
   Flexoresin B1
              No. 2
Cello
   Flexoresin B1   2
   Flexoresin L1   1
              No. 3
Bass Viol
   Flexoresin B1   1
   Flexoresin L1   1

Flesh-Colored Latex Compound
Make a solution of 3 parts of
casein in 100 parts of water and
1 part of 29% ammonia. Add 1
part of Darvan.
The solution is prepared best at
60°C.

To a pebble mill of twice the volume as the material to be ground, add the solution to the powder in equal quantities. Mill for 48 hours. The pigment (powder) is titanium dioxide.

Mix with a stirrer 7 parts of the titanium dioxide dispersion (a 50% dispersion) and 0.02 parts of Rubber Red VD paste and 0.04 parts of Rubber Orange YOD paste.

Add 7 parts of this dispersion to an amount of latex containing 100 parts latex solids, on dry basis.

Preserving Latex
U. S. Patent 2,327,940

Adjust pH to 9.8–10, with ammonia, and add 0.01–0.2% quinoline.

Heat Sterilizing Fresh Latex
U. S. Patent 2,327,939

Adjust pH to 9.8–10, with ammonia, and then heat for 10 hrs. at 45–75°C.

Solvent for Chlorinated Rubber
U. S. Patent 2,338,948

Methylisopropylbenzene	92
Ethyl Lactate	8

Surgeon's Gloves Dusting Powder
Finely powdered potassium bitartrate acts as a bactericidal dusting powder.

Tire-Puncture Sealing Composition
U. S. Patent 2,286,963

Dissolve

Gelatin	3
Water	70

by heating to 90–95°C. then add in the following order:

Calcium Chloride	5½
Fullers' Earth	6
Asbestos Fibers	12
Starch	2½
Cresol	1

Gasolineproof Synthetic Rubber Composition
U. S. Patent 2,138,192

A composition resistant to swelling in gasoline, comprises polymerized chloroprene (3–4 pts.) and chlorinated paraffin wax containing 52–62% chlorine (1 pt.). Softeners, antioxidants, etc. may also be added.

Vulcanization of Butadiene Polymers

1. Rubber or Polymer	100
2. Channel Black	50
3. Zinc Oxide	5
4. Stearic Acid	2
5. Sulfur	2
6. MBT (Captax)	2

Mill 1, 2, 3, and 4 first. Then, mill in 5 and 6.

Synthetic Rubber Emulsion Polymerization
U. S. Patent 2,281,613

Butadiene	7.5
Styrene	2.5
Isohexyl Mercaptan	0.05
Water	18.0
Sodium Oleate	2.0
Ammonium Persulfate	0.03
Temperature	30°C.
Time	Several days

Gas Expanded Rubber
U. S. Patent 2,268,621

A rubber mix comprising:

Rubber, Crepe or Smoked	100
Sulfur	3–50
Light Magnesium Oxide	6

Gilsonite	25
Diphenylguanidine	2
Asphalt	25

It is mixed with a blowing agent, e.g., sodium nitrite and ammonium chloride or diazoaminobenzene, and then gassed, using, carbon dioxide at 1000 lb./sq. inch and heated to evolve gas from the blowing agent to produce a closed-cell, gas-expanded rubber, which is finally vulcanized.

Reviving "Dead" Rubber Balls

Introduce, by a hypodermic syringe, enough rubber cement (1 cc. for a handball) followed by nitrogen or carbon dioxide gas to give a final pressure of 2.6–3 kg./cm².

Reclaiming Scrap Rubber
U. S. Patent 2,325,289

Ground scrap rubber is mixed with 1–5% abietic acid (in absence of steam or water) to 100–140°C. It may then be used in compounding a rubber batch.

Non-Sticking Rubber
U. S. Patent 2,147,312

Rubber is "painted" with a suspension of powdered soapstone and colloidal clay in an aqueous saponaceous wetting agent. That is, to 20 lb. of bentonite in 100 gal. of water are added 0.2% of naphthyldibutyl sodium sulfonate and then 160 lb. of dry soapstone.

Preventing Adhesion of Rubber
U. S. Patent 2,262,689

Talc, used as lubricant for rubber, is rendered more adherent by admixing with a hygroscopic salt, e.g., 0.2% of calcium chloride, or applying a slurry containing:

Talc	64.0
Water	35.5
Glycerin	0.5

It is then allowed to dry.

Reducing Viscosity of Rubber Solutions

Viscosity reduction in solutions of rubber in gasoline by adding alcohol means a saving in the amount of gasoline needed to produce a solution having a specified viscosity. This is because the lower viscosity in gasoline-alcohol blends permit a higher rubber concentration for a given viscosity. Methyl alcohol has a greater effect on viscosity, and hence saves more gasoline, than ethyl alcohol. In solution of natural rubber only 0.5% of methyl alcohol is needed to save 27 to 32% of the gasoline. In solutions of Buna S rubber the respective savings in gasoline due to adding 0.5% of alcohols (methyl to propyl) are: methyl 31, ethyl 18 to 26, propyl 10 to 25%. With Buna SS rubber 0.5% methyl alcohol saves 13–18% of the gasoline while 2% saves 16–21%, but if much more of the alcohol is added the solution becomes unstable in storage. Another effect of methyl, ethyl or propyl alcohol is to stabilize rubber solutions against increase of viscosity during storage. A similar protective effect is obtained by milling butyl alcohol or glycerol into Buna S rubber before making up the solution.

Gelling Neoprene Latex

Sodium Fluosilicate	20

Bentonite   4
Water (Distilled)   76
Sodium fluosilicate if ball milled in this recipe is much more effective as a gelling agent than if the materials are merely slurried together. Ammonium nitrate or chloride (1 to 5 parts) as 2 to 4% solutions may be used to produce gels. More concentrated solutions cause coagulation unless the neoprene latex mix is specially stabilized. Usually these salts are not satisfactory for causing gelation at room temperature but may be used for gelation with heat. One to 2 parts of boric acid added as 5% aqueous solution may be used for room temperature or heat gelation, but room temperature gelation with these amounts of boric acid is a little slow. Ten parts of nitroethane will produce good gels at room temperature with 100 of neoprene as latex or as latex compound.

Fast-Curing Vulcanization Accelerator
U. S. Patent 2,342,870

Tetrabutylthiuram
    Monosulfide   80
Tetrabutylthiourea   5
Tetrabutylurea   15

GR"S"—Olive Drab Raincoat Material
(Calender or Spread)

GR"S" (Synthetic
    Rubber)   100
SRA #1   2
Sulfur   2
Kalvan   60
Buca Clay   50
Cumar MH 2½   15
Stearic Acid   2

Zinc Oxide   5
Heliozone   2
Neozone D   1
Circo Light Process Oil
    (Sun Oil)   10
Red Iron Oxide   1½
Chrome Yellow   6
Ultramarine Blue   12
Cure—2 hrs. at 260°F. Heater Cure.

Self-Curing Lean Cement for Cured GR"S" Coatings
Part 1

Neoprene Type CG   100
Neozone A (Du Pont)   2
Light Calcined Magnesia   4
Zinc Oxide   5
Dissolve in Solvesso # 1444
Part 2
Litharge   10
Accelerator 833   2
Solvesso #1   25
Mix 1 and 2 together just before using. This cement will now cure at room temperature in about 3 days. It can also be cured in 1 hour at 220°F.

Thermosetting Polyvinyl Butyral Coatings for Raincoats

Polyvinyl Butyral   100
Methyl Ricinoleate   60
Dibutyl Sebacate   25
Acrawax C   1
Chrome Yellow   4
Thermax   3½
Red Iron Oxide   ¾
Titanium Dioxide   3
Zinc Oxide   3
Buca Clay   100
Uformite F 200 E   5–10
Make dry paste of all ingredients except the Uformite. Mill on hot mill at about 220°F. Dissolve

in ethanol-butanol mix. Add Uformite to this mixture.

For calendering, mix Uformite and plasticizers together. Then add rest of ingredients. Mill as before. Use as little heat on calender as possible to prevent setting up. Cure 1¾ hrs. at 260°F.

100% Rubber Reclaim Compounds
(Calender or Spread)
Formula No. 1
Black—General Utility

Reclaim (Alkali whole tire—60% R.H.C.)	167
Reogen	3
Stearic Acid	1
Mineral Rubber	10
Neozone D	1
Zinc Oxide	2
Altax	1
P 33 Black	36
Sulfur	2¾

Cures—15 min. at 307°F.
Tensile 1100 lb. — Elongation 450%.

No. 2
Dark Gray—General Utility

Reclaim (Alkali whole tire—60% R.H.C.)	167
RPA #2	2½
Stearic Acid	1
Catalpo Clay	52
Zinc Oxide	5
Captax	1
Sulfur	2¾

Cures—15 min. at 307°F.
Tensile 950 lb. — Elongation 475%.

Code Wire Insulation

Reclaim (Acid process—Neutral)	100
RPA #2	1.00
Stearic Acid	0.75

Mineral Rubber	25.00
Zinc Oxide	1.50
Catalpo Clay	15.00
Thermax	15.00
Altax	0.25
Methyl Zimate	0.75
Sulfur	1.00
Neozone D	1.00

Optimum Cure—15 seconds at 388°F. open steam cure.

Non-Swelling Thiokol Compound
(For Contact with Petroleum and Aromatic Solvents)

Thiokol FA	100.0
Altax	0.4
D.P.G.	0.1
Stearic Acid	0.5
P 33 Black	60.0
Zinc Oxide	10.0

Optimum—Approx. 45 min. at 300°F.

General-Utility Neoprene Compound
(For Oil and Heat Resistance, Good Low Temperature Characteristics)

Neoprene Type GN	100
Latac	¼
Stearic Acid	1
Neozone D	2½
Extra Light Calcined Magnesia	4
Thermax	75
Zinc Oxide	5
Dibutyl Sebacate	10

Optimum—30 min. at 287°F.

Calendered Raincoat Material

Reclaimed Rubber	87
RPA #2	1
Altax	0.3
Zimate	0.03
Retarder W	0.1
Micronex Beads	8

Zinc Oxide	3
Stearic Acid	1
Sulfur	0.75

Heater Cure—1½ hrs. at 260°F.

GR"S" Olive Drab Raincoat Material—Calendered

GR"S"	100
Cumar MH 2½	10
Sulfur	4
Chrome Yellow	3
Lampblack	1¼
Titanium Dioxide	3
Catalpo Clay	100
Naftolen 550	10
Zinc Oxide	10
Brown Factice	20
Captax	2
Zimate	½

Heater Cure—1½ hrs. at 260°F.

GR"I" Compound for Laminating Fabrics (Olive Drab)

GR"I"	100
Zinc Oxide	5
Zinc Stearate	10
Buca Clay (Moore & Munger)	150
Chrome Yellow	3
Lampblack	1¼
Sulfur	2
Butyl Eight	6½

Heater Cure—1 hr. at 260°F.
In processing this compound, make sure that no other rubber or unsaturated compound contaminating stock is present as this will adversely effect the cure.

GR"S" Heel Compound

GR"S"	100
Naftolen 550	20
Neozone D	1
Zinc Oxide	5
MPC Black	40
SRF Black	100

Altax	1½
Ethyl Zimate	¼
Sulfur	4

Optimum—15 min. at 320°F.

Highly-Loaded GR"S" for Mechanical Goods

GR"S"	100
Sulfur	6
Lode Mineral Rubber	40
Zinc Oxide	5
MPC Black	30
P 33 Black	125
Altax	1½
Ethyl Zimate	⅕

Cures—Optimum—30 min. at 307°F.

Self-Curing, 100% Reclaim Compound (Calender or Spread)

Reclaim (Acid, Whole Tire)	167
RPA #2	1
Mineral Rubber	5
Agerite Resin	1
Zinc Oxide	2
Sulfur	1½
Butyl Eight	3

If an alkali reclaim is used, 1 part of Altax should be added to retard the Butyl Eight.

Neoprene Olive Drab Coating (U. S. Army Requirements)

Neoprene GN	100
Neozone A	2
Stearic Acid	½
Light Calcined Magnesia	4
Buca Clay	20
Titanium Dioxide	3
Red Iron Oxide	1
Chrome Yellow	4
Ultramarine Blue	8
Zinc Oxide	5

Heater Cure—1 hr. at 260°F.

Neoprene Latex Coating,
Olive Drab
(For U. S. Army Raincoats)

Neoprene Latex 571	100
Water Glass	¼
Aquarex D	¼
Casein	¼
Buca Clay	25
Zinc Oxide	25
Neozone D	3
Ultramarine Blue	8
Red Iron Oxide	1
Rubber Yellow HN	$2\frac{1}{10}$

All weights are on a dry basis.
All insoluble ingredients should be
ground in a ball mill for 48 hours.
Suggested formula—60% solids.

Neoprene Latex Coating,
Olive Drab

Buca Clay	25
Zinc Oxide	25
Neozone D	3
Ultramarine Blue	8
Red Iron Oxide	1
Rubber Yellow HN	2.1
*Casein Solution (10%)	6.41
Darvan Solution (15%)	17.1
Water	19.19
	106.70

Bunan Oil-Resistant Coating
(Suitable for hose, gaskets and
diaphragms—low extractable
material)

Buna N	100
Zinc Oxide (Kadox)	5
Stearic Acid	½
Benzothiazyl Disulfide	1

* Casein Solution	
Water	86.75
28% Ammonia	3.0
Preservative	0.25
Casein	10.00

Sulfur	2½
Paraplex G 25	30–40
P 33 Black	100–150

By varying the black and the
Paraplex, desired flexibility is obtained.

Optimum cure, approximately
60 min. at 287°F.

Men's Rubber Belt

Pale Crepe Rubber	20.00
Lithopone	20.00
Zinc Oxide	2.00
Paraffin Oil	3.00
Paraffin Wax	3.00
Magnesium Carbonate	5.00
Sulfur	.70
MBT Type Accelerator (e.g.: L–60)	.20
Diphenylguanidine	.02
Organic Orange Rubber Color (e.g.: Vulcan Orange G Ext. I.G.)	.25
Stearic Acid	.20
Antioxidant (e.g.: Flectol H)	.20
Clay	45.43

This is extruded (tubing machine) and cured in soapstone in
open steam. It has a lovely shiny
finish and is beige-colored. Gray
and tan belts can be easily produced by replacing the orange
color, in above formula, by suitable
amounts of black master-batch, together with iron oxide. White
belts can be made by using a little
titanium dioxide plus a trace of
ultramarine.

Artificial Sponge
U. S. Patent 2,138,712

Viscose (20%)	160
Cotton	16
Sodium Sulfate	1200

A sponge-like mass is developed

on and in close union with a pre-formed carrier (e.g., a wooden handle or a swollen fabric base) by immersion in or coating with a mixture of the above ingredients, followed by exposure to steam at 0.5–10 atm. for 15 minutes to 4 hours, washing to remove sulfate, bleaching and dyeing.

Heat Paper Transfer (Decalcomania)
British Patent 526,623

Tallow	4
Printing Ink	1
Stearin	3
Powdered Dry Color	3

Oil-Resistant Sponge

Buna N	100
Zinc Oxide (Kadox)	5
Stearic Acid	12
Altax	1
Sulfur	2
Gastex (Or Pigment)	25
Age Rite Resin D	2
Sodium Bicarbonate	12
Tricresyl Phosphate	50

Cure—60 min. at 305°F.—Blows to double volume. Firm sponge. Softer sponge can be obtained by using additional tricresyl phosphate.

Nitrocellulose Coatings
(For Rubber and Plastic Goods) Very High Spew Points and Good Low Temperature and Fadeometer Resistance
Formula No. 1

Nitrocellulose (5–6 sec.)	100
Ethyl Cellulose (Med. Viscosity)	100
Raw Castor Oil	245
Pigments and Fillers	145

No. 2

Nitrocellulose (5–6 sec.)	100
Ethyl Cellulose (Med. Viscosity)	100
QAC Castor Oil	245
Pigments and Fillers	145

Above compounds have spew point of more than 195°F. a brittle point of about –5°F. and stand over 150 hours in Fadeometer.

Ceramic Made without Heat
U. S. Patent 2,356,214

A product formed by molding without heat and evaporating to dryness a mixture of 15 ounces of magnesite, 25 ounces of Florida clay, 26 ounces of talc, 4 ounces of bentonite, 15 ounces of asbestos, 5 ounces of silica, 3 ounces of dextrine, and 7 ounces of Portland cement with a solution of 5½ ounces of calcium chloride and water.

Molding Clay

Liquid Fish Glue	6 lb.
Water	1 lb.
Whiting	17 lb.
Boiled Linseed Oil	11 lb.
Varnish	12 lb.
Methyl Salicylate	125 cc.

Glueless Printers' Roller

Glycerin	1 l.
Magnesium Chloride Solution (d. = 1.3)	3 l.
Solid Magnesium Chloride	1½ kg.

Dissolve at 80–90°C. and cool to 18–20°C.

Mix 1 l. of above with 0.6–0.8 kg. dry starch (potato). Add 8–10 g. preservative (e.g., Moldex). Stir well; filter and let stand for 6–12 hrs. Remove foam and fill molds. Keep molds at 85–90°C. for 75–90 min.

These rollers can be cleaned with: -

Soap Solution (10%)	5
Kerosene	3

Mix well.

Coloring Gelatin Masses

Colors are best added to the swollen gelatin before melting, preferably dispersed in the glycerol used. Sulfonated castor oil is a softener, and potassium chloride (200 g. to 10 kg. gelatin) acts as a hydroscopic salt. The gelatin, when formed, can be hardened by immersion for 3 to 4 minutes (not more) in dilute formaldehyde (850 cc. to 70 liters cold water). For solid colors, titanium dioxide, ultramarine and lakes are used. These can be "topped" with basic aniline colors, with eosin instead of rhodamine.

Chapter XIX

SOAPS AND CLEANERS

Home Made Soap

Tallow or Grease	10 lb.
Caustic Soda (Lye)	22 oz.
Water	6 lb.
Powdered Borax	½ lb.

Dissolve the caustic soda in the water. Heat the tallow or grease and stir in the caustic soda solution slowly. Stir until all caustic is added and continue stirring and cooking until the mixture takes on a dark transparent appearance and thickens up considerably. Then stir in the borax and when thoroughly mixed, use as a paste or press the soap into molds and dry at ordinary temperature.

Rosin Soap

A vessel of suitable capacity is provided with a steam coil for heating purposes. The following proportions of the ingredients are used: 100 parts rosin, 25 parts caustic soda, and 1,000 parts water. The caustic soda is added to the water in the vessel, and brought into solution by heating; the scum is then removed from the surface of the liquid by any suitable means. The rosin is then added slowly and in such a way that it does not fall in a heap to the bottom, and agitation is kept up during the boiling for 4 to 5 hours. At the end of this time a sample is taken out for examination. The soap should be quite transparent and free from lumps.

Rosin-Tallow Soap

Inasmuch as plain tallow soap does not lather, it is always necessary to add something to make it do so. In laundry soap, rosin is used. For a cheaper and harder product, silicate of soda is sometimes employed. The following formula will serve to illustrate a soap of this character:

Tallow	150
Rosin	75
Sodium Silicate	15

Put the silicate of soda into a small kettle and add water, boiling until it registers 5° Bé. (hot); then add soda ash until it registers 7° Bé., then salt until it makes 8° Bé., while boiling hot. Now run about 70 parts of the soap into the crutcher and add 30 parts of the hot silicate of soda solution. Crutch well, add the perfume, and continue crutching until the soap starts to form. Dump into the frame and treat as usual.

Non-Efflorescing Bar Soap
U. S. Patent 2,278,352

Soap base (75–85 parts) consisting of:

Fatty Acid	66

Rosin Soap	33
Water	31

and an aqueous alkaline solution containing:

Sodium Carbonate	0–10
Sodium Ortho- phosphate	10–30
Sodium Silicate	35–40

Soap Substitute
(Laundry)

A washing agent suitable for application at 50–60°C. comprises:

Xylene	100
Infusorial Clay	100
Kaolin	100
Soda	50

The soda-soaked laundry is washed with 1% of this powder.

Lemon Soap

Sodium Laurylsulfonate	850
Citric Acid	50
Sodium Citrate	10
Lecithin	20
Glycerin	20
Lanolin	30
Mineral Oil	20

Transparent Glycerin Soap

Bleached Tallow	134 lb.
Cochin Cocoanut Oil	88 lb.
Castor Oil	20 lb.
W. W. Rosin	7 lb.
Cane Sugar	64 lb.
Water	32 lb.
Glycerin	34 lb.
Soda Lye (38° Bé.)	135 lb.
Alcohol	16 gal.

Fungicidal Borax Soap
(Washing Mixture)

A washing composition containing approximately 75 per cent of borax and 25 per cent of dry soap possesses considerable fungicidal properties, but is so mild in action as to reduce any tendency to dermatitis.

In its finely divided form, the borax does not cause gumming or caking of the dry powdered soap. Described as a good cleanser, one that will economically and effectively remove dirt from the skin, the borax-soap mixture has a temporary, mild abrasive action that facilitates cleansing. In addition to the benefits imparted by the soap, the borax also contributes detergent and water-softening properties.

Powdered Cleaner
U. S. Patent 2,257,545

Free-flowing non-caking, domestic cleansing powders comprise a mixture of:

Silicon Dioxide Aerogel (d. 0.1–0.3; 80-mesh)	100
Sodium Carbonate	65
Borax	10
Soap	25

Detergent Cake
British Patent 557,593

Sodium Metasilicate	60
Soda Ash	40
China Clay	40

Mix to a paste with 20–30% water; put in molds and allow to harden.

Cleaning Paste

Stearic Acid	75
Refined Tall Oil	25
Sodium Hydroxide (30° Bé.)	55
Sodium Carbonate	10
Water	800

Industrial Hand Soaps and Cleaners

Formula No. 1

Neutral Toilet Soap	30
Colloidal Clay (Bentonite or Kieselguhr)	30
Synthetic Detergent (Santomerse, Sulfatate Nacconol, Igepon or Duponol)	10
Hydrous Wool Fat	5
Perfume	1

No. 2

Sulfonated Neat's Foot Oil	45
Light Liquid Petrolatum	45
Gelatin (25% Aqueous Solution)	10

Heat the oil mixtures on a water bath until clear; then add the gelatin and stir until solution is effected.

No. 3

If a lathering effect is desired, this action can be attained by mixing two parts of Formula 2 with one part of a 20 per cent solution of sodium lauryl sulfate. As a substitute for the popular mechanic abrasive soaps, white granulated corn meal is combined with Formula 2 in the proportion of 1½ parts, by weight, of corn meal and 1 part of the above.

No. 4

(For soap and alkali sensitive skins)

Synthetic Detergent (Sulfatate)	20
Hydrous Lanolin	3
Bentonite	76
Lavender Perfume Oil	1

No. 5

White Soap Chips	35
Naccanol NR	35
Water	500
Glycerin	20
Wood Flour	25
Water-Soluble Perfume Oil	To suit

No. 6

A liquid soap substitute, one that makes a good cleanser and does not defat the skin, is a mixture of sulfonated castor oil having a pH of 7.2 and an oil content of 50 per cent, with 2 per cent of one of the synthetic wetting agents (e.g. Santomerse, Sulfatate Duponol or Igepon). The cleansing power of this mixture can be increased without materially increasing its irritating properties by adding 1 to 2 per cent of an alkali such as trisodium phosphate or sodium hexametaphosphate.

No. 7

Corn Meal, White	50
Butyl Stearate	22
Diglycol Laurate	23
Linseed Fatty Acids	5

No. 8

Tall Oil	12
Mineral Oil (Very Low Viscosity)	77.3
Pine Oil	5.0
Diethylene Glycol	2.0
Caustic Potash (45%)	3.7

Abrasive Hand Cleaning Powder

Naccanol NR	50 g.
Trisodium Phosphate	50 g.
Wood Flour	25 g.
Antiseptic Oil B-5671 (Givaudan)	1 cc.

Triturate the Naccanol and trisodium phosphate together, then mix thoroughly with the wood flour, adding the oil in small quantities during the mixing. Then pass the entire mixture through a 100-mesh sieve.

Hand Cleaner Requiring
No Water
· Formula No. 1
Mixture below makes a gallon.

A

Glyceryl Mono-stearate	1.75	oz.
Sodium Hydroxide	2.50	oz.
Water	3.00	qt.

Heat to boiling.

B

Stearic Acid	2.50	oz.
Mineral Oil	2.50	oz.
Paraffin Wax	11.50	oz.
Spermaceti	7.00	oz.
Lemenone (Perfume)	0.20	oz.

Heat in a double boiler until all is melted.

Pour solution B into solution A with constant stirring, until cold.

No. 2

White Soap Chips	40
Naccanol NR	10
Water	500
Water-Soluble Perfume Oil	1

This is a soft paste suitable for filling into collapsible tubes. It may be used directly on the hands and then wiped off with a rag, without the need of using any water.

Scouring Powder
Formula No. 1

Naccanol NR	25
Trisodium Phosphate	25
Volcanic Ash	10

Triturate the Naccanol and trisodium phosphate together, then mix thoroughly with volcanic ash and pass through a 100-mesh sieve.

No. 2

Yellow Soap Chips	18
Coarse Pumice	40

Sodium Silicate (40°)	3
Glycerin	2
Water	35

Bleaching Soap

The customary procedure is to add the bleaching agent to the kettle soap at the end of the saponification after the alkalinity has been adjusted, but before the separation of the soap. Two pounds of hydrosulfite concentrated, per ton of fats taken for saponification, is sprinkled into the soap kettle, the contents violently agitated for 10–15 minutes by means of live steam, after which the soap is allowed to separate at the top, leaving the water layer at the bottom.

Mercury Fulminate Detector Soap

In an effort to reduce mercury fulminate dermatitis in the explosives industry, a liquid soap can be made, which by a change of color, shows the presence of traces of this dangerous chemical upon the skin.

Diphenylthiocarbazone	0.18	g.
Triethanolamine	250.00	cc.
Liquid Soap	750.00	cc.
Hydroquinone	0.015	g.

The soap, as produced, is orange in color, but in the presence of mercury salts it changes rapidly to a deep, easily recognized purple. The reagent soap solution is said to be so sensitive that one drop will indicate the presence of a minute quantity of mercury, from 0.000002 to 0.00001 g., per square cc.

Antiseptic Liquid Soap

Refined Tall Oil	275	g.
Ethyl Alcohol	200	cc.

Water   425 cc.
Sodium Hydroxide   50 g.
Sodium Carbonate   10 g.
Phenol   25 cc.
Ether   15 cc.

Add 100 cc. water and 200 cc. alcohol to the tall oil, mixing thoroughly. Dissolve the sodium carbonate and sodium hydroxide in remaining 325 cc. of water. To this add the tall oil-alcohol-water mixture and stir well. Finally add the ether and phenol, stirring them well into the mass.

Disinfecting Detergent
U. S. Patent 2,242,315

Calcium Hypochlorite   14
Sodium Hexameta-
  phosphate   71
Sodium Chloride   15

The above gives a clear solution with water and detergent solutions.

Fish and Cannery Factory Antiseptic Wash
(Non-Corrosive)

Formalin   50
Sodium Nitrate   1
Water   2000

Detergent Paper Towel
U. S. Patent 2,333,919

Suitable paper is impregnated with a solution of tetrasodium pyrophosphate so that it contains 0.01–0.2% of latter after drying.

Laundry Blue

Indigo   4 kg.
Oleum   8 l.
Sulfuric Acid   8 l.

Warm to 50°C. and stir until uniform; then add:

Water   600 l.

Laundry Sour and Bluing
U. S. Patent 2,141,589

Ammonium Silico-
  fluoride   159
Sodium Silicofluoride   19
Glucose   72½
Bluing   10

Mix well.

Stabilized Laundry Sour
U. S. Patent 2,331,396

Sodium Acid Fluoride   99
Sodium Hexameta-
  phosphate   1–5

Spectacle Lens Cleaner and Mist Preventer
U. S. Patent 2,333,794

Lens or tissue paper is impregnated with ½% sodium stearate. The lens is wetted slightly before cleaning.

Cleaning Solution for Lenses

Tincture of Green Soap   8
Ammonia   12
Distilled Water   80

Cleaning Bottles

Dirty, small-necked bottles can be cleaned quickly by using copperplated BB shot. The shot is placed in the bottle together with some powdered soap and warm water and shaken thoroughly until the bottle is clean.

Another method, familiar to laboratory workers, uses the same principle for cleaning larger bottles. A short length of chain together with a warm soapy solution is placed in the bottle and shaken vigorously. The chain loosens the hardened, clinging dirt or chemical film to permit removal through the detergent action of the soap.

Non-Corrosive Laboratory Glass Cleaning Solution

To prepare a non-corrosive, alkaline cleaning solution, dissolve the following in water and then make up to one gallon by dilution with water:

Aerosol O. T. (Wetting Agent)	0.95
Calgon or Similar Material	2.6
Dreft (Surface Active Agent)	0.85
Sodium Metasilicate	8.9
Tetrasodium Pyrophosphate	10.7
Trisodium Phosphate	32.0

The best results are obtained in cleaning glassware with a hot solution. The solution may be used repeatedly.

Window and Glassware Cleaners
Formula No. 1
U. S. Patent 1,764,392

A mixture of 100 parts petroleum oil and 33 parts carbon tetrachloride is poured into 3000 parts of 5% soap solution with very good stirring.

No. 2

Methanol	200
Epersol-Y (Hydrolyzed Protein)	10
Water	300
Soluble Perfume Oil	1

No. 3
Canadian Patent 411,330

Isopropyl Alcohol	20–30
Lactic Acid	0.1
Water	To make 100

No. 4

Nekal BX (Wetting Agent)	20
Denatured Alcohol	100
Water	880

No. 5

Borax	1½
Sodium Acetate	1½
Ammonium Chloride	¾
Sodium Sesquicarbonate	10
Sulfatate (Wetting Agent)	1

Dissolve the above solid materials in about 1 gallon of pure hot water, cool, pour into a 5-gallon bottle, add 4 oz. commercial water-glass (sodium silicate) and fill bottle with pure water. This makes the stock solution ready for addition of alcohol solvent and water as needed.

To prepare solution for spray pump bottles place 13 oz. of stock solution and 20 oz. of alcohol in a 5-gallon bottle and dilute to 5 gallons with pure water.

Substitute for Household Ammonia

Use a solution of 3 oz. trisodium phosphate and 5 g. wetting agent (wetanol) per gallon of water.

Saddle Soap

Igepon T (Emulsifying Agent)	15
Water	225
Beeswax	30
Neetsfoot Oil (Cold Pressed)	15
Castor Oil	15

Heat together carefully on a steam bath or a double boiler with good stirring and allow to cool.

Leather Cleaners
Formula No. 1

Castile Soap	6
Water	100

Heat until dissolved, cool and add:

Ammonia Water (26°)   6
Glycerin   14
Ethylene Dichloride   7
No. 2
Powdered Castile Soap   6
Water   160
Boil together until the soap is dissolved, cool and then add:
Ammonia   6
Glycerin   14
Ethylene Dichloride   7
No. 3
Soap Powder   60 g.
Ethyl Ether   60 cc.
Water   1880 cc.
Keep well stoppered in cool place and keep away from open flames. Shake well before using.

Antiseptic Dry Cleaning Fluid
U. S. Patent 2,348,795
Methyl Alcohol   4
Mercuric Chloride or   ⅕
2-Benzyl-4 Chlorophenol   ⅕
Carbon Tetrachloride
                        To make 100

Dry Cleaners' Soap
Soft Soap   60
Diglycol Laurate   7
Ethylene Dichloride   33

Rug Dry Cleaner
Formula No. 1
Water   9
Varsol   27
Diglycol Laurate   1
Diatomaceous Earth   63
To be spread on by sprinkling and then taken up with vacuum cleaner or broom.
No. 2
Water   9
Varsol   25
Sulfatate   3
Diatomaceous Earth   63

Pile Fabric Cleanser
U. S. Patent 2,344,268
Bentonite (200 Mesh)   65
Petroleum Naphtha
  (Light)   25
Wood Flour (60 Mesh)   25

Rug and Textile Cleaning Powder
U. S. Patent 2,344,268
Bentonite (200
  Mesh)   60 lb.
Wood Flour (60
  Mesh)   40 lb.
Stoddard Solvent   4–7 gal.

Jewelry Cleaning Fluid
Ammonium Linoleate S   40
Acetone   60
Ammonia   10
Lavender Oil   1
Water   880

Metal Degreasing Compound
Formula No. 1
Trisodium Phosphate   20
Sodium Metasilicate   10
Sodium Carbonate   10
Sodium Hydroxide   5
Sodium Tall Oil Soap   55
No. 2
British Patent 549,375
Oleic Acid   233 cc.
Sulfonated Castor
  Oil   472 cc.
Sodium Orthosilicate   26 oz.
Water   1 gal.
Sodium Cyanide   2 oz.

Drawing Pen Cleaner
Prepare a "concentrate" consisting of:
Naccanol NR (Wetting
  Agent)   5
Water   95
To make the pen cleaner, as needed, use:

Above Concentrate	2
Water	2
Ammonium Hydroxide	1

Non-Corrosive Tin Plate Cleaner
U. S. Patent 2,285,676

Soda Ash	19.5
Trisodium Phosphate	45.0
Sodium Silicate	25.0
Sodium Silicofluoride	10.0
Sodium Chromate	0.5

Aluminum Cleaner

Trisodium phosphate, 6 lb.; disodium phosphate, 6 lb.; sodium metasilicate, 6 lb.; sodium carbonate, anhydrous, 6 lb.; soft soap, 4 lb.; water, 100 gallons; temperature, 210 to 212°F.

Pre-Anodizing Metal Cleaner
British Patent 549,375

Oleic Acid	233 cc.
Sulfonated Castor Oil	472 cc.
Sodium Orthosilicate	26 oz.
Water	1⅛ gal.

Universal Cleaner

Linseed Oil Fatty Acids	5
Ammonia (26° Bé.)	5
Ethyl Acetate	20
Acetone	10
Tetralin	5
Diglycol Laurate or Butyl Lactate	5
Carbon Tetrachloride	20
Xylol	20
Alcohol	10

Metal-Machinery Cleaner
Formula No. 1
U. S. Patent 2,356,747

Coal Tar Oil, Neutral	40
Monoethanolamine	15
Oleic Acid	15

Ethylene Glycol	15
Ethyl Silicate	½
Phosphoric Acid	½

No. 2

Sodium Hydroxide	30
Sodium Ash	30
Sodium Silicate	20
Yellow Soap	10
Rosin	5
Salt	1½
Water	3½

The above mixture works very well when used in the proportion of 2–5 ounces per gallon of boiling water on dirty, grimy machinery.

Motor Cylinder Cleaner

Concentrated Ammonium Hydroxide	300
Creosote (5%)	100
Alcohol	1600

Remove half of opposite spark plugs. Squirt cleaner in openings, replace spark plugs and run engine until all cylinders fire. Repeat for other cylinders.

Type Cleaner

"Cellosolve"	30
Water	18
Methanol	52

This material can be applied with a small brush or cotton pad, but should be kept sealed when not in use.

Printing-Plate Cleaner

To two quarts each of benzol and alcohol add one-half pound of paraffin wax. Warm this mixture until the paraffin melts and dissolves. When cool, the paraffin will come out of solution giving a rather mushy mixture which may be spread on old engravings with a brush and will not dry for several hours. It is an excellent softener of old ink, and, since it is slow drying, it can be used on press rollers and other articles which are too large to be easily soaked in a solution.

Linotype Matrix Cleaner

Trichloroethylene alone or mixed with benzine in equal proportions is a satisfactory degreasing agent for linotype matrixes. An apparatus for this purpose employs sievelike baskets.

Prior to solvent treatment, the matrixes are first immersed in a soap solution. Following this soap treatment, the matrixes are repeatedly immersed in trichloroethylene or its mixture. For drying the matrixes, the temperature may be increased up to 80–90°C. The trichloroethylene may be recovered, making the soap-solvent degreasing process a fairly economical one.

Rust Remover
Formula No. 1

Hydrochloric Acid (34%)	85
Water	14
Pyridine	1

No. 2
U. S. Patent 2,135,066

The composition consists of a diethylene dichloride-kerosene mixture (70:30) to which are added 50 g. of sodium phosphate per 12 fluid oz. of the first mixture.

Brass Cleaner

Ammonia (28%)	1
Water	1
Denatured Alcohol	5
Lacquer Thinner	3

Brass and Bronze Cleaner

Acetic Acid	6% Solution
Sodium Chloride, to saturate the acid with salt.	

Use soft cloth, brush or cotton soaked in above solution, wash with water after cleansing and dry with dry rag.

Removing Dye Stains from the Hands

First remove all grease by cleaning the hands with an organic solvent, or soap and hot water. Next stain the hands with a strong solution of potassium permanganate. After several minutes, treat the hands with a strong solution of sodium hypochlorite acidified with acetic acid. This will effectively remove both the potassium permanganate stain and most dyestuffs.

Removing Ink Stains from Wall Paper

A solution of equal parts of ammonia and hydrogen peroxide is applied.

Stain Removers

Coffee or Beer

Ammonium Chloride	2
Glycerin	2
Alcohol	2
Water	7

Rinse spot with above mixture, then wash well with water or a 20% soap solution.

Burnt Sugar

Glycerin	10
Water	10
Isopropyl Alcohol	20

After applying above rinse well with water.

Blood

Sodium Persulfate	2
Trisodium Phosphate	2
Hydrogen Peroxide (3%)	3
Water	7

Cadmium, Cobalt, Mercury or Nickel Stains

Potassium
Cyanide   5% Solution
Caution: The above solution is a deadly poison which must be handled very cautiously. Be extremely certain that the soiled garment is washed completely free of the potassium cyanide solution.

Chromium

Sodium Bisulfite 3% Solution
Apply to stain and wash freely with water. This solution will also bleach certain colored materials. It is advisable to test on a small piece of material if the color will be removed, prior to applying in a noticeable spot.

Copper

Potassium
Iodide   25% Solution
Warm the solution slightly, apply and wash freely with water.

Manganese

Soak spot in a 10% solution of ammonium sulfate. Rinse with 1% hydrochloric acid and wash well with water.

Milk

Soak in ether or chloroform, then rinse with a warm 4% solution of borax. Wash well with water.

Mildew

Hydrogen Peroxide (3%)	16
Ammonium Chloride	4
Ethyl Alcohol	10
Water	70

Follow above solution with copious amounts of water.

Picric Acid

Soak in a 20% solution of sodium sulfate. Wash with soap and water.

Urine

Wash in 10% citric acid followed by water.

Scorch

Soak in 1% hydrogen peroxide then rinse in water for one hour.

Egg Yolk

Wash with glycerin followed by an alcoholic soap solution and plenty of water.

Grass Stains

Wash with alcohol or chloroform, let dry, then wash with water to remove residual salts. Use chloroform in well ventilated areas as its fumes are toxic.

Iodine

Apply 10% potassium iodide solution to the stain, followed by a 10% solution of sodium thiosulfate until stain is removed. Wash freely with water.

Iron

Wash stain with an 8% solution of potassium persulfate, followed by water.

Lead

Apply tincture of iodine and allow spot to dry, add 25% solution of potassium iodide until spot is removed. Wash well with water.

Floor-Sweeping Compound
Formula No. 1

Sawdust	20 lb.
Fine Sand	6 lb.
Paraffin Wax	1½ oz.
Light Mineral Oil	10 fl. oz.

Melt the paraffin and add the oil. Work this into the sawdust, then add the sand and mix well.

No. 2

Fine Sand	35
Pine Sawdust	40
Paraffin Oil	15
Water, and Dye	
(If coloring is desired)	10

Some commercial compounds are colored with iron oxide or other pigments and some contain naphthalene flakes, paraffin wax, etc. Essential oils, such as oil of eucalyptus, oil of sassafras, etc., are sometimes added to impart a pleasant odor to the compound or to mask any unpleasant odor of the ingredients used. Pine oil, a small amount of creosote oil, and probably other materials can be used as disinfectants.

The water-wax emulsion type of sweeping compound is an outgrowth of the development of the water-emulsion floor waxes. In this type, the mineral oil is replaced by waxes, resins, water, and emulsifying agents. Instead of a thin film of oil on the floor, a film of wax is deposited. The wax type of sweeping compound is intended for use on floorings, such as linoleum, rubber, asphalt tile, mastic, and polished wood, that may be affected by oils.

Coloring Sweeping Compound

Water- or oil-soluble red or green is used. The water-soluble green is malachite green. The water-soluble red is croceine scarlet. The best oil-soluble green is an alizarine oil green, the red is an azo oil red. Where the water-soluble colors are used, the color is dissolved in water and the sawdust is colored first. Then the oil and sand are added afterwards. If the oil-soluble colors are used, the oil is colored first and then mixed thoroughly with the sawdust and the sand is added afterwards. The whole mass is then thoroughly mixed.

The Cotton Research Foundation, Memphis, Tenn., has developed a sweeping compound consisting of about 95½ parts by weight of cottonseed hull bran and 4½ parts by weight of paraffin oil.

CHAPTER XX

TEXTILES AND FIBERS

Chlorinating Wool
Formula No. 1

For 100 pounds wool: Enter wool for half an hour in cold bath containing 1½ pounds concentrated hydrochloric acid per 10 gallons water. Squeeze gently and work in a bath of the following: For hard wool, 15 pounds bleaching powder and 350 gallons water, and for soft wools 20 pounds bleaching powder and 475 gallons water. Work for half hour, add 3 ounces hydrochloric acid for each 10 gallons of bath and work for ten minutes. Re-enter the wool into the first acid bath to which 8 ounces hydrochloric acid per 10 gallons of water have been added, work fifteen minutes, wash thoroughly in cold water and treat for fifteen minutes in a bath containing 5% of the weight of the wool of sodium thiosulfate at 86°F., and rinse thoroughly.

No. 2

Fifty kg. of a fine wool cloth are treated at ordinary temperature with a filtered solution of 40 kg. of chloride of lime in 1500 liters of water to which an equivalent quantity of hydrochloric acid has been previously added, until the developed hydrochloric acid disappears, which is generally the case after half an hour. The cloth is then abundantly rinsed with cold water. Afterward it is bleached by dipping it into a solution of sodium hydrosulfite or of sulfurous acid and rinsed. Then the bleached fiber is boiled in a solution of 3 kg. of wax soap in 1500 liters of water and rinsed in cold water. The wax soap employed is prepared by saponifying 3 parts of beeswax with 3 parts of solid soda lye. The cloth is then treated for a relatively short time, varying from a few minutes to one-quarter of an hour, according to the thickness of the fiber or other reasons, with a solution of 15 kg. of solid soda lye in 1500 liters of water, wrung out and again copiously rinsed with water. Finally the cloth is boiled in a solution of castile soap to which at the end some acetic acid has been added, dried and calendered.

Lusterizing Wool

To give wool the luster feel of silk, it is entered at 158°F. in a bath containing 1½ to 3 volumes of hydrochloric acid per 1,000 parts of water. A milk of bleaching powder (½ to 1 lb. of bleaching powder to every 5 lb. of yarn) is added, and the yarn is worked about in the bath for three quarters of an hour at the above temperature, after which it is rinsed, a little soda being added, if desired, to neutralize the acid. For light colors, the bleaching powder

(1 lb. per 5 lb. of yarn) should be in the state of a clear solution. Dyeing is best performed in a soap bath, acidified with sulfuric acid; but the wool may also be dyed first in an acid bath, then soaped and scoured.

Textile Dullers (Delusterants) for Hosiery

Formula No. 1

Barium Chloride	40
Dextrin	3
Starch	16
Hydrous Aluminum Sulfate	36
Anhydrous Sodium Acetate	5

No. 2

Barium Chloride	40
Zinc Sulfate (Anhydrous)	16
Dextrin	16
Starch	14
Aluminum Sulfate	14

No. 3

Barium Chloride	48.0
Starch	2.5
Aluminum Sulfate	43.5
Sodium Acetate (Anhydrous)	6.0

All ingredients should be powdered finely and intimately mixed.

No. 4

Sodium Stannate	58.0
Sodium Aluminate	8.0
Titanium Dioxide	2.5
Water	18.5
Glucose	13.0

No. 5

Sulfuric Acid	4	g.
Water	17	l.
Bentonite	3	kg.
Neutral Soap	300	g.
Water	2.7	l.

Finest Titanium Dioxide	6	kg.
Water	1	l.

Mix the first three ingredients, working to a smooth paste. Now mix the next two ingredients separately, heating and stirring until all the soap is dissolved. Finally work the soap solution together with the titanium dioxide into the bentonite paste. Then add the remaining water.

Delustering Shiny Fabric
U. S. Patent 1,942,523

A solution of ammonium-acetate in alcohol is applied to the fabric by means of brush or sponge. The solution is compounded in such proportions or strengths as will best produce the desired result. After application of the mixture, the fabric may be subjected to hot pressing without destroying the newly acquired nap of the fiber.

Delustering Acetate Silk

Treat for ½–1 hour at 90–95°C. in water containing about 1% soap (or a cation-active compound such as cetyl pyridinium bromide) and about 0.2% phenol or cyclohexanol.

Boil-Off of Nylon Fabrics

Method consists of entering the goods into a bath at 170°F. containing about a 20 to 1 ratio of bath to goods, 5% olive oil soap, 2% "Modinal" ES Paste, and one-quarter % trisodium phosphate, all percentages being based upon the weight of the goods. The goods are allowed to soak for 5–6 hours while the temperature of the bath is allowed to slowly drop of its own accord toward room temperature. At the end of this soaking period, the goods are agitated for a few minutes, then enough cold water is added to overflow the tub for a few minutes, thereby floating off any loose solid particles which might otherwise be redeposited on the goods. The bath is then dropped and the goods given a regular rinsing procedure.

Thin Boiling Starch
Formula No. 1
U. S. Patent 2,276,984

Starch is heated at 40–50°C. for 3 hours in aqueous suspension with:

Hydrogen Peroxide (100 Volume)	1.0
Sodium Carbonate	0.2
Copper or Manganese (or Their Salts)	0.1

With suitable pH adjustment other peroxides may be used. The thin-boiling product requires substantially no washing or neutralization.

No. 2
U. S. Patent 2,283,044

Tapioca Flour	247.5
Sodium Borate (1% Solution)	2.5
Sulfonated Castor Oil (1% Solution)	2.5
Water	1000.0

Warm together until gelatinization.

Textile Finish and Size
U. S. Patent 2,275,845

Pearl Maize Starch	20 lb.
Tapioca	80 lb.
Starch-Liquefying Enzyme	1 lb.
Water	60 gal.

The above is heated by live steam to 65–68°C., and again, after keeping for 15–20 minutes, at 100° for 1 hour; softeners and antiseptics are added, and the whole is diluted to 100 gallons.

Sizing Viscose Ribbon

Treat viscose ribbon with an 8% solution of aldehyde-urea resin prepared by adding borax 1 g. and urea 16 g. to 40 g. of neutral 40% formalin, heating to 40°C., letting it stand for 16–18 hrs. at room temperature and bringing the condensate by means of ammonium sulfate to pH 6.6. The sizing is effected at a steam pressure of 1.3 atmospheres and the ribbon is then passed through a drying drum at steam pressures of 2.0–2.5 atmospheres for 7 minutes, and finished in the calender. The ribbon thus treated resembles natural silk.

Sizing Glazed Cotton Yarn

Make a paste with corn starch in the usual manner, afterwards adding to every 20 gal. of paste:

Tallow	7½ lb.
Soft Soap	7½ lb.

together with 5 lb. of Japan wax, and 3 lb. of good clear glue, previously dissolved in a suitable quantity of water. To increase the luster, solid stearine may be substituted for the tallow. For a hard finish, potato starch is to be used in place of the corn starch. Instead of tallow, paraffin wax can be used; in fact, this is preferable, since it obviates any possibility of the size or dressing becoming rancid.

Rayon Size

Sulfonated Castor Oil	15
Glue	30
Sulfonated Wetting Agent (Sulfatate)	2
Sodium Sulfate	8
Water	45

Cover glue with water, heat and stir. When solution is homogeneous add other ingredients, dissolve completely. For sizing use 2–8% of above compound dissolved in water according to working conditions.

Rayon Warp Sizing
U. S. Patent 2,142,801

Water-Miscible Binder (e.g. Dextrin)	3.9–4.6
Softener (Sorbitol)	1.0–1.4

An aqueous mixture of the above is applied; it dries rapidly, and is non-hygroscopic.

Setting Twist in Rayon Knitting Yarns
U. S. Patent 2,340,051

Paraffin wax and glyceryl monostearate in the proportions of 3 parts to 1 by weight are mixed together in the melted state. The required quantity of 0.4% soap solution is then weighed out and heated to approximately 80°C. The hot melted wax-glyceride mixture is then added slowly to the soap solution, vigorous stirring being maintained during this operation. The concentration of the wax-glyceride mixture in the aqueous emulsion is 5% by weight.

Skeins of rayon yarn prior to twisting are immersed in the emulsion for a period of two hours, the ratio of emulsion liquor to weight of rayon to be treated being six to one. The temperature of the liquor is kept at 120°F. during the period of immersion of the skeins. The rayon is then removed, centrifuged and dried. Operations of winding and twisting the yarn then follow. The bobbins or packages of twisted yarn are subjected to a treatment of high humidity at 150°F. for one to two hours. The yarn is then in a condition suitable for knitting. The emulsion remaining in the soaking bath may be re-used by addition of such quantity of fresh emulsion as is required to produce the necessary volume for a subsequent batch of yarn.

Creping Fabrics
British Patent 540,226

Cellulose acetate fabrics containing highly twisted yarns are treated with following at 98–99°C. for 1 hr.:

Soyabean Protein	2–5
Pine Oil	0.2
Tetrahydronaphthalene	0.2
Turkey-Red Oil	0.2
Soap	0.5
Water	To make 100

Scrooping Mercerized Cotton

Any means of imparting scroop to mercerized cotton must be done without oil or soap. Soaking the hosiery in a 5% solution of tartaric acid for half an hour, and then drying, will impart a fine "scroop." A process said to be very good is based upon the use of boric acid. The goods are passed through a bath containing 10 pounds of boric acid, squeezed and dried. It is said that results obtained by this method are quite comparable to silk. Also that colors are not affected if tartaric acid is used.

Silky Scroop on Bleached Cotton Yarn

The dyed yarn, after being rinsed and whizzed, is turned for a few minutes in a cold to lukewarm bath containing ½ to 1 lb. olive oil soap per 10 gal., lifted, and the liquor allowed to drain off a little. It is then entered into a second cold bath containing ½ to 2 lb. acetic acid of 30% or 3 to 8 oz. formic acid of 85% per 10 gal.; or, if a specially permanent scroop be desired, 3 to 8 oz. lactic or tartaric acid per 10 gal. The yarn is worked in this bath for about 10 minutes, whizzed, and dried as hot as possible.

In the case of light shades produced with diamine colors, the soap is frequently added to the dyebath, the goods being then, without rinsing, soured off with acetic or formic acid.

Scroop on Cotton Yarn
Formula No. 1

The dyed yarn after being rinsed and whizzed, is entered into a cold to lukewarm bath containing

½–1 lb. Olive-Oil Soap
                per 10 gallons
treated for a few minutes in this bath, lifted, and allowed to drain.

It is then worked for 10 to 15 minutes in a fresh, cold bath charged per 10 gallons with:

Formic Acid
  (85%)   3–8   oz.
Lactic or Tartaric
  Acid or 5 oz.

Lactic or Tartaric acid for producing a specially good handle   1½–4½ oz.
Glue   1½–2¼ oz.
Potato Starch   1½–2¼ oz.

whizzed, and dried as hot as possible.

The potato starch is first stirred up with cold water and the glue soaked in cold water, whereupon the two are boiled together.

The scroopy feel may be further improved by boiling the yarns with about 5% soda ash and 5% soap previous to dyeing; on bleached yarn a stronger silky scroop is always obtained than on raw yarn.

No. 2

Magnesium Chloride	15.0
Magnesium Sulfate	20.0
Glycerin	5.0
Corn Syrup	42.0
Water	17.8
Formaldehyde	0.2

Dissolve hot; mix well; let cool slowly. This is intended for cotton materials to give a fuller scroop. For application 8–15% of above combination is dissolved in water heated to 120–150°F., and the gravity determined. The cotton material is soaked in this solution. It will take it up in a relatively short time. Then the material is drained, the water pressed out and dried at moderate heat, not exceeding 150°F. For continuous working the operation tank is butted up with the material, preferably dissolved, first in a concentrated stock solution and the concentration of the working tank is frequently checked as to specific gravity.

Scroop Effect on Wool
Formula No. 1

A scroop effect is produced on wool by treatment with a solution of hypochlorite. In practice it is the custom to soap the goods thoroughly, and wash well before chloring. A liquor is made up in a vessel with a capacity of 1000 liters, with 3½ kg. bleaching powder previously dissolved, and passed through a fine sieve. The temperature of the liquor should be about 30°C. The wool is treated in it for half an hour, then washed, and soured in a liquor containing 2½ liters of hydrochloric acid to 1000 liters of water, and again washed. The chloring and the souring are repeated, after which the material is washed and soaped with 3 kg. Marseilles soap to 1000 liters of water, at about 60°C. By now the wool has lost its felting properties and acquired a slightly harsh handle. The dyeing of chlored wool must be carried out with care because of its increased affinity for colored matters, only one-third the amount of the dyestuff required for ordinary wool being needed; with the exception of the amount of dyestuff, the dyeing is accomplished as ordinarily with the acid dyes from a liquor heated to 40°C. The somewhat yellowish tone imparted to the wool by the operation of chloring may be dispelled by treating with bisulfate. After the dyeing, the wool is soaped warm, brightened in a strong solution of acetic acid and dried.

Wool Finish (Scroop)

Magnesium Sulfate	20
Magnesium Chloride	20
Carbamide	35
Water	25

This is a filling for wool materials to give them a fuller scroop.

Emulsion for Finishing Dyed Rayon

Paraffin Wax	10
Stearin	10
Colophony	6

To a fused mixture of the above is added a solution of:

Sodium Hydroxide (40° Bé.)	2
Ammonia (25%) Diluted 1:2.5 with water	6

The mixture is then heated to 80° and water is added to make 100 parts.

Thickening Satin Finish for Calicos

Potato Starch (Farina)	2 kg.
Borax	250 g.
Sulfate of Soda	750 g.
*Soap Solution (See Below)	3 kg.
Water	40 l.

Mercerizing Bath
U. S. Patent 2,345,036

Caustic Soda (32° Bé.)	98–99.7

---

* Soap Solution		
Paraffin Wax	1	kg.
Paraffin Oil	3½	kg.
White Soap	2	kg.
Castor Oil	2½	kg.
Water	6	l.

*Mercerizing
Penetrant   2–0.3

Unkinking Natural Wool
U. S. Patent 2,348,602
Wool is treated with formalde-
hyde in acid medium.

Softener and Lubricant for
Artificial Fibers
U. S. Patent 2,160,458

Acetamide	25
Glycerin	25
Sulfonated Castor Oil	25
Triethanolamine	1
Dextrin	2
Soap	1
Water	23

Softening and Lubricating Wool
U. S. Patent 2,285,357

Methyl Oleate	90
Teaseed Oil	10

This is sprayed on prior to card-
ing or spinning.

Stearic Acid Textile Softener

Stearic Acid	1 lb.
Soluble Oil (Sulfonated)	7 lb.
Ammonia	1 qt.
Water	100 gal.

Melt the acid and pour into hot
water. Then add the ammonia
through a funnel and pipe; mix
thoroughly and add the soluble oil.
If desired, varying amounts of
chip soap may be added to this prepared softener.

Textile Weighter

Sodium Sulfate (Anhydrous)	22

---

*Mercerizing Penetrant

Cresylic Acid	90–97
Diamylphenol	10–3

Epsom Salts	13
Urea	36
Gum Karaya	3
Preservative (Moldex)	$\frac{1}{10}$
Perfume	To suit
Water	To make 100

Wool Oiling Emulsions
Formula No. 1

Oleic Acid	150
Ammonium Hydroxide	5
Water	350

Add ammonium hydroxide to
oleic acid under constant stirring,
add water slowly keeping on stirring until satisfactorily emulsified.

No. 2

Oleic Acid	50
Water	147
Sulfatate (Wetting Agent)	3–5

Dissolve Sulfatate in 10 parts of
hot water under constant stirring,
add oleic acid slowly into solution,
continue stirring until emulsion is
homogeneous. The balance of the
water is added cold. The yellowish white emulsion is fairly stable,
and can be used also after longer
standing, when a slight remixing
is advisable.

The wool should be sprinkled,
before spinning, with this emulsion in the usual way. After spinning it is easily removable.

The use of a stock solution
makes working easier. For this
purpose only one part of the water
is added to the oleic acid, when
emulsified, and the emulsion diluted further just prior to use.

The oleic acid must be carefully
chosen for the above purpose. It
should have a low titer, an average
iodine value (82–84 Wy's) and a
minimum capacity of auto-oxida-

tion to exclude any possible fire hazards.

No. 3

Sulfonated Castor Oil	30
Sulfonated Olive Oil	20
Refined Mineral Oil (Viscosity: 100 S.U.S. at 100°F.)	50

No. 4

Sulfonated Castor Oil	45
Sulfonated Olive Oil	25
Refined Light Mineral Oil	30

No. 5

Sulfonated Castor Oil	30
Sulfonated Olive Oil	20
Olive Oil	50

The olive oil may be replaced with a less expensive vegetable oil with similar characteristics, if it is to be used for cheaper textiles.

These soluble oils give smooth white emulsions with water and are completely removable from the textile material by washing.

The soluble oil is mixed into one part of the water under constant stirring. As soon as a homogeneous emulsion is formed, the rest of the water may be added without stirring. The whole liquid is to be mixed lightly when ready for use.

Textile Softening Cream

Stearine	20
Tallow	24
Coconut Oil	16
Caustic Potash (10% by Weight)	18
Water	22

Melt fats, saponify with caustic potash with constant stirring by using adequate heat; dilute with water while hot. If softer consistency is desired, additional water may be used. It gives a soft finish to all kinds of textiles when used in 1-5% concentration.

This should not separate while standing overnight, and must keep its milky consistency. A gelatinous, soapy compound indicates the use of too much alkali. It must give a neutral reaction to phenolphtalein when tested cold.

For use as a softening agent soak textiles into this textile cream solution at approximately 100°F. for ½-2 hours. Thereafter the water is drained from the goods. Do not rinse. Dry at moderate heat.

Textile Scouring Agents
Formula No. 1

Sulfatate	10
Sulfonated Castor Oil (75%)	20
Oleic Acid	34
Caustic Potash (30° Bé.)	6
Water	30

Mix sulfonated castor oil and oleic acid, add caustic solution under constant stirring. Dissolve Sulfatate in water and blend together.

No. 2

Sulfonated Castor Oil (75%)	13
Oleic Acid	40
Caustic Potash (30° Bé.)	7
Water	40

Blend sulfonated castor oil and oleic acid, add ½ of caustic, dissolve in water, add second half of caustic, stir well, do not heat.

No. 3

Add to 2 parts of sulfonated castor oil and 1 part of sulfonated olive oil, 1 part of a petroleum solvent. Mix in cold, add ammonium hydroxide (sp. gr. 0.910) until it changes to a gelatinous consistency.

Prefabricating Silk Treatment
U. S. Patent 2,290,503

Water	375
Soluble Oil	7
Sodium Bicarbonate	1
Soda Ash	1
Soap	1
Urea	87½
Sodium Nitrate	35

Soak 100 lb. silk in above at 35°C. for 1–2 hours. Centrifuge to rough dry. Silk increases 20–30% in weight on drying.

Improving the Wearing Qualities of Silk Stockings
U. S. Patent 2,295,429

A method for treating fine knitted fabrics, such as silk stockings or the like, to improve wearing qualities and resistance to "runs."

Paraffin	1363	g.
Triple Pressed		
Stearic Acid	227	g.
Lanolin	85	g.
Soluble Oil Base	340	g.
Gelatin	283	g.
Acetanilid	227	g.
Glycerin	453	g.
Technical Dextrin	227	g.
Aluminum Acetate		
(20%)	3620	g.
Hexamethylene-		
tetramine	14.2	g.
Acetic Acid	10.0	cc.
Diastase	7.4	cc.
Water	To make 24.6	l.

Bleaching Jute
Soak the jute for two hours in a bath containing ¾ oz. of water-glass per gallon of water, main-taining the bath at a temperature of 140°F.; then rinse, and bleach at 86°F. in alkaline sodium chloride solution which contains about 1% of chlorine. When taken from the latter bath, rinse thoroughly, sour in a cold hydrochloric acid bath at ¼ to ⅓° Bé., add a small quantity of sulfurous acid, and after half an hour rinse thoroughly.

White Bleach on Union Goods
The best and most permanent white is obtained with peroxide of hydrogen or peroxide of sodium. Prepare the bleaching bath with four to five parts cold water and one part peroxide of hydrogen, adding a little ammonia or silicate of soda to render it slightly alkaline. Enter the well cleaned material, give a few turns, raise the temperature gradually to 40–50°C. (105–120°F.), and leave the goods standing for 6–8 hours, or overnight, care being taken that they are well covered by the bleaching liquor all the time. Then lift the goods and sour off weakly, adding a little bisulfite of potash if necessary, then rinse, and dry slowly.

Peroxide of sodium may be used in the place of peroxide of hydrogen, as follows:

For every 10 gallons cold water add:

Sulfuric Acid	10½ oz.

and after stirring well add gradually:

while stirring continually.

Peroxide of Sodium	8 oz.

Bleaching Textiles
U. S. Patent 2,173,474

Water	1000 gal.

Hydrogen Peroxide
(100 Vol.)   111 gal.
Sulfuric Acid
(96%)   1½ lb.

Immerse goods in the above at 38°C. and squeeze out excess liquor. Allow to remain moist for 2 hr. until bleaching is complete.

Heat Stabilization of Yarn
U. S. Patent 2,278,284

Yarn is impregnated with 8–9% of biuret (on weight of yarn) by passage, as web or thread, through 2% aqueous biuret at 50–60°C.

Crease-Preserving Composition
Swiss Patent 162,987

Paraffin Wax	80
Ceresin Wax	20
Beeswax	14
Montan Wax	5

Melt the paraffin and introduce the three other waxes with further heating. The fabric is impregnated with this composition on its left side and pressed on its right side.

Jute Twine Polishing Size

Starch	8
Glucose	3½
Water	60

Place the starch in a large tub and add cold water gradually. Next add the glucose, stir it and boil for 20 minutes. Then make up:

Borax	1½
Talc	2
Lithopone	3
Gelatin	3
Water	50

Heat the mixture in order to dissolve the borax gelatin. Then combine the two mixtures well, stirring the while, and boil the whole for twenty minutes. Use when cold. Add salicylic acid, 1%, if it is to be kept.

Jute Yarn Dressing

The usual dressing consists of:

Farina or Wheat Flour	90
Tallow or Lard Oil	8
Zinc Chloride	2

When farina alone is used, the mixture should not be boiled but simply raised to boiling point. A very good dressing for jute warps may be made as follows: Steep together for 3 days American sour flour and water in the proportion of 2 lb. of flour to 1 gallon of water. Add 1 lb. of alum and 2 lb. of lard oil for every 20 gallons of steep, and boil the whole together for 1 hour, keeping the mixture in motion while boiling. Another useful dressing for jute warps may be prepared as follows: Place 280 lb. of flour in a 100 gallon cask, and fill up two-thirds with water. Allow the whole to stand for from 8 to 10 days so that fermentation may set in and break up any lumps. This quantity should make three and a half boilings. To each boiling add 3 lb. of bar tallow and boil from 20 to 30 minutes.

Wool-Like Jute

The jute is first treated for from ten minutes to four hours at ordinary temperatures with a hypochlorite solution of 7.4° Tw. or stronger, according to the nature of the material and the degree of bleaching desired. After removing the hypochlorite, as much as possible by pressure, the material is treated at ordinary temperatures for five minutes with caustic soda, 66° Tw. and again pressed. It is then treated for five minutes in a soap foam bath at 97–99.5°C., containing 10 g. soap per liter, washed in water at 50°C., wrung out and dried. The treatment with hypochlorite may be also effected after the caustic treatment or after the soap bath.

Dyeing Sisal and Manila Hemp

When dyeing fiber materials to be used for the manufacture of brushes, etc., necessitating the material being dyed through well, it is best to use a combination of about 2–3% Direct Black and 2–4% logwood extract.

Charge the starting bath with 2% ammonia and ¼–½% soda ash, add 2–3% dyestuff previously well dissolved in condensed water and then about 5% cryst. Glauber's salt; boil up well, enter the material, work for 5–10 minutes, cover with a lattice frame weighted with stones, boil for 2–3 hours, and allow to feed for ¼–½ hour in the cooling bath. Then lift the material, allow it to lie exposed to the air for several hours, and enter into a fresh bath heated to 30–40°C. (85–105°F.) containing pyrolignite of iron of 4–7° Tw.; leave in this bath for ½–1 hour, throw out, and leave exposed to the air for several hours, rinse well, and dry.

If so-called patent or luster-fiber is to be produced, the method of working is exactly as described above; only the fiber is finally taken through a bath of 40–50°C. (105–120°F.) charged as follows:

Soft Soap	2	lb.
Liquor	10	gal.
Gelatin Glue	2	lb.
Logwood Extract	2	lb.
Fustic Extract	½	lb.
Pyrolignite of Iron	½	lb.

Treat the goods in this bath for 30 minutes, allow to drain, and brush dry with suitable brushing machines. If the fiber is not lustered, 8 oz. of whitening per 10 gallons liquor are added to the bath of pyrolignite of iron.

Dyeing Straw Hat Plaits

The plait before dyeing is boiled in water for 1–2 hours: the addition of a little acetic or formic acid improves the color of the straw.

The material is dyed with basic dyes at the boil with addition of 2–5% acetic acid 9° Tw. or 1–3% formic acid 80% until sufficiently level and penetrated, which usually takes about 1–3 hours. It is then left for some time in the bath as it cools down of its own accord. This is particularly necessary with dark shades.

Direct dyestuffs are dyed in the usual way with 1% ash and 20% Glauber's salt fused.

The straw is bleached by steeping the soaked or boiled plait in a bath at about 120°F. into which about 1 lb. Blankit 1 for 10 gallons water has been dredged while stirring. The bath is then slowly heated up to 160–180°F. and the goods left in it for several hours, best overnight. The plait is then rinsed, soured and dyed. The bleaching is done in wooden tubs.

Dyeing Brush Bristles

When dyeing fiber materials to be used for the manufacture of brushes, etc., and necessitating the material being dyed through well, it is best to use a combination of about 2–3% of a Direct Black and 2–4% logwood extract.

Charge the starting bath with 2% ammonia and ¼–½% soda ash, add 2–3% dye previously well dissolved in condensed water, and then about 5% cryst. Glauber's salt; boil up well, enter the material, work for 5–10 minutes, cover with a lattice frame weighted with stones, boil for 2–3 hours, and allow to feed for ½–1 hour in the cooling bath. Then lift the material, allow it to lie exposed to the air for several hours, and enter into a fresh bath heated to 30–40°C. (85–105°F.) containing pyrolignite of iron of 4–7° Tw.; leave in this bath for ½–1 hour, throw out, and leave exposed to the air for several hours, rinse well and dry.

If so-called patent or luster-fiber is to be produced, the method of working is exactly as described above; only the fiber is finally taken through a bath of 40–50°C. (105–120°F.) charged as follows:

Liquor	10 gal.
Gelatin Glue	2 lb.
Soft Soap	2 lb.
Logwood Extract	2 lb.
Fustic Extract	½ lb.
Pyrolignite of Iron	½ lb.

Treat the goods in this bath for thirty minutes, allow to drain, and brush dry with suitable brushing machines. If the fiber is not lustered, 8 oz. of whitening per 10 gallons liquor are added to the bath of pyrolignite of iron.

The dye liquors may be used repeatedly; dyeing in the standing bath requires about ½–⅔ of the stated quantities of dye and logwood extract, equal quantities of soda and ammonia, and about 3% salt calculated on the weight of the goods.

Dyeing Grass

Before dyeing, soak the grass in boiling water to soften it. Then rinse in cold water and squeeze out all the surplus water. Dye the grass in a solution of basic dyes with the addition of Glauber's salt. Rinse well and dry. The grass thus dyed will be brittle, and to overcome this and to soften it and make it pliable, soften it for about one-half hour in a bath composed of two-thirds water and one-third glycerin. Remove the grass, squeeze out the surplus liquor and dry.

Mordanting Rayon

In mordanting either viscose or cuprammonium silks for the basic dyes, the rayon should remain for 2 or 3 hours in a bath containing 2 to 5% of tannin and 1% of hydrochloric acid, on the weight of the goods, at 50°C. (122°F.). The material is then removed, the excess of liquor removed (but not rinsed), and treated for about 20 minutes in a fresh cold bath containing 1 to 2.5%, or about half of the percentage of tannin used, of tartar emetic. More even shades are obtained on viscose mordanted with Katanol than on tannin-antimony mordanted viscose. If particularly fast dyeings are wanted, such as for cross-dyeing, the basic dyes should also have a top mordant by repeating the above process after dyeing. Basic dyes are often used to top the substantive dyes on rayons, thus brightening the shade.

Dyeing Logwood and Acid Colors in Same Bath

Charge the dyebath first with 4% sulfate of iron, 3% sulfate of copper, 15–30% logwood extract, then add sufficient oxalic acid (about 1½–2% of the weight of the goods) to dissolve the precipitate formed in the bath to make the liquor assume a yellowish color, hereafter adding the requisite quantity of acid color in solution.

Enter the wetted out goods at about 60°C. (140° F.), raise in ½ hour to the boil, continue boiling for 1 hour, and exhaust, if necessary, with the addition of ½–1% oxalic acid well diluted with cold water.

After dyeing, rinse very thoroughly, or, if necessary, wash with Fuller's earth and the addition of some acetic acid.

Should a subsequent shading with acid colors be required, the dyebath must first be cooled off somewhat, and then be heated up again gradually after the addition of the dyes.

Aged Black on Loose Cotton

The following solutions are mixed together: 120 parts aniline salt, 40 parts sodium chlorate, 5 parts ammonium chloride, 3 parts copper sulfate. Then 120 parts aluminum acetate, 216.6° Tw. is added and the whole diluted to a density of 12° Tw.

Impregnate the thoroughly wetted-out cotton with this solution, hydro-extract, (the liquor coming from the hydro-extractor is kept), place on hurdles, and dry at 95 to 104°F.

The cotton must be well shaken before it is placed into the stove as well as during ageing. When the cotton is perfectly dry, steam is allowed to pass into the ager for the purpose of completing the aging.

Aging should be completed in from 18 to 24 hours.

Develop a bath, at 104 to 122°F., containing 6% sodium bichromate, 9.5% aniline salt, 2% sulfuric acid, 168° Tw., rinse thoroughly and treat with an emulsion of oil.

Dyeing Wool Bunting Cloth for Flags
Formula No. 1
(Red Shade)
Per 100 lb. Cloth

Du Pont Milling Red SWB Conc. 125%	30 oz.
Du Pont Milling Red SWG Conc. 125%	27 oz.
Glauber's Salt	30 lb.
"Modinal" D Paste (Pat.)	4 oz.
Acetic Acid (56%)	1 qt.

Raise slowly to boil in 1½ hours. Boil ½ hour.

No. 2
(Blue Shade)
Per 100 lb. Cloth

Pontacyl Fast Blue 5R Conc.	13 oz.

Glauber's Salt   30 lb.
"Modinal" D Paste   4 oz.
Raise to the boil in 1 hour.
Shade with Pontacyl Violet 4BL.

Dyeing Aralac
(Casein Fibers)
Formula No. 1
Acid Colors
    Per 100 lb. of Rawstock
Glauber's Salt   10 lb.
Acetic Acid (28%)   2–10 lb.
Volume 500 gal.
Raise slowly to 180–190°F. in 45 minutes.
Run 45 minutes at 180–190°F.
The amount of acid is varied directly as the increase in the depth of shade.
Colors suitable for dyeing by this process:
Milling Yellow 5G Conc.
Milling Yellow GN Conc. 250% (Pat.).
Milling Orange R Conc.
Milling Orange RN Conc. 125% (Pat.).
Milling Red B Conc.
Milling Red 3B Conc.
Milling Red SWB Conc. 125%.
Milling Red SWG Conc. 125%.
Pontacyl Scarlet R Conc.
Pontacyl Violet 4BL Conc. 125%.
Pontacyl Violet S4B.
Brilliant Milling Blue B Conc. 200%.
Pontacyl Fast Blue GB Extra Conc. 125%.
Pontacyl Fast Blue GRB Extra Conc. 125%.
Pontacyl Fast Blue 5R Conc.
Pontacyl Brilliant Blue RR Conc. 200%.
Pontacyl Wool Blue BL Conc. 200%.

Pontacyl Wool Blue GL Conc. 250%.
Brilliant Milling Green B Conc.
Pontacyl Fast Black BBN.
Pontacyl Fast Black BBO.
Pontacyl Fast Black N2B Conc. 200%.
(Use 4 oz. per 100 gallons Modinal DN Paste in the dyebath.)

No. 2
(Chrome Colors)
Topchrome Method
    Per 100 lb. of Rawstock
Glauber's Salt   10 lb.
Acetic Acid (28%)   2–10 lb.
Volume 500 gal.
Raise slowly to 180–190°F. in 45 minutes. Run 45 minutes at this temperature. Add sodium bichromate, ½% for light shades; not more than 1% for heavy shades.

No. 3
Chromate Method
    Per 100 lb. of Rawstock
Sodium or Potassium
    Chromate   ½–¾ lb.
Ammonium Sulfate
    or Ammonium
    Acetate   2–5 lb.
Raise slowly to 180–190°F., run 15 minutes at this temperature, add:
Ammonium Sulfate
    or Ammonium
    Acetate   5–10 lb.
Run ½ hour at 180–190°F.
Colors suitable for these methods:
Pontachrome Fast Yellow R Conc.
Pontachrome Yellow GR.
Pontachrome Yellow GS.
Pontachrome Yellow 3RN.
Pontachrome Yellow SW Conc. 150%.

Pontachrome Orange RL.
Chromate Brown EBN.
Pontachrome Brown G.
Pontachrome Brown HN Conc.
Pontachrome Brown MW Powder.
Pontachrome Brown PG Conc. 125%.
Pontachrome Brown RH Conc.
Pontachrome Fast Red 2RL (Pat.).
Pontachrome Red B.
Pontachrome Azure Blue BR Conc. 200%.
Pontachrome Blue ECR Conc. 200%.
Pontachrome Green G.
Alizarine Blue Black B and BG.
Pontachrome Black F Conc. 125%.
Pontachrome Black FB Conc. 125%.
Pontachrome Black TA.
Pontachrome Blue Black BB Conc. 125%.
Pontachrome Blue Black R Conc.
Pontachrome Blue Black ZF Conc.

At the present time Aralac is largely used in admixture with other fibers.

No. 4

For the piecegoods dyeing of Aralac-wool mixtures:
Heavy Shades

Per 100 lb. of Cloth	
Glauber's Salt	10 lb.
Sulfuric Acid	3 lb.
Modinal DN Paste	10 oz.

No. 5

Light Shades

Per 100 lb. of Cloth	
Glauber's Salt	10 lb.
Acetic Acid (28%)	5–10 lb.
Modinal DN Paste	10 oz.

Volume 500 gal.
Raise slowly to 180–190°F. in 45 minutes, run 45 minutes at this temperature.
The following colors are suitable for dyeing by this process:
Tartrazine Conc.
Pontacyl Light Yellow GG Conc. 125%.
Pontacyl Light Yellow 3G Conc. 150%.
Pontacyl Light Yellow GX.
Orange G.
Pontacyl Carmine 2B.
Pontacyl Carmine 6B Extra Conc. 125%.
Pontacyl Carmine 2G Conc. 150%.
Pontacyl Light Red 4BL Conc. 175%.
Pontacyl Light Red BL Conc. 175%.
Pontacyl Light Scarlet EG.
Pontacyl Fast Violet 10B Conc. 175%.
Pontacyl Violet 4BSN Conc. 125%.
Pontacyl Violet RL.
Anthraquinone Blue B and BN.
Anthraquinone Blue SEN.
Anthraquinone Blue SWB.
Anthraquinone Blue WSA.
Pontacyl Brilliant Blue A Conc.
Pontacyl Brilliant Blue E.
Pontacyl Brilliant Blue V.
Pontacyl Green BL Extra Conc. 200%.
Pontacyl Green NV Extra Conc. 200%.
Pontacyl Green SN Extra.
Pontacyl Black BX.
Pontacyl Black GRF Conc. 125%.
Pontacyl Black RW.
Pontacyl Blue Black RC.
Pontacyl Blue Black SX.

The same procedure and the same list of colors as given under Formula No. 1, are also suitable for the dyeing of mixtures of Aralac and wool.

No. 6

For the Dyeing of Mixtures of Aralac and Cellulose Fibers.

Per 100 lb. of Cloth

Prepare a buffer which is a mix of:

Disodium Phosphate	6
Monosodium Phosphate	1
Volume 250 gal.	
Salt	5–20 lb.
Buffer (Prepared as indicated above)	1–3 lb.

Raise temperature of bath slowly to 180–190°F. in ½ hour, run ½ hour at 180–190°F.

Colors suitable for application by Formula No. 6 are:

Pontamine Fast Yellow 4GL Conc.

Pontamine Yellow CH Conc.

Pontamine Fast Orange RGL Conc. 200% (Pat.).

Pontamine Fast Orange 2GL (Pat.).

Pontamine Fast Orange S Conc. 175%.

Pontamine Fast Orange WS Conc. 175%.

Pontamine Orange DB Conc. 175%.

Pontamine Orange PG Extra Conc. 125%.

Pontamine Orange R Conc.

Pontamine Brown BCW Conc.

Pontamine Brown CG Conc. 150%.

Pontamine Brown D3GN Conc. 125%.

Pontamine Brown NCR Conc. 150%.

Pontamine Brown N3G Conc. 250%.

Pontamine Brown RMR Extra Conc. 125%.

Pontamine Brown XR.

Pontamine Catechu 3G Conc. 200%.

Pontamine Fast Brown 4GL (Pat.).

Pontamine Fast Brown RKL Conc. 125%.

Pontamine Fast Brown SKRL.

Pontamine Fast Brown 3YL Conc. 200% (Pat.).

Pontamine Fast Pink EB Extra.

Pontamine Fast Pink G Conc. 200%.

Pontamine Fast Pink GGN Conc. 125%.

Pontamine Pink 2B Conc. 200%

Purpurine 4B Conc.

Purpurine 10B.

Pontamine Bordeaux B Conc. 150%.

Pontamine Fast Red 8BL Conc. 125%.

Pontamine Fast Red F Conc. 125%.

Pontamine Fast Red FCB Conc. 150%.

Pontamine Garnet R.

Pontamine Red 12B Extra Conc. 200%.

Pontamine Scarlet B.

Pontamine Brilliant Violet B Conc. 200%.

Pontamine Brilliant Violet BN Conc. 200%.

Pontamine Brilliant Violet RN Conc. 150%.

Pontamine Fast Heliotrope B Conc. 200%.

Pontamine Violet N Conc. 150%.

Pontamine Blue RW Conc. 200%.

Pontamine Blue RWG Extra Conc. 200%.

Pontamine Deep Blue BH Conc.

Pontamine Navy Blue DB Conc. 175%.

Pontamine Steel Blue G Extra Conc. 250%.

Pontamine Green BX Conc. 150%.

Pontamine Green 2GB Extra Conc. 150%.

Pontamine Green GX Conc. 125%.

Pontamine Green S Extra Conc. 125%.

Pontamine Green 2Y Conc. 150%.

Pontamine Black BCN Conc. 150%.

Pontamine Black E Double.

Pontamine Black EBN Double 200%.

Pontamine Black EG Extra Conc. 200%.

Pontamine Black RR Conc. 200%.

Pontamine Fast Black GCW Conc. 160%.

Pontamine Fast Black 2GCW Conc. 200%.

Pontamine Fast Black FF Conc. 200%.

Pontamine Fast Black L Conc. 150%.

Pontamine Fast Black LCW Conc. 150%.

No. 7

Dye by the same method as given under Formula No. 6. Rinse, treat in a bath (volume 250 gal.) containing 3 lb. sodium nitrite, 5 lb. sulfuric acid for 15 minutes cold. Rinse and treat in a bath (volume 250 gal.) containing 1½ lb. beta naphthol which has been previously dissolved in a small amount of hot water using 1½ lb. of caustic soda to effect solution.

Colors suitable for application by Formula 7:

Pontamine Diazo Yellow 4G Conc. 200% (Pat.).

Pontamine Diazo Yellow 2GL (Pat.).

Pontamine Diazo Yellow 2GL Conc. 200% (Pat.).

Pontamine Diazo Yellow GM (Pat.).

Pontamine Diazo Orange Conc. 250%.

Pontamine Diazo Orange G Conc. 200% (Pat.).

Pontamine Diazo Orange 3G Conc. 200% (Pat.).

Pontamine Diazo Orange GR Conc. 150% (Pat.).

Pontamine Diazo Orange R Conc. 200%.

Pontamine Diazo Orange RFW.

Pontamine Diazo Orange 2R Conc. 150% (Pat.).

Pontamine Diazo Orange WD (Pat.).

Pontamine Diazo Brown 6G.

Pontamine Diazo Brown R (Pat.).

Pontamine Diazo Bordeaux 7B Extra Conc. 150%.

Pontamine Diazo Bordeaux 2BL Conc. 175%.

Pontamine Diazo Bordeaux RB Conc. 300%.

Pontamine Diazo Red 5BL Conc. 200%.

Pontamine Diazo Red 7BL Conc. 150%.

Pontamine Diazo Red BFW Conc. 175% (Pat.).

Pontamine Diazo Scarlet A Conc. 200%.

Pontamine Diazo Scarlet 2BL.

Pontamine Diazo Scarlet FW Conc. 200% (Pat.).

Pontamine Diazo Scarlet GFW.

Pontamine Diazo Scarlet N Conc.

Pontamine Diazo Scarlet R Conc. 200%.

Pontamine Diazo Violet BL Conc. 250% (Pat.).

Pontamine Diazo Violet RR (Pat.).

Pontamine Diazo Violet RR Conc. 250% (Pat.).

Pontamine Diazo Blue BR Conc. 125%.

Pontamine Diazo Blue 3G (Pat.).

Pontamine Diazo Blue 6G Conc. 200%.

Pontamine Diazo Blue 5GL Conc. 140%.

Pontamine Diazo Blue NA Conc. 200%.

Pontamine Diazo Green BL (Pat.).

Pontamine Diazo Green BL Conc. 150% (Pat.).

Pontamine Diazo Green 3G.

Pontamine Diazo Green 2GL Conc. 200%.

Pontamine Diazo Black BHSW Conc.

Pontamine Diazo Black OB Conc. 150%.

Pontamine Diazo Black RS Conc. 150%.

Pontamine Diazo Black ZV Extra Conc. 150%.

Pontamine Diazo Black ZVN Extra Conc. 150%.

Aralac will frequently be found mixed with various combination fibers such as:

1. Aralac-rayon-cotton
2. Aralac-rayon-wool

3. Aralac-rayon-wool-cotton

When three or more fibers are present in the cloth there is a problem of color selection which requires preliminary experiment using appropriate dyes from the lists previously given. These experiments are conducted in accordance with the method given as formula No. 6.

Basic Chrome Nitrate-Acetate

Potassium Bichromate	300
Boiling Water	300
Nitric Acid (64° Tw.)	360

Add with continual stirring:

Glucose (33%)	90

and then

Acetic Acid (9° Tw.)	500

Allow to stand until the potassium nitrate has crystallized out and dilute to 4° Tw.

Printing Calico with Catechu

Catechu Solution (10%)	75
Starch	10
Gum Tragacanth Solution (6%)	7
Cotton-Seed Oil	2
Sodium Chlorate	2

Boil, cool and add 4 parts acetate of chrome 32° Tw.

The catechu solution is made as follows:

Catechu Cubes	10
Acetic Acid (9° Tw.)	45
Water	45

Textile Printing Paste
U. S. Patent 2,346,041

Ethyl Cellulose	5.00
Naphtha	77.00
Ammonium Oleate	12.18
Oleic Acid	0.70
Water	5.12

The above is mixed with 1.2 times its weight of water.

Silk-Printing Discharge Formula No. 1

Stannous Chloride	15
Methylated Spirits	10
Urea	30
Sodium Thiocyanate	5
Citric Acid	1
Gum Arabic (50%) or GumTragacanth Solution (10%)	29
Water	10

No. 2

Stannous Chloride	10
Methylated Spirits	10
Sym.-Dimethyl Urea	20
Sodium Thiocyanate	5
Gum Arabic Thickening	55

Textile-Printing Resist
Australian Patent 114,954

Albumin	50
Titanium Dioxide	160
Barium Chloride	75
Sodium Sulfate	50
Glycerin	50
Castor Oil	30
Gum Tragacanth Solution (6%)	260
Water	325

The addition of a suitable discharging agent, such as hypo-sulfite, to the resistant produces white effects on colored backgrounds.

Improved Textile Dyeing
British Patent 549,214

The affinity of the materials for various classes of dyes (acid, direct, chrome, S, vat, and dispersed aminoanthraquinone dyes) is improved by the following pretreatment:

Viscose is padded with an aqueous solution (1000 parts, pH 6.4) containing guanidine adipate (30 parts) and 40% aqueous formaldehyde (160 parts) to give a 90–110% by weight increase, dried at about 100°C., and heated at 140°C. for 15 minutes.

Brightening Black Dyeings

This brightening is applied chiefly for blacks; the dyeings not only gain thereby considerably in fullness and depth of shade, but also acquire a much softer handle.

This brightening is also useful if the shades have been dyed too deep and for that reason appear bronzy; in such cases 1½–3 oz. glue previously soaked in cold water are added to the bath in addition to the following weights:

The ingredients per 10 gallons liquor are approximately:

Neutral Soap	3–8	oz.
Olive Oil	1½–4½	oz.
Soda	¾–1½	oz.

To commence with, they are boiled well for 20 or 30 minutes with 1–2 gallons of water as free from lime as possible, and are then added to the bath for which as soft water as possible should also be used.

In this bath the yarns are treated for 15 to 20 minutes, whereupon they are whizzed without rinsing, and dried.

Acetate Rayon "Burnt-Out" Fabrics

First a suitable fabric is chosen. It should contain both acetate rayon and another fiber such as cotton or viscose rayon, these being resistant to the treatment. The resistant fiber should preferably be present in both weft and warp so as to give the final fabric suitable strength. The acetate rayon can be in either warp or weft or both according to the nature of the effect desired.

Such fabric is printed with a pattern using a printing paste which contains not less than 10% of the solvent or disintegrating substance, for example, resorcinol. Two suitable printing pastes are given below:

Formula No. 1

Resorcinol	20
Water	20
Gum Tragacanth Thickening (6%)	60

No. 2

Resorcinol	15
Bentonite	15
Water	70

The printed fabric is dried and then steamed for 10 minutes in a cottage steamer under 10 lbs. steam pressure. The fabric is then run through a brushing machine to remove the disintegrated printed parts of the acetate rayon. It is then found that a voile pattern is produced in the printed parts.

There are further possibilities in this special method of processing. For instance, the fabric may first be dyed and then a discharging substance be applied in the resorcinol printing paste. Alternatively, a suitable color may be added to the printing paste. These variants can be illustrated by the following example:

The acetate rayon mixture fabric is first dyed blue, using dyes capable of being discharged by hydrosulfites. It is then printed in pattern with the following paste:

No. 3

Resorcinol	20
Sodium Sulfoxylate Formaldehyde	10
Water	10
Gum Tragacanth Thickening (6%)	60

Dry and steam as above.

Zinc Dust Resist for Wool

Dyestuff	6	oz.
Water	1½	pt.
Artificial Gum (1:1)	2½-1¼	pt.
Zinc Dust	3¼-6½	lb.
China Clay (1:1)	1¼-2¼	pt.
Make to	1	gal.

This resist, to which all dyestuffs withstanding the action of hydrosulfite may be added, is printed on the white material, dried and covered, generally in blotch work, with colors which can be discharged with zinc-dust. After being well dried again, the material is steamed for one hour with moist steam, then soured cold with dilute hydrochloric acid, 20 cc. hydrochloric acid (36° Tw.) per gallon and well washed.

Reserve for Wool Yarn

Treat 100 pounds yarn in a bath containing 300 gallons water and 10 pounds tannin. Then treat in a hot bath containing 1 pound oxalic acid and 5 pounds tartar emetic for half an hour. Then treat in a bath containing 3 pounds tin salts and 3 pounds hydrochloric acid and dry. Yarn thus treated has little affinity for the usual dyes.

Stripping Dye from Wool
Per 100 lb. of Wool
Formula No. 1

Sulfoxite S Concentrated (Zinc Sulfoxylate Formaldehyde)	1½ lb.
Acetic Acid (28%)	5 lb.

No. 2

Sulfoxite C (Sodium Sulfoxylate Formaldehyde)	3 lb.
Acetic Acid 28%	5 lb.

In either case the chemicals are dissolved in 250 gallons of water. The cloth, previously wetted-out is entered into this bath, the temperature raised gradually to the boil in ½ hour, and boiling continued for ½ hour.

Stripping Dyed Cotton
British Patent 548,490

Cotton dyed with Para Red is stripped by:

Glucose, or Sodium Sulfide	2.0
Sodium Carbonate	1.0
2-Sulfonic Acid Anthraquinone	0.1

at 80° in ¼ hour.

Flameproof, Waterproof and Mildewproof Textile Coating

	Per Cent
Non-Volatile Vehicle	29–32
Pigment and Fillers	28–31
Volatile Vehicle	43–37
Non-Volatile Vehicle:	
Chlorinated Paraffin	22–24
Resin	7–8
Pigment and Fillers:	
Antimony Oxide	7–8
Calcium Carbonate	8–8.5
Iron Oxide	8–9
Coloring Pigments	4–4.5
Mildewproofing Agent	1–1
Volatile Vehicle: (Choice of)	
Ketones	
Mineral Spirits	
Varsol	
Stoddard Solvent	
Xylol	

A formulation such as outlined is usually prepared by grinding the pigment part of the formula in chlorinated paraffin. The grinding may be done either on paint rolls or in a pebble mill, the former seemingly preferred. Not all of the chlorinated paraffin is required to wet down the pigments, the balance of which may be added before cutting the mix with solvent. It is sometimes preferred to withhold the calcium carbonate until after the grinding operation is completed as it is easily mixed into the compound by agitation.

The resin is dissolved in a portion of the solvent. Heat may be used for this operation when a solvent of reasonably high flash point is used. The resin cut is then added to the pigment-chlorinated paraffin mix and this mixture is then cut with solvent to the desired solid content.

The fabric to be treated is passed through a trough containing the impregnating solution and thence through squeeze rolls or over knife edges which remove the excess treating solution. The pickup of compound by the fabric may be controlled by adjustment of the rolls or knives. It is usually desired to obtain a pickup of approximately 50 per cent.

The drying operation consists of removing the solvent by passing the treated fabric through an oven with forced draft, by festooning or over dry cans. The heat treatment also sets up any heat-reactive resins, this way providing the fabric with a dry finish.

Textile Flameproofing
Formula No. 1

Chlorinated Paraffin	15–30
Plasticizer (Aryl Phosphate)	1–5
Zinc or Manganese Borate	10–15
Pigments and Fillers	5–25
Solvent (Naphtha)	25–50

No. 2

Alkyd Resin Binder	10–15
Ethyl Cellulose	2–5
Antimony Trioxide	2–5
Chlorinated Hydrocarbon	10–15
Zinc Oxide	0–2
Magnesium Carbonate	5–10
Pigments (Olive Drab Earth Colors)	10–15
Naphtha	25–50
Aluminum Stearate	5–10
Water	10–15

Such mixtures may be applied to the fabric on a padder. The treatment gives to the fabric both a fire-resistant and a water- and weatherproof finish. The fabric also has, to some extent, resistance to mildew. Mildewproofing agents such as pentachlorophenol may also be added to the mixture.

No. 3

Borax	7 oz.
Boric Acid	3 oz.
Water (Hot)	2 qt.

Note: A suitable wetting agent ("Sulfatate") should be added if new fabrics containing sizing are to be treated.

Articles such as curtains, rugs, fabric towels, draperies, ornamental decorations, pot holders, cotton insulation, etc. are readily rendered fire retardant by treating with this solution. The dry articles may be soaked in the solution, passed through a wringer and hung up to dry. The solution may be sprayed on articles which can not be dipped or soaked. This treatment does *not* resist washing or exposure to rain.

Wood is quite effectively treated by soaking in a hot solution of the above plus a wetting agent.

Fireproof Cloth

Fabric is fireproofed by saturating with a 20% solution of ammonium phosphate, drying, and coating with 3 layers of a mixture containing milk casein 100 parts, water 400 parts, glycerol 130 parts, 25% ammonia solution 10 parts, with intermediate drying periods of 8–12 hours, if drying takes place at room temperature, and shorter periods if higher temperatures are used. If 500 parts of water glass (sp. gr. 1.35–1.40) is added to the above mixture after homogenizing, the preliminary saturation with ammonium phosphate can be omitted. Thin or transparent fabrics can be rendered opaque by the addition of 30–50 parts of zinc oxide or of soot to the above mixture.

Fireproofing Cotton Goods

Unbleached material is first raised, then treated to remove

starchy matter, then washed and dried. The fabric is treated with an aluminate, preferably sodium aluminate, having a sp. gr. of about 1.15 and at ordinary temperature. The cloth is then dried and subjected to the action of more or less pure carbonic acid gas in the presence of moisture, and at about 50°C. The required comparatively small amount of moisture may be left in the cloth when drying it. The time of treatment for a large batch of cloth is about one hour or longer according to the supply of carbonic acid gas. The cloth is then acted upon with a solution which furnishes a supply of carbonic acid. Acid sodium carbonate solution of about 1.18 sp. gr. at about 90 to 100°C. and at ordinary or at a pressure of about 10 pounds per square inch may be used, and the treatment continued for about two or three hours. While the fabric is undergoing treatment with the acid sodium carbonate, liquor carbon dioxide gas may be passed in to replace the carbon dioxide taken up by the material during the treatment. The cloth is then washed and treated with a solution of sodium hypochlorite of 1.015 sp. gr. and at ordinary temperature and pressure. The treatment should be continued for about one hour or longer if necessary to obtain the desired bleaching effect. There may be added to the bleaching solution a proportion of sodium bicarbonate. The cloth may now be washed and dried or dyed the usual way, or treated in any other desired manner.

Nitrated Non-Inflammable
Lace Fabric
U. S. Patent 2,302,107

Cotton fabric is immersed in:

Nitric Acid (97%)	1
Phosphoric Acid (75%)	3

at 38°C. for 1 hr. Neutralize by rinsing in diluted soda ash and wash with water.

Wetting and Conditioning
Agent for Yarns

Formaldehyde	7
Sodium Benzoate	3
Sodium Formate	5
Ethylene Glycol	6
Sulfonated Castor Oil	18
Sulfonated Wetting Agent (Sulfatate)	8
Water	53

Pour all ingredients, except sulfonated castor oil, into the water, stir well to dissolve; the use of moderate heat is permitted. Add sulfonated castor oil to the clear solution.

For wetting and conserving cotton, rayon or wool yarn, use a solution containing 0.5–2% of above combination.

This material serves double purpose by replacing water in the yarn, lost partly at the spinning process, and conserving it by preventing the formation of molds.

It wets the yarn throughout if properly applied by any customary spraying system.

Before its use, the water content of the yarn has to be determined and the required amount of the wetting agent—water solution added. Care shall be taken that the water content of the material after treatment does not exceed the admitted limits for the specific yarn.

Mothproofing Solution
Formula No. 1

A

Santochlor	16
Stearic Acid	1
Carbon Tetrachloride	4

B

Carbowax 4000	6
Triethanolamine	3
Water	5

Solution A, prepared by mixing Santochlor, stearic acid and carbon tetrachloride is added with rapid agitation to solution B, prepared by dissolving the carbowax and triethanolamine in water.

This solution may be brushed on to rugs or upholstery or may be thinned with water and sprayed. Articles properly treated will be mothproof for long periods of time.

No. 2

Citronella Oil	10
Dwarf Pine Oil	3
Wintergreen Oil	1
Monopol Soap	76
Guaiacum Wood Oil	10

The textiles are impregnated by producing a luke warm emulsion consisting of 95 parts of water and 5 parts of the above solution. Textiles are steeped in this emulsion, centrifuged without rinsing and dried under moderate heat.

Mothproofing Emulsion
U. S. Patent 2,351,359

Salicylic Acid	8 g.
Boric Acid	2 g.
Gum Kauri	4 g.
Triphenyl Phosphate	1 g.

14 g. of above mixture is dissolved in:

Cellosolve	100 cc.
Morpholine	10 cc.
Tergitol Penetrant	20 cc.

Then add to it, while mixing:

Water	3900 cc.

Mothproofing for Piano Key Felts

Carbon Tetra- chloride	4	gal.
Paradichlorbenzene	9	lb.
Arsenic Trioxide	1.5	lb.

Rotproofing Fibers
British Patent 551,081

Cotton, wood pulp, paper, wool or silk is treated with 15–20% aqueous sodium naphthenate solution then with 3–7% zinc sulfate solution.

Rotproofing Fabrics
British Patent 557,375

Cotton fabric is impregnated with a hot ¼% water solution of phenyl mercuric acetate to give a 50% weight increase, then treated with a 2½% water solution of salt to give an insoluble precipitate in the fabric.

Improving Absorbency of Towels
U. S. Patent 2,319,822

Towel fabric is treated with:

Slaked Lime	1
Sodium Dihydrogen Phosphate	4
Water	1000

pH not over 6.8.

Porous Waterproofing for Textiles

Sulfate of Alumina	665

Dissolved in:

Water	600

Sugar of Lead 945
Dissolved in:
Water 900

Dissolve each by itself hot, precipitate cold, draw the clear solution off and make to Twaddle 15°. In this manner a standard alumina sulfate-acetate is obtained of which the greater part is deposited on the fiber in drying.

As woolen and half-wool goods still contain some soap from the milling process, a soap passage is as a rule not necessary before impregnating with alumina; otherwise the goods are passed through a weak soap solution (3:1000), squeezed and dried without rinsing.

The goods are impregnated on a hank washing or open width washing machine provided with pressure rollers, by passing the dry goods for one hour through the diluted acetate-sulfate of alumina of 3¾° Tw. (undried goods at 7½° to 15° Tw.). The goods are then slightly centrifuged without rinsing, or squeezed and then dried.

For wool and half-wool goods a single impregnation will suffice in most cases; if a higher grade of waterproof finish is desired, the treatment is repeated, inserting a soap passage if necessary.

Waterproofing Cotton Duck Formula No. 1

	Pounds
Amorphous Mineral Wax or Crude Petrolatum	7½
Yellow Beeswax	1
Refined Bermudez Lake Asphalt	1½

Solvent: 3 gal. Gasoline
and 2 gal. Kerosene

No. 2

Petroleum Asphalt (Medium Hard) Bermudez Asphalt	6
Neutral or Extracted Wool Grease	2½
Lead Oleate, Technical	1½

Solvent: 3 gal. Gasoline
and 2 gal. Kerosene

No. 3

Amorphous Mineral Wax or Crude Petrolatum	8½
Yellow Beeswax	1½

Solvent: 3 gal. Gasoline
and 2 gal. Kerosene

Textile Waterproofing Formula No. 1

Ozokerite	150
Waste Fat	100
Caustic Soda (32.5% Solution)	16–20
Ammonia	15–20
Water	180

Warm together until wax is melted; mix until emulsified. Run cloth through above, then through a solution of ferrous sulfate; then through water and dry.

No. 2
U. S. Patent 2,277,788

An aqueous dispersion which has an affinity for textile fibers is prepared by heating paraffin wax (m.p. about 55°C.) 25 parts at 100°C. for 3 minutes in presence of an aqueous multivalent, water-soluble salt (e.g., aluminum acetate) and a 2.5% aqueous polyvinyl alcohol derivative (sap. val. 80–245, viscosity 20–40 centipoises at 4% concentration at 20°C.).

No. 3
U. S. Patent 2,344,926

Paraffin Wax	3.70
Aluminum Stearate	3.40
Urea-Formaldehyde	
Butanol Ether	2.30
Butanol	3.90
Ethyl Cellulose	0.11
Xylol	0.25
Solvent Naphtha	29.80
Acetic Acid	0.30
Water	To make 100.00

No. 4
U. S. Patent 2,345,142

Basic Aluminum	
Formate	30 kg.
Water, Hot	60 l.
Paraffin Wax	20 kg.
Mineral Oil	10 kg.
Oleic Acid	3 kg.

Mix until uniform.

For use dilute 1 of above with 25 hot water.

No. 5

(1) Double-boiled linseed oil. 100 parts is boiled with 5 parts of shellac for about 10 min. The filtrate is brushed onto the cotton fabric which is then dried in the sun. When dry a second coating is applied. (2) The fabric is left for 24 hours in a saturated solution of aluminum acetate. It is then treated for 3 hours in a steam chamber or boiled 2 hours in the same bath. The fabric is dried in the air and kept immersed for 1 hour in a solution of 8% soap and 2% glue or gum at 80–90°. The fabric is rinsed in water, dried in a drying chamber and calendered.

Water-Repellent Finish for Ribbons

Sago Starch	9
Powdered Aluminum	
Formate	1

Making Casein Fiber Resistant to Hot Water

The fiber is soaked in aqueous formaldehyde at pH 4–5 for 8 to 10 hours, washed and then soaked in 0.9% chromic sulfate at 45°C. for 45 min. The strength of the resulting fiber is 70% that of wool, and it is highly resistant to the action of hot water. The fiber may be dyed in the same way as wool.

Waterproofing Surgical Dressings

A waterproofing process for cotton dressings which can withstand sterilizing by treatment with superheated steam.

The cloth is soaked 24 hours in a saturated aluminum acetate solution, then treated in a steam chamber for 3 hours. After drying in air it is immersed for one hour in a very hot solution of soap (8%) and glue or gum (2%). The dressing is then rinsed with water, dried in a drying chamber, and finally calendered.

Increasing Wet Strength of Yarn
British Patent 550,458

Saponified acetate yarn is scoured, dried, impregnated at 40°C. for ½ hour with a solution of:

Beeswax	10
Acetic Anhydride	5
Benzene	85

Squeezed to 100% by weight increase, dried, and heated for ½ hour at 100–110°C.

Manufacture of Organdie Cloth
U. S. Patent 2,150,825

Cotton goods are treated for

18–20 seconds in a bath containing:

Sulfuric Acid (d. 1.570) 88
Phosphoric Acid (d. 1.575) 12

As retarder; wash, and mercerized with aqueous sodium hydroxide.

Metallized Yarn

The yarn is passed through a bath consisting of:

Gelatin 25
Metallic Powder 25
Water 25

After drying for about twenty minutes the yarn is passed through a bath made up as follows:

Casein 15
Borax 5
Water 80
Metallic Powder 30

After drying a second time, very rapidly, the yarn is passed through a second bath or in a bath of the same composition. The weight of the metallic powder used varies according to the specific gravity and the nature of the material. The effect can be varied by adding different colors to the last bath.

Drum Head and Banjo Cloth
U. S. Patent 2,330,441

Silk is impregnated with a thin solution of rubber cement and dried. One side is then coated with the following solution:

Turpentine 10
Ether 2
Alcohol 5
Beeswax 5
Camphor 5

Then dried.

Asbestos Cloth Substitute

Fabric is fireproofed by saturating it with a 20% solution of ammonium phosphate, drying and coating with three layers of a mixture containing:

Milk Casein 100
Water 400
Glycerin 130
Ammonia (25%) 10

Intermittent drying periods are necessary between the application of each layer. Periods of 8 to 12 hours are needed if drying is done at room temperature, but the time may be shortened if higher temperatures are used. If 500 parts of water glass (sp. gr. 1.35–1.40) is added to the above mixture after homogenizing, the preliminary saturation with ammonium phosphate can be omitted. Thin or transparent fabrics can be made opaque by adding 30 to 50 parts of zinc oxide or soot to the above mixture.

Chapter XXI

MISCELLANEOUS

Separation of Butadiene
British Patent 547,730

Butadiene is removed from gaseous mixtures by absorption in an absorbent solution, containing copper chloride at 10°C., the solution being afterwards heated to 65.5–82° to expel butadiene, which is reabsorbed in a second tower. Any residual gases are recycled in the first stage of the process and finally recovered by heating the second solution.

Suitable absorbent solutions consist of:

Formula No. 1
Copper Chloride	20.0
Ethanolamine	20.0
Ethanolamine Hydro-chloride	32.5
Water	27.5

No. 2
Copper Chloride	20
Glycol	50
Water	30

No. 3
Copper Chloride	20
Ammonium Chloride	20
Water	30

No. 4
Copper Chloride	16.7
Formamide	50.0
Hydrochloric Acid	12.0
Water	21.3

Non-Caking Cracking Catalyst
U. S. Patent 2,280,060

A gel-type catalyst, for hydrocarbon cracking, which has a diminished tendency to become contaminated with coke contains:

Aluminum Oxide	42.5–88.2
Chromic Oxide	8.5–49.0
Cupric Oxide	2.0–15.0

Deodorizing Sulfurous Petroleum
U. S. Patent 2,160,116

Mix with:

Glucose	2
Hydrochloric Acid (d., 1.19)	2

To precipitate odoriferous mercaptans.

Stabilized Kerosene
U. S. Patent 2,165,261

Add 0.08% by volume of acetone.

Anti-Knock Motor Fuel
U. S. Patent 2,145,889

Each gallon of fuel contains 0.5–15 cc. of a mixture of 40–50% by weight of a carbonyl of iron or nickel and 10–40% by weight of a fatty acid ester of a polyhydric ether soluble in the fuel, that is, diglycol stearate.

Aviation Fuel (Non-Clouding)
British Patent 546,998

Ethanolamine	10
Phenylethanol	20
Oleic Acid	55
Ethyl Alcohol,	

Absolute   300
Gasoline, Aviation   4700

The above mixture can take up 0.4% of water at 15°C. without separation into layers and about 0.2% of water at –40°C. without clouding, crystallization, or any separation.

Stabilizing Tetraethyl
Lead Gasoline
U. S. Patent 2,155,678

A composition (containing tetra-ethyl lead) normally unstable and liable to cloud-formation is stabilized and clouding prevented by addition of lecithin (8–10% by weight of the tetraethyl lead).

Colloidal Fuel

Powdered Coal   40
#6 Fuel Oil   60

Mix well. Use through a stream atomizing type burner.

Refinery Gas Polymerization
Catalyst
U. S. Patent 2,158,154

Natural Phosphate Rock   40
Carnotite Ore   40
Zinc Orthophosphate   10
Barium Chloride   6

Improving Fuel Oil
U. S. Patent 2,146,742

To a petroleum distillate is added 5–11 lb. of 2,4-dinitrophenol per 5000 gallons of distillate.

Easily Inflammable Coke
British Patent 547,451

Coal (60 Mesh)   45
Water   48

Tar   6
Caustic Soda   1

Coke is impregnated with above mixture by subjecting it alternately to a vacuum and atmospheric pressure.

Engine Carbon Solvent
U. S. Patent 2,347,983

Cresol   67–50
Dibutyl Phthalate   33–50

Motor Gum Solvent
U. S. Patent 2,236,590

Butyl Stearate   50–90
Orthodibutyl Phthalate,
  or Ethoxybenzene   50–10

0.1–1.0% of the above is added to a motor fuel. The composition has gum-solvent properties at elevated temperatures.

Brine Cooling Fluid

Sodium Chloride   300
Ammonium Sulfate   55
Water   1000

The above mixture solidifies completely at –25°C.

Plugging of Brine Bearing Earth
U. S. Patent 2,156,219

A 15% aqueous solution of lead nitrate is introduced into brine-bearing strata to precipitate lead chloride within the pores, thereby plugging them and keeping out unwanted brine. A 16.7% by weight aqueous solution of magnesium chloride is forced under pressure into water-bearing strata and this is followed by a 33% by weight solution of sodium hydroxide, whereby magnesium hydroxide is precipitated and the precipitate coagulates in the pores, thus blocking the flow of water.

Petroleum Desalting and Demulsifying Composition
U. S. Patent 2,153,560

Napthenic Soap	50–80
Formaldehyde	10–20
Glycerin	10–20

Removing Metal Obstructions from Oil Wells
U. S. Patent 2,152,306

The following solution acts quickly:

Hydrochloric Acid	28
Nitric Acid	5
Copper Chloride	2
Water	To make 100

Freeing Stuck Tools in Oil Wells
Calcium chloride (up to 4 lb. per gal.) or hydrochloric acid is used.

Preserving Gelatin Solutions
Depending upon the hydrogen-ion concentration of the solution, different agents may be used as preservatives. For acid gelatin solutions use:

Sodium Benzoate	0.1
Thymol	0.1
Sodium Salicylate	0.1
Cresol	0.4
p-Chloro-m-xylenol	0.1
Oxyquinoline Sulfate	0.1
Alcohol	8.0
Ethyl Oxy-Benzoate	0.15
Propyl Oxy-Benzoate	0.15
Butyl Oxy-Benzoate	0.15

For gelatin solutions of the alkaline type use:

Thymol	0.1
Chloro-Thymol	0.1
Chloro-Butanol	0.5
Beta-Naphthol	0.2
Phenol	0.5
p-Chloro-m-Xylenol	0.1

Alcohol	8.0
Ethyl Oxy-Benzoate	0.15

Deodorizing Isopropyl Alcohol
Add 1% decolorizing carbon and warm gently for an hour. Allow to stand overnight, add a little magnesium carbonate or talc and filter through filter paper. Allow to age with added ingredients before shipping.

Filter Aids
Filter aids are used to prevent clogging of filter pores with colloidal or other fine particles which slow up or stop filtration.

In filtering oils or fruit juices ½–1% of any of the following serves as filter aids:

Kieselguhr
Bleaching Earth
Fullers' Earth
Florida Earth
Kaolin
Asbestos

Heat-Generating Composition
Formula No. 1
U. S. Patent 2,261,221

Pumice	6.0
Kaolin	4.0
Potassium Chlorate	1.5
Aluminum	3.0
Brass	2.0
Cuprous Oxide	1.0
Linseed Oil	0.1
Toluene Sulfonic Acid	0.6

Each of the ingredients should be in finely powdered form and they should be mixed together carefully with a minimum of friction. This mixture generates heat when moistened with water.

No. 2

Aluminum Powder	20–40

Sodium Thiosulfate	50–500
Maleic Acid	50–150
Copper Oxide	40–100
Fullers' Earth	500–1000

No. 3

Aluminum Powder	17
Copper Carbonate	33
Oxalic Acid	25
Barium Chloride	25

Filler for Taped Electrical Joints
U. S. Patent 2,326,085

Plaster of Paris	25–50
Putty (Linseed Oil and Clay)	25–75

Solid Dielectric
U. S. Patent 2,340,644

Diphenyl Benzene	2
Diphenyl	1
Naphthalene	1

Dielectric (Electrical Insulating)
U. S. Patent 2,341,760–1
Formula No. 1

Chlorinated Diphenyl Benzene	97–40
Hydrogenated Castor Oil	3–60

No. 2

Chlorinated Diphenyl Benzene	60–99
Chlorinated Ortho-nitrodiphenyl	40–1

Electrical Capacitator
(Condenser) Impregnant
U. S. Patent 2,168,156

Tetrahydrofurfuryl Alcohol	64
Ammonium Borate	36

If desired add:

Acetamide	16

Temperature Compensating
Resistor
U. S. Patent 2,264,073

Compensating resistors of high negative temperature coefficient comprise:

A. Tellurium containing silver, 15%, made by adding silver to molten tellurium, cooling, and annealing the alloy at 115–125°C. for about 15 hours.

B. A cadmium-antimony alloy containing equal atomic proportions of cadmium and antimony.

Fluorescent Lamp Coating
U. S. Patent 2,176,151

Cadmium Silicate	50.0
Zinc Beryllium Silicate	50.0
Manganese	0.8

Vacuum Tube Getter
U. S. Patent 2,167,762

Potassium Uranium Fluoride	95–94
Phosphorus	5–6

Hydraulic Fluid
Formula No. 1
U. S. Patent 2,345,586

Glycerin	30
Propylene Glycol	9
Isobutyl Alcohol	61
Borax	0.5–3

No. 2
German Patent 711,381

Mineral oil containing 2% stearic acid, wool fat or its acids or alcohols or fatty acid acids of the latter is used where it must be exposed to air and where rapid flow is necessary.

No. 3
U. S. Patent 2,337,650

Butyl Alcohol	50
Glycerin	5

2-Methyl 2,4-Penta-
nediol   35–40
Castor Oil   <10

Door Check Fluid

Glycerin   67.0
Potassium Linoleate   1.2
Water   31.8

Magnetizable Fluid
U. S. Patent 2,149,782

Mercury   83 cc.
Steel Dust   14 cc.
Graphite   3 cc.

This is used as an indicating liquid that can be reset by a magnet.

Manometer Fluid
British Patent 547,447

Diethyl phthalate with or without an oil soluble dye, to color it.

Anti-Incendiary Material
British Patent 545,514

Comminuted cellulosic material (sawdust) is impregnated with an aqueous solution of:

Calcium Chloride   18.0
Zinc Chloride   5.0
Ammonium Chloride   4.0
Borax   1.0
Naphthol Green   0.3

Magnesium and Aluminum Powder Fire Extinguisher
U. S. Patent 2,307,083

Tricresyl Phosphate   2–20
Powdered Graphite   80–98

Dry Fire Extinguisher

Fine Granular Silica   70
Sodium Bicarbonate   25
Armenian Bolus   5

Fire Foam Extinguisher
U. S. Patent 2,289,688

A

Aluminum Sulfate   46.5
Water   53.5

B

Sodium Bicarbonate   10.00
Sodium Lauryl Sulfate 0.17
Borax   0.15
Water   89.68

Fire Extinguisher Foam Stabilizer
Formula No. 1
British Patent 560,354

Beet or Turnip Roots 100
Caustic Soda Solution
(66–76%)   7.5

Boil for thirty minutes; press out liquid and use dry residue for spraying.

No. 2
Canadian Patent 417,315

Neutralized Hydrolyzed
Fish Scale Proteins
(35% Solution)   3683
Monobutylether of
Hydroxybiphenyl   414
Butyl "Cellosolve"   250
Water   1700
Cresylic Acid   25
Sodium Dichromate   12½

Bonded Fibrous Insulation
U. S. Patent 2,163,567

Insulating fibers, e.g., asbestos are impregnated with a hot solution of:

Paraffin Wax   5
Mineral Oil   20
Rosin   70
Slaked Lime   5

Heat-Exchange Medium
U. S. Patent 2,276,120
Formula No. 1

Stannous Chloride 56.1 mols

Zinc Chloride   43.9 mols
(Melting Point 171°C.)
No. 2
Cuprous Chloride   21.7 mols
Stannous Chloride 78.3 mols
(Melting Point 172°C.)

Diamond Dust Brooch
U. S. Patent 2,344,024
Diamond Dust   25–50
Alumina   36.8–75
Bentonite   0–10
Zinc Silicate   0–6
Make into a heavy paste with water, mix and dry in several stages. Extrude into rods and fire at 2400°F. for 1–2 hrs.

Clutch Plate Studs
U. S. Patent 2,284,785
Formula No. 1
Molybdenum Sulfide   69
Graphite   14
No. 2
Graphite   10
Silica   10
Felspar   10
To either of above add 25 of the following binder:
Boiled Linseed Oil   32
Rosin   67
Mineral Oil   1
Bake at 345–400°C., reaching this temperature in three stages.

Brake and Clutch Frictional
Material
U. S. Patent 2,284,785
Molybdenum Sulfide   50–80
Graphite   10–20
Linseed Oil   25

Sewage Treatment
U. S. Patent 2,268,647
Precipitation of suspended and colloidal matter from sewage is effected by making the sewage slightly alkaline, then adding 8 parts per million of a mixture comprising approximately:
Mercurous Oxide   4.0
Potassium Chromate   11.0
Zinc Oxide   8.5
Acid Sodium Carbonate   68.0
Sodium Carbonate   8.5

Water Disinfectant
Calcium Chlorite   10 g.
Hydrochloric Acid
(25%)   3 g.
Use this amount per cubic meter of water to kill bacillus coli.

Carbonaceous Zeolite
(Ion-Exchange Composition)
U. S. Patent 2,170,065
Lignite (powdered) is mixed with an equal weight of 70% sulfuric acid and heated gently for 1 hr.; then more strongly. Wash out acid and dry.

Cation Exchange Water Softener
U. S. Patent 2,294,764
Semi-Hard Coal   100
Anhydrous Iron Chloride 80
Heat at 280–315°C. Cool when reaction is complete. Wash with water; then with 2% caustic soda solution, with 2% sulfuric acid; then with water and dry.

Water Impurity Coagulant
U. S. Patent 2,284,827
Sodium Aluminate   1
Bentonite   4
Mix and add to water at rate of 0.05–0.5 lb./1000 gal.

Coagulant for Hard Water
U. S. Patent 2,152,942
Water having more than 100 p.p.m. calcium carbonate is treated in order indicated, with:

1 Calcium Oxide   5   g./gal.
2 Ferrous Sulfate 0.7 g./gal.
3 Aluminum
    Sulfate   0.4 g./gal.
As primary coagulants
4 Ferric Sulfate   0.22 g./gal.
As a secondary coagulant.

Water Purifying Compound
U. S. Patent 2,245,495
Aluminum Sulfate   666
Anhydrous Barium
    Peroxide   169
Manganese Dioxide   90
The manganese dioxide, which catalyzes the liberation of oxygen from hydrogen peroxide formed, may be replaced by potassium permanganate.

Water Softener Briquette
British Patent 548,290
Soda Ash   80.20
Sodium Aluminate   10.26
Trisodium Phosphate   5.58
Water   57.00
Make into a slurry; run into forms and dry gently.

Water Corrective
Phosphoric Acid
    (75%)   3200
Sulfuric Acid (95%)   276
Lactic Acid (50%)   403

Boiler Scale Compound
Formula No. 1
U. S. Patent 2,156,173
Disodium Phosphate   55
Sodium Bisulfite   45
Fuse together and form into bricks.
No. 2
U. S. Patent 2,291,146
Soda Ash   7
Trisodium Phosphate   5

Borax   3
Water   85
This will maintain constant alkalinity in boiler water.
No. 3
U. S. Patent 2,291,146
Soda Ash   7
Trisodium Phosphate   5
Borax   3
Black Molasses   1–2
Water   To make 100
This will keep alkalinity almost constant at all temperatures.

Processing Tannin for Boiler
Water Treatment
As the tannin is delivered it is not very soluble for use as a water treatment chemical. Hence, it should be processed, whether it is Cutch, Quebracho, Chestnut Tannin or a mixture. The process involved, merely consists of adding caustic soda with a small amount of water, then diluting the mass after the reaction has subsided.
The proportions are approximately as follows:
Powdered Tannin
    Extract (Cutch, Que-
    bracho, or Chestnut
    Tannin)   50
Flake Caustic Soda   25
Water   100
First obtain a strong metal oil drum with only one head, clean it thoroughly with hot water, or steam, and then add the 50 parts of tannin extract and half the water (50 parts). Mix to a thick mass with a wooden paddle and then add the caustic soda, about 1 to 2 parts at one time. Very soon the mass will begin to get warm on stirring. Continue to add the caustic soda a little at a time until

all of it has been added, stirring constantly. Perhaps the solution will bubble up violently and if so add some of the remaining water to cool the mass. When all the caustic has been added, pour in the rest of the water, stir very strongly for five minutes, and let stand and cool 24 hours.

This material can be used in conjunction with soda ash to treat boiler water, and the tannin liquid should be used at the rate of about ¼ lb. per thousand gallons of water.

If on testing the boiler water which contains this amount of tannin one finds it a deeper color than a light amber, decrease the amount of tannin liquor to about ⅕ or even ⅙ lb. per thousand gallons of make up water.

Locating Leaks in Pipe Lines Under Water

Make up 0.1% Uranine dye solution in water.

Pump into pipe line.

Characteristic fluorescent green appears in the water over the leaks.

Radiator Rusting Inhibitor
Canadian Patent 411,765

Sodium Nitrite	0.1–1.0
Sodium Arsenite or	
Sodium Phosphate	0.1–1.0

Radiator and Water Boiler
"Stop-Leak"
U. S. Patent 2,293,546
Formula No. 1

Soya Bean Oil	1.0
Glycerol	0.2
Rosin	7.5
Monoethanolamine	1.2
Water	90.1

The soya bean oil and glycerol are heated together at about 235°C. to 245°C. for such a time as is required to insure complete reaction to form a product which is substantially the mono-ester. To this product is then added the rosin and the temperature raised to 255 to 265°C., where it is held for such a time as is necessary to insure substantially complete reaction between the rosin and the free hydroxyl groups of the mono or diglycerides. The common practice of adding about 0.3% of lime as a catalyst for the above reactions aids in shortening the cooking time to about 20 minutes for the initial reaction, 15 to 30 minutes for the final one. The progress of the final reaction is readily followed by a determination of the acid number of the product which, in this example, should be between 120 and 130. The molten resin may then be dispersed with stirring in hot water (between 70 to 95°C.) containing the monoethanolamine.

The resulting concentrate may be added to a leaky cooling system in such quantity as to give a preferred final concentration of around 0.2 to 1% of the resinous material in the system. Through evaporation of the water and the presence of heat a firm, tight seal is gradually built up at the point where leakage occurred. Smaller or greater percentages of stop-leak may be used; smaller percentages merely lengthen the time necessary to seal the leak, while larger percentages provide faster action, but are usually unnecessary.

No. 2

The composition of No. 1 to which is added 1 to 5% of wood flour.

No. 3

The composition of No. 2 in which finely divided asbestos is substituted for wood flour.

No. 4

The composition of No. 2 in which aluminum powder is substituted for wood flour.

Auto Anti-Freeze
Formula No. 1
British Patent 554,607

Powdered Glue	60 g.
Sodium Nitrate	6½ lb.
Water	1 gal.

No. 2
U. S. Patent 2,176,492

Sodium Nitrite	½–3½
Butyl Stearate	½–5
Alcohol	To make 100

No. 3

This antifreeze will not corrode the cooling system, and is perfectly safe all winter unless the motor becomes overheated.

Triethanolamine	5 fl. oz.
Light Mineral Oil	15 fl. oz.
Carbitol	15 fl. oz.
Methanol	192 gal.
Isopropyl Alcohol	797 fl. oz.

Yield 5 gal.

Three parts of this anti-freeze and four parts of water will protect the cooling system to minus 10°F.

Non-Foaming Anti-Freeze
U. S. Patent 2,298,465

Ethyl Oleate or Phenyl Stearate	0.3–0.8
Ethylene Glycol	To make 100

Airplane D-Icing Coating
U. S. Patent 2,346,891

Exposed aircraft surfaces are first painted with a solution of polymerized glyceryl phthalate in acetone or other solvent, then coated with a solution of:

Glycol Stearate	4 lb.
Ethylene Glycol	1 gal.

This solution is made by warming at 70°C. It is then cooled to 45°C. and stirred intermittently until cold. The surfaces of the plane must be repainted with the latter solution from time to time, as needed.

Windshield Defroster
Formula No. 1

Ethylene Glycol	35.0
Glycerol	35.0
Ethyl Alcohol	20.0
Water	9.9
Aerosol (10% Solution)	0.1

No. 2

This material is a liquid which is applied with a cloth to the inside surfaces of windshields, windows, etc. to prevent fogging. Spray or wipe on and polish dry with a clean, dry cloth.

Aerosol OT (Aqueous 10%)	1	fl. oz.
Glycerin	1	pt.
Alcohol	4½	pt.
Acetone	½	pt.
Water	8	pt.

Mix together.
Yield 1¾ gal.

Cathode or X-Ray Tube Screen Coating
U. S. Patent 2,169,046

Zinc Oxide	35
Silica	19
Magnesia	1

Barium Carbonate   2
Mix together with solution of:
Manganese Chloride   ½
Water   100
Then add:
Hydrofluoric Acid (48%)   20
Evaporate mixture to dryness
and fire at 1000°C.

X-Ray Protective Composition
U. S. Patent 2,315,061

To liquid latex or compounded
rubber is added a composition comprising the following:

Lead Monoxide   78
Barium Sulfate   14
Bismuth Oxychloride   8

Sufficient water to produce a
paste.

When. mixing the above composition, the water should have a
temperature ranging from 50 to
65°F. Very thorough agitation is
essential to insure a smoooth product and any mechanical mixer
may be employed.

When liquid latex is used to receive the composition, produced
from lead monoxide, barium sulfate, bismuth oxychloride and
water, 3 parts of liquid latex
should be used to 1 part of the
composition. The . composition
should be added to the liquid latex
very slowly and with gentle rather
than violent stirring, to the end
that bubbles may be avoided and
that no gases are liberated from
the latex.

The lead monoxide, barium sulfate and bismuth oxychloride is
dispersed throughout the solution
after which it is spread to form a
sheet or employed in the production of gloves, for example, by the
application to a form through dip-

ping. Relatively thin coats of the
composition establishes a protective covering and when the material is from 0.03 to 0.04 inch in
thickness, its repulsive effect is the
same as ½ millimeter of lead.

When the above ingredients are
mixed with compounded rubber,
2 parts of compounded rubber in
liquid form, to 1 part of the mixture will produce a heavier material capable of withstanding
shearing and bending stresses so
that aprons, gloves or gowns made
of compounded rubber and the
said mixture, may be employed
commercially in the handling of
machine parts, castings, tires or
other products where X-rays are
used for inspection or the like.

When the latex and mixture
composition of lead monoxide, barium sulfate and bismuth oxychloride is spread to produce a sheet
or placed upon a mold, curing is
accomplished by confining the substance in a compartment having a
temperature of from 100 to 125°F.
for at least 15 min., after which
tempering in a water bath at
160°F. should take place. When
"building up" a glove, for example, the mold is first dipped into
the material and then cured as just
mentioned, after which additional
dippings and curings may take
place until the desired thickness
is obtained.

When using compounded rubber, the coagulation method of
curing is preferable—the dipped
form or sheet of material is submerged in a solution composed of
1 part of methyl alcohol to 9 parts
glacial acetic acid. The material
is left in this coagulation solution

from 2 to 3 min., after which the excess moisture is evaporated by the application of warm air. The material is next placed in boiling water or confined in an oven having a temperature of in excess of 210°F. for not less than 40 min.

Polarity Test Paper

Dissolve 1 g. phenolphthalein in a small quantity of alcohol. Add this solution to a 10% solution of potassium chloride in water. Soak filter paper in this solution and dry. A strip of this paper, moistened in water and placed in contact with the two terminals will show red at the negative terminal. Caution: Do not use with high voltages.

Animal Tissue Fixative

Picric Acid	5
Isopropanol	55
Acetone	30
Acetic (Glacial)	5
Formaldehyde (40%)	5

The length of fixation depends, as with other fixatives, on the size and nature of the tissues involved. From two hours to four days is recommended. Tissues have been left in this fixative for several days without apparent harm.

The tissues that are not imbedded in paraffin are stored in 70% isopropanol.

Since this solution fixes and dehydrates at the same time, it permits a direct transfer from the fixative to isopropanol. In general practice, tissues are trimmed and placed in the labeled cheesecloth "tea" bags in which they are transferred from one solution to another and through the paraffins until imbedded.

After fixation the tissues are washed in two changes of isopropanol (nearly absolute), one to two hours in each change. Then they are passed through three changes of dioxane, one to two hours in each change. The tissues are usually left overnight in the third change of dioxane. Infiltration is begun with two hours in a ⅓ dioxane-⅔ paraffin mixture and completed in three changes of pure paraffin, one half to one hour for each, in a vacuum oven.

Tissues are sectioned from 4 to 7 microns thick. The picric acid is removed from the mounted sections with a 1.5% solution of ammonia hydroxide in 95% ethanol prior to staining.

Regenerating Cuprous Chloride Absorbent Solution

For the regeneration of a solution of cuprous chloride that has been used for the absorption of carbon monoxide:

First, pass a current of air through the solution till the CO-complex has been broken up and the carbon monoxide removed. The aeration is continued till the copper has all been oxidized to the cupric state. This stage is detected by diluting a small volume of the solution with water. If no white precipitate appears the oxidation is complete.

The solution is then poured into a bottle containing clean scrap copper, together with sufficient 60% hydrochloric acid to fill the bottle. The latter is securely stoppered and allowed to stand two or three days or until the solution becomes straw yellow, when it is again ready for use.

Solvent-Resistant Pump Packings
Use potassium oleate for the impregnation of the packing.

Settling Fumes
Use a mixture of about equal parts of barium sulfate and ethyl acetate and spray in room. Caution: This is inflammable.

Activated Carbon (Charcoal)

Sawdust	500 g.
Conc. Sulfuric Acid	150 cc.
Phosphoric Acid (85%)	600 cc.

Heat to 120–150°C., while mixing. Drive off sulfuric at 380–400°C. Heat to 950–1000°C. for 2 hrs. Cool, wash acid-free, dry at 120°C., grind and sift to 1000 mesh.

Alpha Cellulose
U. S. Patent 2,301,314
α-Cellulose of high quality and uniform viscosity is made by treating comminuted hard wood with nitric acid of 5% or less strength for about 24–30 hours at a temperature of 60–80°C., then steaming the treated material for about 45 minutes, rinsing and adding 2% sodium hydroxide and cooking for about 2 hours; then, preliminarily bleaching by use of chlorine in an amount of about 0.6–2% of the weight of the air-dried pulp, then treating the pulp with about 9–10 times its dry weight of 6–10% sodium hydroxide for 10–20 minutes at room temperature and finally bleaching

for approximately 2 hours by using sodium hypochlorite about 0.15–0.3% of the dry weight of the pulp.

Lycopodium Substitute

Marble Dust	97–96½
Stearin	3–3½

The above must be finely ground so that 80% is under 0.053 mm. particle size.

Extracting Saponin from Soap Bark
U. S. Patent 2,172,265

Soap Bark	1 lb.
Methanol	6 lb.
Ammonia	1 oz.

Boil, under reflux, and filter. Evaporate filtrate to $\frac{1}{5}$th volume. At 49°C. acetone is added to precipitate the saponin. Wash precipitate with acetone and dry in an atmosphere of CO_2.

Cuprous Oxide
U. S. Patent 2,280,168
Copper sulfate (2 mols) is treated with aqueous sodium sulfite (9 mols) to form a clear solution to which sodium hydroxide (4 mols) is added to precipitate cuprous oxide and regenerate sodium sulfite for re-use.

Hydrogen Peroxide
U. S. Patent 2,153,658
Barium dioxide is added to phosphoric acid at 0–4°C., slowly with good mixing. Do not add all the barium dioxide necessary to neutralize the acid. Heat to 20°C., cool to 0–4°C., add more barium dioxide until faintly acid. Heat to 35°C. to precipitate acid phosphate. Add barium carbonate un-

til neutral. Filter and add just enough sulfuric acid to precipitate any barium compounds in the filtrate. Filter off this precipitate.

Potassium Chlorate

Quicklime	1000 g.
Chlorine Gas	1000 g.
Muriate of Potash Crystals (Commercial grade)	350 g.

Yield: Varies from 250 to 270 grams.

A two gallon enamelled pail is placed under a mechanical agitator. 1000 cc. of cold water is added and the quicklime is fed in while the mass is being agitated at 40 to 50 r.p.m. When the heat of reaction, produced by the slaking, has fallen to 130°F., chlorine gas is passed in from an ordinary 5 lb. chlorine cylinder. A rubber tube is slipped over the outlet of the cylinder, and a glass tube of ¼ inch or ⅜ inch bore is inserted into the other end of the rubber tube. The glass tube should extend to within ½ inch of the bottom of the pail and be clamped securely to the edge of the pail.

Chlorine gas is then passed into the warm lime water at such a rate that no chlorine gas escapes. After 2½ to 3 hours, there will be a sudden rise in temperature, the latter may go to 200°F. and a sudden evolution of foam may rise to within three or four inches of the top of the pail. This indicates the end of the reaction and the chlorine valve should be immediately shut off. Stirring may be continued for ten minutes. The lime suspension has become quite clear, except for a small quantity of sandy or gritty material which was present in the lime used. The calcium chlorate solution is filtered from its mechanical impurities, heated to 190°F., and converted to chlorate of potash by the addition of muriate of potash crystals. The pail and contents are now set in a larger pail or box and cracked ice is packed around it. When the temperature of the solution has fallen to 35 or 40°F., the crop of chlorate of potash crystals is filtered off, washed with ice water, and dried carefully on a glass plate in an oven or in the air.

Removing Glass Tubes and Rods from Rubber Stoppers

Soak in methyl alcohol and gently work glass free from stopper. After a short time, even very firmly adhering glass tubes and rods will come free.

Removing Frozen Stoppers and Ground Glass Joints

Keep frozen parts completely immersed in concentrated sulfuric acid and heat the acid to 100°C. or higher, as necessary. After some time, the parts will separate unless the glass is actually sintered together or etched together by caustic.

Glass Syringe Aids

If glass surfaces of syringes have become "frozen" or locked these valuable instruments need need not be discarded. Very often the plunger can be separated by merely boiling the frozen syringes in an aqueous solution containing 25 per cent of glycerin.

Glycerin is an excellent preventive against syringe "freezing" during storage and it has the advantage that it can be washed away with water before sterilization or use. Indeed, lubrication and sterility can be achieved in one simple method.

A mixture of equal parts of 90 per cent alcohol and glycerin of phenol (equivalent to Glycerite of Phenol U.S.P.) forms a safe and effective means of storing syringes that have been sterilized by boiling. Tests have confirmed both the safety and efficacy of this procedure. When the syringe is taken out of this mixture, the alcohol evaporates quickly, leaving behind a thin film of glycerin of phenol which serves not only to keep the interior of the barrel sterile, but also prevents the piston from sticking.

Restoring Fraudulent or Faded Documents

Moisten, blot and cover with a few drops of 8-hydroxyquinoline in 6% acetic acid; wash with water after a few minutes and dry. The iron (from ink) becomes visible on exposure to day—or ultra-violet light.

Electrochemical (Facsimile) Recording Solution

A good solution for facsimile work is made by adding to 100 ml. of silver nitrate solution sufficient sodium thiosulfate solution just to dissolve the silver thiosulfate. Add to this resulting solution an excess of ¼ of its volume of the sodium thiosulfate solution. This final solution is good for facsimile work with currents as high as 10 milliamperes on a good rag or duplicating paper mounted on a nickel or nickel-plated drum. The stylus should be of tungsten and negative to the drum. Stock solutions consisting of 525 grams/liter sodium thiosulfate in 1000 ml. water; 139 grams/liter sodium thiosulfate; lead acetate 400 grams/liter; silver nitrate 100 grams/liter; and sodium potassium tartrate-saturated water solution at 10°C. After treatment the paper may be dried and stored. Before use it is moistened and after recording may be washed free of chemicals, dried and kept indefinitely.

Dyestuffs Recommended for the Coloring of Trichlorethylene

The following colors are soluble in this solvent; the amount to be used being dependent upon the intensity of color desired.

Oil Yellow.
Oil Yellow N.
Oil Fast Yellow EG.
Oil Orange.
Oil Brown N.
Anthraquinone Violet Base.
Anthraquinone Blue SKY Base.
Anthraquinone Blue AB Base.
Anthraquinone Iris R Base.
Anthraquinone Green G Base.
Oil Black BG.

TABLES

Weights and Measures
Troy Weight
24 grains = 1 pwt.
20 pwts. = 1 ounce
12 ounces = 1 pound

Apothecaries' Weight
20 grains = 1 scruple
3 scruples = 1 dram
8 drams = 1 ounce
12 ounces = 1 pound
The ounce and pound are the same as in Troy Weight.

Avoirdupois Weight
27$\frac{11}{32}$ grains = 1 dram
16 drams = 1 ounce
16 ounces = 1 pound
2000 lb. = 1 short ton
2240 lb. = 1 long ton

Dry Measure
8 quarts = 1 peck
2 pints = 1 quart
4 pecks = 1 bushel
36 bushels = 1 chaldron

Liquid Measure
4 gills = 1 pint
2 pints = 1 quart
4 quarts = 1 gallon
31½ gals. = 1 barrel
2 barrels = 1 hogshead
1 teaspoonful = ⅛ oz.
1 tablespoonful = ½ oz.
16 fluid oz. = 1 pint

Circular Measure
60 seconds = 1 minute
60 minutes = 1 degree
360 degrees = 1 circle

Long Measure
12 inches = 1 foot
3 feet = 1 yard
5½ yards = 1 rod
5280 feet = 1 stat. mile
320 rods = 1 stat. mile

Square Measure
144 sq. in. = 1 sq. ft.
9 sq. ft. = 1 sq. yard
30¼ sq. yds. = 1 sq. rod
43,560 sq. ft. = 1 acre
40 sq. rods = 1 rood
4 roods = 1 acre
640 acres = 1 sq. mile

Metric Equivalents
Length
1 inch = 2.54 centimeters
1 foot = 0.305 meter
1 yard = 0.914 meter
1 mile = 1.609 kilometers
1 centimeter = 0.394 in.
1 meter = 3.281 ft.
1 meter = 1.094 yd.
1 kilometer = 0.621 mile

Capacity
1 U. S. fluid oz. = 29.573 milliliters
1 U. S. liquid qt. = 0.946 liter
1 U. S. dry qt. = 1.101 liters
1 U. S. gallon = 3.785 liters
1 U. S. bushel = 0.3524 hectoliter
1 cu. in. = 16.4 cu. centimeters
1 milliliter = 0.034 U. S. fluid ounce
1 liter = 1.057 U. S. liquid qt.
1 liter = 0.908 U. S. dry qt.
1 liter = 0.264 U. S. gallon
1 hectoliter = 2.838 U. S. bu.
1 cu. centimeter = 0.061 cu. in.
1 liter = 1000 milliliters or 700 cu. c.

Weight
1 grain = 0.065 gram
1 apoth. scruple = 1.296 grams
1 av. oz. = 28.350 grams
1 troy oz. = 31.103 grams
1 av. lb. = 0.454 kilogram
1 troy lb. = 0.373 kilogram
1 gram = 15.432 grains
1 gram = 0.772 apoth. scruple
1 gram = 0.035 av. oz.
1 gram = 0.032 troy oz.
1 kilogram = 2.205 av. lb.
1 kilogram = 2.679 troy lb.

Approximate pH Values
The following tables give approximate pH values for a number of substances such as acids, bases, foods, biological fluids, etc. All values are rounded off to the nearest tenth and are based on measurements made at 25°C.

pH Values of Acids
Hydrochloric, N	0.1
Hydrochloric, 0.1N	1.1
Hydrochloric, 0.01N	2.0
Sulfuric, N	0.3
Sulfuric, 0.1N	1.2
Sulfuric, 0.01N	2.1
Orthophosphoric, 0.1N	1.5

Sulfurous, 0.1N	1.5
Oxalic, 0.1N	1.6
Tartaric, 0.1N	2.2
Malic, 0.1N	2.2
Citric, 0.1N	2.2
Formic, 0.1N	2.3
Lactic, 0.1N	2.4
Acetic, N	2.4
Acetic, 0.1N	2.9
Acetic, 0.01N	3.4
Benzoic, 0.1N	3.1
Alum, 0.1N	3.2
Carbonic (saturated)	3.8
Hydrogen Sulfide, 0.1N	4.1
Arsenious (saturated)	5.0
Hydrocyanic, 0.1N	5.1
Boric, 0.1N	5.2

pH Values of Bases

Sodium Hydroxide, N	14.0
Sodium Hydroxide, 0.1N	13.0
Sodium Hydroxide, 0.01N	12.0
Potassium Hydroxide, N	14.0
Potassium Hydroxide, 0.1N	13.0
Potassium Hydroxide, 0.01N	12.0
Lime (saturated)	12.4
Sodium Metasilicate, 0.1N	12.6
Trisodium Phosphate, 0.1N	12.0
Sodium Carbonate, 0.1N	11.6
Ammonia, N	11.6
Ammonia, 0.1N	11.1
Ammonia, 0.01N	10.6
Potassium Cyanide, 0.1N	11.0
Magnesia (saturated)	10.5
Sodium Sesquicarbonate, 0.1N	10.1
Ferrous Hydroxide (saturated)	9.5
Calcium Carbonate (saturated)	9.4
Borax, 0.1N	9.2
Sodium Bicarbonate, 0.1N	8.4

pH Values of Foods

Apples	2.9–3.3
Apricots	3.6–4.0
Asparagus	5.4–5.8
Bananas	4.5–4.7
Beans	5.0–6.0
Beers	4.0–5.0
Beets	4.9–5.5
Blackberries	3.2–3.6
Bread, white	5.0–6.0
Butter	6.1–6.4
Cabbage	5.2–5.4
Carrots	4.9–5.3
Cheese	4.8–6.4
Cherries	3.2–4.0
Cider	2.9–3.3

Corn	6.0–6.5
Crackers	6.5–8.5
Dates	6.2–6.4
Eggs, fresh white	7.6–8.0
Flour, wheat	5.5–6.5
Gooseberries	2.8–3.0
Grapefruit	3.0–3.3
Grapes	3.5–4.5
Hominy (rye)	6.8–8.0
Jams, fruit	3.5–4.0
Jellies, fruit	2.8–3.4
Lemons	2.2–2.4
Limes	1.8–2.0
Maple Syrup	6.5–7.0
Milk, cows	6.3–6.6
Olives	3.6–3.8
Oranges	3.0–4.0
Oysters	6.1–6.6
Peaches	3.4–3.6
Pears	3.6–4.0
Peas	5.8–6.4
Pickles, dill	3.2–3.6
Pickles, sour	3.0–3.4
Pimento	4.6–5.2
Plums	2.8–3.0
Potatoes	5.6–6.0
Pumpkin	4.8–5.2
Raspberries	3.2–3.6
Rhubarb	3.1–3.2
Salmon	6.1–6.3
Sauerkraut	3.4–3.6
Shrimp	6.8–7.0
Soft Drinks	2.0–4.0
Spinach	5.1–5.7
Squash	5.0–5.4
Strawberries	3.0–3.5
Sweet Potatoes	5.3–5.6
Tomatoes	4.0–4.4
Tuna	5.9–6.1
Turnips	5.2–5.6
Vinegar	2.4–3.4
Water, drinking	6.5–8.0
Wines	2.8–3.8

pH Values of Biologic Materials

Blood, plasma, human	7.3–7.5
Spinal Fluid, human	7.3–7.5
Blood, whole, dog	6.9–7.2
Saliva, human	6.5–7.5
Gastric Contents, human	1.0–3.0
Duodenal Contents, human	4.8–8.2
Feces, human	4.6–8.4
Urine, human	4.8–8.4
Milk, human	6.6–7.6
Bile, human	6.8–7.0

INTERCONVERSION TABLES AND CHART
for Units of Volume and Weight, and Energy

MULTIPLY BY

TO CONVERT FROM	To Cu. In.	To Cu. Ft.	To Cu. Yd.	To Fl. Oz.	To Pint	To Quart	To Gallon	To Grain	To Oz. Troy	To Oz. Av.	To Lb. Troy	To Lb. Av.	To CC. or G.	To Lit. or Kg.	To Cu. M.
Cu. in.	1.00000	0_35787	0_42143	.554112	.034632	.017316	.004329	252.891	.526857	.578037	.043905	.036127	16.3871	.016387	0_41639
Cu. Ft.	1728.00	1.00000	.037037	957.505	59.8442	29.9221	7.48052	438996	910.408	998.848	75.8674	62.4280	28316.9	28.3169	.028317
Cu. Yd.	46656.0	27.0000	1.00000	25852.6	1615.79	807.896	201.974	117990_3	24581.0	26968.9	2048.42	1685.56	764556	764.556	.764556
Fl. Oz.	1.80469	.001044	0_43868	1.00000	.062500	.031250	.007813	456.390	.950813	1.04318	.079234	.065199	29.5736	.029573	0_42957
Pint	28.8750	.016710	0_36189	16.0000	1.00000	.500000	.125000	7302.23	15.2130	16.6908	1.26775	1.04318	473.177	.473177	0_34732
Quart	57.7500	.033420	.001238	32.0000	2.00000	1.00000	.250000	14604.5	30.4260	33.3816	2.53550	2.08635	946.354	.946354	0_39463
Gallon	231.000	.133681	.004951	128.000	8.00000	4.00000	1.00000	58417.9	121.704	133.527	10.1420	8.34541	3785.42	3.78542	.003785
Grain	.003954	0_52288	0_78475	.002191	0_31369	0_46850	0_41712	1.00000	.002083	.002286	0_31736	0_31428	.064799	0_46479	0_76479
Oz. Troy	1.89805	.001098	0_44068	1.05173	.065733	.032867	.008217	480.000	1.00000	1.09714	.083333	.068571	31.1035	.031104	0_43110
Oz. Av.	1.72999	.001001	0_43708	.958608	.059913	.029957	.007489	437.500	.911457	1.00000	.075955	.062500	28.3495	.028350	0_42835
Lb. Troy	22.7766	.013181	0_34882	12.6208	.788800	.394400	.098600	5760.00	12.0000	13.1657	1.00000	.822857	373.242	.373242	0_33732
Lb. Av.	27.6799	.016018	0_35933	15.3378	.958611	.479306	.119826	7000.00	14.5833	16.0000	1.21528	1.00000	453.593	.453593	0_44536
CC. or Gram	.061024	0_43531	0_51308	.033814	.002113	.001057	0_32642	15.4323	.032151	.035274	.002679	.002205	1.00000	.001000	.000001
Liter or Kg.	61.0237	.035315	.001308	33.8140	2.11337	1.05669	.264172	15432.3	32.1507	35.2739	2.67923	2.20462	1000.00	1.00000	.001000
Cu. M.	61023.7	35.3146	1.30795	33814.0	2113.37	1056.69	264.172	154320_3	32150.7	35273.7	2679.23	2204.62	1000000	1000.00	1.00000

Note. The small subnumeral following a zero indicates that the zero is to be taken that number of times; thus, 0_31428 is equivalent to .0001428.

Values used in constructing table:

1 inch = 2.540001 cm.

1 cu. in. = 16.387083 cc. = 16.387083 g H_2O at 4°C. = 39°F.

1 lb. av. = 453.5926 g.

∴ 1 gal. = 8.34541 lb.

∴ 1 lb. av. = 27.679886 cu. in. H_2O at 4°C.

1 lb. av. = 7000 grains.

∴ 1 gallon = 58417.87 grains.

231 cu. in. = 1 gallon = 3785.4162 g.

TO CONVERT FROM — MULTIPLY BY

TO CONVERT FROM	B. T. U.	P. C. U.	Cal.	Ft. Lb.	Ft. Tons	Kg. M.	HP Hrs.	KW Hrs.	Joules	Lb. C	Lb. H₂O
B. T. U.	1.00000	.555556	.251996	778.000	.389001	107.563	$.0_33929$	$.0_32931$	1055.20	$.0_46876$	.001031
P. C. U.	1.80000	1.00000	2.20462	1400.40	.700202	193.613	$.0_37072$	$.0_35276$	1899.36	$.0_31238$	.001855
Calories	3.96832	2.20462	1.00000	3087.40	1.54368	426.844	.001559	.001163	4187.37	$.0_32729$	.004089
Ft. Lb.	.001285	$.0_37141$	$.0_33239$	1.00000	.000500	.138255	$.0_65050$	$.0_63767$	1.35625	$.0_78838$	$.0_51325$
Ft. Tons	2.57069	1.42816	.647804	2000.00	1.00000	276.511	.001010	$.0_37535$	2712.59	$.0_31768$	.002649
Kg M.	.009297	.005165	.002343	7.23301	.003617	1.00000	$.0_53653$	$.0_52725$	9.81009	$.0_66394$	$.0_59580$
HP Hrs.	2544.99	1413.88	641.327	1980000	990.004	273747	1.00000	.746000	2685473	.175044	2.62261
KW Hrs.	3411.57	1895.32	859.702	2654200	1327.10	366959	1.34041	1.00000	3599889	.234648	3.51562
Joules	$.0_39477$	$.0_35265$	$.0_32388$	.737311	$.0_33687$	.101936	$.0_63724$	$.0_62778$	1.00000	$.0_76518$	$.0_69766$
Lb. C	14544.0	8080.00	3665.03	11315_3	5657.63	1564396	5.71434	4.26285	15347_3	1.00000	14.9876
Lb. H₂O	970.400	539.111	244.537	754971	377.487	104379	.381270	.284424	1023966	.066744	1.00000

"P. C. U." refers to the "pound-centigrade unit." The ton used is 2000 pounds. "Lb. C" refers to pounds of carbon oxidized, 100% efficiency equivalent to the corresponding number of heat units. "Lb. H₂O" refers to pounds of water evaporated at 100°C. = 212°F. at 100% efficiency.

By the use of the foregoing table about 330 interconversions among twenty-six of the standard engineering units of measure can be directly estimated from the alignment chart to three significant figures or calculated by simple multiplication to six significant figures. The multiplier factor given in the table is located on the center scale "A" giving the point which when aligned with any number point on "C1" determines the product on "C." Imperfections in the scale due to lack of precision in printing should be checked at intervals along "A" scale by actual division of "C" by "C1," the lines being left out so that the reader can do this. A line scratched on a transparent celluloid triangle gives the best medium for making alignments.

When volume and weight interconversions are given, water is the medium the calculations are based upon. By the introduction of specific gravity factors the medium can be changed, giving the weight of any volume of any material.

CONVERSION OF THERMOMETER READINGS

F°	C°	F°	C°	F°	C°	F°	C°	F°	C°	F°	C°
—40	—40.00	30	—1.11	80	26.67	250	121.11	500	260.00	900	482.22
—38	—38.89	31	—0.56	81	27.22	255	123.89	505	262.78	910	487.78
—36	—37.78	32	0.00	82	27.78	260	126.67	510	265.56	920	493.33
—34	—36.67	33	0.56	83	28.33	265	129.44	515	268.33	930	498.89
—32	—35.56	34	1.11	84	28.89	270	132.22	520	271.11	940	504.44
—30	—34.44	35	1.67	85	29.44	275	135.00	525	273.89	950	510.00
—28	—33.33	36	2.22	86	30.00	280	137.78	530	276.67	960	515.56
—26	—32.22	37	2.78	87	30.56	285	140.55	535	279.44	970	521.11
—24	—31.11	38	3.33	88	31.11	290	143.33	540	282.22	980	526.67
—22	—30.00	39	3.89	89	31.67	295	146.11	545	285.00	990	532.22
—20	—28.89	40	4.44	90	32.22	300	148.89	550	287.78	1000	537.78
—18	—27.78	41	5.00	91	32.78	305	151.67	555	290.55	1050	565.56
—16	—26.67	42	5.56	92	33.33	310	154.44	560	293.33	1100	593.33
—14	—25.56	43	6.11	93	33.89	315	157.22	565	296.11	1150	621.11
—12	—24.44	44	6.67	94	39.44	320	160.00	570	298.89	1200	648.89
—10	—23.33	45	7.22	95	35.00	325	162.78	575	301.67	1250	676.67
— 8	—22.22	46	7.78	96	35.56	330	165.56	580	304.44	1300	704.44
— 6	—21.11	47	8.33	97	36.11	335	168.33	585	307.22	1350	732.22
— 4	—20.00	48	8.89	98	36.67	340	171.11	590	310.00	1400	760.00
— 2	—18.89	49	9.44	99	37.22	345	173.89	595	312.78	1450	787.78
0	—17.78	50	10.00	100	37.78	350	176.67	600	315.56	1500	815.56
1	—17.22	51	10.56	105	40.55	355	179.44	610	321.11	1550	843.33
2	—16.67	52	11.11	110	43.33	360	182.22	620	326.67	1600	871.11
3	—16.11	53	11.67	115	46.11	365	185.00	630	332.22	1650	898.89
4	—15.56	54	12.22	120	48.89	370	187.78	640	337.78	1700	926.67
5	—15.00	55	12.78	125	51.67	375	190.55	650	343.33	1750	954.44
6	—14.44	56	13.33	130	54.44	380	193.33	660	348.89	1800	982.22
7	—13.89	57	13.89	135	57.22	385	196.11	670	354.44	1850	1010.00
8	—13.33	58	14.44	140	60.00	390	198.89	680	360.00	1900	1037.78
9	—12.78	59	15.00	145	62.78	395	201.67	690	365.56	1950	1065.56
10	—12.22	60	15.56	150	65.56	400	204.44	700	371.11	2000	1093.33
11	—11.67	61	16.11	155	68.33	405	207.22	710	376.67	2050	1121.11
12	—11.11	62	16.67	160	71.11	410	210.00	720	382.22	2100	1148.89
13	—10.56	63	17.22	165	73.89	415	212.78	730	387.78	2150	1176.67
14	—10.00	64	17.78	170	76.67	420	215.56	740	393.33	2200	1204.44
15	— 9.44	65	18.33	175	79.44	425	218.33	750	398.89	2250	1232.22
16	— 8.89	66	18.89	180	82.22	430	221.11	760	404.44	2300	1260.00
17	— 8.33	67	19.44	185	85.00	435	223.89	770	410.00	2350	1287.78
18	— 7.78	68	20.00	190	87.78	440	226.67	780	415.56	2400	1315.56
19	— 7.22	69	20.56	195	90.55	445	229.44	790	421.11	2450	1343.33
20	— 6.67	70	21.11	200	93.33	450	232.22	800	426.67	2500	1371.11
21	— 6.11	71	21.67	205	96.11	455	235.00	810	432.22	2550	1398.89
22	— 5.56	72	22.22	210	98.89	460	237.78	820	437.78	2600	1426.67
23	— 5.00	73	22.78	215	101.67	465	240.55	830	443.33	2650	1454.44
24	— 4.44	74	23.33	220	104.44	470	243.33	840	448.89	2700	1482.22
25	— 3.89	75	23.89	225	107.22	475	246.11	850	454.44	2750	1510.00
26	— 3.33	76	24.44	230	110.00	480	248.89	860	460.00	2800	1537.78
27	— 2.78	77	25.00	235	112.78	485	251.67	870	465.56	2850	1565.56
28	— 2.22	78	25.56	240	115.56	490	254.44	880	471.11	2900	1593.33
29	— 1.67	79	26.11	245	118.33	495	257.22	890	476.67	2950	1621.11

ALCOHOL PROOF AND PERCENTAGE TABLE

U. S. Proof at 60° F.	Per cent Alcohol by Volume at 60° F.	Per cent Alcohol by Weight	U. S. Proof at 60° F.	Per cent Alcohol by Volume at 60° F.	Per cent Alcohol by Weight
0	0.0	0.00	57	28.5	——
1	0.5	——	58	29.0	23.82
2	1.0	0.80	59	29.5	——
3	1.5	——	60	30.0	24.67
4	2.0	1.59	61	30.5	——
5	2.5	——	62	31.0	25.52
6	3.0	2.39	63	31.5	——
7	3.5	——	64	32.0	26.38
8	4.0	3.19	65	32.5	——
9	4.5	——	66	33.0	27.24
10	5.0	4.00	67	33.5	——
11	5.5	——	68	34.0	28.10
12	6.0	4.80	69	34.5	——
13	6.5	——	70	35.0	28.97
14	7.0	5.61	71	35.5	——
15	7.5	——	72	36.0	29.84
16	8.0	6.42	73	36.5	——
17	8.5	——	74	37.0	30.72
18	9.0	7.23	75	37.5	——
19	9.5	——	76	38.0	31.60
20	10.0	8.05	77	38.5	——
21	10.5	——	78	39.0	32.48
22	11.0	8.86	79	39.5	——
23	11.5	——	80	40.0	33.36
24	12.0	9.68	81	40.5	——
25	12.5	——	82	41.0	34.25
26	13.0	10.50	83	41.5	——
27	13.5	——	84	42.0	35.15
28	14.0	11.32	85	42.5	——
29	14.5	——	86	43.0	36.05
30	15.0	12.14	87	43.5	——
31	15.5	——	88	44.0	36.96
32	16.0	12.96	89	44.5	——
33	16.5	——	90	45.0	37.86
34	17.0	13.79	91	45.5	——
35	17.5	——	92	46.0	38.78
36	18.0	14.61	93	46.5	——
37	18.5	——	94	47.0	39.70
38	19.0	15.44	95	47.5	——
39	19.5	——	96	48.0	40.62
40	20.0	16.27	97	48.5	——
41	20.5	——	98	49.0	41.55
42	21.0	17.10	99	49.5	——
43	21.5	——	100	50.0	42.49
44	22.0	17.93	101	50.5	——
45	22.5	——	102	51.0	43.43
46	23.0	18.77	103	51.5	——
47	23.5	——	104	52.0	44.37
48	24.0	19.60	105	52.5	——
49	24.5	——	106	53.0	45.33
50	25.0	20.44	107	53.5	——
51	25.5	——	108	54.0	46.28
52	26.0	21.28	109	54.5	——
53	26.5	——	110	55.0	47.24
54	27.0	22.13	111	55.5	——
55	27.5	——	112	56.0	48.21
56	28.0	22.97	113	56.5	——

U. S. Proof at 60° F.	Per cent Alcohol by Volume at 60° F.	Per cent Alcohol by Weight	U. S. Proof at 60° F.	Per cent Alcohol by Volume at 60° F.	Per cent Alcohol by Weight
114	57.0	49.19	158	79.0	72.38
115	57.5	——	159	79.5	——
116	58.0	50.17	160	80.0	73.53
117	58.5	——	161	80.5	——
118	59.0	51.15	162	81.0	74.69
119	59.5	——	163	81.5	——
120	60.0	52.15	164	82.0	75.86
121	60.5	——	165	82.5	——
122	61.0	53.15	166	83.0	77.04
123	61.5	——	167	83.5	——
124	62.0	54.15	168	84.0	78.23
125	62.5	——	169	84.5	——
126	63.0	55.16	170	85.0	79.44
127	63.5	——	171	85.5	——
128	64.0	56.18	172	86.0	80.62
129	64.5	——	173	86.5	——
130	65.0	57.21	174	87.0	81.90
131	65.5	——	175	87.5	——
132	66.0	58.24	176	88.0	83.14
133	66.5	——	177	88.5	——
134	67.0	59.28	178	89.0	84.41
135	67.5	——	179	89.5	——
136	68.0	60.32	180	90.0	85.69
137	68.5	——	181	90.5	——
138	69.0	61.38	182	91.0	86.99
139	69.5	——	183	91.5	——
140	70.0	62.44	184	92.0	88.31
141	70.5	——	185	92.5	——
142	71.0	63.51	186	93.0	89.65
143	71.5	——	187	93.5	——
144	72.0	64.59	188	94.0	91.02
145	72.5	——	189	94.5	——
146	73.0	65.67	190	95.0	92.42
147	73.5	——	191	95.5	——
148	74.0	66.77	192	96.0	93.85
149	74.5	——	193	96.5	——
150	75.0	67.87	194	97.0	95.32
151	75.5	——	195	97.5	——
152	76.0	68.92	196	98.0	96.82
153	76.5	——	197	98.5	——
154	77.0	70.10	198	99.0	98.38
155	77.5	——	199	99.5	——
156	78.0	71.23	200	100.0	100.00
157	78.5	——			

Buffer Systems

The following table gives some common buffer systems and the approximate pH of maximum buffer capacity. The zone of effective buffer action will vary with concentration but the general average will be ± 1.0 pH from the value given, for concentrations approximately 0.1 molar.

Glycocoll-Sodium Chloride-Hydrochloric Acid	2.0
Potassium Acid Phthalate-Hydrochloric Acid	2.8
Primary Potassium Citrate	3.7
Acetic Acid-Sodium Acetate	4.6
Potassium Acid Phthalate-Sodium Hydroxide	5.0
Secondary Sodium Citrate	5.0
Carbonic Acid-Bicarbonate	6.5
Primary Phosphate-Secondary Phosphate	6.8
Primary Phosphate-Sodium Hydroxide	6.8
Boric Acid-Borax	8.5
Borax	9.2
Boric Acid-Sodium Hydroxide ..	9.2
Bicarbonate-Carbonate	10.2
Secondary Phosphate-Sodium Hydroxide	11.5

Courtesy of W. A. Taylor & Company

REFERENCES AND ACKNOWLEDGMENTS

Abrasive & Cleaning Methods
Agr. Gaz. N. S. Wales
Allg. Oes. v. Gettzeitung
Aluminum Co. of Amer.
American Colloid Co.
Amer. Cyanamid & Chem. Corp.
Amer. Druggist
Amer. Dry Milk Inst.
Amer. Dyestuff Reporter
Amer. Electrop. Society
Amer. Gum Importers' Ass'n
Amer. Paint Jol.
Amer. Perfumer
Amer. Photography
Amer. Wool & Cotton Reporter
Analyst
Anal. Fis. Quim.
Army Ordnance
Ault & Wiborg Varnish Wks. Handbook

Baker's Helper
Bakers Review
Baker's Weekly
Behr Manning Corp.
Better Enameling
Boonton Molding Co.
Bottler & Packer
Boyce Thompson Inst.
Brewers' Tech. Review
Brick & Clay Record
Br. Jol. Dent. Science
Brit. Jol. of Photography
Brit. Medical Jol.
Bull. Imp. Hyg. Lab.
Bulletin of Imperial Institute
Bull. Soc. Franc. Phot.

Camera
Camera (Luzern)
Canadian Jol. of Med. Technology
Canadian Textile Jol.
Canner
Cement & Cement Mfr.
Ceramic Age
Chemical Abstracts
Chemical Analyst
Chemical Industries
Chemical Products
Chemical Weekblad
Chem. Zent.
Chemist & Druggist
Chr. Hansen's Lab.
Cleaning & Dyeing World
Combustion
Confectioner's Jol.
Consumers' Guide
Cowles Laundry Tips
Cramer's Manual

Dairy World
Damsk. Tids. Farm
Dental Items
Dental Lab'y Review
Devt. Part. Zeitung
Diamant
Drug & Cosmetic Industry
Druggists Circular
Drugs, Oils & Paints
DuPont Rubber Bulletins

Eastman Kodak Co.
Electric Journal
Electrochemical Society

Farbe u. Lacke
Farben Zeitung
Farm. Tid.
Farming S. Africa
Fein Mechanic v. Prezision
Fettchem. Umschan.
Fils & Tissus
Flavors
Focus
Food Manufacture
Fruit Products Jol.

Gelatin, Leim, Klebstoffe
General Abrasive Co.
Glass Industry
Graphic Arts Monthly

Hawaiian Planters' Record
Hercules Powder Co. Bulletins
Hide & Leather
Ice Cream Review
India Rubber World
Indian Lac Research Inst.
Indian Soap Jol.
Indiana Acad. of Sciences
Indiana Farmer's Guide
Industrial & Eng. Chemistry
Industrial Chemist
Industrial Finishing
Instruments
Intern'l Salt Co.
Int'l Tin Res. & Dev. Council
Iowa State College Bull.

J. Amer. Dental Assn.
J. Amer. Medical Assn.
J. Chem. Eng.
J. Chinese Chem. Soc.
J. Federation Curriers
J. Federation Light Leather Tanners
J. Res. Nat. Bur. Standards
J. Rubber Industry
J. Russ. Rubber Ind.

Jol. Soc. Leather Trades
Jol. Soc. Rubber Ind. Japan
Jol. Tech. Physics
Jol. of Technical Methods (I.A.M.M.)

Keram Steklo
Khimstroi
Kozhevenna-Obuvnaya Prom.
Kunstdunger, Und Leim

Lakokras, Ind.
Leather Trades Review
Leather Worker
Les Mat. Grasses
Lithographic Tech. Foundation

Malayan Agric. Jol.
Manufacturing Chemist
Manufacturing Confectioner
Meat
Meat Merchandising
Melliand
Metal Industry
Metall und Erz
Metallurg
Metallurgist
Metals & Alloys
Mich. Agric. Exp. Sta.
Milk Dealer
Mineralogist
Monatschr. Textil-Ind.
Monsanto Chem. Co.
Munic. Eng. San. Record

Nat'l Butter & Cheese Jol.
Nat'l Provisioner
New York Physician
Nickelsworth
Nitrocellulose
Nord. Tid. Fot.

Ober Flachen Tech.
Oil & Color Trades Jol.
Oil & Soap

Pacific Plastics
Pacific Rural Press
Paint Technology
Paper Trade Jol.
Parfum Mod.
Peinture, Pigments, Vernis
Phar. Acta Helva
Pharmaceutical Jol.
Phila. Quartz Co.

Phot. Abstracts
Photo. Chronik
Photo Art Monthly
Phot. Ind.
Phot. Korr.
Photog. Kronik
Phot. Rev.
Photo Rundschau
Physics
Phytopathology
Pix
Plater's Guide Book
Portland Cement Assn.
Power
Practical Druggist
Practical Everyday Chemistry
Printing Industry
Prob. Edelmetalle
Process Engr. Mo.
Proc. World Petroleum Congress

Rayon & Mell. Tex. Monthly
Refiner & Nat. Gas Mfr.
Rev. Aluminum
Rev. Amer. Electro Society
Rev. Trimest & Can.
Rock Products

Science
Sharpless Solvents Corp.
Silver Technologist
Shoe and Leather Journal
Soap
Soap Gazette & Perfumer
Solvent News
Sovet-Sakhar
Spirits
Steel
Synthetic & Applied Finishes

Textile Colorist
Textile Mfr.
Textile Recorder

Univ. Nebr. Agric. Coll. Bull.
U. S. Department of Agriculture
U. S. Bureau of Mines
U. S. Bureau of Standards

Veneers and Plywood

Z. Elektrochem.
Zeit. Unters. Lebensm.

TRADE-NAME CHEMICALS

During the past few years, the practice of marketing raw materials, under names which in themselves are not descriptive chemically of the products they represent, has become very prevalent. No modern book of formulae could justify its claims either to completeness or modernity without numerous formulae containing these so-called "Trade Names."

Without wishing to enter into any discussion regarding the justification of "Trade Names," the editors recognize the tremendous service rendered to commercial chemistry by manufacturers of "Trade Name" products, both in the physical data supplied and the formulation suggested.

Deprived of the protection afforded their products by this system of nomenclature, these manufacturers would have been forced to stand helplessly by while the fruits of their labor were being filched from them by competitors who, unhampered by expenses of research, experimentation and promotion, would be able to produce something "just as good" at prices far below those of the original producers.

That these competitive products were "just as good" solely in the minds of the imitators would only be evidenced in costly experimental work on the part of the purchaser and, in the meantime, irreparable damage would have been done to the truly ethical product. It is obvious, of course, that under these circumstances, there would be no incentive for manufacturers to develop new materials.

Because of this, and also because the "Chemical Formulary" is primarily concerned with the physical results of compounding rather than with the chemistry involved, the editors felt that the inclusion of formulae containing various trade name products would be of definite value to the producer of finished chemical materials. If they had been left out many ideas and processes would have been automatically eliminated.

As a further service the better known "Trade Name" products are included with the list of chemicals and supplies.

CHEMICALS AND SUPPLIES:
WHERE TO BUY THEM*

Numbers on right refer to list of suppliers on pages directly following this list. Thus to find out who supplies borax look in left hand column, alongside borax, on page 432. The number there is 34. Now turn to page 444 and find number 34. Alongside is the supplier, American Potash & Chemical Corp., New York, N. Y.

Product	No.
A	
A.A.P. Naphthols	15
A-Syrup	423
Aacagum	251
Abalyn	276
Abietic Acid	276
Abopon	251
Accelerator 808	195
Accelerator 833	195
Accelerators, Vulcanization	195
Acceloid	245
Accroides	463
Acelose	18
Acetaloid	2
Acetamide	20
Acetic Acid	140
Acetic Anhydride	16
Acetoin	335
Acetol	445
Acetone	150
Acetphenetidin	360
Acetyl Cellulose..see Cellulose Acetate	
Acetyl Salicylic Acid	368
Acidolene	349
Acids, Fatty	540
Acimul	251
Acrawax	251
Acriflavine	1
Acrolite	157
Acrxyeol	442
Acrylic Resins	450
Acryloid	450
Acrysol	450
Acto	507
Adeps Lanae..............see Lanolin	
Adheso Wax	251
Adipic Acid	195
A.D.M. No. 100 Oil	44
Aerogel	368
Aerosol	23
Agar	14
Agene	399
Agerite Powder	559
Akcocene	23
Aktivin	8
Albacer	251
Alba-Floc	550

Product	No.
Albalith	387
Albasol	381
Albatex	137
Alberit	471
Albertol	471
Albinol	464a
Albolit	53
Albolith	387
Albone C	195
Albron	11
Albumen	515
Albusol	344
Alcohol, Denatured	449
Alcohol, Pure	551
Aldehol	312
Aldol	392
Aldydal	294
Alfalate	322
Alframine	365
Alginic Acid	313
Alizarin	595
Alkalies	148
Alkaloids	360
Alkanet	285
Alkanol	195
Alkyd Resins	409
Alloxan	88
Almond Oil	342
Aloes	413
Aloin	398
Aloxite	109
Alperox	335
Alpha Naphthol	286
Alphasol	23
Altax	559
Alugel	376
Alumina	11
Aluminum	11
Aluminum Acetate	392
Aluminum Bronze Powder	549
Aluminum Chloride	99
Aluminum Hydrate	121
Aluminum Oleate	479
Aluminum Silicate	580
Aluminum Stearate	229
Aluminum Sulfate	513
Alums	256
Alundum	397

* Please see addenda p. 453.

Product	No.	Product	No.
Alvar	478	Aqualube	251
Amandol	195	Aquamel	251
Amberette	522	Aquapel	408
Amberlac	442	Aquaplex	442
Amberol	442	Aquaresin	251
Ambreno	195	Aquarex	195
Amco Acetate	316	Aquarome	216
Amerine	26	Aquasol	23
Ameripol	254	Arachis Oil	see Peanut Oil
Amerith	117	Arapali	450
Ameroid	33	Araskleen	368
Amidine	103	Aratone	50
Aminomethylpropanediol	150	Archer-Daniels No. 635	44
Aminostearin	251	Archer-Daniels-Midland Oil	44
Aminox	384	Arctic Syntex	144
Ammonia	377	Areskap	368
Ammoniac Resin	23	Aresklene	368
Ammonium Alginate	313	Aridex	195
Ammonium Bichromate	431	Arlacel	52
Ammonium Bifluoride	269	Arlex	52
Ammonium Carbonate	586	Arochlor	521
Ammonium Chloride	420	Aroflex	519
Ammonium Laurate	251	Arolite	519
Ammonium Linoleate	251	Arosol	239
Ammonium Nitrate	234	Arsenic	35
Ammonium Oleate	251	Arsenious Oxide	see White Arsenic
Ammonium Persulfate	92	Artisil	462
Ammonium Phosphate	521	Asbestine	299
Ammonium Stearate	251	Asbestos	429
Ammonium Sulfamate	195	Ascarite	534
Ammonium Sulfate	60	Asetoform	258
Ammonium Sulfite	344	Aseptex	251
Ammonium Sulforicinoleate	251	Asphalt	65
Ammonium Thiocyanate	603	Asphaltum	9
Ammonium Thioglycollate	600	Astrinite	246
Amsco Solvent	32	Astrulan	23
Amyl Acetate	96	Atrapol	406
Anchoracel 2p	38	Aurosal	58
Angstrol	190	Avenex	602
Anhydrone	62	Avirol	379
Aniline Chloride		Avitex	195
see Aniline Hydrochloride		Avonac	379
Aniline Dyes	240	Azo	36
Aniline Hydrochloride	99		
Aniline Oil	188	**B**	
Ansol	552	Badex	515
Anthraquinone	412	Bakelite	56
Antidolorin	227	Bakers' Plasticizer	57
Antimony	530	Balsams	288
Antimony Chloride	475	Barak	195
Antimony Oxide	283	Bardal	68
Antimony Sulfide	223	Bardex	68
Antimony Trioxide		Barite	See Barytes
see Antimony Oxide		Barium Carbonate	66
Anti-Oxidants	248	Barium Nitrate	102
Anti-scorch	195	Barium Peroxide	66
Antox	195	Barium Silico Fluoride	25
Apco	39	Barium Sulfate	66
Apocthinner	39	Barium Sulfide	173
Appramine	574	Baroid	382
Aquadag	3	Barretan	68
Aqualoid	309	Bartyl	248

Product	No.
Barytes	382
Basic Colors	15
Batavia Dammar	23
Bay Rum	451
Bayberry Wax	84
Beckacite	440
Beckamine	440
Beckolin	440
Beckophen	440
Beckosol	440
Beeswax	192
Beetle Resin	23
Bellaphan	311
Belro	275
Bensapol	588
Bentonite	22
Benzaldehyde	279
Benzidine	236
Benzine	32
Benzocaine	1
Benzoic Acid	111
Benzol	68
Benzoyl Peroxide	335
Benzyl Cellulose	6
Bergamot Oil	407
Beryllium	72
Beryllium Salts	223
Beta Naphthol	99
Beutene	384
Bicarbonate of Soda	136
Bismuth	122
Bismuth Subnitrate	388
Bitumen	see Asphaltum
Black Leaf	539
Blanc Fixe	291
Blandol	494
Blankit	240
Bleaching Powder	208
Blendene	251
Blood Albumen	370
Blood, Dried	46
Bludtan	123
Boea	297
Bole	580
Bonderite	411
Bondogen	559
Bone Ash	176
Bone Black	484
Bone Glue	170
Bone Meal	60
Bone Oil	532
Borax	34
Bordeaux Mixture	358
Bordow	188
Boric Acid	83
Borol	193
Botanical Products	414
Bresin	276
Brisgo	276
Bromine	181
Bromo "Acid"	436
Bromo-Fluorescein	251

Product	No.
Bronze Powder	189
Brosco	466
B.R.T. No. 7	68
Burgundy Pitch	331
Butacite	195
Butalyde	149
Butex	402
Butoben	360
Butyl Acetate	149
Butyl Acetyl Ricinoleate	150
Butyl Alcohol (Normal)	434
Butyl Aldehyde	150
Butyl Amine	150
Butyl Carbitol	106
Butyl Cellosolve	106
Butyl Lactate	150
Butyl Propionate	96
Butyl Stearate	316
Butyric Ether	396

C

Product	No.
Cadalyte	256
Cadmium	555
Cadmolith	128
Cajuput Oil	292
Calagum	251
Calcene	148
Calcium Arsenate	85
Calcium Carbonate	330
Calcium Carbonate (Precipitated)	360
Calcium Chloride	366
Calcium Chloride (Anhydrous)	213
Calcium Cyanamid	23
Calcium Fluoride	238
Calcium Hydroxide	see Lime, Hydrate
Calcium Oxide	see Lime
Calcium Phosphate	433
Calcium Polysulfide	98
Calcium Propionate	251
Calcium Stearate	524
Calcium Sulfate	see Plaster of Paris
Calcium Sulfide (Luminous)	29
Calcocid	99
Calcolac	99
Calcoloid	99
Calcozine	99
Calgon	94
Calgonite	100
Calomel	590
Calorite	505
Camphor	69
Camphor Oil	342
Candelilla Wax	298
Capillary Syrup	see Glucose Syrup
Captax	563
Caramel Color	231
Caraway Oil	336
Carbitol	106
Carbofrax	109
Carbolac	97
Carbolic Acid	see Phenol
Carbolic Oil	441

Product	No.
Carbolineum	108
Carbon, Activated	305
Carbon Bisulfide	62
Carbon Black	547
Carbon, Decolorizing	169
Carbon Tetrachloride	395
Carbonex	68
Carboraffin	13
Carboseal	106
Carbowax	106
Carboxide	106
Cardamom Seed	390
Carmine	444
Carnauba Wax	452
Carragheen	see Irish Moss
Casco	112
Casein	112
Castile Soap	156
Castor Oil	57
Castor Oil, Blown	57
Castor Oil, Sulfonated	588
Castrolite	454
Catalin	17
Catalpo	369
Catylon	270
Caustic Soda	351
CCH	351
Celascour	15
Celeron	157
Celite	306
Cellit	294
Cellon	198
Cellosolve	106
Celluloid	117
Celluloid Scrap	476
Cellulose Acetate	114
Cellulose Nitrate	361
Cement	417
Censteric	150
Ceraflux	251
Cercon	153
Cerelose	162
Cereps	575
Ceresalt	203
Ceresin Wax	482
Cerol	462
Cetamin	251
Cetec	241
Cetyl Alcohol	290
Cetyl Trimethyl Ammonium Bromide	406
Chalk, Precipitated	135
Charcoal	439
China Clay	525
China Wood Oil	63
Chloramine	1
Chlorasol	106
Chlorex	106
Chlorinated Rubber	276
Chlorine (Liquid)	208
Chloroform	188
Chlorophyll	21

Product	No.
Chloropicrin	41
Chlorosol	106
Cholesterin	360
Cholesterol	see Cholesterin
Chremnitz White	217
Chrome Alum	564
Chrome Green	315
Chrome Yellow	40
Chromic Acid	374
Chromic Oxide	see Chromium Oxide
Chromium Oxide	283
Chromium Sulfate	360
Cinchophen	99
Cinelin	138
Citral	248
Citrene	248
Citric Acid	422
Citronella Oil	458
Clarinol	276
Clay	547a
Clay No. 33	369
Clovel	251
Coal Tar	165
Coal Tar Colors	318
Coal Tar Oil	68
Cobalt Acetate	90
Cobalt Chloride	354
Cobalt Driers	354
Cobalt Linoleate	354
Cobalt Naphthenate	401
Coblac	79
Cocoa Butter	10
Cocoa Butter Oxanhydride	54
Coconut Butter	432
Coconut Oil	59
Coconut Oil Fatty Acid	5
Cod Liver Oil	451
Codite	157
Collodion	160
Collodion Wool	see Pyroxylin
Colloresin	240
Colophony	see Rosin
Color L34	251
Colors, Dry	282
Colors, Oil Soluble	151
Colza Oil	see Rapeseed Oil
Cominol	151
Condensite	56
Congo Resin	see Resins, Natural
Conoco	158
Convertit	400
Copal	23
Copal, Esterified	226
Copper Carbonate	160
Copper Chloride	62
Copper Cyanide	267
Copper Nitrate	62
Copper Oxides	283
Copper Sulfate	64
Coppercide	298
Corn Oil	30
Corn Sugar	500

Product	No.
Corn Syrup	162
Cosmic Black	544
Cotton Seed Oil	71
Cotton Seed Oil, Hydrogenated	432
Coumarin	352
Coumarone Resin	68
Cream of Tartar	269
Creosote	319
Creosol Sulfonic Acid	368
Cresols	161
Cresophan	258
Cresylic Acid	68
Cromodine	19
Cryolite	568
Cryptone	387
Crysalba	368
Cumar	68
Cupric Chloride...see Copper Chloride	
Curbay Binder	552
Curgon	471
Cuttle Fish Bone	228
Cyclamal	239
Cycline	368
Cyclohexylamine	195
Cyclohexanol	195
Cyclonol	195
Cymanol	296

D

Product	No.
Dammar Gum	331
Dapol	247
Darco	169
Darvan	559
Daxad	178
Deceresol	23
Degras	27
Deo-base	494
Deramin	251
Derris Extract	472
Derris Root	73
Devolite	471
Dextrins	370
Dextrose	162
Diacetin	316
Diafoam	442
Diakonn	295
Diamond K Linseed Oil	499
Diamyl Phthalate	552
Diastafor	504
Diastase	526
Diatol	552
Diatomaceous Earth	586
Dibutyl Cellosolve Phthalate	368
Dibutyl Tartrate	312
Dibutylphthalate	316
Dicalite	180
Dichlorbenzol	284
Dichlorethylene	195
Dichlorethylether	106
Diethanolamine Lactate	251
Diethylcarbonate	552
Diethylene Glycol	106

Product	No.
Diethyl Phthalate	560
Digestase	345
Diglycol Laurate	251
Diglycol Oleate	251
Diglycol Stearate	251
Dilecto	157
Dimethylolurea	195
Dinitrophenol	379
Diolin	195
Dionin	360
Dioxan	106
Dipentene	276
Diphenyl	521
Diphenyl, Chlorinated	368
Diphenyl Oxide	248
Diphenyl Phthalate	368
Discolite	454
Disodium Phosphate	565
Disperso	586
Distoline	587
Dowicide	188
Dow-metal	188
Dow plasticizers	188
D. P. G.	368
Drierite	265
Driers (Paint and Oil)	401
Driers, Varnish	401
Drop Black	582
Dry Ice	351
Dulux	195
Duolith	320
Duponol	195
Duphax	501
Duphonol	195
DuPont Rubber Red	195
Duprene	195
Duraplex	442
Durez	245
Durite	516
Durocer	251
Durophene	471
Dutox	195
Dyestuffs	379
Dynax	195

E

Product	No.
East-India Gum	23
Eastman Products	202
Egg, Dried	430
Egg Yolk	515
Elaine	210
Elemi	552
Emulgor A	251
Emulphor	240
Emulsifier L83A	251
Emulsifier A3076A	251
Emulsifier S489	251
Emulsone	251
Emulsone B	251
Eosin	440
Epersol-Y	598a
Ephedrine	360

Product	No.
Epsom Salt	238
Erinoid	33
Erio Chrome Dyes	235
Escolite	163
Essential Oils	152
Esso	507
Ester Gum	329
Esterol	409
Estersol	552
Ethavan	368
Ether	106
Ethocel	188
Ethox	561
Ethyl Acetate	361
Ethyl Cellulose	276
Ethyl Lactate	23
Ethyl Parasept	279
Ethyl Protol	187
Ethylamine	75
Ethylene Diamine	75
Ethylene Dichloride	188
Ethyleneglycol	106
Eucalyptus Oil	220
Eugenol	248
Eulan	240
Euresol	78
Exton	195

F

Product	No.
Fabroil	241
Factolac	285
Falba Absorption Base	421
Falkide	214
Falkovar	214
Feectol	455
Feldspar	155
Fer-ox	586
Ferric Chloride......see Iron Chloride	
Ferrisul	368
Ferro Chrome	209
Ferro Manganese	209
Ferro Silicon	209
Ferrous Chloride....see Iron Chloride	
Ferrox	513
Fiberlon	218
Filac	287
Fillers	584
Film Scrap	287
Filter-Cel	306
Filtrol	219
Fish Glue	278
Fish Oil	214
Fixalt	368
Fixtan	256
Flavine	593
Flavors	293
Flaxseed	80
Flectol	455
Flexalyn	276
Flexoresin	251
Flexowax C	251
Fluorspar	281

Product	No.
Foamapin	251
Foamex	251
Formaldehyde	279
Formalin	see Formaldehyde
Formamide	195
Formica	224
Formic Acid	565
Formvar	478
Freon	317
Friction Black Pulp	200
Fuller's Earth	460
Fusel Oil	552
Fyrex	565

G

Product	No.
G Protein	251
Gallagum	251
Gallic Acid	595
Gamboge	452
Gardinol	195
Gastex	237
Gelatin	48
Gelloid	588
Gelowax	251
Gelozone	581
Gelva	478
Gelva Resin	478
Geraniol	312
Geranium Lake	301
Geranium Oil	464
Gilsonite	331
Ginseng	328
Glandular Products	585
Glassheen	251
Glauber's Salt	302
Glaurin	251
Glucarine	251
Glucose	463
Glue	167
Glumide	251
Glutrin	448
Glycerawax	251
Glycerin	145
Glycerol	see Glycerin
Glyceryl Monooleate	251
Glyceryl Mono Stearate	251
Glyceryl Phthalate	251
Glyceryl Sebacate	450
Glyceryl Stearate	251
Glyceryl Tristearate	251
Glyco Wax A	251
Glyco Wax B	251
Glycol	see Ethylene Glycol
Glycol Bori-Borate	251
Glycol Distearate	251
Glycol Oleate	251
Glycol Phthalate	251
Glycol Stearate	251
Glycolac	251, 321
Glycomel	251
Glycomine	251
Glycopon	251

Product	No.
Glycoride	251
Glycosterin	251
Glyptal	242
Gold Chloride	344
Graphite	291
Graphite, Colloidal	4
Grapeseed Oil	168
Green Soap	355
Ground Nut Oil	see Peanut Oil
Guai-a-phene	146
Guantal	455
Gum Arabic	194
Gum Batu	536
Gum Benzoin	413
Gum Copal	331
Gum Dammar	536
Gum Elemi	226
Gum Karaya	228
Gum, Locust Bean	298
Gum Manila	297
Gum Mastic	463
Gum Myrrh	285
Gum Sandarac	331
Gum Tragacanth	359
Gums, Varnish	297
Gutta Percha	191
Gypsum	553

H

Halex	89
Halowax	262
Harcol	251
Harveg	272
Harvite	484
Haskelite	271
Heliozone	195
Hemlock Bark	528
Henna Leaves	414
Herbs	139
Hercolyn	276
Hercusol	276
Herkolite	241
Hevealac	251
Heveatex	277
Hexamethylene Diisocyanate	196
Hexamethylenetetramine	279
Hexone	106
Hexyl Alcohol	106
Hormones	604
Hyamine	450
Hycar	598
Hydralite C	240
Hydrated Lime	see Lime
Hydristear	583
Hydrochloric Acid	238
Hydrofluoric Acid	250
Hydrogen Peroxide	573
Hydrogenated Castor Oil	195
Hydrogenated Cottonseed Oil	432
Hydrogenated Fish Oil	492
Hydromalin	251
Hydromel	251

Product	No.
Hydronol	195
Hydropel	151
Hydroquinone	202
Hydroresin	251
Hydrowax	251
Hygropon	251

I

Ichthyol	360
Idalol	274
IG Wax	240
IG Wax O	240
Igepon	240
Ignex	251
Indian Red	79
Indigisols	105
Indigo	437
Indium	72
Indolylacetic Acid	360
Indur	441
Indusoil	296
Infusorial Earth	see Diatomaceous Earth
Insect Wax, Chinese	451
Invadine	137
Invert Sugar	400
Iodine	388
Iridium	58
Irish Moss	414
Iron Ammonium Citrate	468
Iron Chloride	133
Iron Oxide	79
Iron Sulfate	238
Isco	298
Isocholesterol	see Cholesterin
Isohol	251
Isolene	383
Isoline	587
Isomerpin	195
Isophan	115
Isopropyl Acetate	264
Isopropyl Alcohol	503
Isopropyl Ether	106
Ivory Black	79

J

Japan Wax	491
Jasmogene	563
Jointex	251

K

K. D. Gum	331
Kainite	230
Kalite	559
Kaolin	465
Kapsol	561
Karo	162
Kellin	499
Kellogg KVO	499
Kellogg Varnish Oil	499
Kellsey	499
Keltone	313

Product	No.
Kerol	91
Kerosene	147
Kerosene, Deodorized	482
Ketanol	195
Ketonone	23
Kilfoam	20
Kolineum	319
Kopol	71a
Koreon	374
Kronisol	561
Kronitex	561
Kryocide	420

L

Product	No.
Laboratory Equipment, 118, 129, 132, 206, 221, 386, 470	
Lacquer Blue	40
Lacquers	338
Lactac	195
Lactic Acid	43
Lactoid	89
Lactol Spirits	128
Lake Colors	315
Lamp Black	79
Lanette Wax	195
Lanolin	360
Lard Oil	212
Latac	195
Latex	277, 384
Laurex	384
Lauric-Ol	249
Lauryl Alcohol	195
Lauryl Sulfonate	195
Lavekol	240
Lavender Oil	558
Lavene	248
Lead Acetate	238
Lead Arsenate	238
Lead Borate	354
Lead and Its Oxides	199
Lead Oleate	479
Lead Stearate	601
Lecithin	28
Lecivit	309
Lemenone	251
Lemon Juice, Concentrated	375
Lemon Oil	292
Lemonal	558
Leonil	23
Le Page's Cement	457
Leptyne	425
Lerbinia	216
Lethane	450
Leucosol	195
Leucotrope W.	588
Leukanol	450
Leukonin	269
Levelene	15
Lewisol	329
Licorice	339
Lime	417
Lime, Hydrated	298

Product	No.
Limestone	467
Lindol	116
Lingasan	275
Linoleic Acid	251
Linseed	see Flaxseed
Linseed Oil	80
Litex	529
Litharge	199
Lithium Chloride	352
Lithium Salts	352
Lithoform	19
Lithopone	320
Locust Bean Powder	194
Lognite	593
Logwood Extract	24
Lorinol	81
Lorol	195
Lubrisol	494
Lucidol	335
Lufax	450
Lumarith	116
Luminol	202
Luxene	56
Lycopodium	355
Lynsol	275
Lysol	326

M

Product	No.
Mabelite	23
Mafos	351
Magnesia	110
Magnesite	243
Magnesium Carbonate	360
Magnesium Chloride	586
Magnesium Hydroxide	348
Magnesium Powder	72
Magnesium Silicate	580
Mahogany Soaps	494
Makalot	141
Maleic Acid	379
Manganar	195
Manganese	7
Manganese Dioxide	189
Manhaden Oil	55
Manila Gum	297
Manjak	596
Mannitan Monostearate	52
Mannitol	52
Manol	482
Manox	346
Mapico	79
Mapromin	446
Mapromol	446
Marble Dust	263
Marcol	502
Marlite	432
Marseilles Soap	see Castile Soap
Mazola	162
Melamac	23
Mellittis	248
Menthol	289
Menthone	248

Product	No.
Menthyl Salicylate	258
Mercurous Nitrate	590
Mercury	289
Mercury Compounds	590
Merpentine	195
Merpol	195
Mersol	368
Mertanol	368
Merusol	506
Mesityl Oxide	149
Methacrylic Acid	195
Methalate C	233
Methanol	257
Methasol	295
Methenamine	279
Methicol	247
Methocel	188
Methox	561
Methyl Abietate	276
Methyl Acetate	106
Methyl Acetone	175
Methyl Amyl Ketone	106
Methyl Anthranilate	222
Methyl Cellosolve	106
Methyl Cellosolve Acetate	227
Methyl Cellulose	188
Methylcyclohexanone	68
Methyl Ethyl Ketone	106
Methyl Hexalin	68
Methyl Isobutyl Ketone	258
Methyl Orange	143
Methyl Parasept	279
Methyl p-Hydroxybenzoate	279
Methyl Salicylate	188
Methylanol	195
Methylate	233
Methylene Chloride	567
Metol	202
Metro-Nite	362
Metso	423
Mica	495
Micoid	364
Migasol	137
Milcol	251
Milk Gloss	251
Milk Sugar	344
Mineral Oil	see Paraffin Oil
Mineral Rubber	65
Mineral Seal Oil	507
Mineral Spirits	32
Minium	see Lead Oxide
Mirasol	107
Moldex	251
Moldslip	251
Molybdenum	141
Molybdenum Compounds	141
Molybdenum Oxide	141
Monex	384
Monoamylamine Oleate	251
Monoethanolamine Lactate	251
Monolite	241
Montan Color	251

Product	No.
Montan Wax	517
Morpholine	106
Mowilith	294
Mulsene	251
Multifex	599
Muriatic Acid	see Hydrochloric Acid
Musk Xylol	248
Mycoban	195

N

Naccolene	379
Naccon	379
Nacconol	379
Naphtha	172
Naphthalene	68
Naphthalene, Chlorinated	262
Napthenic Acid	251
Napoleum Spirits	172
Narobin	251
National Oil Red	379
Neatsfoot Oil	381
Nekal	240
Nelgin	251
Neolan	137
Neomerpin	195
Neoprene	295
Neozone D	195
Neutroleum	233
Neville Resin	385
Nevindene	269
Nevinol	385
Nevtex	385
N-Glo-5	391
Nickel Chloride	160
Nickel Sulfate	269
Nicotine	539
Nicotine Sulfate	323
Nicotinic Acid	99
Nigrosine	99
Nipabenzyl	461
Nipagen	253
Nipasol	461
Nitramon	195
Nitre Cake	541
Nitric Acid	368
Nitrobenzol	99
Nitrocellulose	196
Nitro-cotton	see Nitrocellulose
Nitroethane	150
Nonaethylene Glycol	106
Nonaethylene Glycol Laurate	251
Nopco	381
Novolak	56
Nuad	401
Nuba Resin	385
Nu-Char	296
Nulomoline	400
Nuodex	401
Nusoap	401

O

Ocenol	196
Ochres	490

Product	No.
Octyl Acetate	106
Octyl Alcohol	106
Oil, Citronella	292
Oil, Mineral	505
Oil, Olive	325
Oil, Red	see Oleic Acid
Oil Root Beer C	474
Oilate	401
Oildag	3
Oilsolate	442
Oiricuri Wax	87
Oiticica Oil	303
Olate	432
Olein	119
Oleoresins	474
Oleyl Alcohol	195
Olive Oil	518
Olive Oil Substitute	251
Olive Oil, Sulfonated	454
Ondulum	251
Opal Wax	195
Orange Oil	454
Oroco	453
Ortho Dichlorbenzene	284
Ortho-phenylphenate	188
Orthosil	420
Osmo-Kaolin	225
Ouricuri Wax	186
Oxalic Acid	374
Oxgall	385
Oxygen	131
Oxynone	455
Oxyquinoline Sulfate	74
Ozokerite Wax	518

P

Product	No.
Palm Kernel Oil	59
Palm Oil	586
Palmitic Acid	46
Panoline	507
Para Aminophenol	564
Parachlormetacresol	368
Parachol	251
Paracide	284
Para-dor	188
Paradura	409
Paraffin, Chlorinated	20
Paraffin Oils	469
Paraffin Wax	403
Para-flux	261
Paralac	295
Paraldehyde	279
Paramet	409
Paranol	409
Para-Phenylenediamine	37
Parapont	195
Parasept	279
Paris Black	79
Paris Green	177
Paris White	497
Parlon	276
Paroil	12

Product	No.
Peachol	251
Peanut Oil	207
Pearl Essence	357
Pectin	101
Peerless Clay	559
Pegopren	252
Pentacetate	477
Pentaerythritol	279
Pentaerythritol Abietate	276
Pentalyn	276
Pentasol	477
Pentrol	312
Peppermint Oil	342
Peptone	183
Perchloron	420
Perfume Bases	293
Perilla Oil	308
Permosalt	251
Permosalt A	251
Perone	195
Perrol	558
Perspex	295
Petrex	275
Petrobenzol	39
Petrohol	502
Petrolatum	419
Petrolene	39
Petroleum Ether	489
Petroleum Jelly	494
Petroleum Spirits	520
Petromix	494
Petropol 2138	32
Pharmasol	105
Phenac	23
Phenol	16
Phenol-Formaldehyde Resins	197
Phenolic Resin	440
Phenothiazine	99
Phenyl Cellosolve	106
Phenyl Chloride	284
Phenyl Mercuric Nitrate	598
Phisol	406
Phloroglucinol	598
Phobophene	579
Phosphoric Acid	565
Phosphorus	300
Phosphotex	368
Phthalic Anhydride	368
Piccolyte	418
Pigment 725	387
Pigment Colors	40
Pine Oil	244
Pine Tar	496
Pitch	438
Plaskon	426
Plastacele	195
Plaster of Paris	580
Plastogen	559
Plastopal	294
Plaz	471
Plexigum	450
Plexite	450

Product	No.
Plextol	471
Plioform	255
Pliolite	255
Pliowax	225
Pollopas	428
Polyacrylic Esters	195
Polycol	251
Polyisobutylene	507
Polymerized Glycol Stearate	251
Polyrin	251
Polystyrene	368
Polyvinyl Acetate	106
Polyvinyl Alcohol	392
Polyvinyl Butyral	368
Polyvinyl Butyral Resin	368
Polyvinyl Chloride	368
Polyzime	526
Ponolith	320
Pontalite	195
Pontianak Resin	363
Pontol	195
Potash, Caustic	393
Potassium Abietate	251
Potassium Carbonate	542
Potassium Chlorate	542
Potassium Chloride	34
Potassium Dichromate	431
Potassium Hydroxide	360
Potassium Iodine	388
Potassium Metabisulfite	344
Potassium Nitrate......see Saltpeter	
Potassium Oleate	251
Potassium Palmolate	251
Potassium Perchlorate	542
Potassium Permanganate	111
Potassium Silicate	423
Potassium Tartrate	360
Prague Powder	259
Prestabit	240
Proflex	251
Proofit	251
Propylene Glycol	188
Propylene Glycol Stearate	251
Propylene Oxide	106
Protectoid	116
Protoflex	251
Protovac	112
Provatol	204
Proxate	332
Prussian Blue	315
Prystal	17
Puerine	349
Pumice	166
Psyllium Seeds	324
Pylam Red	436
Pyrax	559
Pyrefume	414
Pyrethol	353
Pyrethrum	414
Pyrethrum Extract	356
Pyridin	319
Pyro............see Pyrogallic Acid	

Product	No.
Pyrogallic Acid	595
Pyrolusite	189
Pyroxylin Solutions	205

Q

Quakersol	415
Quebracho	24
Quince Seed	285
Quinine Bisulfate	258
Quinine Hydrochloride	360
Quiholine	68

R

Raisin Seed Oil	444
Rancidex	251
Rapeseed Oil	63
Rapidase	572
Rauzene	438
Rayox	559
Red Oil	119
Red Squill	285
Redmanol	56
Reogen	559
Resin DA 1	251
Resin R-H-35	195
Resinox	442a
Resins, Natural	23
Resins, Synthetic	440
Resipon	45
Resoglaz	6
Resorcin	416
Revertex	443
Rezidel	251
Rezinel	251
Rezyl	23
Rheolan	251
Rhodium	58
Rhonite	450
Rhoplex	450
Rhotex	450
Ricinoleic Acid	57
Rochelle Salts	422
Rodo	559
Rose Water	336
Roseol	342
Rosin	244
Rosin Oil	383
Rosin, Polymerized	276
Rosoap A	251
Rotenone	535
Rubber	201
Rubber Hydrochloride	347
Rubber Latex	334
Rubber Resin	251
Rubber, Synthetic........195, 254, 255	
Rubidium Salts	340

S

"S" Syrup	423
Saccharine	279
Sal Soda	136

Product	No.	Product	No.
Salicylic Acid	188	Sodium Metasilicate	423
Salt	371	Sodium Naphthalene Sulfonate	270
Salt Cake	23	Sodium Nitrate	70
Saltpeter	164	Sodium Nitrite	493
Santicizers	368	Sodium Oleate	251
Santobane	368	Sodium Oxalate	565
Santolite	368	Sodium Perborate	195
Santomask	368	Sodium Phosphate	521
Santomerse	368	Sodium Propionate	251
Santowax	368	Sodium Pyrophosphate	565
Santox	368	Sodium Resinate	408
Sapamine	137	Sodium Silicate	423
Saponin	309	Sodium Silico Fluoride	256
Savolin	251	Sodium Stannate	269
Schultz Silica	124	Sodium Stannite	354
Selenium	31	Sodium Sulfate	238
Sellatan A	235	Sodium Sulfite	358
Sepia	see Cuttle Fish Bone	Sodium Thiocyanate	603
Serinol	310	Sodium Thioglycollate	606
Serrasol	130	Sodium Tungstate	62
Shellac	595	Soligen	6
Shellac Wax	291	Solox	552
Sherpetco	482	Solozone	195
Sicapon	251	Solux	195
Siennas	217	Solvent Naphtha	68
Silex	580	Solvesso	507
Silica	67	Solwax	441
Silica Black	304	Sorbitol	251
Silicon	72	Sorbitol Laurate	251
Silvatol	137	Sorbitol Oleate	251
Silver	266	Sorbitol Stearate	251
Silver Cyanide	160	Soya Protein	249
Silver Nitrate	202	Soybean Flour	500
Slaked Lime	see Lime	Soybean Oil	499
Soap	511	Span	52
Soda Ash	179	Sperm Oil	159
Soda, Caustic	351	Spermaceti	517
Soda, Sal	154	Speron	97
Sodium Acetate	542	Spindle Oil	507
Sodium Alginate	313	Squill	414
Sodium Aluminate	376	SRA Black	15
Sodium Arsenite	268	Stacol	251
Sodium Benzoate	284	Stago CS	597
Sodium Bicarbonate	136	Standard Viscous Oil	505
Sodium Bichromate	431	Stannic Chloride	see Tin Chloride
Sodium Bisulfite	256	Stannous Chloride	see Tin Chloride
Sodium Borate	see Borax	Starch	512
Sodium Borophosphate	251	Staybellite	276
Sodium Carbonate	493	Stearacol	251
Sodium Chlorate	404	Stearic Acid	119
Sodium Chlorite	351	Stearin	577
Sodium Choleate	183	Stearin Pitch	260
Sodium Cyanide	195	Stearite	586
Sodium Fluoride	23	Stearol	427
Sodium Hexametaphosphate	94	Stearoricinol	251
Sodium Hydrosulfite	454	Stearyl Alcohol	444
Sodium Hydroxide	360	Stetsol	599
Sodium Hypochlorite	174	Stoddard Solvent	172
Sodium Hypochlorite Liquid	447	Storax	342
Sodium Hyposulfite	256	Stramonium	414
Sodium Lauryl Sulfate	251	Stripolite	454
Sodium Metaphosphate	94	Stripper, T. S.	45

Product	No.
Stroba Wax	251
Strontium Nitrate	256
Strychnine	422
Styrax	see Storax
Styrene	188
Styrex	188
Sublan	251
Sucrose Octoacetate	392
Sulfadiazine	99
Sulfanilamide	99
Sulfanole	574
Sulfatate	251
Sulfathiazole	99
Sulfo Turk A.	251
Sulfo Turk B.	251
Sulfo Turk C.	251
Sulfogene	195
Sulfonated Castor Oil	93
Sulfonated Coconut Oil	454
Sulfonated Fatty Alcohol	251
Sulfonated Mineral Oil	251
Sulfonated Olive Oil	588
Sulforicinol	77
Sulfur	513
Sulfur Chloride	513
Sulfur Dioxide	567
Sulfuric Acid	361
Sulfurized Oils	501
Sunoco Spirits	520
Superphosphate	566
Surfex	435
Suspendite	251
Suspensone	251
Syncrolite	309
Syntex	144
Synthane	523
Synthenol	499

T

Talc	166
Talcum	see Talc
Talloil	123
Tallow	575
Tamol	450
Tanax	23
Tannic Acid	595
Tantalum	215
Tar Acid Oil	68
Tartar Emetic	43
Tartaric Acid	258
Tea Seed Oil	337
Teglac	23
Tegofan	471
Telloy	559
Tellurium	559
Tellurium Oxide	35
Tenex	244
Tenite	531
Tepidone	195
Tergitol	106
Terpesol	276
Terpineol	292

Product	No.
Tetrachlorethane	188
Tetrachlorethylene	195
Tetrahydronaphthalene	195
Tetralin	195
Tetrone	195
Textac	276
Textone	351
Thallium Sulfate	309
Theop	251
Thermoplex	442
Thiamin Hydrochloride	360
Thiocarbamilid	368
Thionex	195
Thiourea	309
Thorium Salts	576
Thymol	480
Ti-Cal	320
Tidolith	548
Timonex	533
Tin	546
Tin Chloride	475
Tin Oxide	354
Tinctures	410
Tintite	251
Ti-Sil	320
Titanium Dioxide	185
Titanium Tetrachloride	513
Titanox	538
Ti-Tone	320
Ti-Tree Oil	51
Toluene	68
Toluol	307
Toners	510
Tonsil	460
Tornesit	275
Triacetin	392
Triamylamine	477
Tributyl Citrate	150
Trichlorethylene	195
Triclene	195
Tricresyl Phosphate	258
Triethanolamine	106
Triethanolamine Lactate	251
Triethanolamine Naphthenate	251
Triethanolamine Oleate	251
Triethanolamine Phthalate	251
Triethanolamine Stearate	251
Trigamine	251
Trigamine Stearate	251
Trihydroxyethylamine	see Triethanolamine
Trikalin	251
Triphenylguanidine	195
Triphenylphosphate	368
Tripoli	527
Trisodium Phosphate	565
Triton	450
Troluoil	39
Tuads	559
Tung Oil	see China Wood Oil
Tungsten	215
Tunguran, A	6

Product	No.
Turkelene	251
Turkerol	251
Turkey Red Oil	381
Turmeric	414
Turpentine	42
Turpentine Substitute	39
Turpentine (Venice)	383
Turtle Oil	473
Tween	52
Twitchell Base	210
Typaphor Black	238
Tysenite	559

U

Product	No.
Uformite	442
Ultramarine Blue	510
Ultranate	49
Ultrasene	49
Ultravon	137
Umbers	217
Unilith	548
Union Solvent	545
Unyte	426
Uranium Nitrate	269
Urea	480
Ureka C	455
Ursulin	23
Uversol	269

V

Product	No.
Valex	95
Vandex	559
Vanilla Beans	536
Vanillal	487
Vanillin	474
Vanzyme	559
Varcrex	104
Varcum	562
Varnish	372
Varnish Gums & Resins	23
Varnolene	507
Varsol	507
Vaso	567
Vat Colors	15
Vatsol	23
Vegetable Colors	437
Vegetable F Wax	517
Vermiculite	280
Vermilion	217
Victron	554
Vinapas	6
Vinsol	275
Vinyl Acetate	392
Vinyl Chloride	106
Vinylite	106
Virifoam	251
Viscogum	251
Viscoloid	195
Vistanex	6

Product	No.
Vitamins	360
Vitriol	see Sulfuric Acid
V. M. P. Naphtha	507
Volclay	22
Vultex	570

W

Product	No.
Water Glass	see Sodium Silicate
Wax L33	251
Wax, Microcrystalline	507
Wax, Synthetic	251
Wetanol	251
Wetting Out Agents	251
White Arsenic	424
White Lead	380
Whiting	148
Witch Hazel Extract	181
Witco Yellow	586
Wood Flour	333
Wood Oil	see China Wood Oil
Wool Wax	82
Wyo-Jel	591

X

Product	No.
X-13	238
Xerol	232
Xylene	see Xylol
Xylerol	251
Xylol	68
Xynomine	406

Y

Product	No.
Yeast	504
Yelkin	453
Yumidol	251

Z

Product	No.
Zein	30
Zelan	195
Zenite	195
Zikol	391
Zimate	559
Zinc	273
Zinc Carbonate	586
Zinc Chloride	586
Zinc Chromate	571
Zinc Lactate	43
Zinc Oleate	269
Zinc Oxide	360
Zinc Resinate	391
Zinc Silicofluoride	269
Zinc Stearate	360
Zinc Sulfate	456
Zinc Sulfide	387
Zinol	391
Zirconium	537
Zirconium Oxide	223
Zirex	391
Zopaque	127

SELLERS OF CHEMICALS AND SUPPLIES

No.	Name	Address
1.	Abbott Laboratories	North Chicago, Ill.
2.	Acetate Products Corp.	London, England
3.	Acheson Colloids Corp.	Port Huron, Mich.
4.	Acheson Graphite Corp.	Niagara Falls, N. Y.
5.	Acme Oil Corp.	Chicago, Ill.
6.	Advance Solvents & Chem. Corp.	New York, N. Y.
7.	Ajax Metal Co.	Philadelphia, Pa.
8.	Aktivin Corp.	New York, N. Y.
9.	Allied Asphalt & Mineral Corp.	New York, N. Y.
10.	Alpha Lux Co., Inc.	New York, N. Y.
11.	Aluminum Co. of America	Pittsburgh, Pa.
12.	Amecco Chemicals, Inc.	Rochester, N. Y.
13.	American Active Carbon Co.	Columbus, O.
14.	American Agar Co., Inc.	San Diego, Calif.
15.	American Aniline Products, Inc.	New York, N. Y.
16.	American-Brit. Chem. Supplies, Inc.	New York, N. Y.
17.	American Catalin Corp.	New York, N. Y.
18.	American Cellulose Co.	Indianapolis, Ind.
19.	American Chemical Paint Co.	Ambler, Pa.
20.	American Chemical Products Co.	Rochester, N. Y.
21.	American Chlorophyll, Inc.	New York, N. Y.
22.	American Colloid Co.	Chicago, Ill.
23.	American Cyanamid & Chem. Co.	New York, N. Y.
24.	American Dyewood Co.	New York, N. Y.
25.	American Fluoride Corp.	New York, N. Y.
26.	American Insulator Corp.	New Freedom, Pa.
27.	American Lanolin Corp.	Lawrence, Mass.
28.	American Lecithin Corp.	New York, N. Y.
29.	American Luminous Products Co.	Huntington Park, Calif.
30.	American Maize Products Co.	New York, N. Y.
31.	American Metal Co.	New York, N. Y.
32.	American Mineral Spirit Co.	New York, N. Y.
33.	American Plastics Corp.	New York, N. Y.
34.	American Potash & Chem. Corp.	New York, N. Y.
35.	American Smelting & Refining Co.	New York, N. Y.
36.	American Zinc Co.	New York City
37.	Amido Products Co.	New York, N. Y.
38.	Anchor Chemical Co.	Manchester, England
39.	Anderson Prichard Oil Corp.	Oklahoma City, Okla.
40.	Ansbacher-Siegle Corp.	Rosebank, New York
41.	Ansul Chem. Co.	Marinette, Wis.
42.	Antwerp Naval Stores Co., Inc.	Boston, Mass.
43.	Apex Chem. Co.	New York, N. Y.
44.	Archer-Daniels-Midland Co.	Minneapolis, Minn.
45.	Arkansas Co.	New York, N. Y.
46.	Armour & Co.	Chicago, Ill.
47.	Asbury Graphite Mills	Asbury Park, N. J.
48.	Atlantic Gelatine Co.	Woburn, Mass.
49.	Atlantic Refining Co.	Philadelphia, Pa.
50.	Atlantic Research Associates	Newtonville, Mass.
51.	Atlas Import Co.	Chicago, Ill.
52.	Atlas Powder Co.	Wilmington, Del.
53.	Augsburger, Kunst Fabrik	Augsburg, Germany

No.	Name	Address
54.	Autoxygen, Inc.	New York, N. Y.
55.	Badcock, Robert & Co.	New York, N. Y.
56.	Bakelite Corp.	New York, N. Y.
57.	Baker Castor Oil Co.	Jersey City, N. J.
58.	Baker & Co., Inc.	Newark, N. J.
59.	Baker, Franklin Co.	Hoboken, N. J.
60.	Baker, H. J. & Bro.	New York, N. Y.
61.	Baker, J. E., Co.	York, Pa.
62.	Baker, J. T. Chem. Co.	Philipsburg, N. J.
63.	Balfour, Guthrie & Co., Ltd.	New York, N. Y.
64.	Barada & Page, Inc.	Kansas City, Mo.
65.	Barber Asphalt Co.	Philadelphia, Pa.
66.	Barium Reduction Corp.	Charleston, W. Va.
67.	Barnsdall Tripoli Corp.	Seneca, Mo.
68.	Barrett Co.	New York, N. Y.
69.	Barry, E. J., Inc.	New York, N. Y.
70.	Battelle & Renwick	New York, N. Y.
71.	Battleboro Oil Co.	Battleboro, N. C.
71a.	Beck, Koller & Co.	Detroit, Mich.
72.	Belmont Smelting & Refining Wks.	Brooklyn, N. Y.
73.	Benkert, W. & Co., Inc.	New York, N. Y.
74.	Benzol Products Co.	Newark, N. J.
75.	F. C. Bersworth Labs.	Framingham, Mass.
76.	Beryllium Corp. of America	New York, N. Y.
77.	Bick & Co., Inc.	Reading, Pa.
78.	Bilhuber-Knoll Corp.	New York, N. Y.
79.	Binney & Smith	New York, N. Y.
80.	Bisbee Linseed Co.	Philadelphia, Pa.
81.	Bohme, A. G., H. Th.	Chemnitz, Germany
82.	Bopf-Whittam Corp.	Linden, N. J.
83.	Borax Union, Inc.	San Francisco, Calif.
84.	The W. H. Bowdlear Co.	Syracuse, N. Y.
85.	Bowker Chem. Corp.	New York, N. Y.
86.	Bradley & Baker	New York, N. Y.
87.	Brazil Oiticica, Inc.	New York, N. Y.
88.	British Drug Houses, Ltd.	London, England
89.	British Xylonite Co.	London, England
90.	Brooke, Fred L., Co.	Chicago, Ill.
91.	Bud Aromatic Chemical Co., Inc.	New York, N. Y.
92.	Buffalo Electro Chem. Co., Inc.	Buffalo, N. Y.
93.	Burkard-Schier Chem. Co.	Chattanooga, Tenn.
94.	Buromin Corp.	Pittsburgh, Pa.
95.	Bush, W. J. & Co., Inc.	New York, N. Y.
96.	C. P. Chemical Solvents, Inc.	New York, N. Y.
97.	Cabot, Godfrey L., Inc.	Boston, Mass.
98.	Calcium Sulphide Corp.	Damascus, Va.
99.	Calco Chemical Co.	Bound Brook, N. J.
100.	Calgon, Inc.	Pittsburgh, Pa.
101.	Calif. Fruit Growers' Exchange	Ontario, Calif.
102.	Campbell, C. W. Co., Inc.	New York, N. Y.
103.	Campbell, John & Co.	New York, N. Y.
104.	Campbell Rex & Co.	London, England
105.	Carbic Color & Chemical Co.	New York, N. Y.
106.	Carbide & Carbon Chem. Corp.	New York, N. Y.
107.	C. I. Osborn Co.	New York, N. Y.
108.	Carbolincum Wood Preserving Co.	Milwaukee, Wis.
109.	Carborundum Co.	Niagara Falls, N. Y.
110.	Carey, Philip Co.	Lockland, Ohio
111.	Carus Chem. Co., Inc.	La Salle, Ill.
112.	Casein Mfg. Co.	New York, N. Y.
113.	The Casein Mfg. Co. of Amer., Inc.	New York, N. Y.
114.	Celanese Corp. of America	New York, N. Y.

No.	Name	Address
115.	Cellonwerke	Charlottenburg, Germany
116.	Celluloid Corp	Newark, N. J.
117.	Celluloid Corp	New York, N. Y.
118.	Central Scientific Co	Chicago, Ill.
119.	Century Stearic Acid Wks	New York, N. Y.
120.	Century Stearic Acid & Candle Wks	New York, N. Y.
121.	Ceramic Color & Chem. Mfg. Co	New Brighton, Pa.
122.	Cerro de Pasco Copper Corp	New York, N. Y.
123.	Champion Paper & Fibre Co	Canton, N. C.
124.	Chaplin-Bibbo	New York, N. Y.
125.	Chazy Marble Lime Co., Inc	Chazy, N. Y.
126.	Chesebrough Mfg. Co	New York, N. Y.
127.	Chemical & Pigment Co	Baltimore, Md.
128.	Chemical & Pigment Co., Inc	Scranton, Pa.
129.	Chemical Publ. Co., Inc	Brooklyn, N. Y.
130.	Chemical Solvents, Inc	New York, N. Y.
131.	Cheney Chem. Co	Cleveland, Ohio
132.	Chicago Apparatus Co	Chicago, Ill.
133.	Chicago Copper & Chem. Co	Blue Island, Ill.
134.	Chipman Chem. Co., Inc	Bound Brook, N. J.
135.	Chrystal, Charles B. Co., Inc.	New York, N. Y.
136.	Church & Dwight Co., Inc.	New York, N. Y.
137.	Ciba Co., Inc	New York, N. Y.
138.	Cinelin Co	Indianapolis, Ind.
139.	Clarke, John & Co.	New York, N. Y.
140.	The Cleveland-Cliffs Iron Co.	Cleveland, Ohio
141.	Climax Molybdenum Co	New York, N. Y.
142.	Clinton Co	Clinton, Ia.
143.	Coleman & Bell Co	Norwood, Ohio
144.	Colgate-Palmolive-Peet Co	Chicago, Ill.
145.	Colgate-Palmolive-Peet Co	Jersey City, N. J.
146.	Colledge, E. W., Inc.	Cleveland, Ohio
147.	Colonial Beacon Oil Co.	Everett, Mass.
148.	Columbia Alkali Corp	New York, N. Y.
149.	Commercial Solvents Corp	New York, N. Y.
150.	Commercial Solvents Corp.	Terre Haute, Ind.
151.	Commonwealth Color & Chem. Co	Brooklyn, N. Y.
152.	Compagnie Duval	New York, N. Y.
153.	Conewango Refining Co	Warren, Pa.
154.	Consolidated Chem. Sales Corp	Newark, N. J.
155.	Consolidated Feldspar Corp	Trenton, N. J.
156.	Conti Products Corp	New York, N. Y.
157.	Continental Diamond Fibre Co.	Bridgeport, Pa.
158.	Continental Oil Co	Ponca City, Okla.
159.	Cook Swan Co., Inc	New York, N. Y.
160.	Cooper, Charles & Co.	New York, N. Y.
161.	Coopers Creek Chem. Co.	W. Conshohocken, Pa.
162.	Corn Products Refining Co	New York, N. Y.
163.	Cowles Detergent Co.	Cleveland, Ohio
164.	Croton Chem. Corp	Brooklyn, N. Y.
165.	Crowley Tar Products Co.	New York, N. Y.
166.	Crystal, Charles B. Co., Inc.	New York, N. Y.
167.	Cudahy Packing Co	Chicago, Ill.
168.	Danco, Gerard J.	New York, N. Y.
169.	Darco Sales Corp	New York, N. Y.
170.	Darling & Co	Chicago, Ill.
171.	Davison Chem. Corp.	Baltimore, Md.
172.	Deep Rock Oil Corp	Chicago, Ill.
173.	C. P. De Lore Co	St. Louis, Mo.
174.	Delta Chem. Mfg. Co.	Baltimore, Md.
175.	Delta Chem. & Iron Co	Wells, Mich.
176.	Denver Fire Clay Co	Denver, Colo.

No.	Name	Address
177.	Devoe & Reynolds Co	New York, N. Y.
178.	Dewey & Almy Chem. Co.	Boston, Mass.
179.	Diamond Alkali Co.	Pittsburgh, Pa.
180.	Dicalite Co	New York, N. Y.
181.	Dickinson, E. E. Co.	Essex, Conn.
182.	Dickinson, J. Q. & Co	Malden, W. Va.
183.	Difco Laboratories, Inc	Detroit, Mich.
184.	Digestive Ferments Co	Detroit, Mich.
185.	Marshall Dill	San Francisco, Calif.
186.	Distributing & Trading Co	New York, N. Y.
187.	Dodge & Olcott Co	New York, N. Y.
188.	Dow Chemical Co	Midland, Mich.
189.	Drakenfeld, B. F. & Co	New York, N. Y.
190.	Dreyer, P. R. Co	New York, N. Y.
191.	Dreyfus Co., L. A.	Rosebank, N. Y.
192.	Drury, A. C. & Co., Inc	Chicago, Ill.
193.	Ducas, B. P. Co	New York, N. Y.
194.	Duche, T. M. & Sons	New York, N. Y.
195.	DuPont, E. I., de Nemours & Co	Wilmington, Del.
196.	E. I. DuPont de Nemours & Co., Inc.	Parlin, N. J.
197.	Durite Plastics	Philadelphia, Pa.
198.	Dynamit, A. G.	Troisdorf, Germany
199.	The Eagle-Picher Lead Co	Cincinnati, Ohio
200.	Eakins, J. S. & W. R., Inc	Brooklyn, N. Y.
201.	Earle Bros	New York, N. Y.
202.	Eastman Kodak Co	Rochester, N. Y.
203.	Economic Materials Co	Chicago, Ill.
204.	Eff Laboratories, Inc	Cleveland, Ohio
205.	Egyptian Lacquer Co	Kearney, N. J.
206.	Eimer & Amend	New York, N. Y.
207.	Elbert & Co.	New York, N. Y.
208.	Electro Bleaching Gas Co.	New York, N. Y.
209.	Electro-Metallurgical Co.	New York, N. Y.
210.	Emery Industries, Inc	Cincinnati, Ohio
211.	Empire Distilling Corp	New York, N. Y.
212.	Enterprise Animal Oil Co	Philadelphia, Pa.
213.	Fales Chem. Co., Inc	Cornwall Landing, N. Y.
214.	Falk & Co	Pittsburgh, Pa.
215.	Fansteel Metallurgical Corp	No. Chicago, Ill.
216.	Felton Chemical Co	Brooklyn, N. Y.
217.	Fezandie & Sperrle, Inc	New York, N. Y.
218.	Fiberloid Corp	Indian Orchard, Mass.
219.	Filtrol Co	Los Angeles, Calif.
220.	Fishbeck, Chas. Co	New York, N. Y.
221.	Fisher Scientific Co	Pittsburgh, Pa.
222.	Florasynth Laboratories	New York, N. Y.
223.	Foote Mineral Co.	Philadelphia, Pa.
224.	Formica Insulation Co	Cincinnati, Ohio
225.	Fougera, E. & Co	New York, N. Y.
226.	France, Campbell & Darling	Kenilworth, N. J.
227.	Franco-American Chemical Wks	Carlstadt, N. J.
228.	Frank-Vliet Co.	New York, N. Y.
229.	Franks Chem. Products Co., Inc.	Brooklyn, N. Y.
230.	French Potash Co	New York, N. Y.
231.	Alex Fries & Bro.	Cincinnati, Ohio
232.	Fries Bros.	New York, N. Y.
233.	Fritzchie Bros	New York, N. Y.
234.	Garrigues, Stewart & Davies, Inc.	New York, N. Y.
235.	Geigy Co., Inc.	New York, N. Y.
236.	General Aniline Works, Inc	New York, N. Y.
237.	General Atlas Carbon Co	New York, N. Y.
238.	General Chemical Co	New York, N. Y.

No.	Name	Address
239.	General Drug Co.	New York, N. Y.
240.	General Dyestuffs Corp.	New York, N. Y.
241.	General Electric Co.	Pittsfield, Mass.
242.	General Electric Co.	Schenectady, N. Y.
243.	General Magnesite & Magnesia Co.	Philadelphia, Pa.
244.	General Naval Stores Co.	New York, N. Y.
245.	General Plastics Corp.	London, England
246.	General Plastics, Inc.	No. Tonawanda, N. Y.
247.	Girdler Corp.	Louisville, Ky.
248.	Givaudan-Delawanna, Inc.	New York, N. Y.
249.	Glidden Co.	Cleveland, Ohio
250.	Globe Chem. Co.	Cincinnati, Ohio
251.	Glyco Products Co., Inc.	Brooklyn, N. Y.
252.	Goldschmidt, A. G., Th.	Essen, Germany
253.	Goldschmidt Corp.	New York, N. Y.
254.	Goodrich, B. F., Co.	Akron, Ohio
255.	Goodyear Tire & Rubber Co.	Akron, Ohio
256.	Grasselli Chemical Co.	Cleveland, Ohio
257.	W. S. Gray Co.	New York, N. Y.
258.	Greeff, R. W. & Co.	New York, N. Y.
259.	Griffith Laboratories	Chicago, Ill.
260.	Gross, A. & Co.	New York, N. Y.
261.	Hall, C. P. & Co.	Akron, Ohio
262.	Halowax Corp.	New York, N. Y.
263.	Hammil & Gillespie, Inc.	New York, N. Y.
264.	Hamilton, A. K.	New York, N. Y.
265.	Hammond Drierite Co.	Yellow Springs, Ohio
266.	Handy & Harman	New York, N. Y.
267.	Hardy, Charles, Inc.	New York, N. Y.
268.	Harrison Mfg. Co.	Rahway, N. J.
269.	Harshaw Chemical Co.	Cleveland, Ohio
270.	Hart Products Corp.	New York, N. Y.
271.	Haskelite Mfg. Corp.	Chicago, Ill.
272.	Haveg Corp.	Newark, Del.
273.	Hegeler Zinc Co.	Danville, Ill.
274.	Heine & Co.	New York, N. Y.
275.	Hercules Powder Co.	New York, N. Y.
276.	Hercules Powder Co.	Wilmington, Del.
277.	Heveatex Corp.	Melrose, Mass.
278.	C. B. Hewitt & Bro.	New York, N. Y.
279.	Heyden Chemical Works	New York, N. Y.
280.	Hill Bros. Chem. Co.	Los Angeles, Calif.
281.	Hillside Fluor Spar Mines	Chicago, Ill.
282.	Holland Aniline Dye Co.	Holland, Mich.
283.	O. Hommel Co.	Pittsburgh, Pa.
284.	Hooker Electro-Chemical Co.	New York, N. Y.
285.	Hopkins, J. L. & Co.	New York, N. Y.
286.	Hord Color Products	Sandusky, Ohio
287.	Horn Jefferys & Co.	Burbank, Calif.
288.	Horner, James B., Inc.	New York, N. Y.
289.	Huisking, Chas. L. & Co., Inc.	New York, N. Y.
290.	Hummel Chemical Co., Inc.	New York, N. Y.
291.	Hurst, Adolph & Co., Inc.	New York, N. Y.
292.	D. W. Hutchinson & Co., Inc.	New York, N. Y.
293.	Hymes, Lewis Associates	New York, N. Y.
294.	I. G. Farbenindustrie	Frankfurt, Germany
295.	Imperial Chem. Industries	London, England
296.	Industrial Chem. Sales Co.	New York, N. Y.
297.	Innes, O. G., Corp.	New York, N. Y.
298.	Innis Speiden Co.	New York, N. Y.
299.	International Pulp Corp.	New York, N. Y.
300.	International Selling Corp.	New York, N. Y.

No.	Name	Address
301.	Interstate Color Co., Inc.	New York, N. Y.
302.	Iowa Soda Products Co.	Council Bluffs, Ia.
303.	Jackson, L. N. & Co.	New York, N. Y.
304.	Jacobson, C. A.	W. Va. University, Morgantown, W. Va.
305.	The Jennison-Wright Co.	Toledo, Ohio
306.	Johns-Manville Corp.	New York, N. Y.
307.	Jones & Laughlin Steel Corp.	Pittsburgh, Pa.
308.	Jones, S. L. & Co.	San Francisco, Calif.
309.	Jungmann & Co.	New York, N. Y.
310.	Kali Mfg. Co.	Philadelphia, Pa.
311.	Kalle & Co.	Wiesbaden Bierich, Germany
312.	Kay Fries Chem., Inc.	New York, N. Y.
313.	Kelco Co.	San Diego, Calif.
314.	Kentucky Clay Mining Co.	Mayfield, Ky.
315.	Kentucky Color & Chem. Co.	Louisville, Ky.
316.	Kessler Chem. Corp.	Philadelphia, Pa.
317.	Kinetic Chem., Inc.	Wilmington, Del.
318.	H. Kohnstamm & Co.	New York, N. Y.
319.	Koppers Products Co.	Pittsburgh, Pa.
320.	Krebs Pigment & Color Corp.	Newark, N. J.
321.	Kuhlman, Etabls.	Paris, France
322.	Kurt, Albert, G. M. B. H.	Amoneburg, Germany
323.	Lattimer-Goodwin Chem. Co.	Grand Junction, Ohio
324.	Laxseed Co.	New York, N. Y.
325.	Leghorn Trading Co., Inc.	New York, N. Y.
326.	Lehn & Fink Corp.	New York, N. Y.
327.	Theo. Leonhard Wax Co., Inc.	Haledon, Paterson, N. J.
328.	Lewis, C. H. & Co.	New York, N. Y.
329.	Lewis, John D., Inc.	Providence, R. I.
330.	Limestone Products Corp. of Amer.	Newton, N. J.
331.	Lincks, Geo. H.	New York, N. Y.
332.	Liquid Carbonic Corp.	Chicago, Ill.
333.	Litter, D. H., Co.	New York, N. Y.
334.	Littlejohn & Co., Inc.	New York, N. Y.
335.	Lucidol Corp.	Buffalo, N. Y.
336.	Geo. Lueders & Co.	New York, N. Y.
337.	Lundt & Co.	New York, N. Y.
338.	Maas & Waldstein	Newark, N. J.
339.	MacAndrews & Forbes Co.	New York, N. Y.
340.	Mackay, A. D.	New York, N. Y.
341.	Magnetic Pigment Co.	New York, N. Y.
342.	Magnus, Mabee & Reynard, Inc.	New York, N. Y.
343.	Makalot Corp.	Boston, Mass.
344.	Mallinckrodt Chemical Works	St. Louis, Mo.
345.	Malt Diastase Co.	Brooklyn, N. Y.
346.	Manchester Oxide Co.	Manchester, England
347.	Marbon Corp.	Gary, Ind.
348.	Marine Magnesium Prod. Corp.	S. San Francisco, Calif.
349.	Martin, Dennis Co.	Newark, N. J.
350.	Martin, L., Co.	New York, N. Y.
351.	Mathieson Alkali Co.	New York, N. Y.
352.	Maywood Chem. Works	Maywood, N. J.
353.	McCormick & Co.	Baltimore, Md.
354.	The McGean Chem. Co.	Cleveland, Ohio
355.	McKesson & Robbins, Inc.	New York, N. Y.
356.	McLaughlin, Gormley, King & Co.	Minneapolis, Minn.
357.	Mearl Corp.	New York, N. Y.
358.	Mechling Bros. Chem. Co.	Camden, N. J.
359.	E. Meer & Co., Inc.	New York, N. Y.
360.	Merck & Co.	Rahway, N. J.
361.	Merrimac Chemical Co.	Boston, Mass.
362.	Metro-Nite Co.	Milwaukee, Wis.

No.	Name	Address
363.	Meyer & Sons, J.	Philadelphia, Pa.
364.	Mica Insulator Co.	New York, N. Y.
365.	Michel Export Co.	New York, N. Y.
366.	Michigan Alkali Co.	New York, N. Y.
367.	Miller, Carl F., Co.	Seattle, Wash.
368.	Monsanto Chem. Works	St. Louis, Mo.
369.	Moore-Munger	New York, N. Y.
370.	Morningstar, Nicol, Inc.	New York, N. Y.
371.	Morton Salt Co.	Chicago, Ill.
372.	Murphy Varnish Co.	Newark, N. J.
373.	Mutual Chem. Co. of Amer.	New York, N. Y.
374.	Mutual Chem. Co. of America	New York, N. Y.
375.	Mutual Citrus Products Co.	Anaheim, Calif.
376.	National Aluminate Corp.	Chicago, Ill.
377.	Nat'l Ammonia Co., Inc.	Philadelphia, Pa.
379.	Nat'l Aniline & Chem. Wks.	New York, N. Y.
380.	National Lead Co.	New York, N. Y.
381.	National Oil Products Co.	Harrison, N. J.
382.	Nat'l Pigments & Chem. Co.	St. Louis, Mo.
383.	National Rosin Oil & Size Co.	New York, N. Y.
384.	Naugatuck Chem. Co.	Naugatuck, Conn.
385.	Neville Co.	Pittsburgh, Pa.
386.	N. J. Laboratory Supply Co.	Newark, N. J.
387.	N. J. Zinc Co.	New York, N. Y.
388.	The N. Y. Quinine & Chem. Wks., Inc.	Brooklyn, N. Y.
390.	Newmann-Buslee & Wolfe, Inc.	Chicago, Ill.
391.	Newport Industries, Inc.	New York City
392.	Niacet Chem. Co.	Niagara Falls, N. Y.
393.	Niagara Alkali Co.	New York, N. Y.
394.	Niagara Chemicals Corp.	Niagara Falls, N. Y.
395.	Niagara Smelting Corp.	Niagara Falls, N. Y.
396.	The Northwestern Chem. Co.	Wauwatosa, Wis.
397.	Norton Co.	Worcester, Mass.
398.	Norwich Pharmacal Co.	Norwich, N. Y.
399.	Novadel-Agene Corp.	Newark, N. J.
400.	Nulomoline Co.	New York, N. Y.
401.	Nuodex Products, Inc.	Elizabeth, N. J.
402.	Ohio-Apex, Inc.	Nitro, W. Va.
403.	Oil States Petroleum Co.	New York, N. Y.
404.	Oldbury Electro-Chem. Co.	New York, N. Y.
405.	Olive Branch Minerals Co.	Cairo, Ill.
406.	Onyx Oil & Chem. Co.	Passaic, N. J.
407.	Orbis Products Corp.	New York, N. Y.
408.	Papermakers' Chem. Corp.	Wilmington, Del.
409.	Paramet Chem. Corp.	Long Island City, N. Y.
410.	Parke, Davis & Co.	Detroit, Mich.
411.	Parker Rust Proof Co.	Detroit, Mich.
412.	Patent Chemicals, Inc.	New York, N. Y.
413.	Peek & Velsor, Inc.	New York, N. Y.
414.	Penick, S. B. & Co.	New York, N. Y.
415.	Penn. Alcohol Corp.	Philadelphia, Pa.
416.	Penn. Coal Products Co.	Petrolia, Pa.
417.	Penn.-Dixie Cement Corp.	New York City
418.	Penn. Industrial Chem. Corp.	Clairton, Pa.
419.	Penn. Refining Co.	Butler, Pa.
420.	Penn. Salt Mfg. Co.	Philadelphia, Pa.
421.	Pfaltz-Bauer, Inc.	New York, N. Y.
422.	Pfizer, Chas. & Co., Inc.	New York, N. Y.
423.	Phila. Quartz Co.	Philadelphia, Pa.
424.	Philipp Bros.	New York, N. Y.
425.	Pittsburgh Plate Glass Co.	Pittsburgh, Pa.
426.	Plaskon Corp.	Toledo, Ohio

No.	Name	Address
427.	Plymouth Organic Labs.	New York, N. Y.
428.	Pollopas, Ltd.	London, England
429.	Powhatan Mining Corp.	Woodlawn, Baltimore, Md.
430.	Pray, W. P.	New York, N. Y.
431.	Prior Chem. Corp.	New York, N. Y.
432.	Procter & Gamble Co.	Cincinnati, Ohio
433.	Provident Chem. Wks.	St. Louis, Mo.
434.	Publicker, Inc.	Philadelphia, Pa.
435.	Pure Calcium Products Co.	Painesville, Ohio
436.	Pylam Products Co.	New York, N. Y.
437.	Ransom, L. E., Co.	New York, N. Y.
438.	Robert Rauh, Inc.	Newark, N. J.
439.	Read, Chas. L. & Co., Inc.	New York, N. Y.
440.	Reichhold Chemicals, Inc.	Detroit, Mich.
441.	Reilly Tar & Chem. Corp.	Indianapolis, Ind.
442.	Resinous Prod. & Chem. Co.	Philadelphia, Pa.
442a.	Resinox Corp.	New York City
443.	Revertex Corp.	Brooklyn, N. Y.
444.	Revson, R. F., Co.	New York, N. Y.
445.	Rhone-Poulene, Inc.	Paris, France
446.	Richards Chem. Works	Jersey City, N. J.
447.	Riverside Chem. Co.	No. Tonawanda, N. Y.
448.	Robeson Process Co.	New York, N. Y.
449.	Rogers & McClellan	Boston, Mass.
450.	Rohm & Haas	Philadelphia, Pa.
451.	Rosenthal, H. H., Co.	New York, N. Y.
452.	Ross, Frank B., Co., Inc.	New York, N. Y.
453.	Ross-Rowe, Inc.	New York, N. Y.
454.	Royce Chem. Co.	Carlton Hill, N. J.
455.	Rubber Service Labs. Co.	Akron, Ohio
456.	Russell, W. R. & Co.	New York, N. Y.
457.	Russia Cement Co.	Gloucester, Mass.
458.	Ryland, H. C., Inc.	New York, N. Y.
459.	Saginaw Salt Products Co.	Saginaw, Mich.
460.	Salomon, L. A. & Bro.	New York, N. Y.
461.	Samuelson & Co., P.	London, England
462.	Sandoz Chem. Works	New York, N. Y.
463.	Scheel, Wm. H.	New York, N. Y.
464.	Schimmel & Co.	New York, N. Y.
464a.	Schliemann Co., Inc.	New York City
465.	Schofield-Daniel Co.	New York City
466.	Scholler Bros., Inc.	Philadelphia, Pa.
467.	F. E. Schundler & Co.	Joliet, Ill.
468.	Schuylkill Chem. Co.	Philadelphia, Pa.
469.	Schwabacher, S. & Co., Inc.	New York, N. Y.
470.	Scientific Glass Apparatus Co.	Bloomfield, N. J.
471.	Scott, Bader & Co.	London, England
472.	Seacoast Laboratories	New York, N. Y.
473.	Edwin Seebach Co.	New York, N. Y.
474.	Seeley & Co., Inc.	New York, N. Y.
475.	Seldner & Enequist, Inc.	Brooklyn, N. Y.
476.	Serinsky, Moses, Co.	Indianapolis, Ind.
477.	Sharples Solvents Corp.	Philadelphia, Pa.
478.	Shawinigan, Ltd.	New York, N. Y.
479.	Shepherd Chem. Co.	Norwood, Cincinnati, Ohio
480.	Sherka Chem. Co., Inc.	Bloomfield, N. J.
482.	Sherwood Petroleum Co.	Englewood, N. J.
483.	Thomas J. Shields Co.	New York, N. Y.
484.	Siemon Colors, Inc.	Newark, N. J.
485.	Siemon & Co.	Bridgeport, Conn.
486.	Silica Products Co.	Kansas City, Mo.
487.	Silver, Geo., Import Co.	New York, N. Y.

No.	Name	Address
488.	Sinclair Refining Co.	Olmstead, Ill.
489.	Skelly Oil Co.	Chicago, Ill.
490.	Smith Chem. & Color Co.	Brooklyn, N. Y.
491.	Smith & Nickols, Inc.	New York, N. Y.
492.	Smith, Werner G., Co.	Cleveland, Ohio
493.	Solvay Sales Corp.	New York, N. Y.
494.	Sonneborn, L., Sons	New York, N. Y.
495.	Southern Mica Co.	Franklin, N. C.
496.	Southern Pine Chem. Co.	Jacksonville, Fla.
497.	Southwark Mfg. Co.	Camden, N. J.
498.	Sparhawk Co.	Sparkhill, N. Y.
499.	Spencer Kellogg & Sons Sales Corp.	Buffalo, N. Y.
500.	A. E. Staley Mfg. Co.	Decatur, Ill.
501.	Stamford Rubber Supply Co.	Stamford, Conn.
502.	Stanco Distributors	New York, N. Y.
503.	Standard Alcohol Co.	New York, N. Y.
504.	Standard Brands, Inc.	New York, N. Y.
505.	Standard Oil Co. of Calif.	San Francisco, Calif.
506.	Standard Oil Co. of Indiana	Chicago, Ill.
507.	Standard Oil Co. of N. J.	New York, N. Y.
508.	Standard Oil Co. of N. Y.	New York, N. Y.
509.	Standard Silicate Co.	Pittsburgh, Pa.
510.	Standard Ultramarine Co.	Huntington, W. Va.
511.	Stanley Co., John T.	New York City
512.	Starch Products Co.	New York, N. Y.
513.	Stauffer Chem. Co.	New York, N. Y.
514.	Stauffer Chem. Co. of Texas	Freeport, Texas
515.	Stein, Hall & Co.	New York, N. Y.
516.	Stokes & Smith Co.	Philadelphia, Pa.
517.	Strahl & Pitsch	New York, N. Y.
518.	Strohmeyer & Arpe Co.	New York, N. Y.
519.	Stroock & Wittenberg Corp.	New York, N. Y.
520.	Sun Oil Co.	Philadelphia, Pa.
521.	Swann Chemical Co.	New York, N. Y.
522.	Synfleur Scientific Labs.	Monticello, N. Y.
523.	Synthane Corp.	Oaks, Pa.
524.	The Synthetic Products Co.	Cleveland, Ohio
525.	Taintor Trading Co.	New York, N. Y.
526.	Takamine Laboratory, Inc.	Clifton, N. J.
527.	Tamms Silica Co.	Chicago, Ill.
528.	Tanners Supply Co.	Grand Rapids, Mich.
529.	Tannin Corp.	New York, N. Y.
530.	C. Tennant & Sons Co. of N. Y.	New York, N. Y.
531.	Tenn. Eastman Corp.	Kingsport, Tenn.
532.	Texas Chem. Co.	Houston, Texas
533.	Texas Mining & Smelting Co.	Laredo, Texas
534.	Thomas, Arthur H., Co.	Philadelphia, Pa.
535.	Thorocide, Inc.	St. Louis, Mo.
536.	Thurston & Braidich	New York, N. Y.
537.	Titanium Alloy Mfg. Co.	Niagara Falls, N. Y.
538.	Titanium Pigments Co.	New York, N. Y.
539.	Tobacco By-Products & Chem. Corp.	Louisville, Ky.
540.	Trask, Arthur C., Co.	Chicago, Ill.
541.	Trojan Powder Co.	Allentown, Pa.
542.	Turner, Joseph & Co.	Ridgefield, N. J.
543.	Uhe, George Co.	New York, N. Y.
544.	Uhlich, Paul Co.	New York, N. Y.
545.	Union Oil Co.	Los Angeles, Calif.
546.	Union Smelting & Refining Co., Inc.	Newark, N. J.
547.	United Carbon Co.	Charleston, W. Va.
547a.	United Clay Mines Corp.	Trenton, N. J.
548.	United Color & Pigment Co.	Newark, N. J.

No.	Name	Address
549.	U. S. Bronze Powder Works, Inc.	New York, N. Y.
550.	U. S. Gypsum Co.	Chicago, Ill.
551.	U. S. Industrial Alcohol Co.	New York, N. Y.
552.	U. S. Industrial Chem. Co.	New York, N. Y.
553.	U. S. Phosphoric Prod. Corp.	New York, N. Y.
554.	U. S. Rubber Products, Inc.	New York, N. Y.
555.	U. S. Smelting, Refining & Mining Co.	New York, N. Y.
556.	Utah Gilsonite Co.	St. Louis, Mo.
557.	Van Allen, L. R. & Co.	Chicago, Ill.
558.	Van-Ameringen Haebler, Inc.	New York, N. Y.
559.	Vanderbilt, R. T., Co.	New York, N. Y.
560.	Van Dyk & Co., Inc.	Jersey City, N. J.
561.	Van Schaack Bros. Chem. Co.	Chicago, Ill.
562.	Varcum Chem. Corp.	Niagara Falls, N. Y.
563.	Verley, Albert & Co.	Chicago, Ill.
564.	Verona Chem. Co.	Newark, N. J.
565.	Victor Chem. Works	Chicago, Ill.
566.	Virginia-Carolina Chem. Corp.	Richmond, Va.
567.	Virginia Smelting Works	W. Norfolk, Va.
568.	Vitro Mfg. Co.	Pittsburgh, Pa.
570.	Vultex Chem. Co.	Cambridge, Mass.
571.	Waldo, E. M. & F., Inc.	Muirkirk, Md.
572.	Wallerstein Co., Inc.	New York, N. Y.
573.	The Warner Chem. Co.	New York, N. Y.
574.	Warwick Chem. Co.	West Warwick, R. I.
575.	Welch, Holme & Clark Co.	New York, N. Y.
576.	Welsbach & Co.	Gloucester, N. J.
577.	Werk, M., Co.	Cincinnati, Ohio
578.	Western Charcoal Co.	Chicago, Ill.
579.	Westinghouse Elec. & Mfg. Co.	E. Pittsburgh, Pa.
580.	Whittaker, Clark & Daniels	New York, N. Y.
581.	Wiffen & Co., Sons, Ltd.	London, England
582.	Wilckes-Martin-Wilckes Co.	New York, N. Y.
583.	Will & Baumer Candle Co.	New York, N. Y.
584.	C. K. Williams & Co.	Easton, Pa.
585.	The Wilson Laboratories	Chicago, Ill.
586.	Wishnick-Tumpeer, Inc.	New York, N. Y.
587.	Woburn Degreasing Co.	Harrison, N. J.
588.	Wolf, Jacques & Co.	Passaic, N. J.
589.	Wood Flour, Inc.	Manchester, N. H.
590.	Wood Ridge Mfg. Co.	Wood Ridge, N. J.
591.	Wyodak Chem. Co.	Cleveland, Ohio
593.	Young, J. S. & Co.	Hanover, Pa.
595.	Zinsser, Wm. & Co.	New York, N. Y.
596.	Zophar Mills, Inc.	Brooklyn, N. Y.

ADDENDA

597.	Borne-Scrymser Co.	New York, N. Y.
598.	Edwal Labs.	Chicago, Ill.
598a.	J. Einstein	Flushing, N. Y.
599.	Hycar Corp.	Akron, O.
600.	Martin Laboratories	New York, N. Y.
601.	Metasap Chem. Co.	Harrison, N. J.
602.	Quaker Oats Co.	Chicago, Ill.
603.	Rochester Gas & Elec. Corp.	Rochester, N. Y.
604.	Schering Corp.	Bloomfield, N. J.
605.	Sherwin-Williams Co.	Cleveland, O.
606.	Stanton Lab.	Wyncote, Pa.

WHERE TO BUY CHEMICALS OUTSIDE
THE UNITED STATES

Argentina, Buenos Aires—M. Goetz, Rincón 332

Australia, Adelaide—Robert Bryce & Co., Pty., Ltd., 73–75 Wakefield Street

—————, Melbourne, C1—Robert Bryce & Co., Pty., Ltd., 526 Little Bourke Melbourne Street

—————, Sydney—Robert Bryce & Co., Pty., Ltd., 188–190 Kent Street

Bolivia, La Paz—M. Romulo Vildoso, Calle Potosi 137

Brazil, Sao Paulo—Empresa Commercial Mercur Ltda., Caixa Postale 4232

Canada, Montreal—Chemicals, Ltd., 384 St. Paul St., W.

—————, Toronto 2—Canada Colors & Chemicals, Ltd., 1090 King St., West

—————, Vancouver, B. C.—Shanahan's Ltd., Foot of Campbell Avenue

Chili, Santiago—Bernardo Dornblatt, Classificador 195B

Cuba, Havana—Raul Guillent, 215 Bank of Nova Scotia Bldg., P. O. Box 1133

England, London—Rex Campbell & Co., Ltd., 7, Idol Lane, Eastcheap, E.C. 3

India, Calcutta—Kaisers Trading Company, 159 Lower Chitpore Rd.

Mexico, Mexico D. F.—R. Koestinger, J. M. Velasco, 119, Insurgentes, Mixcoac

New Zealand, Wellington—Robert Bryce & Co., Pty., Ltd., 19 Lower Tory Street

South Africa, Johannesburg—Philip Elzas & Co., 132 London House, Loveday St.

Sweden, Huddinge—E. Landerholm

Switzerland, Zurich—Oswald E. Boll, Dufourstrasse 157

Uruguay, Montevideo—Compania Industrial Alfa, Lda., Porongos 2228

Venezuela, Caracas—A. G. Bulgaris, Apartado 1752

INDEX

A

Abrasive, Diamond 347
  Synthetic 347
  Wheel, Bonding Adhesive for.... 30
Absorbent Solution, Regenerating
  Cuprous Chloride 416
Absorption Base 85
Acetic Acid Tablets 91
Acknowledgments 427
Acne Lotion 92
Adhesive (see also Agglutinant,
  Calking, Cement, Glue, Muci-
  lage, Paste, Putty, Seal)....12, 19
  for Abrasive Wheel, Bonding..... 30
  Calcium Sucrate 33
  Casein 25
  Celluloid to Hard Rubber 38
  for Cellulose Acetate 40
  Cellulose Acetate Foil 40
  Cheap Starch 21
  China 47
  Container-Joint Thermosetting .. 42
  Decalcomania 34
  Denture 48, 75
  Dextrin 23
  Envelope 18
  Ethyl Cellulose 40
  Flexible Thermoplastic 41
  Glass to Metal 43
  High-Melting Thermoplastic 41
  Laminating 27, 41
  Lantern Slide 43
  Latex 34
  Light Polyvinyl 40
  Millboard 44
  Moistureproof Cellophane 32
  Neutral Starch 21
  Non-Setting 27
  Paper 34
  Paper Board 33
  Permanently Tacky Pressure 38
  Plywood 33
  for Polyvinyl Chloride Plastics.. 40
  Potato-Starch 23
  Pressure-Sensitive 45
  for Rectifiers, Conductive 48
  Rubber-Resin 39
  Sodium Silicate 28
  Solution, Rosin 38
  for Sound-Deadening Pads 44
  Starch 20
  Stoneware 47
  Surgical Strapping 45

Adhesive—Continued
  Synthetic Resin 33
  Tape 45
  Tape, Pressure-Sensitive 45
  Thermoplastic 41
  Thermosetting 43
  Transparent Resin 41
  Veneer 21
  Wall-Paper 18
  Water-Insoluble 34
  Waterproof Surgical 45
  Waxed-Paper 30
  Wood Glue 33
Agglutinant, Water-Soluble 33
Agricultural Insecticide (see also
  Insecticide) 121
Air Deodorant 110
  Odor Neutralizer 109
  Sterilization 110
Alberene Cement 47
Alcohol Proof and Percentage Table.425
Alloy for Brass, Brazing 268
  Brazing 268
  Brazing, for Welding to Iron....268
  Cadmium Bearing 261
  Cleaning and Covering for Molten
    Copper 262
  Corrosion-Resistant Denture.....264
  Dental Casting 263
  Dental Clasp 264
  Dental Filling 263
  Dental Inlay 264
  Dental Model 264
  Electric Contact 262
  Electric Fuse 261
  Electric Resistance 262
  Electric Wire Resistance 262
  Hard Platinum 263
  Hardened-Lead Bearing 261
  Heat-Hardenable Copper 262
  Iron Bearing 261
  Jewelry 263
  Lead 261
  Non-Corrosive Lead 261
  Refining White Metal 263
  Resistor 262
  Spark Plug Electrode 262
Allspice, Imitation 54
Almond Extract 54
Alumina Sol 325
Aluminum Alkyd, Ready Mixed....292
  Alloy Casting, Hardened 260
  Coating 252
  Plating 252

Aluminum—Continued
Pre-Painting Treatment 249
Stripping Oxide Film from 249
Vehicle 292
Ammonia, Substitute for Household.373
Ammunition Primer 349
Animal Tissue Fixative 416
Ant Poison 8
Antacid Mixture 91
Anthelmintic Emulsion, Tetrachloro-
ethylene 129
Sheep 129
Tablet 104
Anti-Acid Powder 91
Anti-Freeze 15
Auto 414
Liquid 16
Non-Foaming 414
Anti-Incendiary Material 410
Anti-Mist Paper for Glass 326
Anti-Perspirant 87
Liquid 87
Anti-Perspiration Soap 87
Antipruritic Lotion 95
Antiseptic 86
Dental 76
Dusting Powder 70
Dye 93
Hand Soap 87
Military Foot Powder 89
Soap 87
Solution 86
Telephone Mouthpiece 110
Wash 93
Wash, Fish and Cannery Factory.372
Apparatus 3
Where to Buy 5
Apple Confection 150
Leather, Old-Fashioned 151
Preventing Dropping of 118
Asbestos Cloth Substitute 405
Roofing Paper 240
Aseptic Wax 97
Asphalt Aggregate, Light-Weight...240
Emulsion 44
Improving Adhesion of 240
Asthma Cigarette 83
Inhalant, Hay-Fever 83
Nose Drops 83
Smoke Powder 83
Spray 82
Astringent Dusting Powder 70
Athlete's Foot Powder 89
Foot Remedy 90
Foot, Soaking Bath for 90
Rub 89
Auto Rubbing Compound 347

B

Baking Powder 174
Bandoline 78

Barbecue Sauce for Pumping Barbe-
cue Hams 199
Barium Sulfate Suspension, Medic-
inal 91
Basic Chrome Nitrate—Acetate 396
Bath, Bubble 109
Bathing Composition 79
Bay Rum 89
Bearing, Non-Corrosive 260
Bedbug Exterminator 8
Beetles, Control of Hide and Skin..218
Beverage, Prepared Lemon 58
Syrup, Quick-Dissolving 54
Flavors and 51
Bird Food, Canary 9
Bitter-Tasting Product, Non-Toxic..105
Bituminous Composition, Tough....358
Blackboard, White 235
Blasting Powder 349
Primer, Non-Combustible 349
Bleach on Union Goods 387
Wood Floor 12
Bleaching Jute 387
Nut Shells 143
Textiles 387
Blueing Steel 250
Blueprint, Correction Fluid for 340
Lacquer for 340
Paper 339, 340
Red Writing Fluid for 340
Bockwurst 209
Boiler Compound 15
Scale Compound 412
Water Treatment, Processing Tan-
nin 412
Bologna, Ring 211
with Soya Flour 212
Bonding Aluminum to Steel 49
Bordeaux Mixture 127
Bow Rosins 359
Boxwood Leaf Miner Spray 122
Brake and Clutch Frictional Mate-
rial 411
Brazing Wire, Copper 268
Breaking Emulsion 114
Brick, Hard Refractory Non-Acid..237
Tile, High-Strength 236
Brine Bearing Earth, Plugging of..407
Brine Cooling Fluid 407
Bristle, Bactericidal 221
Broiler All-Mash 132
Feed 132
Bronze, Tin-Free Gear 262
Bubble Bath 109
Buffing Compound 347, 348
Wax, Cream 348
Bug Control 124
Bunan Oil-Resistant Coating.......365
Burn Cream 97
Emulsion, Tannic Acid 97
Film Treatment, Wound and..... 96
Ointment 97
Butadiene, Polymerizing 358

Butadiene Polymers, Vulcanization of 360
Butadiene, Separation of 406
Butter, Non-Melting 181
  Production, Reducing Churning Time in 181
  Spread, Stabilized Homogenized..181
  Substitute, Fat-Free 181
Buttercream 161
Buttermilk, Cultured 182
Butterscotch Sauce 152
Buyer's Guide 430
Buying Chemicals and Apparatus... 5

C

Cadmium, Zinc and Tin Coating, Identifying 257
Cake, Coffee 155
  Strawberry and Other Short 155
Calculating Costs 5
Calf Feed, Dry 137
  Meal 136
  Starter Mixture, Reinforced 138
Calking Compound 48
  Compound, Metal Filler and...... 48
  Wax, Boat 48
Callous Skin Remover 91
  Softener and Remover 90
Calorizing Metal 259
Camouflage Coating 281
  Face Cream 61
Can Joint Seal 34
  Seal 32
Canary Bird Food 9
Candle 12
  Coating 353
Candy with Chocolate Flavor 148
  Fig-Flavored 150
  Grapefruit-Flavored 149
  Jellied Tart 147
  Jelly 147
  Licorice-Flavored 148
  Orange-Flavored 149
  Pectin 146
Canned Fruit Salad 139
  Jellied Consommé 140
  Tomato Aspic 141
Canning, Precautions in 139
Canvas, Fireproofing 15
Capsule Filling Hints 106
Caramel Wafer Shell 173
Carbon, Activated 417
  Paper, Transfer 217
Carob Beans, Dehusking 132
Case-Hardening Compound 259
  Iron or Steel 259
  Small Articles 259
Casein Fiber Resistant to Water, Making 404
Cattle Grub Control 129
  Louse Dust 129
  Salt Iodine Blocks 138

Cedar Chest Compound 98
Celery Salt 184
Cellulose Acetate, Improving Stretch of 358
  Alpha 417
  Foil, Moistureproofing for 328
Cement. See also Adhesive, Agglutinant, Calking, Glue, Mucilage, Paste, Putty, Seal.
  Alberene 47
  Aquarium 12
  Brown 38
  Building 234
  Buna N 35
  Buna S 35
  Butyl 35
  for Ceramics 30
  Chlorinated Rubber 37
  for Crockery, Hard 39
  for Cured GR"S" Coatings, Self-Curing Lean 362
  Fine Plastic Building 234
  Floor Hardener 12, 235
  Floor, Preparing for Painting...235
  Flooring, Waterproofing 235
  for Glass, Rubber 38
  for High Temperatures, Iron ... 30
  High-Vacuum 43
  Hycar 35
  Hycar O.R. 37
  Hydraulic 234
  Insole 37
  Lacquer 44
  for Leather Driving Belts 37
  Lens 43
  Low-Temperature-Curing 35
  for Marble and Alabaster 30
  Metal to Metal 38
  Mix, Increasing Fluidity of 234
  Mixed 36
  Neoprene 35
  Oil-Well Water-Sealing 49
  Oxychloride 234
  Pettman 44
  Photo-Mounting 35
  Pipe-Joint 42
  Quick-Setting Waterproof 44
  Red Ebonite 36
  Refractory 48
  Rosin Linoleum 39
  Rubber to Metal 37
  Rubber Resin 39
  Semi-Active Black 35
  Soft, Black, Stable 36
  Stainless-Steel Pipe-Joint 41
  Stoneware 47
  and Stucco, Waterproofing for ...287
  Synthetic Rubber 34
  Tacky Channel Black 35
  Thiokol 35
  for Uncured Vinylite-Bakelite Coated Cloth 39
  Universal 44

Cement—Continued
Vinylite 284
Vistanex 35
Watch Crystal 43
Waterproofing 14
White 36
Cementing Leather with Gutta-Per-
cha 49
Form for Slip Casting 238
Ceramic Made with Heat 366
Switch Composition 237
Chapped Lip Stick 109
Charcoal, Activated 417
Cheese Coating, Non-Molding 182
Coating Wax 183
Reducing Blowing in 182
Chemicals and Supplies, Sellers of..444
and Supplies, Where to Buy...5, 430
Cherry Fruit Fly Control 122
with Other Natural Flavors 52
Chest Rub 7
Rub Salve 8
Chick Starter 133
Starting and Growing Mash 132
Chicken, Smoked Roast 202
Chigger Bites Treatment 98
Chili Con Carne 201
Chloracne Treatment 92
Chocolate, Reducing Viscosity of...153
Chromate Skin Injuries, Salve for.. 93
Chromium-Plating Bath 251
Cast Iron 251
Clarification, Filtering and 4
Clay, Molding 366
Cleaner (see also Detergent, Soap,
Spot Remover, Stain Remover) 13
Aluminum 375
Brass 376
Drawing Pen 374
Household 13
Leather 373, 374
Linotype Matrix 376
Metal Machinery 375
and Mist Preventer, Spectacle
Lens 372
Motor Cylinder 375
Non-Corrosive Tin Plate 374
Paint-Brush 14
Powdered 369
Pre-Anodizing Metal 375
Printing Plate 375
Rug Dry 374
Straw-Hat 13
Type 375
Universal 375
Wall-Paper 13
Window 13
Window and Glassware 373
Cleaners' Soap, Dry 374
Cleaning Bath, Electrolytic Metal..254
Bath, Non-Aqueous Metal 249
Bottles 372

Cleaning—Continued
Eggs, Solution for 130
Fluid, Antiseptic Dry 374
Fluid, Dry 13
Fluid, Jewelry 374
Paste 369
Powder, Rug and Textile 374
and Protecting Metal 244
Solution, Non-Corrosive Labora-
tory Glass 373
Cleanser, Liquid Skin 59
Pile Fabric 374
Cloth, Drum Head and Banjo......405
Fireproof 400
Manufacture of Organdie 404
Substitute, Asbestos 405
Clutch Frictional Material, Brake
and 411
Plate Stud 411
Coal-Tar Skin Treatment 92
Coating. See also Lacquer, Paint,
Varnish.
for Acetate Sheet, Moistureproof.357
Acoustic Sound-Deadening 310
Airplane Deicing 414
for Aluminum, Protective 310
for Asphalt Containers, Non-Stick.311
for Asphalt Drums 310
Base, Artificial Leather.318, 319, 320
Biscuit 162
Bunan Oil-Resistant 365
Candle 353
Cathode or X-Ray Tube Screen..414
for Cellophane, Moistureproof....358
for Cloth, Flexible 285
Composition, Fusible 312
Composition for Ice Trays 310
Dark Chocolate 162
Fire-Resistant 315
Flameproof, Waterproof and Mil-
dewproof Textile 399
Fluorescent Lamp 409
Foundry Mold 310
Fruit Protective 312, 313, 314
Lemon 162
or Lining for Iron Pipes 310
for Lithographic Plates, Protec-
tive 310
Malt Milk 162
Milk Bottle Cap 328, 358
Moistureproof 285
Moistureproof Hot Melt 310
Neoprene Latex 365
Neoprene Olive Drab 364
Nitrocellulose 366
Oilproof Barrel 311
Orange 162
Paint or Lacquer "Stop-Off"....311
for Paper Containers, Oilproof...311
for Petroleum Refinery Equip-
ment 310
Special 309

Coating—Continued
of Stainless Steel, Black 250
Steel, Black 278
for Stone Top Tables, Protective.312
Vehicle, Wall 292
White 163
Wine Tank 312
Cocoa Coating, Heat-Resistant.....154
Coating, Heat-Resistant Candy
and Cake 153
Malt Powder 9
Powder, Sweet 9
Codling Moth Ovicide and Larvi-
cide 123
Moth Spray 123
Cod-Liver Oil, Orange Juice, Malt
Extract Emulsion 107
Coke, Easily Inflammable 407
Colloid 111
Colloidal Graphite, Aqueous 114
Sulfur 113
Cologne, Cream 72
Eau de 72
Coloring. See also Dyeing.
Cadmium Brown 250
Stainless Steel 250
Compounding, Containers for 3
Concrete. See also Cement.
Made with Sea Water 234
Plasticizing 234
Road Protective 234
Consultants 5
Containers for Compounding 3
Conversion Table 422, 423
of Thermometer Readings 424
Copper, Antique Coloring for 9
Deplating of Steel 254
Plating, Simple 253
Powdered 258
Cork, Glue Binder for 20
Substitute 240
Corn Collodion 90
Remover 90
Corrosion Inhibitor, Lubricating
Oil 230
Inhibitor, Pickling 244
of Metal in Contact with Chloro-
form, Preventing 246
Preventing Moisture Absorption
and 246
Resisting Coating for Copper.....262
Silica Gel for Preventing 246
Corrosionproofing Magnesium Al-
loys 247
Corrosive Chemicals 4
Effect of Brine on Refrigerating
Machinery, Preventing 244
Cosmetic Stocking 108
and Drug Products 59
Costs, Calculating 5
Cotton Duck, Waterproofing 403
Crab Meat, Preservation of 213

Crayon, Hot Metal Marking 217
Luminous 217
Marking 9
Red Marking 217
Cream, Acid 62
Artificial 175
Brushless Shaving 7, 80
Burn 97
Camouflage Face 61
Cleansing 5, 59
Cold 5, 6, 63
Cologne 72
Concentrated Day 62
Cosmetic Base 63
Deodorant 88
Four-Purpose 62
Germicidal 86
Hair 79
Hand 67
Hand Protective 67
Hormone 61
Insect-Repellent 98
Medicinal Vanishing 103
with Milk, Diluting 176
Modern Tissue 63
Nail 67
Neutral Cleansing 2
Non-Greasy Cold 6
Shoe 343, 344
Skin-Treatment 92
Substitute for Coffee 175
Sulfathiazole—Urea 61
Suntan 64
Tissue 63
Toilet 60
Under Arm 87
Vanishing 6, 67
Waterproofing Skin 92
Whipped 160
Whipping Aid 176
Crease Preserving Composition.....388
Crucible, Refractory 237
Cuprous Oxide 417
Curd Milk, Soft 175
Curry Powder 184
Cut-Flower Preservative 118

D

Decalcomania 366
Adhesive 34
Decolorizing 4
Defoamer for Pulp and Paper
Stock 325
Defroster, Windshield 414
Dehusking Carob Beans 132
Deicing Coating, Airplane 414
Deinking Newspaper 328
Delustering Acetate Silk 380
Shiny Fabric 380
Delustrant. See Duller.

Dental Anesthetics, Topical 75
  Antiseptic 76
  Casting Alloy 263
  Cavity Cement Lining 75
  Clasp Alloy 264
  Copper Amalgam 263
  Cotton, Styptic 75
  Desensitizer 74
  Duplicating and Impression Com-
    pound 354, 355
  Epithelial Solvent 74
  Filling Alloy 263
  Inlay Alloy 264
  Instrument Sterilizing Fluid 75
  Investment 237, 267
  Model Alloy 264
  Paste, Pumice-Bentonite 73
  Plaster of Paris Cast, Separating
    Fluid for 236
  Pulp Devitalizer 75
Dentifrice, Antiseptic 73
  Salt-Lime 74
  Sodium Oleate 74
Dentin 74
Denture Alloy, Corrosion-Resistant.264
  Base, Ceramic 238
  Cleaner 76
Deodorant, Air 109, 110
  Body 87
  Cream 88
  Fungicide 88
  Spray 8
Deodorizer, Cooking (Fish) 214
Depilatory 108
  Feather and Animal Hair 130
Dermatitis Lotion 93
  Salve for Metal Poisoning 94
Detergent. See also Cleaner.
  Cake 369
  Disinfecting 372
  Paper Towel 372
Developer. See Photographic.
Diamond Dust Brooch 411
  Embedding Composition 258
Dielectric, Solid 409
Disinfectant 86
  Cresol 8
  Pine Oil 121
  Poultry House 129
  Spray 120
  Water 411
Disinfecting Detergent 372
  Seeds 128
  Shoes 86
Dissolving, Mixing and 4
Document, Detecting Falsified 339
  Restoring Faded 339, 419
Dope, Airplane 323
Doughnuts with Potatoes 171
Dressing, Belt 232
  Fine Leather 219
  Flavor for French 183

Dressing—Continued
  French 183
  GR"S" Rubber Belt 232
Drug Penetrating Aid 104
  Products, Cosmetics and 59
Dry Cleaning. See Cleaning.
Duller for Hosiery 380
Dyeing Aged Black on Loose Cot-
    ton 391
  Aralac 392–396
  Aralac and Cellulose Fiber Mix-
    tures 394
  Aralac-Rayon-Wool-Cotton 396
  Brightening Black 397
  Brush Bristles 390
  Casein Fibers 392–396
  Grass 390
  Improved Textile 397
  Logwood and Acid Colors in Same
    Bath 391
  Sisal and Manila Hemp 389
  Straw Hat Plaits 389
  Wool Bunting Cloth 391
Dyestuff for the Coloring of Tri-
    chlorethylene 419

E

Eau de Cologne 72
Eggs, Solution for Cleaning 130
Electric Capacitor 409
  Insulation. See Insulation.
  Joint, Filler for Taped 409
  Shaving Aid 80
Electrical Insulating Filling Com-
    pound 358
  Sealing Compound 39
Electroplating, Brush 253
Elementary Preparations 5
Elixir, Vitamin B 105
Embalming Fluid 109
Emulsion 111
  Asphalt 114
  Breaking 114
  Castor Oil 111
  Ceresin Wax 111
  Cetyl Alcohol 111
  Cod-Liver Oil 112
  Cod-Liver Oil, Orange Juice Malt
    Extract 107
  Cottonseed Pitch 114
  Cumar 112
  for Finishing Dyed Rayon 384
  Fruit Protective 142
  Hop Extract 55
  Kerosene Spray 126
  Lemon Oil 56
  Liquid Petrolatum 106
  Mothproofing 402
  Neat's Foot Oil 111
  Ointment 100
  Orange Oil 56
  Ouricury Wax 111

Emulsion—Continued
Paint, Congo Resin 303
Paint for Exterior Use, Water...285
Paint Paste 303
Paint Tint 304
Paraffin Wax 111
Pentaerithritol Abietate 113
Peppermint Oil 112
Polyisobutylene 112
Polymerization, Synthetic Rub-
ber 360
Polymerization of Vinyl Acetate..357
Polyvinyl Chloride Polymeriza-
tion 112
Spray, Dichloroethylether 111
"Stabelite" Ester 113
Stone Paint 286
Sulfur 114
Tetrachloroethylene Anthelmintic.129
Type Finish, Interior 300
Waterproofing Leather 218
Wool Oiling 385
Enamel. See also Paint.
Architectural White 308
Base for Tints, Quick Dry 298
Black Baking 279
Black Wrinkle 316
Boro-Silicate 240
Brown Floor 300
Floor 300
Four-Hour Brushing Black......299
Gray Floor 300
Green Floor 300
Heat-Resisting Black 278
Interior Quick-Dry 298
Liquid 279
Liquid, White Architectural.....308
Lusterless Olive Drab 281
Lusterless Sanding 281
Marine Interior 281
for Metal, Clear Vitreous 239
Non-Yellowing White Quick-Dry..298
Olive Drab Wrinkle 316
Quick-Dry Blue 299
Quick-Dry Green 299
Quick-Dry Red 299
Quick-Drying 281
Wagon and Tractor 274
Enamelled Iron, Vitreous Undercoat
for 239
Enteric-Pill Coating 106
Erasing Fluid for Tracing Cloth....217
Etch, Glass 216
Non-Ferrous Metallographic 256
Etchant for Silver Solders, Metallo-
graphic 256
Etching Fluid, Glass 10
Macroscopic Brass 256
Magnesium Metallographic Speci-
mens 257
Resist, Acid 257
Solution for Beryllium 256

Ethyl Cellulose 356
Experiments 2
Explosive Black Powder, Stable....349
Charge, Oil Well 349
Demolition 349
Exterminator, Bedbug 8
Vermin 120
Extra Reading 5
Extract, Pure Lemon 9
Eye Burn Treatment 84
Drops 93
Drops, Non-Irritating 83
Wash 93
Wash, Hay Fever 84
Wash, Neutral 84
Eyebrow Black 84
Dressing 80
Eyelids, Treating Hives on 84

F

Fabric, Acetate Rayon "Burn-
Out" 397, 398
Crêping 382
Fireproofing Light 15
GR"I" Compound for Laminat-
ing 364
Rotproofing 402
False Teeth Plate. See Denture.
Farm Specialties 115
Fatty Acids, Preventing Discolora-
tion of Higher 233
Feed, Calf 136
Dry Calf 137
Poultry 132
Preserving Green 138
Feminine-Hygiene Preparation 108
Fertilizer, Granulated 118
Fiber 379
Rotproofing 402
Softener and Lubricant for Arti-
ficial 385
Fiberboard, Oilproofing 15
Figuring 2
Filling, Buttercream 161
Chocolate Pie 168
French Custard Cream 164
Fruit-Flavor Pie 168
Lemon Pie 169
Powdered Pie 165
for Sugar Wafer 172
Film or Prints, Copper Sulfide In-
tensification of 331
Filter Aid 408
Filtering and Clarification 4
Finish for Calicos, Thickening
Satin 384
Interior Emulsion Type 300
Interior Flat Wall 295
Interior Semi-Gloss 296
for Laboratory Table Tops 324

Fire Extinguisher, Dry 15, 410
  Extinguisher, Foam 410
  Extinguisher Foam Stabilizer....410
  Extinguisher, Magnesium Alumi-
    num Powder 410
  Extinguishing Liquid 15
  Kindler 15
Firebrick 237
Fireproof Paper 15
Fireproofing Canvas 15
  Light Fabrics 15
Fish Preserving Ice 213
Flavor and Beverages 51
  Emulsifier and Stabilizer for..... 55
  Rye-Bread 55
  Vanilla (Artificial) 9
Flavoring from Orange Peels, Mar-
  malade 55
  Vehicle, Pharmaceutical 105
Floor Sweeping Compound 378
Flotation Frother, Fluorspar 114
  Reagent for Oxide Ores 114
Flour, Bleaching and Maturing.....174
  Prepared Biscuit 173
  Prepared Pancake 173
  Prepared Waffle 174
Flower Preservative, Cut 118
Fluid, Door Check 410
  Hydraulic 409
  Magnetizable 410
  Manometer 410
Fluorescent Lamp Coating 409
Flux, Aluminum or Magnesium
  Welding 270
  Aluminum Scrap Melting 270
  Brazing 269
  Copper Welding 269
  for Galvanized Iron, Soldering...266
  Gold Melting 269
  Hard Soldering 266
  Magnesium Refining 270
  for Melting Non-Ferrous Metals..270
  Multiple Ceramic 238
  Nickel Chrome Alloy Welding....269
  Non-Corrosive Magnesium 270
  Soldering 16, 265
  for Welding Aluminum 269
  for Welding Iron Base Alloys....270
  for Welding Steel Contaminated
    with Sulfur 270
Fly Control, Cherry Fruit 122
  Paper 8
  Spray 8
  Spray, D.D.T. 120
Fondant Icing Sugar 152
Food, Canary Bird 9
  Color, Liquid Green 175
  Color, Liquid Red 175
  Color, Liquid Yellow 175
  Products 139
  Protein 213
Foot Bath 89
  Perspiration, Control of 87
  Powder, Antiseptic Military 89

Foundry Mold Coating 310
  Sand, Magnesium Alloy 266
Frankfurter, High-Grade 207
  Skinless 207
Fruit Concentrates, Pure 51
  Dessert, New 141
  Ice Cream or Soda Fountain
    Crushed 57
  Preventing Discoloration of Cut..142
  Protective Coating 142
  Syrup for Pickling 142
Fuel, Anti-Knock Motor 406
  Colloidal 407
  Non-Clouding Aviation 406
  Oil, Improving 407
Fudge, Chocolate 162
Fumes, Settling 417
Fungicide, Deodorant 88
Fungous Diseases in Pet Fish,
  Treating 130
Fur Bleaching 223, 226
  Carroting Solution 222
  Dyeing 223
  Felt Hat, After-Chrome for 223
  Wax Finish for 223
Furnace Luting Lining, Electric....237
Furniture Filler, Wax 46, 315
  Filling Scratches on Varnished...315

G

Galvanized Coating, Repairing Dam-
  aged 249
Garden Specialties 115
Gargle Powder 73
Gasoline, Solidified 15
  Stabilizing Tetraethyl Lead 407
Gel, Glycerin Face 62
Gelatin Composition for True-to-
  Scale Prints 329
  Masses, Coloring 366
  Solution, Preserving 408
Germicidal Bombs 86
  Cream 86
Germicide 86
  Anaerobic Wound Infection 97
Getter, Vacuum Tube 409
Gilder's Size 311
Gingivitis Mouth Wash 74
Glass Batch, Clear Colorless 239
  Batch, High-Silica 240
  Batch, Optical 239
  Etching Fluid 10
  Frosting 322
  Low-Expansion 239
  Ruby 239
  Setting Composition, Plate 43
  Silver Mirror on 253
  "Glassine" Paper 14
Glaze. See also Icing.
  Bakers' 154
  Fruit and Berry 155
  High-Temperature-Resistant Ce-
    ramic 238
  Pecan Bun 156

Glazed Topping for Coffee Cake....155
Glove, Invisible 67
Glue. See also Adhesive, Aggluti-
  nant, Calking, Cement, Muci-
  lage, Paste, Putty, Seal.
  Binder for Cork 20
  Bookbinders' Flexible 32
  Coating, Grease-Resistant 327
  Dried-Blood 28
  Flexible 2
  Gum Arabic 30
  "Iceproof" 18
  Magazine and Catalog 31
  Marine 46
  Non-Warping Paper 19
  Plywood 41
  Size 19
  Soybean 28
  from Sweet Potatoes, Salt 23
  Veneer 41
  Water-Resistant 18
Glycerin Gelatin Base 153
  Substitute in Cosmetics 107
Grafting Wax 11, 45, 119
  Wax Salve 119
Grain and Mash 132
Grape with Other Natural Flavors. 52
  Shipment Protection 142
Graphite, Aqueous Colloidal 114
  Grease 11
Grease, High-Temperature Stop-
  cock 232
Greaseproofing Paper and Fiber-
  board 15
Grinding, Pulverizing and 4
  Wheel. See Abrasive Wheel.
GR"S" Compounds 362, 364
Gum. See Adhesive, Agglutinant,
  Calking, Cement, Glue, Mu-
  cilage, Paste, Putty, Seal.
Gypsum, Waterproofing Cast 235

H

Hair. See also Permanent.
  Bluing for White 80
  Cream 79
  Gloss 78
  Lotion 76
  Pomade 78
  Pomade, Wavy 79
  "Tonic" 76
  Waving Fluid 77
  Waving Solutions, Cold 77
Ham, New England Style Pressed..185
Hamburger Style Patties 186
Hand Cleaner without Water 371
  Cleaning Powder, Abrasive 370
  Lotion 65
  Wash, Lithographer's Protective. 67
Hardening Bath, High Speed Steel.259
  Copper 262
Hay-Fever Asthma Inhalant 83
  Eye Wash 84
  Treatment 83

Head Cheese, Country Style 201
  Lice Lotion 80
Heat Exchange Medium 410
  Generating Composition.....408, 409
  Paper Transfer 366
Heating 3
  Powder, Chemical 77
Heel Compound, GR"S" 364
Hemorrhoidal Suppository 99
Hide Beetle Control 129
  Dehairing 218
  and Skin Beetles, Control of.....218
  Softening Dry 218
Hives on Eyelids, Treating 84
Honeycomb, Artificial 132
Hop Extract Emulsion 55
Horticulture. See Garden.
Household Products 8
Hydrogen Peroxide 417
Hygiene Preparation, Feminine 108

I

Ice Cream Bar Coating 180
  Cream, Chocolate 178
  Cream or Soda Fountain Crushed
    Fruit 57
  Cream, Vanilla 178
  Stabilizer for Water 179
  Tangerine Water 179
  Water 178
Icing 155
  Bakers' 154
  or Base, Boiled 163
  Boiled Chocolate 157
  Buttercream 161
  Chocolate 156
  Coffee Cake 158
  Cookie Flat 161
  Cream 156
  Custard 163
  Doughnut 161
  Fondant-Type 157
  Fruit, Nut, etc. 157
  Home-Made Boiled 162
  Malted Chocolate 160
  Marshmallow 158
  Peach Fluff 160
  Pineapple Fluff or Fruit 157
  Special 163
  White 158
Impetigo and Wound Powder 96
Ingredients 1
Inhalant, Hay-Fever Asthma 83
  Nasal 81
  Vapor 82
Ink, Black Marking 215
  Black Stencil 215
  Black Writing 215
  Blue-Black Writing 9
  Ceramic Stenciling 216
  China and Glass 216
  for Clock Numerals 216
  Duplicating 215

Ink—Continued
  Fluorescent 215
  Hectograph 215
  Invisible, Detecting 215
  Laboratory 216
  Laundry-Marking 9
  for Marking Glass 2
  Printing 216
  for Ruling on Glass 216
  Semi-Gallate Writing 215
  for Stamp Pads 215
  Sympathetic 215
  Thinner, India 217
Insect Powder, D.D.T. 120
  Repellent 8, 98
  Repellent Cream 98
Insecticidal Aerosol 119
Insecticide (see also Larvicide)....119
  Agricultural 121
  Dry 120
  Nicotine Bentonite 127
  for Poultry Houses 129
  Shade Tree 125
  Slug 121
Insulating Compound for Molten
    Cast Steel 267
  Filling Compound, Electrical.....358
Insulation, Bonded Fibrous 410
  Code Wire 363
  Electrical 358, 359
  Fireproof Thermal 236
  High-Frequency Ceramic 238
  Sound 236
Insulator, Ceramic Electrical 238
Ion Exchange Composition 411
Iron, Powdered 258
Itch Lotion, Grain 96
  Remedy 95

J

Jam, Strawberry 144
Japanese Beetle Larvae Control....121
Javelle Water 14
Jelly for Bakers, Firm 170
  for Bakers, Spreading 169
  Candy 147
  Cocoa 145
  Fig Slab 170
  Grape 145
  Paste, Orange 171
  Piping 156
  Powder, Bakers' Flavored Pectin.164
  Surgical Lubricating 105
  Washable 100
  Wine 145
Jointing Compound, Plastic Heat-
    proof 42

K

Kallitype Silver Printing 336
Kerosene, Stabilized 406

Ketchup, Spice for 184
Killing Bath 224
Killing, Brush 225
Killing with Catalyst 225
Kohl 84

L

Label Base, Heat-Sealable 48
Lace Fabric, Nitrated Non-Inflam-
    mable 401
Lacquer 317
  Base, Perspirationproof 322
  for Blueprint, Direct Print, Map,
    etc. 340
  Book 320
  Brushing Linoleum 320
  Cellulose Acetate 320
  Cheese Wrapper Foil 320
  Clear Furniture 318
  Coated Paper, Avoiding Stickiness
    of 320
  Coating, Flame Resistant 322
  Crystallizing 322
  Exterior   Clear   Transparent
    Metal 323
  Filling and Binding 322
  Flat 321
  for Glass Identification, Dipping.322
  Gold 321
  Imitation Shellac 324
  for Lenses, Protective 322
  for Moistureproofing Cellophane..321
  Prolamine (Zein) Plastic 323
  Thinner 321, 322
  White Automobile or Refrigera-
    tor 318
  Wood 320
Laminated Wood 41
Lanolin Substitute 85
Lard, Preserving 213
Larvicide, Codling Moth 123
  Mosquito 119
Latex, Heat-Sterilizing Fresh.....360
  Preserving 360
Laundry Bleach (Javelle Water)... 14
  Blue 14, 372
  Sour and Bluing 372
  Sour, Stabilized 372
Lawn Grub Control 121
Laxative, Vegetable 107
Lead Coated Iron, Powdered 258
  Coating Alloy, Hot 251
Leak Sealer. See also Anti-Leak.
Leak-Sealing Composition 49
Leather Belt Stuffing 219
  Coating for Artificial...318, 319, 320
  Dressing, Fine 219
  Dye Solution, Black 220
  Emulsion, Waterproofing 218
  Goods, Oil for Softening 219
  Preservative 10
  Preventing Mold Growth on.....219

Leather—Continued
Strips, Wax for 220
Waterproofing Sole 218
Lemon Beverage, Prepared 58
Cheese 167
Extract 54
Extract, Pure 9
Oil Emulsion 56
Lining, Asphalt Drum 312
Liniment 89
Methol 89
Sore Muscle 7
Lip Stick, Chapped 109
Lithographer's Protective Hand
Wash 67
Lithographic Albumin Substitute...337
Image Coating 337
Paper Coating 338
Plate Desensitizing Etch 337
Printing Plate Treatment 338
Liverwurst 210
Marble 209
Loaf, Baked Hamburger 186
Baked Veal and Pork 189
Barbecue-Style Pork 195
Blood and Tongue 185
Celery Meat 193
Cooked Corn Beef and Spaghetti.188
Cooked Ham 187
Corned Beef 187
Corsica-Style Meat 192
Defense Meat 190
Delectable Meat 194
Dutch Meat 193
Glaze for Meat 199
Hamburger and Cheese 199
Liver 195
Liver and Bacon 189
Liver and Cheese 198
Luncheon Meat 189
Luxury Meat 191
Meat, Macaroni and Cheese.....198
Meat, Pickle and Pimento 192
Minced Luncheon 196
Mosaic Liver 194
Pork Luncheon 185
Roast-Beef 186
Skim-Milk Meat 197
Southern Meat 191
Southern Peanut Meat 194
Tongue and Cheese 197
Translucent Liver 196
Utility Meat 190
Veal 188
Loss, Spoilage and 4
Lotion, Acne 92
After-Shaving 81
Antipruritic 95
Base, Alginate Hand 67
Dermatitis 93
Face 65
Grain Itch 96
Hair 76

Lotion—Continued
Hand 6, 65
Head Lice 80
Suntan 64
Vinegar Face 3
Lubricant (see also Grease) 11
Airplane Parts 229
Air-Pump 230
Anti-Corrosive Spindle 230
Anti-Size Thread 230
Ceramic Binder 238
Clock 230
Corrosionless Bearing 230
Emulsive 231
Ethyl Cellulose Molding 231
Forging Tool and Die 231
Gasoline Line 229
Gun 11, 229
Hydrocarbon Resistant Stopcock.232
Leather Belt 232
Leather Packing and Gasket 231
Metal Can 229
Non-Flowing 230
Organic Vapor Resistant Stop-
cock 232
Plastic Molding 230
Rubber Mold 231
Solid 230
Stopcock 232
Wire Drawing 229
Lubricating Oil Corrosion Inhibitor.230
Paste, Anti-Size 230
Lubrication, Powder Metallurgy
Die 229
Lute, Pipe-Joint 42
Lycopodium Substitute 417

M

Magnesium Alloy, Casting and Roll-
ing 261
Finish, Bright 248
Gray Finish for 248
Pre-Painting Treatment for.....248
Magnet Alloy, Permanent 260
Malt Powder, Cocoa 9
Malted Milk Powder 9
Malto-Dextrin 174
Manganese-Plating 253
Marking. See also Ink.
Glass 216
Porcelain 216
Marmalade, Orange 143
Marshmallow, Starch-Albumen 152
Mascara, Cake 84
Mash, Breeding 133
Chick Starting and Growing.....132
Grain and 132
Poultry 131
Turkey 136
Turkey Growing 135
Masking Tape 45
Mayonnaise 183

Measures 420
Measuring, Weighing and 4
Meat, Cure for Sausage 212
Curing Salt 212
Dried Salted 213
Preservative 212
Meatwurst, Smoked 209
Mechanical Goods, Highly Loaded
GR"S" for 364
Melamine Resin 356
Mercerizing Bath 384
Meringue 167
Metal Carbide 258
Cleaning and Protecting 244
Coating on One Side, Hot 251
Degreasing Compound 374
Filler and Calking Compound.... 48
Protection 276
Protective Synthetic Coating.....278
Metallic Spray Coating 251
Methods 3
Metol Poisoning Salve 94
Metric Equivalents 420
Mica, Permanent Marking of 239
Milk of Almond 65
Bottle Cap Coating 328, 358
Mixing and Dissolving 4
Mold Coating, Casting 266
Coating, Ingot 266
Composition, Foundry 266
Composition for Precision Cast-
ings 267
and Core, Foundry 266
Facing Sand 266
Wash, Foundry 267
Molded Cork and Cane Stalk Prod-
uct 355
Molding Clay 366
Material 11
Mordanting Rayon 390
Mortar, Mason's 235
Mosquito-Repelling Oil 8
Mothproofing Emulsion 402
Fluid 8
for Piano Key Felts 402
Solution 402
Mouth Rinse, Germicidal Dental... 73
Treatment, Trench 74
Wash 7, 73
Wash, Gingivitis 74
Wash, Glycerin Thymol 73
Mucilage (see also Adhesive, Agglu-
tinant, Calking, Cement, Glue,
Paste, Putty, Seal) 31
of Quince Seed 93
Mustache Wax 109

N

Nasal Constrictor 82
Inhalant 81
Ointment 93
Spray 81

Nasal—Continued
Spray, Benzedrine 81
Spray, Throat and 81
Nectar, Imitation Cherry 53
Imitation Grape 53
Imitation Raspberry 53
Imitation Strawberry 53
Lime 52
Orange 52
Neoprene Compound, General Util-
ity 363
Latex Coating 365
Latex, Gelling 361
Olive Drab Coating 364
Newspaper, Deinking 328
Nickel-Plating Bath 252
on Magnesium 252
Nicotine Bentonite Insecticide 127
Decoction 128
Dust 128
Nitrocellulose Coating 366
Non-Staling Baked Products 172
Non-Toxic Bitter-Tasting Product..105
Nose Drops, Asthma 83
Drops, Isotonic 82
Nut Shells, Bleaching 143
Skins, Removal of Filbert 143
Skins, Removing 143
Nutmeg, Imitation 54
Nylon Fabric, Boil-Off of 380

O

Offset Composition, Printing 216
Oil, Baby Skin 97
Base Well Drilling Fluid 230
Box Calf 219
Extracting Fish Liver.........233
and Fat, Bleaching 233
Floor 11
Flushing 231
General-Purpose Grinding 309
Metal-Cutting 231
Metal-Quenching 231
Milky Suntan 65
Mixing Varnish and Paint 309
Mosquito-Repelling 8
Non-Rusting Turbin 230
Penetrating 11, 232
for Shampoo, Crude 80
Soluble 231
Substitute for Coconut 175
Suntan 64
Well, Freeing Stuck Tools in 408
Well, Removing Metal Obstruct-
tions from 408
Oilproofing Paper and Fiberboard.. 15
Ointment. See also Salve.
Analgesic 85
Antiseptic 2
Anti-Sunburn 65
Base 60, 84
Base, Ophthalmic 103

Ointment—Continued
Base, Stainless 85
Base, Washable 60
Burn 97
Cable Rash 93
Chlorophyll Impetigo 96
Cod-Liver Oil 85
Emulsion 100
Eye 93
Mustard-Gas 97
Skin 93
Stable Zinc Peroxide 92
Sulfa—Zinc Peroxide 96
Sunburn 64
Ulcer 92
Vaginal 100
Oleomargarine, Non-Weeping 181
Orange Oil Emulsion 56
Ovicide, Codling-Moth 123

P

Packing Gasket, Chemically Re-
    sistant 359
    Plastic Stuffing Box 359
    Solvent-Resistant Pump 417
Paint. See also Enamel.
    Anti-Fouling Shipbottom 279
    Base for Tints, Brick and
        Stucco 173
    Base for Tints, House 272
    Black Bridge 278
    Black Hull 280
    Brick and Stucco 172
    Brick and Stucco White 172
    Brushes, Cleaning 314
    in Cans, Preventing Skinning of..315
    Casein Powder 312
    Cement Water 287
    Clear Protective Coat for Phos-
        phorescent 317
    Clear Varnish for Heat Resistant
        Aluminum 292
    Congo Resin Camouflage 287
    Congo Resin Emulsion 303
    Congo Resin Fire-Retardant ...289
    Congo Resin Traffic 275
    Congo Resin Vehicle for Gasoline-
        Soluble 282
    Emulsion Stone 286
    Exterior House 271
    Exterior Metal 275
    Exterior Trim and Trellis ...273
    for Exterior Use, Water Emul-
        sion 285
    Fire-Resisting Canvas Preserva-
        tive 280
    Flame-Resisting 285
    Flat Wall 295
    Interior Gloss 296, 297
    Light Gray Deck 279
    Light Gray Hull 280
    Lusterless Congo Resin Ammuni-
        tion 282

Paint—Continued
    Marine 279
    Marine Interior Flat 281
    Marine Outside White 280
    Non-Leafing Aluminum Paste ...310
    Oleo-Resinous Emulsifiable ...288
    Outside White House 271
    Paste, Emulsion 303
    Phosphorescent 316
    Red Deck 280
    Remover 12, 324
    Semi-Gloss Polishing 311
    Special 309
    for Steel, Carbon Black 278
    for Structural Steel, Anti-Corro-
        sive 278
    for Structural Steel, Red Lead...278
    Tint, Emulsion 304
    Traffic 274
    Water 287
    Waterproof Awning 285
    Water-Reducible Oil Type Paste
        301, 302
    for Wooden Vessels, Anti-Fouling
        Copperbottom 279
Painting Glazed Tile 316
Palladium Plating Bath 251
Paper, Avoiding Stickiness of Lac-
    quer-Coated 320
    Blueprint 339, 340
    Containers, Oilproofing 327
    Fireproof 15
    for Glass, Anti-Mist 326
    "Glassine" 14
    Glaze Coating 326
    Grease-Resistant 327
    Grease-Resistant Glue Coating
        for 327
    Heat-Sealable Waxed 328
    Improving Transparency of
        Waxed 328
    Laminating 327
    Non-Slip, Non-Hygroscopic, Ad-
        hesive-Coated 326
    Non-Tarnish 326
    Oilproofing 15
    Polarity Test 416
    Pulp Oilproofing 327
    Size, Lime in Place of Caustic
        Soda in 325
    Softening 326
    Stock, Defomar for Pulp and ...325
    Towel, Detergent 372
    Transparentizing Solution 326
    Tub Sizing for 325
    Waterproof Cigarette 328
    Waxed 328
Parasiticide. See also Insecticide.
    Spray 123
Paste. See also Adhesive, Agglu-
    tinant, Calking, Cement, Glue,
    Mucilage, Putty, Seal.
    Billposter's 18

Paste—Continued
Cold-Water 18
Labeling 17
Non-Warping 17
Paperhanger's 12
Textile Printing 396
Tin-Labeling 20
Pastry. See Cake.
Pea Aphid Control 123
Soup Cubes 143
Peach Base 160
Pectin Candy 146
Solution 146, 165
Penetrating Aid, Drug 104
Perfume, Solid 71
Spanish "Paste" Leather 72
Permanent Magnet Alloy 260
Waving Composition 76
Waving, Heating Powder for 77
Perspiration, Control of Foot 87
Petrolatum Emulsion, Liquid 106
Petroleum, Deodorizing Sulfurous..406
pH Values 420, 421
Pharmaceutical Tablet Coating 106
Phenothiazine Suspension 129
Phosphatizing 276
Photoengraver's Cold Enamel 337
Cold Enamel Lacquer 337
Photograph, Eliminating Abrasion
Fog on 334
Photographic Acid Hardening Fixer
16, 333
Bleaching Solution 334
Developer, Blue Tone and Blue-
Black Tone 337
Developer, Fine Grain 330
Developer, Hard 330
Developer, Low-Contrast 329
Developer, Non-Dichroic Fog 330
Developer, Reddish-Tone 337
Developer, Soft 330
Developer without Speed Loss,
Fine Grain 331
Developer, Stable 329
Developer, Two-Solution Type ...330
Developer, Universal 329
Developing, Economical 330
Developing Solution 16
Etching Bath 336
Film, Reduction of Overdevel-
oped 334
Film, Restoring Shrunken 335
Fixing Bath, Recovery of Silver
from 340
Fixing and Hardening Bath 333
Flashlight Powder 340
Gas Light Paper, Developing ...336
Gelatin Hardening Solution 332
Intensifiers 332
Negative, Fog Prevention in 335
Negative, Fog Removal from 335
Negative, Intensifying Bath for..331

Photographic—Continued
Negative, Non-Curling 335
Negative, Normalizing Overdevel-
oped 334
Negative, Preventing Drying Spots
on 335
Negative, Rapid Drying of 335
Paper, Home-Sensitized 329
Paper, Sensitized 329
Print-Out Emulsion 336
Print-Out Paper 336
Print on Oxidized Aluminum 338
Print, Preventing Curl in 335
Resist 334
Solution 16
Substitutes 339
Toning Agent 336
Physic (see also Laxative) 107
Pickled Metals, After-Treatment of.244
Pickling Agent for Copper Beryl-
lium 243
Corrosion Inhibitor 244
Fruit, Syrup for 142
Iron and Steel 242
and Rustproofing for Iron 276
Solution for Chrome and Chrome
Nickel Steel 243
Solution, Stainless Steel 243
Stainless Steel 243
Steel 243
Pie Base, Custard 167
Lemon Cream 164
Pigment, Ceramic 238
Pill Coating, Enteric 106
Excipient 106
Pineapple Extract, Imitation 53
Pipe Joint Calking 42
Pipe Lines under Water, Locating
Leaks in 413
Plant Food, Soilless Growth 115
Growth Regulator 117
Nutrient, Tomato 117
Plaster, Catalyst for Quick-Setting
Anhydride 236
Gasoline-Resistant 235
of Paris, Retarding the Harden-
ing of 236
Parting Compound for Pouring ..267
Surgical-Cast 106
Wall-Patching 12
Plastics 351
Injection Molding Powder 356
Molding Composition, Acrylic ...355
Molding Powder 357
Piping, Flexible 358
Polymerization Inhibitor for
Methacrylate 355
Self-Hardening 355
Sulfur 355
Wood Dough 12
Plate Glass Setting Composition .. 43
Poison, Ant 8

Poison—Continued
Bait, Mole Cricket 130
Bait, Sowbug 120
Ivy Killer 130
Ivy Treatment 94
Oak Remedy 94
Silver-Fish 120
Subterranean Grass Caterpillar..121
Tomato Moth 122
Poisonous Chemicals 4
Polish 10
Auto 341, 342
Auto Paste 11
Black Shoe 10
Bright-Drying Floor 345, 346
Chromium 342
and Cleaner, Windshield Glass ..342
Colorless Shoe 343
Floor Paste 11
Furniture 11, 344
Glass 342
for Gold and Soft Metals 342
Leather 343
Liquid Furniture and Floor 344
Liquid Leather 343
Liquid Wax 341
Metal 10
Nail 70
Oven 342
Paste Leather 343
Paste, Shoe 343
Paste Wax 341
Remover, Oily Nail 70
Silver 342
Two-Tone Furniture 344
Wax Shoe 343
Polishing Cloth, Metal 342
Electrolytic Metal 254
Electrolytic Steel 254
Stainless Steel Anodic 254
Wax 11
Polystyrene Sponge 356
Stabilizing 356
Pork and Apple Patties, Breakfast.201
Potassium Chlorate 418
Potato Leafhopper Dust 122
Non-Discoloring Peeled 143
Protecting against Decay 119
Ring Rot and Scab Control 128
Poultry Mashes 131
Powder, Anti-Acid 91
Antiseptic Dusting 70
Antiseptic Military Foot 89
Astringent Dusting 70
Athlete's Foot 89
Baking 174
Curry 184
D.D.T. Insect 120
Douche 70
Face 70
Foot 7
Gargle 73

Powder—Continued
Impetigo and Wound 96
Potato Leafhopper 122
Roach 120
Rug and Textile Cleaning 374
Scouring 371
Surgeon's Gloves Dusting 360
Tooth 7
Pre-Galvanizing Coating 249
Preparations, Elementary 5
Preserve, Strawberry 144
Primer, Adhesive Scratch Resist-
ing 324
Alkyd Resin 277
Galvanized Iron 276
Navy Chromate 280
Red Lead 276
Transparent Flexible Metal 278
Undercoat, White 272
Zinc Chromate 276, 277
Zinc Chromate—Iron Oxide Al-
kyd 277
Zinc Dust—Zinc Oxide 276
Zinc Tetroxychromate 277
Zinc Tetroxychromate—Iron Ox-
ide 277
Print, Copper Sulfide Intensification
of Film or 331
Correction Fluid for Vandyke ...340
Developer for B W 340
Removal of Yellow Spots on 333
Restoring Yellow 333
Printer's Roller, Glueless 366
Protein Food 213
Pudding, Blood 210
Liver 210
Pulverizing and Grinding 4
Putty (see also Adhesive, Agglu-
tinant, Calking, Cement, Glue,
Mucilage, Paste, Seal) 12, 47
Brass to Marble 46
China to China 47
Glass to Glass 46
Glass to Wood 46
for Knife Handles 47
Metal to Celluloid 47
Metal to China 47
Metal to Glass 46
Metal to Horn 46
Metal to Linoleum 46
Metal to Stone 46
Metal to Wood 46
Rubber to Glass 46
Rubber to Metal 46
Rubber to Stone 46
Sticking to Hand, Preventing ... 47
Stone to Hard Rubber 47
Stone to Horn 47
Stone to Stone 47
White Lead 47
Pyrotechnic Flare 349
Torch, Green 349

Q

Quality of Ingredients 1
Quince Seed Mucilage 93

R

Raincoat Material, Calendered 363
  Material, Olive Drab 362, 364
  Thermosetting Polyvinyl Butyral
    Coating for 362
  Thermosetting Waterproof Coat-
    ing for 282
Raspberry with Other Natural Fla-
  vors 52
Rayon Knitting Yarn, Setting Twist
  in 382
Recording Solution, Electrochemi-
  cal 419
Rectal Suppository 99
Red Pepper, Capsicine-Free 184
  Pepper Substitute 184
References 427
Refinery Gas Polymerization Cat-
  alyst 407
Refractory (see also Firebrick)
                          236, 237
  Clay 236
  Coating for Metal or Ceramics...237
  Fusion Cast 237
  Non-Acid Brick, Hard 237
  Zircon 237
Resin 351
  Fine-Melt Congo 293
  Treatment, Congo 292
Resistor, Temperature Compensat-
  ing 409
Resurfacing Cold-Rolled Low-Car-
  bon Steel 260
Rhenium-Plating 252
Ribbons, Water-Repellent Finish
  for 404
Roach Powder 120
Roll, Frankfurter 173
Roofing Paper, Asbestos 240
Root Beer Extract, Concentrated.. 54
Rotenone-Derris Substitute 120
Rub, Athlete's 89
Rubber 351
  Ball, Reviving Dead 361
  Belt, Men's 365
  Compound, 100% Reclaim Self-
    Curing 364
  Emulsion Polymerization, Syn-
    thetic 360
  Gas-Expanded 360
  Non-Sticking 361
  Preventing Adhesion of 361
  Reclaim Compound, 100% 363
  Reclaiming Scrap 361
  Solution, Reducing Viscosity of..361
  Solvent for Chlorinated 360
Rubbing Alcohol, Isopropyl 89

Rust Inhibitor for Engines 244
  Prevention Compound 10
  Remover 376
Rusting Inhibitor, Radiator 413
Rustproofing 245
  for Iron, Pickling and 276

S

Sage Oil, Imitation 54
Salami Cotto (Cooked) 206
Salt, Celery 184
Salt-Water Pit Lining 240
Salve. See also Ointment.
  for Chromate Skin Injuries 93
  Grafting-Wax 119
  Healing 85
  Metol Poisoning 94
  for Metol Poisoning, Dermatitis. 94
Sandwich Special, Pro-Lac 186
Saponin from Soap, Extracting 417
Sausage, Braunschweiger Style
    Liver 208
  Cervelate 203
  Farmer-Style 204
  Liver 208
  Marbled 203
  Mayence-Style 208
  Meat, Cure for 212
  Mortadella-Style 206
  Nutrition 203
  Polish-Style 205
  Seasoning, Pork 212
  Summer 204
  Thueringer-Style 205
  Vienna Style 207
Scabies Preparation 95
Scouring Powder 371
Scrapple 200
Scroop on Bleached Cotton Yarn,
    Silky 383
  on Cotton Yarn 383
  Effect on Wool 384
Scrooping, Mercerized Cotton 382
Seal, Aluminum Thread-Joint Seal. 42
  Can 32
  Can Joint 34
  Can Seam 42
  Elastic Refrigerator 49
  Metal-Joint 42
Sealer, Crack 42
  for Floors, Penetrating 306
Sealing Composition, Leak 49
  Compound, Electrical 39
  Compound, Glass 43
  Expansion Joint Plastic 42
  Tire Puncture 360
Sedative 104
Sellers of Chemicals and Supplies..344
Sewage Treatment 411
Shampoo, Crude Oil for 80
  Soapless 79

Shaving Aid, Electric 80
Sheep Anthelmintic 129
  Tick Dip 129
Sheepskin, Curing Hairy 222
  Washing 221
Sheeting, Thermosetting Waterproof
  Coating for 282
Shellac Compound, Injection Mold-
  ing 358
  Substitute for Records 358
Sherbet 179
  Mix, Pineapple 179
  Stabilizer for 179
Shoe Bottom Filler 219
  Box Toe Stiffening 220
  Coloring Used 220
  Disinfecting 86
  Dressing, White 10
  Sole Stain 220
  Stiffener 220
  Waterproofing for 10
Shortening, Dry Lecithin 174
Silk Printing Discharge 397
  Stocking, Improving the Wearing
    Quality of 387
  Treatment, Prefabricating 387
Silver Fish Poison 120
  Mirror on Glass 253
  Plating without Electricity 256
  Plating Magnesium 255
  Printing, Kallitype 336
  from Stripping Solution, Recover-
    ing 255
  Tarnish Inhibitor 249
Size, Bookbinders' 32
  Gilder's 311
  Glue 19
  Jute Twine Polishing 388
  Lime in Place of Caustic Soda in
    Paper 325
  Rayon 382
  Rosin 325
  Textile Finish and 381
Sizing Glazed Cotton Yarn 381
  for Paper, Tub 325
  Rayon Warp 382
  Viscose Ribbon 381
  and Waterproofing, Paraffin Wax.328
Skin Cleanser, Liquid 59
  Protective 67
  Treatment, Coal-Tar 92
Smoke, Colored 350
  Flare, Colored Rocket 350
Soap (see also Cleaner) 13
  Aluminum Chloride 87
  Anti-Perspiration 87
  Antiseptic Hand 87
  Antiseptic Liquid 371
  Bleaching 371
  Fungicidal Borax 369
  Home-Made 368
  Industrial Hand 370
  Lemon 369

Soap—Continued
  Liquid 13
  Mercury Fulminate Detector 371
  Non-Efflorescing Bar 368
  Paste, Mechanic's Hand 13
  Rosin 368
  Rosin—Tallow 368
  Saddle 13
  Silver Fluoride 87
  Substitute 369
  Transparent Glycerin 369
Softener and Lubricant for Arti-
    ficial Fibers 385
  Stearic Acid Textile 385
Softening Cream, Textile 386
  and Lubricating Wool 385
Soilless-Growth Plant Food 115
Solder, Aluminum 264
  for Aluminum Foil 264
  for Aluminum and Stainless Steel,
    Powdered 264
  Brazing 265
  Gas Meter 265
  Hard 265
  Low-Tin 265
  Soft 265
  Tinless Soft 265
  for Tinplate, Soft 265
  for Zinc-Aluminum Alloys 264
Soldering Flux 16
Solvent, Engine Carbon 407
  Motor Gum 407
Sound-Deadening Pads, Adhesive for 44
Soup Cubes, Oxtail-Type 213
  Cubes, Pea 143
Sowbug Poison Bait 120
Spearmint Extract 54
Spice, Imitation 54
Spoilage and Loss 4
Sponge, Artificial 365
  Iron Reducing Agent 258
  Oil-Resistant 366
  Polystyrene 356
Spot Remover, Grease, Oil, Paint
    and Lacquer 13
  Remover, Rust and Ink 14
Spray, Asthma 82
  Base, Horticultural 124
  Benzedrine Nasal 81
  Boxwood Leaf Miner 122
  Cattle Horn Fly 130
  Codling Moth 123
  Dry Agricultural 127
  Emulsion, Kerosene 126
  General Agricultural 126
  Gladiolus Thrip 124, 125
  Nasal 81
  Ornamental Plant Insecticide ...124
  Parasiticide 123
  Paris Green 124
  Rotenone 123
  Spreader, Agricultural 127
  Thrip 124

Spray—Continued
Throat and Nasal 81
Water Hyacinth 123
Stain from Hand, Removing Dye ..376
Pink Ceramic Underglaze 238
Preventing Silver Nitrate 339
Remover. See Cleaner, Spot Remover.
Removing Beer 376
Removing Blood 377
Removing Burnt Sugar 376
Removing Chromium 377
Removing Cobalt 377
Removing Coffee 376
Removing Copper 377
Removing Egg Yolk 377
Removing Grass 377
Removing Iodine 377
Removing Iron 377
Removing Lead 378
Removing Manganese 377
Removing Mercury 377
Removing Mildew 377
Removing Milk 377
Removing Nickel 377
Removing Picric Acid 377
Removing Scorch 377
Removing Urine 377
from Wall Paper, Removing Ink.376
Wood 317
Stainless Steel, Sea-Water-Resistant 244
Stair Tread Coating, Non-Slip 268
Starch, Thin-Boiling 381
Steel, Blue-Black Finish on 9
Stereotype Surfacing 254
Sterilization, Air 110
Stocking, Cosmetic 108
Stone, Artificial Building 235
Filler, Alberene 47
"Stop Leak," Radiator and Water
Boiler 413, 414
Stoppers and Ground Glass Joints,
Removing Frozen 418
Removing Glass Tubes and Rods
from Rubber 418
Strawberry with Other Natural
Flavors 51
Stripping Dyed Cotton 399
Oxide Film from Aluminum 249
Stud, Clutch Plate 411
Styrene, Inhibiting Polymerization
of 356
Sugar Wafer 172
Wafer, Filler for 172
Wafer Shell 173
Sugarless Syrup Substitute 54
Sulfa Drug Compositions 99
Drug Surgical Film 96
Vaginal Suppository 99
Sulfathiazole Suspension 93
Sulfur, Colloidal 113
Sunburn Ointment 64

Suntan Cream 64
Lotion 64
Oil 64
Oil, Milky 65
Suppository, Cocoa-Butter 99
Hemorrhoidal 99
Mold Lubricant 99
Rectal 99
"Sulfa" Vaginal 99
Vaginitis 99
Surgical Cast Plaster 106
Dressing, Flexible Adhesive 106
Dressing, Waterproofing 404
Film, Sulfa-Drug 96
Lubricating Jelly 105
Suspension, Phenothiazine 129
Sweeping Compound, Coloring ...378
Compound, Floor 378
Sweet Cocoa Powder 9
Synthetic Rubber Composition, Gasolineproof 360
Syringe Aids, Glass 418
Syrup, Chocolate 55
Orange Juice 57
Orangeade 56
Pectin 154
for Pickling Fruit 142
Quick-Dissolving Beverage 54
Substitute, Sugarless 54
Sugarless Pharmaceutical 104

T

Tablet Coating, Pharmaceutical 106
Tannic Acid Burn Emulsion 97
Tanning, Egg-Yolk Substitute for..218
Extract 218
Tape, Masking 45
Taylor's Dummy, Thermoplastic....358
Temperature Compensating Resistor 409
Measurements 3
Tempering Compound, Metal 259
Tent, Thermosetting Waterproof
Coating for 282
Textile 379
Bleaching 387
Coating, Flameproof, Waterproof
and Mildewproof 399
Duller 380
Dyeing, Improved 397
Finish and Size 381
Flameproofing 400
Printing Paste 396
Printing Resist 397
Scouring Agent 386
Softener, Stearic Acid 385
Softening Cream 386
Waterproofing 403, 404
Weighter 385
Thermometer Readings, Conversion
of 424
Thermostat, Electric 261

Thiokol Compound, Non-Swelling...363
Throat and Nasal Spray 81
Tin Foil, Non-Discoloring 261
Tinning Compound, Cold 250
  Copper 250
  Hot 250
Tinplate Protective Coating 250
Tin-Plating of Cast Iron, Immer-
  sion 251
Tire Puncture Sealing 360
Tobacco Blue Mold Control 128
Toilet Water. See Cologne.
Tomato Aspic, Canned 141
  Pinworm Control 122
  Plant Nutrient 117
Toothache Drops 74
Topping, Orange Crumb 163
Towels, Improving Absorbency of..402
Trade Name Chemicals 429
Tree Gall Treatment 119
Trench Mouth Treatment 74
True-to-Scale Prints, Gelatin Com-
  position for 329
Tungsten Carbide, Pulverizing Ce-
  mented 258
Turkey Growing Mash 135
  Mash 136
  Starter 134
Turkish Paste 151
Typewriter Ribbon, Rejuvenating..217

V

Vacuum Tube Getter 409
Vaginal Ointment 100
  Suppository, "Sulfa" 99
Vaginitis Suppository 99
Vapor Inhalant 82
Varnish, Alkyd Blending 309
  for Art Work, Pale Gloss 306
  Chemical-Resistant Spar 290
  Clear Automobile 291
  Congo Resin 293, 294
  Exterior 289
  Floor and Deck 305
  Furniture Rubbing 307
  Insulating Impregnating 308
  Interior 304
  Low-Cost Floor and Trim 305
  Mopping 306
  Sealer, Congo Resin 295
  Spar 289
  Synthetic Resin 291
  Synthetic Rubbing 307
  Traffic Paint 291
  Wall Sizing 308
Vegetable Preservation 143
Vinyl Acetate, Emulsion Polymeriza-
  tion of 357
Vitamin B Elixir 105
Vulcanization Accelerator, Fast Cur-
  ing 362

W

Walnut Extract, Imitation 53
Wash, Bakers' 154
Washable Jelly 100
Washing Composition 79
Water, Coagulant for Hard........411
  Corrective 412
  Disinfectant 411
  Impurity Coagulant 411
  Purifying Compound 412
  Repellent Finish for Ribbons ...404
  Resistant Casein Fiber 404
  Softener Briquette 412
  Softener, Cation Exchange 411
  Soluble Adhesives 17
Waterproof Coating 282
Waterproofing Cement 14
  Cement and Stucco 287
  Heavy Canvas 14
  Liquid 14
  Paper and Fiberboard 14
Wave Set Fluid 77
Wax 351
  Aseptic 97
  Batik 351
  Cable 352
  Cartridge 351
  Cheese Coating 183, 351
  Coopers' 351
  Copper Engraving 352
  Cream Buffing 348
  Dental Adhesive 354
  Dental Base Plate 354
  Dental Carving 354
  Dental Inlay 354
  Dental Investment 354
  Finish for Fur 223
  for Fire-Gilding 352
  Floor 346, 347
  Furniture Filler 46, 315
  Grafting 11, 45, 119, 352
  Gravure 352
  Inhibiting Discoloration of Par-
    affin 353
  Ironing 352
  Laundry 352
  for Leather Strips 220
  Modelling 351, 352
  Modifier for Paraffin 353
  Mustache 109
  Penetration of Modified 125°F
    M.P. Paraffin 353
  Plating, Electrolytic 355
  Polishing 11
  Remover, Floor 345
  Saddler's 353
  Salve, Grafting 119
  Shoemakers' Thread 352, 353
  Ski 351
  Substitute, Carnauba 353
  Substitute, Lanette 353
  Wood Filler 46, 315

Weed Killer 130
Weighing and Measuring 4
Weights and Measures 420
Weld Spatter Coating, Non-Adher-
  ing 270
Welding Composition, White Metal.268
  Electrode Coating 268
  Rod 268
  Rod Coating, Aluminum 269
  Rod Flux Coating 269
  Rod for Red Brass 268
Well Drilling Fluid, Oil-Base 230
Whipped Cream 160
Wood Dough Plastic 12
  Filler 46, 315
  Filler, Wax 46, 315
  Fireproofing 241
  Floor Bleach 12
  Ply Lamination 41
  Preservative 240
  Resin-Impregnated 359
Wool, Chlorinating 379
  Lusterizing 379
  Oiling Emulsion 385
  Softening and Lubricating 385
  Stripping Dye from 399

Wool—Continued
  Unkinking Natural 385
  Yarn, Reserve for 398
  Zinc Dust Resist for 398
Wool-like Jute 388
Wound and Burn Film Treatment.. 96

X

X-Ray Photographs, Improving 335
  Protective Composition 415

Y

Yarn Dressing, Jute 388
  Heat Stabilization of 388
  Increasing Wet Strength of 404
  Metallized 405
  Wetting and Conditioning Agent
    for 401

Z

Zein-Rosin Solution, Stable 323
Zeolite, Carbonaceous 411
Zinc-Plating, Bright 253